# ALGEBRAIC BETHE ANSATZ AND CORRELATION FUNCTIONS

## An Advanced Course

# ALGEBRAIC BETHE ANSATZ AND CORRELATION FUNCTIONS
## An Advanced Course

**Nikita Slavnov**
Steklov Mathematical Institute, Russian Academy of Sciences, Russia

World Scientific

NEW JERSEY • LONDON • SINGAPORE • BEIJING • SHANGHAI • HONG KONG • TAIPEI • CHENNAI • TOKYO

*Published by*

World Scientific Publishing Co. Pte. Ltd.
5 Toh Tuck Link, Singapore 596224
*USA office:* 27 Warren Street, Suite 401-402, Hackensack, NJ 07601
*UK office:* 57 Shelton Street, Covent Garden, London WC2H 9HE

**Library of Congress Cataloging-in-Publication Data**
Names: Slavnov, Nikita, author.
Title: Algebraic Bethe ansatz and correlation functions : an advanced course /
Nikita Slavnov, Steklov Mathematical Institute, Russian Academy of Sciences, Russia.
Description: New Jersey : World Scientific, [2022] | Includes bibliographical references.
Identifiers: LCCN 2022002702 | ISBN 9789811254253 (hardcover) |
ISBN 9789811254260 (ebook for institutions) | ISBN 9789811254277 (ebook for individuals)
Subjects: LCSH: Bethe-ansatz technique. | Correlation (Statistics) | Mathematical physics.
Classification: LCC QC20.7.B47 S53 2022 | DDC 530.13--dc23/eng20220321
LC record available at https://lccn.loc.gov/2022002702

**British Library Cataloguing-in-Publication Data**
A catalogue record for this book is available from the British Library.



For any available supplementary material, please visit
https://www.worldscientific.com/worldscibooks/10.1142/12776#t=suppl

Desk Editors: Soundararajan Raghuraman/Lai Fun Kwong

Typeset by Stallion Press
Email: enquiries@stallionpress.com

# Preface

In 1931, H. Bethe proposed a new method for constructing quantum Hamiltonian eigenfunctions of the Heisenberg spin chain [H. Bethe (1931)]. This method was called the Bethe ansatz. It gave rise to a new approach to studying a broad class of quantum systems.

Although the Bethe ansatz is used to describe low-dimensional systems, this method has played an essential role in such areas of physics as statistical mechanics, quantum field theory, and condensed matter physics. Interest in the Bethe ansatz has significantly increased in recent years. There are at least two reasons that have contributed to this growth. First, it is the experimental creation of optical lattices of ultracold atoms. Thanks to the development of laser cooling technologies, we now can obtain and study quantum systems that previously existed only in the minds of theorists. The second reason is that at the beginning of the 21st century, it was unexpectedly discovered that the Bethe ansatz could be effectively used to solve a number of problems in supersymmetric gauge theories and string theory [J. A. Minahan and K. Zarembo (2003)]. This discovery caused an avalanche of publications on the topic of the Bethe ansatz.

In the late 70s of the 20th century, in the works of Leningrad School, under the leadership of L. D. Faddeev, a Quantum Inverse Scattering Method (QISM) was developed [L. D. Faddeev et al (1979); L. D. Faddeev and L. A. Takhtajan (1979); E. K. Sklyanin (1982a)]. In the framework of this method, it was found that many quantum models solved by the Bethe ansatz and having a completely different physical interpretation can be described by the same algebra of operators, actually being different representations of this algebra. In this case, many important physical systems properties can be established already at the algebraic level without using its concrete representation. This approach was called the *algebraic*

*Bethe ansatz*. The original method developed by H. Bethe is now commonly referred to as the *coordinate Bethe ansatz*.

The algebraic Bethe ansatz solvable models, by construction, have an infinite set of integrals of motion. This is a sign of integrability. Thus, it becomes possible to obtain exact results without the use of perturbation theory. This opportunity is very attractive. Therefore, almost immediately after creating the QISM, attempts were made to apply this method to calculate correlation functions, which are the most important characteristics of quantum models. The results of those studies were subsequently published in the book [V. E. Korepin et al (1993)]. However, a long time has passed since then, during which the algebraic Bethe ansatz has been continuously developed and improved. This book describes new approaches to calculating correlation functions within the algebraic Bethe ansatz framework.

This book grew out of lectures on the algebraic Bethe ansatz the author gave in 2015–2016 at Steklov Mathematical Institute [N. A. Slavnov (2017)]. However, during the work on the book, the source material was largely revised. Let us briefly describe the content.

The first two chapters are devoted to the general settings of the algebraic Bethe ansatz and the solution of the spectral problem. Much of what is described in these chapters can be found in other monographs and review articles. Chapter 1 introduces the concepts of an $R$-matrix, a monodromy matrix, and a transfer matrix. We consider specific examples and show how the transfer matrix generates the quantum Hamiltonian of the $XXX$ Heisenberg chain. Chapter 2 describes a method for finding a transfer matrix eigenvectors in the algebraic Bethe ansatz framework. Here we obtain a system of Bethe equations and study its properties. Much attention is paid to the twisted transfer matrix and its eigenvectors.

In the next two chapters, we talk about constructing local operators within the algebraic Bethe ansatz framework. Chapter 3 provides a detailed solution to the quantum inverse problem for spin chains. Chapter 4 is devoted to the composite model.

In Chapters 5–7, we discuss scalar products of Bethe vectors. Scalar products are the main tool for studying correlation functions. Chapter 5 deals with the general case of the scalar product. Chapter 6 is devoted to scalar products in which one of the vectors is the eigenvector of the quantum Hamiltonian. Finally, Chapter 7 describes new methods for calculating scalar products that have emerged relatively recently.

The next two chapters are devoted to calculating matrix elements of operators in the basis of physical states. We call such matrix elements form factors. In Chapter 8, we calculate the form factors of the monodromy matrix entries. They are directly reduced to scalar products. In Chapter 9, we study the form factors of local operators, which can be considered the simplest correlation functions.

In Chapter 10, we study the thermodynamic limit of the models discussed in the previous chapters. From a formal point of view, this topic goes beyond the algebraic Bethe ansatz. However, such a chapter is necessary for further calculation of correlation functions in the thermodynamic limit.

Finally, the last two chapters are devoted to calculating correlation functions using the algebraic Bethe ansatz. We consider two examples. In Chapter 11, we obtain a multiple-integral representation for the two-point correlation function of the third components of spin in the $XXZ$ Heisenberg chain. In Chapter 12, we use form factor expansion to calculate the long-distance and large-time asymptotics of correlation functions in the model of one-dimensional bosons. Some technical calculations are placed in appendices.

As one can see from the above content, this book focuses on those aspects of the algebraic Bethe ansatz directly related to the calculation of correlation functions. Therefore, this book in no way claims to be a kind of encyclopedia of the algebraic Bethe ansatz. Many important and interesting features of this method are not covered in it. Moreover, even the main topic is far from being fully addressed. In particular, we do not mention such exciting and beautiful approaches as the method of dual fields and the quantum transfer matrix method. We also do not say anything about the very interesting connection between correlation functions of quantum systems and classical exactly solvable equations. Covering these topics would take up too much space, even though they are directly related to the main topic.

Fortunately, the literature on the algebraic Bethe ansatz and related problems is so extensive that making an exhaustive list would be pretty challenging. Let us list nevertheless just a few review articles and books: [L. A. Takhtajan (1985); L. D. Faddeev (1984); V. E. Korepin et al (1993); L. D. Faddeev (1998); A. G. Izergin (1999); J. Plunkett (2009); F. Franchini (2017); Y. Wang et al (2015); F. Levkovich-Maslyuk (2016); H.-P. Eckle (2019)]. The interested reader will easily find in this literature those topics that are not included in this book.

One of the features of this book is that it uses a shorthand notation for the products of some functions and operators. They make formulas much more compact and easy to read. However, one needs to get used to these symbols. The reader may have some difficulties at first. Nevertheless, we hope that in time the reader will appreciate the convenience of this shorthand notation.

There are no special exercises in the book for readers to follow on their own. However, readers will have the opportunity to test their skills continually. For example, most scalar product formulas are given in a universal form that works equally for rational and trigonometric R-matrix models. However, for simplicity, the proofs of these formulas are mainly given for the case of a rational $R$-matrix. The reader can independently prove them in the case of a trigonometric $R$-matrix.

# About the Author

**Nikita Slavnov** heads the Department of Theoretical Physics of Steklov Mathematical Institute, Moscow, Russia. His research work focuses on quantum integrable systems and related topics. He is the author of over 100 scientific publications. He conducted research in theoretical physics at Ecole Normale Superieure de Lyon, France, the University Savoie-Mont Blanc, Annecy, France, and the University of Tours, France. He organizes and teaches a course of lectures on integrable systems at Steklov Mathematical Institute. He is the author of lecture courses on the algebraic Bethe ansatz (Bialowieza, Poland) and the nested Bethe ansatz (Les Houches, France).

# Acknowledgment

My first steps in science took place in the Leningrad School's unforgettable atmosphere, created by Ludwig Faddeev. It is a great honor for me to realize that I am a small part of this school. I am especially grateful to the late Anatoly Izergin and Vladimir Korepin, my first teachers in the algebraic Bethe ansatz. I also remember with great gratitude the numerous scientific discussions with Nikolai Bogolyubov, Petr Kulish, Nikolai Reshetikhin, Evgeny Sklyanin, Fedor Smirnov, and Vitaly Tarasov. My special thanks go to Alexander Its, who introduced me to the wonderful methods of analyzing correlation functions.

A big part of my scientific life is connected with my work in the Lyon Group in collaboration with Nikolai Kitanine, Jean Michel Maillet, and Veronique Terras. A little later, Karol Kozlowski and Giuliano Niccoli joined our group. Our work on correlation functions went on for many years, which I remember with great pleasure. I sincerely thank my colleagues and friends.

I am very grateful to Samuel Belliard, Stanislav Pakuliak, and Eric Ragoucy, with whom we actively worked on the topic of the nested algebraic Bethe ansatz. These works have helped me in many ways to look at the Bethe ansatz from a new perspective. I also thank my young colleagues Arthur Hutsalyuk and Andrey Liashyk, who contributed a lot to the success of our work with their energy.

Many thanks to Jean-Sebastién Caux and Holger Frahm. Thanks to them, I realized that the Bethe ansatz is not only a theory but also real physics.

I thank my colleagues in the Department of Theoretical Physics (former Quantum Field Theory Department) of Steklov Mathematical Institute. Many thanks to the late Mikhail Polivanov, who created a wonderful scientific atmosphere in the department and always paid attention to my work. Many thanks to Irina Arefeva, Gleb Arutyunov, Dmitry Bykov, Leonid Chekhov, Sergey Frolov, Vladimir Pavlov. I especially want to thank Andrei Pogrebkov and Andrei Zotov, together with whom we organized a seminar on integrable systems.

It has always been a great pleasure for me to discuss with my colleagues Sergei Derkachev, Fabian Essler, Frank Göhmann, Michio Jimbo, Sergei Khoroshkin, Andreas Klümper, Ivan Kostov, Barry McCoy, Tetsuji Miwa, Rodrigo Pimenta, Didina Serban, Junji Suzuki, Anton Zabrodin. I express my sincere gratitude to them. I thank for the hospitality of my colleagues from the Theoretical Physics Laboratories in Lyon, Annecy, and Tours.

My special thanks to Samuel Belliard and Karol Kozlowski, who read the initial draft of this book and made many useful and very valuable comments.

Finally, I would like to express my deep gratitude to my father, Andrei Slavnov. His assessment of my work has always been the highest criterion for me.

# Contents

*Preface* v

*About the Author* ix

*Acknowledgment* xi

1. Quantum integrable systems 1

1.1 $XXX$ Heisenberg chain . . . . . . . . . . . . . . . . . . 2
1.2 Construction of integrable systems . . . . . . . . . . . . . 4
1.3 Yang–Baxter equation . . . . . . . . . . . . . . . . . . . 8
1.4 Constructing the monodromy matrix . . . . . . . . . . . . 12
1.5 Hamiltonian of the $XXX$ spin chain . . . . . . . . . . . . 15
1.6 Examples of monodromy matrices . . . . . . . . . . . . . . 17
1.6.1 Inhomogeneous $XXX$ chain . . . . . . . . . . . . 17
1.6.2 Quantum nonlinear Schrödinger equation . . . . . 18
1.6.3 Trigonometric $R$-matrix . . . . . . . . . . . . . . 21
1.7 Hilbert space . . . . . . . . . . . . . . . . . . . . . . . . 23
1.8 Commutation relations . . . . . . . . . . . . . . . . . . . 26
1.8.1 Lattice QNLSE model . . . . . . . . . . . . . . . . 28
1.8.2 Quantum determinant . . . . . . . . . . . . . . . . 29

2. Algebraic Bethe ansatz 31

2.1 Actions of the monodromy matrix entries on vectors . . . 32
2.1.1 Shorthand notation . . . . . . . . . . . . . . . . . 32
2.1.2 Action of operators $A$ and $D$ on Bethe vectors . . 33
2.1.3 Eigenvectors and Bethe equations . . . . . . . . . 37
2.1.4 Example of on-shell Bethe vector . . . . . . . . . . 40
2.1.5 Compatibility of commutation relations . . . . . . 42

2.2 Twisted monodromy matrix . . . . . . . . . . . . . . . . . . 43
2.3 Bethe equations . . . . . . . . . . . . . . . . . . . . . . . 46
2.3.1 Coinciding roots . . . . . . . . . . . . . . . . . . . 46
2.4 Spurious solutions in spin chains . . . . . . . . . . . . . . . 48
2.4.1 Spurious solution of twisted Bethe equations . . . 50
2.4.2 On the completeness of the twisted on-shell Bethe vectors . . . . . . . . . . . . . . . . . . . . . 53
2.5 Dual Bethe vectors . . . . . . . . . . . . . . . . . . . . . . 56
2.6 Generalized $L$-operator . . . . . . . . . . . . . . . . . . . . 57

3. Quantum inverse problem 61

3.1 Cyclic permutations in the monodromy matrix . . . . . . 62
3.2 Quantum inverse problem . . . . . . . . . . . . . . . . . . 65

4. Composite model 69

4.1 Summation over partitions . . . . . . . . . . . . . . . . . . 69
4.2 Definition of composite model . . . . . . . . . . . . . . . . 70
4.3 Multi-composite model . . . . . . . . . . . . . . . . . . . . 75
4.3.1 Spin chains . . . . . . . . . . . . . . . . . . . . . . 76
4.3.2 One-dimensional Bose gas . . . . . . . . . . . . . 78
4.4 Special on-shell Bethe vector . . . . . . . . . . . . . . . . . 82

5. Scalar products of off-shell Bethe vectors 85

5.1 Scalar product of off-shell Bethe vectors . . . . . . . . . . 87
5.1.1 Action of $C(v)$ on Bethe vectors . . . . . . . . . . 88
5.1.2 Generalized model and scalar products . . . . . . 91
5.1.3 Highest coefficient of the scalar products . . . . . 93
5.2 Determinant representation for the highest coefficient . . . 97
5.2.1 Properties of the determinant representation . . . 97
5.2.2 Summation identity . . . . . . . . . . . . . . . . . 99
5.3 Different representations for the scalar product . . . . . . 103

6. Scalar products with on-shell Bethe vectors 107

6.1 Identities for determinants . . . . . . . . . . . . . . . . . . 108
6.2 Determinant formula for the scalar product . . . . . . . . 113
6.3 Orthogonality of the eigenvectors . . . . . . . . . . . . . . 116
6.4 Norm of the twisted on-shell vector . . . . . . . . . . . . . 119
6.4.1 Basis of twisted on-shell Bethe vectors . . . . . . . 123
6.5 Particular cases of scalar products . . . . . . . . . . . . . 125

7. Alternative methods to compute scalar products 129

7.1 Scalar product and multiple action of transfer matrix . . . . . . . . . . . . . . . . . . . . . . . . . 129

7.1.1 Scalar product and completely indirect action . . 130

7.1.2 Completely indirect action . . . . . . . . . . . . . . 135

7.1.3 Multiple action of transfer matrices . . . . . . . . 139

7.2 System of linear equations for scalar products . . . . . . . 142

7.2.1 Cauchy matrix . . . . . . . . . . . . . . . . . . . . 144

7.2.2 Transformation of the system of linear equations . . . . . . . . . . . . . . . . . . . . . . . 147

8. Form factors of the monodromy matrix elements 153

8.1 Form factors of the off-diagonal elements . . . . . . . . . . 154

8.2 Form factors of the diagonal elements . . . . . . . . . . . 155

8.3 Universal form factor . . . . . . . . . . . . . . . . . . . . . 158

8.3.1 Determinant representations . . . . . . . . . . . . 159

8.4 Relationships between universal form factors . . . . . . . . 162

9. Form factors of local operators 165

9.1 Form factors of local operators in the spin chains . . . . . 166

9.1.1 Magnetization . . . . . . . . . . . . . . . . . . . . 168

9.2 Form factors of local operators in the one-dimensional Bose gas . . . . . . . . . . . . . . . . . . . . . . . . . . . 170

9.2.1 Field form factors . . . . . . . . . . . . . . . . . . 172

9.3 Calculation of generating function . . . . . . . . . . . . . 176

9.4 One more determinant representation . . . . . . . . . . . . 182

10. Thermodynamic limit 189

10.1 Bethe equations in the one-dimensional Bose gas . . . . . 191

10.2 Ground state in the thermodynamic limit . . . . . . . . . 195

10.2.1 Ground state density . . . . . . . . . . . . . . . . 195

10.2.2 Ground state counting function . . . . . . . . . . 198

10.3 Excited states . . . . . . . . . . . . . . . . . . . . . . . . 200

10.3.1 Counting function of excited states . . . . . . . . 200

10.3.2 Shift function . . . . . . . . . . . . . . . . . . . . 201

10.3.3 Energy and momentum of the excited states . . . 204

10.4 Chemical potential . . . . . . . . . . . . . . . . . . . . . . 208

10.5 Dressed charge and shift function . . . . . . . . . . . . . . 210

10.6 Thermodynamic limit of the $XXZ$ chain . . . . . . . . . . 213

10.6.1 Zero magnetic field . . . . . . . . . . . . . . . . . . 218
10.6.2 Dressed charge at the Fermi boundary . . . . . . . 219
10.7 Examples of calculating the thermodynamic limit . . . . . 221
10.7.1 Thermodynamic limit of sums and products . . . 221
10.7.2 Definition of Fredholm determinant . . . . . . . . 223
10.7.3 Norm of on-shell Bethe vectors in the thermodynamic limit . . . . . . . . . . . . . . . . 225
10.7.4 Form factors in the thermodynamic limit . . . . . 227

11. Multiple integral representations for correlation functions 231

11.1 Two-point correlation functions via the algebraic Bethe ansatz . . . . . . . . . . . . . . . . . . . . . . . 233
11.2 Two-point functions in the $XXZ$ chain . . . . . . . . . . . 236
11.3 Generating function . . . . . . . . . . . . . . . . . . . . 238
11.4 Multiple integral representation for generating function . . . . . . . . . . . . . . . . . . . . . . . . . . 240
11.4.1 Replacement of sums with integrals . . . . . . . . 240
11.4.2 Sum over partitions of inhomogeneities . . . . . . 241
11.4.3 Sum over partitions of Bethe parameters . . . . . 244
11.5 Generating function via form factor expansion . . . . . . . 251
11.5.1 Sums over solutions of twisted Bethe equations . . 252
11.5.2 Poles of the integrand . . . . . . . . . . . . . . . . 254
11.5.3 Calculating the integral . . . . . . . . . . . . . . . 257
11.5.4 Spurious solutions . . . . . . . . . . . . . . . . . . 260
11.6 Correlation function of free fermions . . . . . . . . . . . . 262
11.7 Emptiness formation probability . . . . . . . . . . . . . . 267
11.7.1 Emptiness formation probability at $\Delta = 1/2$ . . . 270

12. Asymptotics of correlation functions via form factor expansion 279

12.1 General scheme of the form factor approach . . . . . . . . 281
12.2 Generating function . . . . . . . . . . . . . . . . . . . . 285
12.3 Special form factor . . . . . . . . . . . . . . . . . . . . . 289
12.3.1 Smooth part . . . . . . . . . . . . . . . . . . . . . 289
12.3.2 Thermodynamic limit of the discrete part . . . . . 292
12.4 Form factor with particles and holes . . . . . . . . . . . . 300
12.4.1 Smooth part . . . . . . . . . . . . . . . . . . . . . 300
12.4.2 Discrete part . . . . . . . . . . . . . . . . . . . . . 301

12.4.3 Excitations away from the Fermi boundaries . . . 303
12.4.4 Critical states and critical form factors . . . . . . 304
12.4.5 Discrete part of critical form factor amplitude . . 308
12.5 Summation of critical form factors . . . . . . . . . . . . . 312
12.6 Long-distance asymptotics of the correlation function . . . 315
12.6.1 The case $\ell \neq 0$ . . . . . . . . . . . . . . . . . . . . 317
12.6.2 The case $\ell = 0$ . . . . . . . . . . . . . . . . . . . . 318
12.7 Correlation function of fields . . . . . . . . . . . . . . . . . 320
12.8 Large-time asymptotics . . . . . . . . . . . . . . . . . . . 322
12.8.1 Correlation function of fields . . . . . . . . . . . . 329
12.8.2 Density correlation function . . . . . . . . . . . . 331

Appendix A The $\psi$-function and the Barnes $G$-function 335

A.1 The $\psi$-function . . . . . . . . . . . . . . . . . . . . . . . . 335
A.2 The Barnes $G$-function . . . . . . . . . . . . . . . . . . . . 339

Appendix B Finite-size corrections to the excitation energy 341

B.1 Corrections to the shift function . . . . . . . . . . . . . . . 341
B.2 Excitation energy . . . . . . . . . . . . . . . . . . . . . . . 344

Appendix C Identities for Fredholm determinants 347

Appendix D Integrals with Vandermonde determinant 351

*Bibliography* 353

*Index* 375

# Chapter 1

# Quantum integrable systems

This chapter describes a general scheme for constructing quantum integrable models within the Quantum Inverse Scattering Method (QISM) framework. Much of what is described here can be found in original works [L. D. Faddeev et al (1979); L. D. Faddeev and L. A. Takhtajan (1979); E. K. Sklyanin (1982a)] or monographs devoted to this topic [L. A. Takhtajan (1985); L. D. Faddeev (1998); V. E. Korepin et al (1993); F. Franchini (2017); H.-P. Eckle (2019)].

In Section 1.1, we give an example of a model ($XXX$ Heisenberg chain [W. Heisenberg (1928)]) solvable by the algebraic Bethe ansatz. From a formal point of view, the Hamiltonian of the $XXX$ chain is a matrix of finite size. Hence, the spectrum of this matrix can be found via the standard linear algebra methods. However, using standard methods in this particular example is not realistic. The search for other approaches to solving this problem led H. Bethe to create a new method [H. Bethe (1931)], which later became one of the starting points of the QISM.

Section 1.2 introduces such important concepts of the QISM as the monodromy matrix, the $R$-matrix, the transfer matrix, and the $RTT$-relation. We show how the $RTT$-relation gives us the possibility of constructing quantum integrable systems. In Section 1.3, we show that the compatibility condition of the $RTT$-relation implies the Yang–Baxter equation for the $R$-matrix. We also find one of the solutions to this equation. Using the found solution, we construct the monodromy matrix and obtain the Hamiltonian of the $XXX$ chain in Sections 1.4 and 1.5. Section 1.6 provides other examples of the algebraic Bethe ansatz solvable models. These are just the models that will be discussed later. More detailed lists can be found in [L. D. Faddeev (1998); V. E. Korepin et al (1993); H.-P. Eckle (2019)]. In Section 1.7, we discuss the Hilbert space in which the elements of the monodromy matrix act. We also introduce the concept of a vacuum vector. The existence of this vector in the Hilbert space of the Hamiltonian is a key condition for the applicability of the algebraic Bethe ansatz. In Section 1.8, we give the commutation relations between the monodromy matrix elements in models with rational and trigonometric $R$-matrices. Here we also introduce the concept of the quantum determinant.

We have limited ourselves to the basic concepts of the QISM necessary for further reading of the book. Note, however, that the QISM is actually a quantization of the classical inverse scattering method [C. S. Gardner et al (1967)]. The details of this approach can be found in the book [L. D. Faddeev and L. A. Takhtajan (1987)]. It is also necessary to mention the theory of exactly solvable models of two-dimensional statistical physics [R. J. Baxter (1982)], in which the concepts of transfer-matrix and $R$-matrix appeared. Finally, in this book, we almost did not touch on the original method proposed by H. Bethe in 1931 and called the coordinate Bethe ansatz. An excellent description of the coordinate Bethe ansatz can be found in [M. Gaudin (1983); V. E. Korepin et al (1993); H.-P. Eckle (2019)].

## 1.1 $XXX$ Heisenberg chain

Consider a chain of $N$ spin-1/2 particles (see Fig. 1.1) and assume that the nearest neighbors in this chain interact with each other. The Hamiltonian of this quantum system can be written in the form

$$H = \sum_{k=1}^{N}\Big(J_x\sigma_k^x\sigma_{k+1}^x + J_y\sigma_k^y\sigma_{k+1}^y + J_z\sigma_k^z\sigma_{k+1}^z\Big). \tag{1.1}$$

Here $J_x$, $J_y$, and $J_z$ respectively are interaction constants along the axes $x$, $y$, and $z$, while $\sigma_k^{x,y,z}$ are the projections of the spin operators of the $k$th particle in these axes. The model is called the $XYZ$ Heisenberg chain (magnet), if all three interaction constants are different. If two of the interaction constants are equal, say, $J_x = J_y \neq J_z$, the model is called the $XXZ$

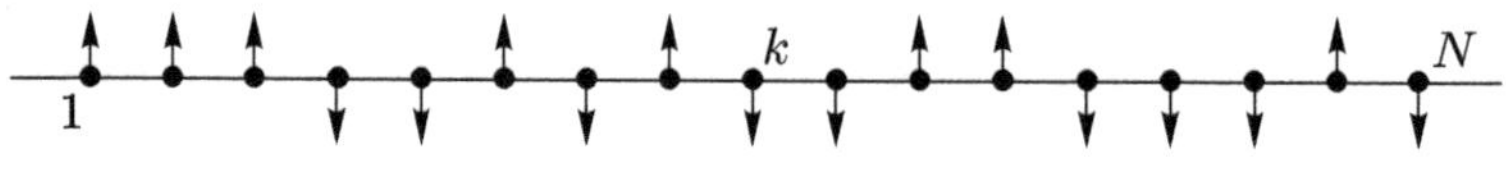


Fig. 1.1 Heisenberg spin chain.

Heisenberg chain. Finally, in the case $J_x = J_y = J_z$ the model is known as the $XXX$ Heisenberg chain. Its Hamiltonian has the form

$$H = J\sum_{k=1}^{N}\Big(\sigma_k^x\sigma_{k+1}^x + \sigma_k^y\sigma_{k+1}^y + \sigma_k^z\sigma_{k+1}^z\Big). \tag{1.2}$$

In what follows, for simplicity, we will set $J = 1$. More often, in the literature, one can find the choice $J = -1$. The common sign of the Hamiltonian affects the properties of the ground state as $N \to \infty$. We will discuss this issue in Chapter 10. For most of this book, the specific choice of the constant $J$ is irrelevant.

**Remark 1.1.** We implicitly assumed the periodic boundary conditions in (1.1), (1.2). Indeed, in the sum over $k$, we have the projections of the operator $\sigma_{N+1}$ for $k = N$. However, there is no such particle in the chain. The situation can be corrected if we set $\sigma_{N+1}^{x,y,z} = \sigma_1^{x,y,z}$. Then the chain turns into the ring. In this case, the model is still one-dimensional in the sense that the coordinate of the particle on the chain is given by one integer $k$. Generally, the requirement of periodicity is not necessary. One can also consider open chains [E. K. Sklyanin (1988); M. Gaudin (1983); F. C. Alcaraz et al (1987); H. Asakawa and M. Suzuki (1996); M. Bortz and J. Sirker (2005); P. A. de Sa and A. M. Tsvelik (1995); S. Fujimoto and S. Eggert (2004); A. Furusaki and T. Hikihara (2004); H. Frahm and A. A. Zvyagin (1997); Y.-S. Wang (2000, 2002); F. Göhmann et al (2005c); N. Kitanine et al (2007, 2008); F. H. L. Essler et al (2005)]. In this case, we simply replace the upper summation limit in the formulas (1.1), (1.2) with $N - 1$. However, due to several reasons, we prefer to deal with closed chains. First, these models are simpler from a mathematical point of view. Secondly, in closed chains, the condition of translational invariance is ensured from the very beginning. Finally, the most interesting are the long chains (formally, we send $N \to \infty$). If we are interested in the processes that occur far from the boundaries of such a long chain, they should not depend on the boundary conditions for physical reasons (see, e.g., [D. Ruelle (1969)]). Therefore, we can choose the model that deals with the simplest mathematical tools to describe these physical processes.

The components of the spin-1/2 operator are proportional to the Pauli matrices $S^{x,y,z} = \sigma^{x,y,z}/2$, where

$$\sigma^x = \begin{pmatrix}0 & 1\\ 1 & 0\end{pmatrix}, \qquad \sigma^y = \begin{pmatrix}0 & -i\\ i & 0\end{pmatrix}, \qquad \sigma^z = \begin{pmatrix}1 & 0\\ 0 & -1\end{pmatrix}. \tag{1.3}$$

From a mathematical point of view, the additional subscript of the $\sigma$-matrices in (1.2) means that these operators act in the tensor product of spaces $V_1 \otimes \cdots \otimes V_N$, where $V_k \sim \mathbb{C}^2$, $k = 1, \dots, N$. Each $\sigma_k^{x,y,z}$ acts non-trivially in $V_k$ only. In other words,

$$\sigma_k^\alpha = \underbrace{\mathbf{1} \otimes \cdots \otimes \mathbf{1}}_{k-1} \otimes \sigma^\alpha \otimes \underbrace{\mathbf{1} \otimes \cdots \otimes \mathbf{1}}_{N-k}, \qquad \alpha = x, y, z, \tag{1.4}$$

where $\mathbf{1}$ is the identity matrix in $\mathbb{C}^2$. Thus, these operators are the matrices of the size $2^N \times 2^N$. Hence, the Hamiltonian also is a $2^N \times 2^N$ matrix.

The standard problem of quantum mechanics is to find the spectrum of the Hamiltonian of a quantum system. Since, in our case, the Hamiltonian is a matrix of finite size, we are dealing with linear algebra. Therefore, we could find the eigenvalues of the matrix $H$ by solving the characteristic equation

$$\det(H - \lambda \mathbf{I}) = 0, \tag{1.5}$$

where $\mathbf{I} = \mathbf{1} \otimes \cdots \otimes \mathbf{1}$. Let us, however, assume that the chain under consideration is a model of a one-dimensional macroscopic crystal. It means that the number of chain sites $N$ is equal to the number of atoms in this macroscopic crystal, say, $N \sim 10^6$. Then the Hamiltonian is a matrix of the size of $2^{10^6} \times 2^{10^6}$, and Equation (1.5) is an algebraic equation of degree $2^{10^6}$. This number is so huge that there is nothing to compare it with. Clearly, no computer can cope with such a task. We conclude that the standard methods of linear algebra, in this case, are useless. Therefore, we need to look for some alternative ways. H. Bethe proposed such an alternative way in 1931 in work [H. Bethe (1931)]. Later on, this method was called Bethe ansatz.

In the late '70s, early '80s of the last century, in the works of the Leningrad School (see, for example, [E. K. Sklyanin (1982a); L. D. Faddeev et al (1979); L. D. Faddeev and L. A. Takhtajan (1979)]), Quantum Inverse Scattering Method (QISM) was developed. One of the tools of this method is an algebraic Bethe ansatz. In order to distinguish this new approach from the original method proposed by H. Bethe, the latter is now called the coordinate Bethe ansatz. We will study the algebraic Bethe ansatz. In Chapter 4, we will establish a link between the two approaches.

## 1.2 Construction of integrable systems

The QISM combines several approaches that have emerged in solving one-dimensional quantum systems: coordinate Bethe ansatz, classical exactly solvable equations, and low-dimensional statistical physics models. A consistent description of all these methods would take up too much space.

We choose another way. Namely, we assume that we are given some abstract algebra with bilinear commutation relations. After that, we will see what consequences follow from this and fill our abstract algebra with concrete content.

The general procedure for constructing quantum integrable systems within the framework of the QISM looks as follows. Suppose we have a $2 \times 2$ matrix $T(u)$. Its matrix elements $T_{ij}(u)$ are operators acting in some Hilbert space $\mathcal{H}$ and depending on a complex variable $u$ called a *spectral parameter*. Following the tradition, we will often denote this matrix as

$$T(u) = \begin{pmatrix} A(u) & B(u) \\ C(u) & D(u) \end{pmatrix}. \tag{1.6}$$

In the framework of the QISM, the matrix $T(u)$ is called the *monodromy matrix*. This term is borrowed from the classical inverse scattering method, where an analog of $T(u)$ consists of scattering data.

The matrix $T(u)$ acts in the tensor product of two spaces $\mathbb{C}^2 \otimes \mathcal{H}$. The space $\mathbb{C}^2$ in which $T(u)$ has the form of a $2 \times 2$ matrix (1.6) is called an *auxiliary linear space*. The space $\mathcal{H}$ in which the matrix elements $A$, $B$, $C$, and $D$ act is called a *quantum space*.

Let the commutation relations of the operators $T_{ij}(u)$ be as follows:

$$R_{12}(u,v)T_1(u)T_2(v) = T_2(v)T_1(u)R_{12}(u,v), \tag{1.7}$$

where $R_{12}(u,v)$ is a $c$-number matrix called an *$R$-matrix*. We call equation (1.7) the *$RTT$-relation*. It holds in the tensor product of three spaces $V_1 \otimes V_2 \otimes \mathcal{H}$, where $V_k \sim \mathbb{C}^2$, $k = 1, 2$. The subscripts show in which of these spaces the matrix acts non-trivially. The $R$-matrix $R_{12}(u,v)$ acts non-trivially in $V_1 \otimes V_2$ and as the identity operator in $\mathcal{H}$. The matrices $T_k$ act non-trivially in $V_k \otimes \mathcal{H}$. Thus, $T_1(u) = T(u) \otimes \mathbf{1}$ and $T_2(v) = \mathbf{1} \otimes T(v)$. Then we can write down equation (1.7) in the form

$$R(u,v)\big(T(u) \otimes \mathbf{1}\big)\big(\mathbf{1} \otimes T(v)\big) = \big(\mathbf{1} \otimes T(v)\big)\big(T(u) \otimes \mathbf{1}\big)R(u,v). \tag{1.8}$$

Sometimes the form (1.8) is more convenient.

As we said above, the $RTT$-relation sets commutation relations between elements of the monodromy matrix. To see this more explicitly, we can decompose the matrices over the standard basis matrices $E^{jk}$:

$$E^{11} = \left(\begin{smallmatrix} 1 & 0 \\ 0 & 0 \end{smallmatrix}\right), \qquad E^{12} = \left(\begin{smallmatrix} 0 & 1 \\ 0 & 0 \end{smallmatrix}\right), \qquad E^{21} = \left(\begin{smallmatrix} 0 & 0 \\ 1 & 0 \end{smallmatrix}\right), \qquad E^{22} = \left(\begin{smallmatrix} 0 & 0 \\ 0 & 1 \end{smallmatrix}\right). \tag{1.9}$$

In other words, the matrix $E^{jk}$ has a unit at the intersection of the $j$th row and the $k$th column. All other entries are zero. Then

$$T_1(u) = \sum_{i,j} T_{ij}(u)E_1^{ij}, \qquad T_2(u) = \sum_{k,l} T_{kl}(v)E_2^{kl}, \tag{1.10}$$

where the subscripts of the standard basis matrices refer to the spaces $V_1$ and $V_2$. Each summation index takes the values 1 and 2. The $R$-matrix acts in the tensor product $V_1 \otimes V_2$. Therefore, it can be decomposed as follows:

$$R(u,v) = \sum_{a,b,c,d} R_{abcd}(u,v) E_1^{ab} E_2^{cd}. \tag{1.11}$$

Then the $RTT$-relation (1.7) takes the form

$$\sum_{a,b,c,d,i,j,k,l} R_{abcd}(u,v) T_{ij}(u) T_{kl}(v) E_1^{ab} E_2^{cd} E_1^{ij} E_2^{kl} \\ = \sum_{a,b,c,d,i,j,k,l} R_{abcd}(u,v) T_{kl}(v) T_{ij}(u) E_1^{ij} E_2^{kl} E_1^{ab} E_2^{cd}. \tag{1.12}$$

Multiplying the standard basis matrices via $E_p^{ij} E_p^{kl} = E_p^{il} \delta_{jk}$, we arrive at

$$\sum_{a,c,i,j,k,l} R_{aick}(u,v) T_{ij}(u) T_{kl}(v) E_1^{aj} E_2^{cl} \\ = \sum_{b,d,i,j,k,l} R_{jb\,ld}(u,v) T_{kl}(v) T_{ij}(u) E_1^{ib} E_2^{kd}. \tag{1.13}$$

Comparing the coefficients for the same products of the $E$-matrices, we finally obtain

$$\sum_{a,b} R_{iakb}(u,v) T_{aj}(u) T_{b\,l}(v) = \sum_{a,b} R_{ajb\,l}(u,v) T_{kb}(v) T_{ia}(u). \tag{1.14}$$

In both sides of this equality, we have bilinear combinations of the monodromy matrix entries. In the lhs, operators with the argument $u$ are on the left, and in the rhs, they are on the right. Thus, we are dealing with bilinear commutation relations.

Let us turn back to Equation (1.7). Multiplying it with $R_{12}^{-1}(u,v)$ from the right

$$R_{12}(u,v) T_1(u) T_2(v) R_{12}^{-1}(u,v) = T_2(v) T_1(u), \tag{1.15}$$

and taking the trace with respect to the space $V_1 \otimes V_2$ we obtain

$$\mathrm{tr}_{12} T_1(u) T_2(v) = \mathrm{tr}_{12} T_2(v) T_1(u). \tag{1.16}$$

Here we used the possibility of cyclic permutation in the trace. Since the trace of the tensor product is equal to the product of traces, we obtain

$$\mathcal{T}(u)\mathcal{T}(v) = \mathcal{T}(v)\mathcal{T}(u), \tag{1.17}$$

where

$$\mathcal{T}(u) = \operatorname{tr} T(u) = A(u) + D(u). \tag{1.18}$$

The trace of the monodromy matrix $\mathcal{T}(u) = \operatorname{tr} T(u)$ is called the *transfer matrix*. In this case, the word 'matrix' should be understood in the sense of an 'operator' (similarly to the case of $S$-matrix).

Let us expand the operator $\mathcal{T}(u)$ into power series over $u$ centered in some point $u_0$

$$\mathcal{T}(u) = \sum_k (u - u_0)^k I_k, \tag{1.19}$$

where the coefficients $I_k$ are some operators acting in the Hilbert space $\mathcal{H}$. In this formula, we do not specify the summation area of $k$. Generally, the sum in (1.19) can be either finite or infinite. In any case, differentiating (1.17) $k$ times over $u$ and $n$ times over $v$ at $u, v = u_0$, we find that all the operators $I_k$ commute

$$[I_k, I_n] = 0, \qquad \forall k, n. \tag{1.20}$$

If we now set one of these $I_k$ to be the Hamiltonian of some quantum system, then we get a model with, generally speaking, an infinite set of integrals of motion, that is, an integrable system. This is the general scheme for constructing integrable models in the framework of the QISM.

**Remark 1.2.** We do not consider the question of the definition of quantum integrability in this book and refer the reader to works [S. Weigert (1992); J. Clémente-Gallardo and G. Marmo (2009); J.-S. Caux, J. Mossel (2011)].

Let us stop here and discuss the results obtained. It should be mentioned that the existence of an infinite set of integrals of motion does not yet guarantee that the model obtained describes any real physical integrable system. The operators $I_k$ are not necessarily independent; generically, they are not self-adjoint, are not local, and so on. All these questions are undoubtedly important, and they require a separate study. The reader can find the details in the already mentioned works [L. A. Takhtajan (1985); L. D. Faddeev (1998); V. E. Korepin et al (1993)]. We only note that it is crucial to choose the decomposition point $u_0$ appropriately to make the $I_k$ operators local. Besides, instead of expanding the operator $\mathcal{T}(u)$ into the series, one can expand a function of this operator. Below, we demonstrate this with a concrete example.

Finally, we note that other choices of the generating function for the integrals of motion are also possible. For example, in the models that we will consider below, a relation

$$[T_{ij}(u), T_{ij}(v)] = 0 \tag{1.21}$$

is fulfilled for any $i$ and $j$. Therefore, instead of taking the trace $\operatorname{tr} T(u)$, we could consider any matrix element of the matrix $T(u)$ and construct a set of commuting operators via exactly the same scheme that we used above. Sometimes, this way is used. We have chosen the trace $\operatorname{tr} T(u)$ as the generating function of the integrals of motion because we want to obtain models with periodic boundary conditions. As we shall see later, these boundary conditions automatically arise due to the cyclicity of the trace.

To conclude this section, we note that we can also consider monodromy matrices of size $n \times n$ with $n > 2$. This allows us to consider physical systems in which several degrees of freedom of fundamental particles interact, for example, the spin and the charge of the electrons. A generalization of the coordinate Bethe ansatz for such models was given in [C. N. Yang (1967); B. Sutherland (1968, 1975)]. An algebraic version of this method was proposed in [P. P. Kulish and N. Yu. Reshetikhin (1981); P. P. Kulish and N. Yu. Reshetikhin (1982, 1983)] and called a *nested algebraic Bethe ansatz*. Various approaches based on this method can be found in [S. Belliard et al (2013a,b); N. A. Slavnov (2020); A. Liashyk and N. A. Slavnov (2018); V. Tarasov and A. Varchenko (1995b, 2013); S. Belliard and E. Ragoucy (2008); L. Frappat et al (2009); S. Khoroshkin and S. Pakuliak (2008a,b); B. Enriquez et al (2007); S. Khoroshkin et al (2007); A. Os'kin et al (2009); A. Hutsalyuk et al (2017b); E. K. Sklyanin (1996); N. Gromov et al (2017); P. Ryan and D. Volin (2019); J. M. Maillet and G. Niccoli (2018)].

## 1.3 Yang–Baxter equation

A question arises whether we can find such matrices $T(u)$ and $R(u,v)$ that satisfy the $RTT$-relation (1.7) and with which we can build a meaningful physical model. We start by searching for possible matrices $R(u,v)$. At first glance, it seems that any invertible matrix can be chosen as the $R$-matrix. However, this is not the case because the commutation relations (1.7) should be consistent.

Consider a product of three monodromy matrices $T_1(u_1)T_2(u_2)T_3(u_3)$. This operator acts in the tensor product of spaces $V_1 \otimes V_2 \otimes V_3 \otimes \mathcal{H}$, where

$V_k \sim \mathbb{C}^2$, $k = 1, 2, 3$. As usual, the subscripts show in which copy of $\mathbb{C}^2$ the matrix acts non-trivially.

Using the $RTT$-relation, we can reorder these matrices to the opposite order $T_3(u_3)T_2(u_2)T_1(u_1)$. For doing this, we can use two different ways: either we move $T_1$ to the right and then we permute $T_2$ and $T_3$, or we can move $T_3$ to the left and then permute $T_1$ and $T_2$:

$$\begin{array}{ccccc} & T_2(u_2)T_1(u_1)T_3(u_3) & \to & T_2(u_2)T_3(u_3)T_1(u_1) & \\ \nearrow & & & & \searrow \\ T_1(u_1)T_2(u_2)T_3(u_3) & & & & T_3(u_3)T_2(u_2)T_1(u_1). \\ \searrow & & & & \nearrow \\ & T_1(u_1)T_3(u_3)T_2(u_2) & \to & T_3(u_3)T_1(u_1)T_2(u_2) & \end{array} \tag{1.22}$$

Each permutation of two monodromy matrices can be done by means of the $R$-matrix

$$R_{jk}(u_j, u_k)T_j(u_j)T_k(u_k)R_{jk}^{-1}(u_j, u_k) = T_k(u_k)T_j(u_j), \qquad j, k = 1, 2, 3. \tag{1.23}$$

The subscripts of the $R$-matrices show in which pair of $\mathbb{C}^2$ this matrix acts non-trivially. The first way of reordering the matrices $T_j$ leads us to

$$\begin{aligned} &R_{23}(u_2, u_3)R_{13}(u_1, u_3)R_{12}(u_1, u_2)T_1(u_1)T_2(u_2)T_3(u_3) \\ &\qquad \times R_{12}^{-1}(u_1, u_2)R_{13}^{-1}(u_1, u_3)R_{23}^{-1}(u_2, u_3) = T_3(u_3)T_2(u_2)T_1(u_1). \end{aligned} \tag{1.24}$$

The second way gives

$$\begin{aligned} &R_{12}(u_1, u_2)R_{13}(u_1, u_3)R_{23}(u_2, u_3)T_1(u_1)T_2(u_2)T_3(u_3) \\ &\qquad \times R_{23}^{-1}(u_2, u_3)R_{13}^{-1}(u_1, u_3)R_{12}^{-1}(u_1, u_2) = T_3(u_3)T_2(u_2)T_1(u_1). \end{aligned} \tag{1.25}$$

If we require that

$$R_{23}(u_2, u_3)R_{13}(u_1, u_3)R_{12}(u_1, u_2) = R_{12}(u_1, u_2)R_{13}(u_1, u_3)R_{23}(u_2, u_3)\,, \tag{1.26}$$

then the results of both reordering methods coincide.

Equation (1.26) is a matrix functional equation. It is called the *Yang–Baxter equation* [R. J. Baxter (1978, 1982); J. B. McGuire (1964); C. N. Yang (1967)] and plays an exceptionally important role in the theory of exactly solvable models of statistical physics [M. Jimbo (1990); J. H. H. Perk and H. Au-Yang (2006); R. J. Baxter (1972a,b, 1973); V. V. Bazhanov and Yu. G. Stroganov (1981); B. Berg et al (1978); M. Jimbo et al (1988); J. H. H. Perk and C. L. Schultz (1981);

N. Yu. Reshetikhin (1987); V. I. Vichirko and N. Yu. Reshetikhin (1984)] and factorized $S$-matrices [V. A. Fateev and A. B. Zamolodchikov (1982); A. B. Zamolodchikov (1977); A. B. Zamolodchikov and Al. B. Zamolodchikov (1977, 1978, 1979); C. N. Yang and M. L. Ge (1989)].

It is worth mentioning that the Yang–Baxter equation is not necessary, but only a sufficient condition of the compatibility of the commutation relations between the monodromy matrix entries. However, an example of an $R$-matrix that, on the one hand, provides compatibility but, on the other hand, does not satisfy the Yang–Baxter equation is not known.

Let us try to find at least one non-trivial solution to the Yang–Baxter equation. First, we observe that if $R(u_1, u_2)$ is a solution to (1.26), then $\varphi(u_1, u_2)R_{12}(u_1, u_2)$ also is a solution, where $\varphi(u_1, u_2)$ is an arbitrary scalar function of two variables. Therefore, we can look for the solutions of equation (1.26) up to multiplication by a scalar function.

One obvious solution to (1.26) is the identity matrix. The second slightly less obvious solution is the permutation matrix $P$. Let us recall the basic properties of the permutation matrix.

Let us have two copies of $\mathbb{C}^2$, say, $V_n$ and $V_m$. The permutation matrix $P_{nm}$ acts in the tensor product $V_n \otimes V_m$. It has the following representation in terms of the standard basis matrices:

$$P_{nm} = \sum_{i,j} E_n^{ij} E_m^{ji}. \tag{1.27}$$

The main property of the permutation matrix is that it exchanges the spaces $V_n$ and $V_m$. Namely, let $X_{nm}$ be a matrix acting in $V_n \otimes V_m$. Then

$$P_{nm} X_{nm} P_{nm} = X_{mn}, \tag{1.28}$$

where

$$X_{nm} = \sum_{i,j,k,l} X_{ijkl} E_n^{ij} E_m^{kl}, \qquad X_{mn} = \sum_{i,j,k,l} X_{ijkl} E_m^{ij} E_n^{kl}. \tag{1.29}$$

It follows from (1.28) that $P_{nm} = P_{mn}$ and $P_{nm}^2 = I$, where $I = \mathbf{1} \otimes \mathbf{1}$.

Using these properties we can easily prove that the permutation matrix satisfies the Yang–Baxter equation. Indeed, multiplying the product $P_{12}P_{13}P_{23}$ from the right by $P_{12}^2$ we obtain

$$P_{12}P_{13}P_{23} = P_{12}P_{13}P_{23}P_{12}P_{12} = P_{23}P_{13}P_{12}. \tag{1.30}$$

Here we took advantage of the fact that $P_{13}P_{23}$ is between two permutation matrices $P_{12}$. Then by virtue of (1.28), we just have to exchange the subscripts 1 and 2: $P_{13}P_{23} \to P_{23}P_{13}$.

Let us try to find a solution that depends on the difference $u_1 - u_2$ and is a linear combination of the identity and permutation matrices. Let $u_1 - u_2 = u$ and $u_2 - u_3 = v$. Then $u_1 - u_3 = u + v$, and the Yang–Baxter equation takes the form

$$R_{12}(u)R_{13}(u+v)R_{23}(v) = R_{23}(v)R_{13}(u+v)R_{12}(u). \tag{1.31}$$

We are looking for the solution to this equation in the form $R(u) = \phi(u)I + \eta P$, where $\phi(u)$ is a smooth function of $u$ and $\eta$ is a constant. The matrix equation (1.31) yields a functional equation for the function $\phi(u)$. Let us derive this equation. To do this, we substitute $R(u) = \phi(u)I + \eta P$ into (1.31) and compare the coefficients for the same powers of $\eta$. For $\eta^0$ and $\eta^3$, we obtain obvious identities. For $\eta^1$, an identity also arises:

$$\begin{aligned}&\phi(v)\phi(u+v)P_{12} + \phi(v)\phi(u)P_{13} + \phi(u)\phi(u+v)P_{23}\\ &\quad = \phi(u)\phi(u+v)P_{23} + \phi(v)\phi(u)P_{13} + \phi(v)\phi(u+v)P_{12}.\end{aligned} \tag{1.32}$$

A non-trivial equation arises for $\eta^2$ only:

$$\begin{aligned}&\phi(u)P_{13}P_{23} + \phi(v)P_{12}P_{13} + \phi(u+v)P_{12}P_{23}\\ &\quad = \phi(u)P_{23}P_{13} + \phi(v)P_{13}P_{12} + \phi(u+v)P_{23}P_{12}.\end{aligned} \tag{1.33}$$

Multiplying (1.33) from the left by $P_{12}$ we obtain

$$\begin{aligned}&\phi(u)P_{12}P_{13}P_{23} + \phi(v)P_{13} + \phi(u+v)P_{23}\\ &\quad = \phi(u)P_{12}P_{23}P_{13} + \phi(v)P_{23} + \phi(u+v)P_{13}.\end{aligned} \tag{1.34}$$

Now we multiply this equation from the right by $P_{13}$:

$$\phi(u)+\phi(v)+\phi(u+v)P_{23}P_{13} = \phi(u)P_{12}P_{23}+\phi(v)P_{23}P_{13}+\phi(u+v), \tag{1.35}$$

where we have omitted the identity matrix $\mathbf{1}\otimes\mathbf{1}\otimes\mathbf{1}$ for brevity. Finally, multiplying (1.35) from the right by $P_{23}$, we eventually obtain

$$\big(\phi(u) + \phi(v) - \phi(u+v)\big)P_{23} = \big(\phi(u) + \phi(v) - \phi(u+v)\big)P_{12}. \tag{1.36}$$

Obviously, this equality is possible if and only if

$$\phi(u+v) = \phi(u) + \phi(v), \tag{1.37}$$

for any $u$ and $v$. It is well known that the linear function $\phi(u) = au$ (where $a$ is an arbitrary constant) is the only smooth solution of the functional Equation (1.37). The constant $a$ can be included in the constant $\eta$. Thus, we find that the matrix

$$R(u_1, u_2) = (u_1 - u_2)I + \eta P, \tag{1.38}$$

solves the Yang–Baxter equation (1.26). This solution depends on the difference of the arguments: $R(u,v) = R(u-v)$. It can be written explicitly as

$$R(u-v) = \begin{pmatrix} u-v+\eta & 0 & 0 & 0 \\ 0 & u-v & \eta & 0 \\ 0 & \eta & u-v & 0 \\ 0 & 0 & 0 & u-v+\eta \end{pmatrix}. \tag{1.39}$$

We also note that in the derivation of this solution, we did not use the fact that the spaces $V_k$ were isomorphic to $\mathbb{C}^2$. Therefore, the solution obtained is also valid for $R$-matrices acting in the space $\mathbb{C}^N \otimes \mathbb{C}^N$ for arbitrary $N$.

Of course, the obtained solution of the Yang–Baxter equation is not the only one. The $R$-matrix (1.38) is a degeneration of the trigonometric $R$-matrix, described in Section 1.6.3. In turn, the trigonometric $R$-matrix is a degeneration of the elliptic $R$-matrix [R. J. Baxter (1972a, 1982); A. A. Belavin (1981b); E. K. Sklyanin (1982b)]. An extensive literature is devoted to the search for solutions to the Yang-Baxter equation [A. A. Belavin (1981a,b); V. V. Bazhanov (1987); I. V. Cherednik (1986); E. Date et al (1988); A. G. Izergin and V. E. Korepin (1981b); M. Jimbo (1986); A. N. Kirillov and N. Yu. Reshetikhin (1987); I. M. Krichever (1981); P. P. Kulish et al (1981); P. P. Kulish and E. K. Sklyanin (1982b); N. Yu. Reshetikhin and P. Wiegmann (1987); V. N. Tolstoy and S. M. Khoroshkin (1992); A. B. Zamolodchikov and Al. B. Zamolodchikov (1977, 1979)]. Each such solution corresponds to some integrable model.

## 1.4 Constructing the monodromy matrix

Now we construct a matrix $T(u)$ satisfying relation (1.7). It is clear that the identity matrix satisfies this equation, but this example is not of much interest. In fact, we already have one non-trivial example: this is the $R$-matrix. Indeed, due to the Yang–Baxter equation

$$R_{12}(u-v)R_{13}(u-w)R_{23}(v-w) = R_{23}(v-w)R_{13}(u-w)R_{12}(u-v), \tag{1.40}$$

we can say that $T(u) = R(u-w)$. We should only agree upon how we understand this equation. We know that $R(u-w)$ is a $4\times 4$ matrix. However, one can understand it as a $2\times 2$ matrix whose entries are operators acting in the Hilbert space $\mathcal{H} = \mathbb{C}^2$. In other words, these operators are also

$2 \times 2$ matrices. Let us set, for instance, $w = \eta/2$. Then we have

$$R(u - \tfrac{\eta}{2}) = \begin{pmatrix} u + \frac{\eta}{2} & 0 & 0 & 0 \\ 0 & u - \frac{\eta}{2} & \eta & 0 \\ 0 & \eta & u - \frac{\eta}{2} & 0 \\ 0 & 0 & 0 & u + \frac{\eta}{2} \end{pmatrix} = \begin{pmatrix} u + \frac{\eta}{2}\sigma^z & \eta\sigma^- \\ \eta\sigma^+ & u - \frac{\eta}{2}\sigma^z \end{pmatrix}, \tag{1.41}$$

where $\sigma^{\pm} = \frac{1}{2}(\sigma^x \pm i\sigma^y)$. We would like to note that the choice of $w = \eta/2$ was made solely for the sake of nice writing the $R$-matrix in terms of $\sigma$-matrices. We can fix $w$ in any other way. We will do this in Section 1.6.1.

Thus, setting $T(u) = R(u - \eta/2)$, that is,

$$T(u) = \begin{pmatrix} u + \frac{\eta}{2}\sigma^z & \eta\sigma^- \\ \eta\sigma^+ & u - \frac{\eta}{2}\sigma^z \end{pmatrix}, \tag{1.42}$$

we obtain a monodromy matrix satisfying the $RTT$-relation due to (1.40).

Although we have managed to find a non-trivial example of the matrix $T(u)$, the quantum model constructed on this monodromy matrix base is still rather poor because

$$\mathcal{T}(u) = (u + \tfrac{\eta}{2}\sigma^z) + (u - \tfrac{\eta}{2}\sigma^z) = 2u\mathbf{1}. \tag{1.43}$$

Nevertheless, this example is not useless because it gives a hint to construct a reasonable model. But first, we will establish one important property of the monodromy matrix.

Consider two $2 \times 2$ matrices $T^{(1)}(u)$ and $T^{(2)}(u)$ depending on the complex argument $u$. Let the entries $T^{(\ell)}_{jk}(u)$ be operators acting in a space $\mathcal{H}_\ell$, $\ell = 1, 2$. It is crucial that the spaces $\mathcal{H}_1$ and $\mathcal{H}_2$ are different. Then $[T^{(1)}_{ij}(u_1), T^{(2)}_{kl}(u_2)] = 0$ for all $u_1$, $u_2$ and all $i, j, k, l$.

Consider the spaces $V_1 \otimes V_2 \otimes \mathcal{H}_1$ and $V_1 \otimes V_2 \otimes \mathcal{H}_2$, where $V_k \sim \mathbb{C}^2$, $k = 1, 2$. As usual we can introduce the matrices $T^{(1)}_k$ and $T^{(2)}_k$ for $k = 1, 2$. They act non-trivially only in $V_k \otimes \mathcal{H}_1$ and $V_k \otimes \mathcal{H}_2$.

**Proposition 1.1.** *Let both matrices $T^{(1)}$ and $T^{(2)}$ enjoy equation* (1.7)

$$R_{12}(u_1, u_2)T^{(\ell)}_1(u_1)T^{(\ell)}_2(u_2) = T^{(\ell)}_2(u_2)T^{(\ell)}_1(u_1)R_{12}(u_1, u_2), \tag{1.44}$$

*for $\ell = 1, 2$. Then their product $T(u) = T^{(1)}(u)T^{(2)}(u)$ also enjoys equation* (1.7).

**Proof.** Let us substitute the product $T^{(1)}(u)T^{(2)}(u)$ in the lhs of (1.7)

$$\begin{aligned} &R_{12}(u_1, u_2)T^{(1)}_1(u_1)T^{(2)}_1(u_1)T^{(1)}_2(u_2)T^{(2)}_2(u_2) \\ &\quad = R_{12}(u_1, u_2)T^{(1)}_1(u_1)T^{(1)}_2(u_2)T^{(2)}_1(u_1)T^{(2)}_2(u_2). \end{aligned} \tag{1.45}$$

Here we have used that $T_1^{(2)}(u_1)$ acts in the space $V_1 \otimes \mathcal{H}_2$, while $T_2^{(1)}(u_2)$ acts in the space $V_2 \otimes \mathcal{H}_1$. Hence, these matrices commute. Now, due to (1.44), we can move the matrix $R_{12}(u_1, u_2)$ to the right

$$\begin{aligned} &R_{12}(u_1,u_2)T_1^{(1)}(u_1)T_2^{(1)}(u_2)T_1^{(2)}(u_1)T_2^{(2)}(u_2) \\ &\quad = T_2^{(1)}(u_2)T_1^{(1)}(u_1)T_2^{(2)}(u_2)T_1^{(2)}(u_1)R_{12}(u_1,u_2). \end{aligned} \tag{1.46}$$

It remains to swap $T_1^{(1)}(u_1)$ and $T_2^{(2)}(u_2)$, and we arrive at

$$\begin{aligned} &R_{12}(u_1,u_2)T_1^{(1)}(u_1)T_1^{(2)}(u_1)T_2^{(1)}(u_2)T_2^{(2)}(u_2) \\ &\quad = T_2^{(1)}(u_2)T_2^{(2)}(u_2)T_1^{(1)}(u_1)T_1^{(2)}(u_1)R_{12}(u_1,u_2). \end{aligned} \tag{1.47}$$

□

Note that proposition 1.1 holds for any $R$-matrix, not just for the $R$-matrix (1.38).

Consider now a space $\mathcal{H}$, that is the tensor product of $N$ spaces $\mathbb{C}^2$: $\mathcal{H} = V_1 \otimes \cdots \otimes V_N$, where every $V_n \sim \mathbb{C}^2$. Let us build matrices

$$L_n(u) = \begin{pmatrix} u + \frac{\eta}{2}\sigma_n^z & \eta\sigma_n^- \\ \eta\sigma_n^+ & u - \frac{\eta}{2}\sigma_n^z \end{pmatrix}, \tag{1.48}$$

where $\sigma_n^\alpha$ is defined by (1.4). We call the matrices $L_n(u)$ *$L$-operators*. Each matrix element of every $L$-operator is, in turn, an operator acting in the space $\mathcal{H}$. Observe that it acts non-trivially only in $V_n$. Hence, every $L_n(u)$ satisfies the $RTT$-relation (1.8)

$$R(u-v)\big(L_n(u)\otimes\mathbf{1}\big)\big(\mathbf{1}\otimes L_n(v)\big) = \big(\mathbf{1}\otimes L_n(v)\big)\big(L_n(u)\otimes\mathbf{1}\big)R(u-v). \tag{1.49}$$

Now we define the monodromy matrix as follows:

$$T(u) = L_N(u)\dots L_2(u)L_1(u). \tag{1.50}$$

Due to proposition 1.1 this matrix enjoys equation (1.7).

Consider an example when $T(u)$ is the product of two $L$-operators: $T(u) = L_2(u)L_1(u)$. It is easy to see that in this case,

$$\mathcal{T}(u) = (u+\tfrac{\eta}{2}\sigma_2^z)(u+\tfrac{\eta}{2}\sigma_1^z)+(u-\tfrac{\eta}{2}\sigma_2^z)(u-\tfrac{\eta}{2}\sigma_1^z)+\eta^2\sigma_2^-\sigma_1^+ +\eta^2\sigma_2^+\sigma_1^-. \tag{1.51}$$

Setting here $u = 0$ we obtain

$$\tfrac{2}{\eta^2}\mathcal{T}(0) = \sigma_2^z\sigma_1^z + 2\sigma_2^-\sigma_1^+ + 2\sigma_2^+\sigma_1^-, \tag{1.52}$$

or, using equations $\sigma_k^\pm = \frac{1}{2}(\sigma_k^x \pm i\sigma_k^y)$,

$$\tfrac{2}{\eta^2}\mathcal{T}(0) = \sigma_2^x\sigma_1^x + \sigma_2^y\sigma_1^y + \sigma_2^z\sigma_1^z. \tag{1.53}$$

We see that we have obtained the Hamiltonian of the $XXX$ spin chain consisting of two sites.

## 1.5 Hamiltonian of the $XXX$ spin chain

We now show how one can use the monodromy matrix (1.50) for obtaining the Hamiltonian of the $XXX$ chain in the general case. It is easy to check that the Hamiltonian (1.2) can be written in the form

$$H = 2\sum_{k=1}^{N} P_{k\,k+1} - N, \tag{1.54}$$

where we omitted the identity operator at $N$ for brevity. Indeed, using

$$\begin{aligned} \sigma_k^x &= E_k^{12} + E_k^{21}, & \sigma_k^y &= -i(E_k^{12} - E_k^{21}), \\ \sigma_k^z &= E_k^{11} - E_k^{22}, & \mathbf{1} &= E_k^{11} + E_k^{22}, \end{aligned} \tag{1.55}$$

we obtain

$$\sigma_k^x\sigma_{k+1}^x + \sigma_k^y\sigma_{k+1}^y + \sigma_k^z\sigma_{k+1}^z = 2\sum_{a,b=1}^{2} E_k^{ab}E_{k+1}^{ba} - \mathbf{1}\otimes\mathbf{1} = 2P_{k\,k+1} - \mathbf{1}\otimes\mathbf{1}, \tag{1.56}$$

which implies (1.54).

We prove that the Hamiltonian of the $XXX$ chain (1.2) can be obtained from the transfer matrix of the monodromy matrix (1.50) by the following formula:

$$H = 2\eta\frac{\mathrm{d}\mathcal{T}(u)}{\mathrm{d}u}\,\mathcal{T}^{-1}(u)\Big|_{u=\frac{\eta}{2}} - N. \tag{1.57}$$

First, let us agree upon notation. The entries of the monodromy matrix (1.50) act in the space $\mathcal{H} = V_1\otimes\cdots\otimes V_N$, where any $V_k \sim \mathbb{C}^2$. The space $V_k$ is called a *local quantum space* associated with the $k$th site. However, the monodromy matrix itself is the $2\times 2$ matrix and acts in the auxiliary linear space $\mathbb{C}^2$. We denote this space by $V_0$. Thus, the monodromy matrix $T(u)$ is an operator acting in the tensor product of $N+1$ spaces $V_0\otimes V_1\otimes\cdots\otimes V_N$. Each $L$-operator (1.48) acts non-trivially in the space $V_0\otimes V_n$ and coincides with the $R$-matrix: $L_n(u) = R_{0n}(u-\frac{\eta}{2})$. Then, due to (1.50), we have

$$T(u) = L_N(u)\cdots L_1(u) = R_{0\,N}(u-\tfrac{\eta}{2})\cdots R_{0\,1}(u-\tfrac{\eta}{2}). \tag{1.58}$$

Thus,

$$\mathcal{T}(u) = \mathrm{tr}_0\big(R_{0\,N}(u-\tfrac{\eta}{2})\cdots R_{0\,1}(u-\tfrac{\eta}{2})\big), \tag{1.59}$$

where the trace is taken over the auxiliary space. Since we can make cyclic permutations within the trace, we see that the transfer matrix (1.59) automatically provides periodic boundary conditions.

Let us calculate the derivative of $\mathcal{T}(u)$ at $u = \frac{\eta}{2}$. We also take into account that $R_{0\,k}(0)$ coincides up to the factor $\eta$ with the permutation matrix $P_{0\,k}$. Then

$$\begin{aligned}\left.\frac{\mathrm{d}\mathcal{T}(u)}{\mathrm{d}u}\right|_{u=\frac{\eta}{2}} &= \frac{\mathrm{d}}{\mathrm{d}u}\mathrm{tr}_0\big(R_{0\,N}(u-\tfrac{\eta}{2})\cdots R_{0\,1}(u-\tfrac{\eta}{2})\big)\Big|_{u=\frac{\eta}{2}}\\ &= \eta^{N-1}\sum_{k=1}^{N}\mathrm{tr}_0\big(P_{0\,N}\cdots R'_{0\,k}(0)\cdots P_{0\,1}\big)\\ &= \eta^{N-1}\sum_{k=1}^{N}\mathrm{tr}_0\big(R'_{0\,k}(0)P_{0\,k-1}\cdots P_{0\,k+1}\big).\end{aligned} \tag{1.60}$$

Here we have used the cyclicity of the trace. Multiplying the obtained product from the right with the identity matrix $P^2_{0\,k+1}$ and using cyclic permutation again, we find

$$\begin{aligned}\left.\frac{\mathrm{d}\mathcal{T}(u)}{\mathrm{d}u}\right|_{u=\frac{\eta}{2}} &= \eta^{N-1}\sum_{k=1}^{N}\mathrm{tr}_0\big(R'_{0\,k}(0)P_{0\,k-1}\cdots P_{0\,k+1}P^2_{0\,k+1}\big)\\ &= \eta^{N-1}\sum_{k=1}^{N}\mathrm{tr}_0\big(P_{0\,k+1}\big[R'_{0\,k}(0)P_{0\,k-1}\cdots P_{0\,k+1}\big]P_{0\,k+1}\big).\end{aligned} \tag{1.61}$$

According to the general rule, we should replace the space $V_0$ by the space $V_{k+1}$ in the matrix product between two permutation matrices $P_{0\,k+1}$, that is

$$\left.\frac{\mathrm{d}\mathcal{T}(u)}{\mathrm{d}u}\right|_{u=\frac{\eta}{2}} = \eta^{N-1}\sum_{k=1}^{N}\mathrm{tr}_0\big(R'_{k+1\,k}(0)P_{k+1\,k-1}\cdots P_{k+1\,k+2}P_{k+1\,0}\big). \tag{1.62}$$

Now, there is only one matrix $P_{k+1\,0}$ acting in the space $V_0$. It is easy to check that the trace of the permutation matrix with respect to one of the two spaces in which it acts gives the identity matrix in the other space. Hence, we finally obtain

$$\begin{aligned}\left.\frac{\mathrm{d}\mathcal{T}(u)}{\mathrm{d}u}\right|_{u=\frac{\eta}{2}} &= \eta^{N-1}\sum_{k=1}^{N}R'_{k+1\,k}(0)P_{k+1\,k-1}\cdots P_{k+1\,k+2}\\ &= \eta^{N-1}\sum_{k=1}^{N}P_{k+1\,k-1}\cdots P_{k+1\,k+2}\,,\end{aligned} \tag{1.63}$$

because $R'_{k+1\,k}(0) = I$.

Let us calculate now the transfer matrix at $u = \frac{\eta}{2}$. We have

$$\mathcal{T}(\tfrac{\eta}{2}) = \eta^N\mathrm{tr}_0\big(P_{0\,N}\cdots P_{0\,1}\big). \tag{1.64}$$

To compute the trace, we fix an arbitrary $k$ and repeat all the calculations that we did to derive the transfer matrix derivative. Then we obtain

$$\mathcal{T}(\tfrac{\eta}{2}) = \eta^N P_{k+1\,k} P_{k+1\,k-1} \cdots P_{k+1\,k+2}. \tag{1.65}$$

Then, using $P_{ab}^2 = I$, we find

$$\mathcal{T}^{-1}(\tfrac{\eta}{2}) = \eta^{-N} P_{k+1\,k+2} \cdots P_{k+1\,k-1} P_{k+1\,k}. \tag{1.66}$$

It is important to stress that the subscript $k$ in (1.65) and (1.66) is arbitrary from the set $\{1, \dots, N\}$. Therefore, when we multiply the matrix $\mathcal{T}'(\frac{\eta}{2})$ from the right by $\mathcal{T}^{-1}(\frac{\eta}{2})$, we can always choose for each term in the sum over $k$ in (1.63) the same subscript $k$ in (1.66). Hence,

$$2\eta \frac{\mathrm{d}\mathcal{T}(u)}{\mathrm{d}u} \mathcal{T}^{-1}(u)\Big|_{u=\frac{\eta}{2}} = 2\sum_{k=1}^{N} P_{k+1\,k}, \tag{1.67}$$

and we do reproduce (1.54) up to the term proportional to the identity matrix.

Observe that in the example considered, we expanded the logarithm of $\mathcal{T}(u)$ into a series, but not the transfer matrix itself. It was also crucial that the expansion was carried out at the point where the $L$-operator turns into the permutation matrix. It is these two conditions that have provided the locality of the resulting Hamiltonian. The latter turned out to be a sum of operators describing the interaction between the nearest neighbors only. Other methods for obtaining local Hamiltonians from the transfer matrix can be found in [A. G. Izergin and V. E. Korepin (1981a,c); T. Oota (2004)].

It should also be noted that although the $R$-matrix depends on the parameter $\eta$, the Hamiltonian no longer depends on $\eta$. Indeed, in the $XXX$ model, this parameter does not play any role. Therefore, it is often chosen for convenience reasons, such as $\eta = 1$ or $\eta = i$. However, we will keep this parameter for the sake of generality.

## 1.6 Examples of monodromy matrices

### 1.6.1 *Inhomogeneous XXX chain*

In Section 1.4, we constructed the $L$-operator of the $XXX$ chain by fixing one of the arguments of the $R$-matrix equal to $\eta/2$. However, as we have already noted, we could choose this argument in any other way. Moreover, we could make different choices at different sites.

Let us return once again to the quantum space $\mathcal{H} = V_1 \otimes \cdots \otimes V_N$, where any $V_k \sim \mathbb{C}^2$. Let, as before, $V_0$ denote the auxiliary linear space, and let

$R_{0k}(u-v)$ act in $V_0 \otimes V_k$. We now define the $L$-operator corresponding to the $k$th site ($k = 1, \ldots, N$) as $L_k(u) = R_{0k}(u - \xi_k)$. Here $\xi_k$ are complex numbers such that generically $\xi_k \neq \xi_n$ for $k \neq n$. We can write down this $L$-operator as a $2 \times 2$ matrix as follows:

$$L_k(u) = \begin{pmatrix} u - \xi_k + \eta E_k^{11} & \eta E_k^{21} \\ \eta E_k^{12} & u - \xi_k + \eta E_k^{22} \end{pmatrix}, \tag{1.68}$$

where the standard basis matrices $E_k^{ab}$ act in the space $V_k$. Obviously, these $L$-operators satisfy the $RTT$-relation due to the Yang–Baxter equation for all $k = 1, \ldots, N$. Then a matrix

$$T(u) = L_N(u) \ldots L_1(u) \tag{1.69}$$

should satisfy the $RTT$-relation due to Proposition 1.1.

It is worth mentioning that the parameter $u$ is shifted by different $\xi_n$ at different chain sites. The parameters $\xi_n$ are called *inhomogeneities*, and the model constructed via the monodromy matrix (1.69) is called the *inhomogeneous XXX chain*. Following the standard procedure, one can construct a set of commuting operators by expanding the transfer matrix of this model in a series. However, we will not be able to obtain integrals of motion with the interaction of several nearest neighbors. Indeed, as we have already noted, the locality of interaction in the homogeneous chain was achieved because *all* $L$-operators turned into a permutation matrix at the point $u = \eta/2$. In the case of the inhomogeneous chain, by setting $u = \xi_n$, we can guarantee that *only one* $L$-operator $L_n$ becomes a permutation matrix. This ultimately leads to the loss of locality.

Nevertheless, this model often appears to be very useful in intermediate computations when working with the usual (homogeneous) $XXX$ chain. The latter can be treated as a particular case of the inhomogeneous chain where $\xi_j = \xi_k$ for $j, k = 1, \ldots, N$. Note also that the inhomogeneous $XXX$ chain is closely related to the Gaudin model [M. Gaudin (1976, 1983)]. The latter occurs in the limit $\eta \to 0$ [P. P. Kulish and E. K. Sklyanin (1982b); L. D. Faddeev (1984); E. K. Sklyanin (1989)].

### 1.6.2 *Quantum nonlinear Schrödinger equation*

Let us give one more example of the $L$-operator that satisfies the $RTT$-relation but generates a completely different physical model. Consider a set of operators $\psi_n$ and $\psi_n^\dagger$ ($n = 1, \ldots, N$) obeying the following commutation relations:

$$[\psi_n, \psi_m^\dagger] = \frac{\delta_{nm}}{\Delta}, \tag{1.70}$$

where $\Delta$ is some number. Generically, this number can be complex. However, for physical applications, it is real and positive. One can consider the operators $\psi_n$ and $\psi_n^\dagger$ as a discretization of Bose fields.[1] Indeed, let us consider Bose fields $\Psi(x)$ and $\Psi^\dagger(x)$ depending on a one-dimensional real variable $x$ and obeying canonical commutation relations

$$[\Psi(x), \Psi^\dagger(y)] = \delta(x-y). \tag{1.71}$$

Consider also an interval $[0, L]$ and choose $N$ points $x_1, \dots, x_N$ in this interval, such that $x_n = \Delta n$, and $x_N = L$. Then we can set

$$\psi_n = \frac{1}{\Delta}\int_{x_{n-1}+0}^{x_n-0} \Psi(x)\,\mathrm{d}x, \qquad \psi_n^\dagger = \frac{1}{\Delta}\int_{x_{n-1}+0}^{x_n-0} \Psi^\dagger(x)\,\mathrm{d}x. \tag{1.72}$$

It is easy to check that the operators $\psi_n$ and $\psi_n^\dagger$ constructed in this way do satisfy relations (1.70).

We allow the operators considered above to act in a Fock space with vacuum $|0\rangle$ in such a way that

$$\Psi(x)|0\rangle = 0, \qquad \psi_n|0\rangle = 0. \tag{1.73}$$

Consider an $L$-operator of the form

$$L_n(u) = \begin{pmatrix} 1 - \frac{iu\Delta}{2} + \frac{c\Delta^2}{2}\psi_n^\dagger\psi_n & -i\Delta\psi_n^\dagger\rho_n^+ \\ i\Delta\rho_n^-\psi_n & 1 + \frac{iu\Delta}{2} + \frac{c\Delta^2}{2}\psi_n^\dagger\psi_n \end{pmatrix}, \tag{1.74}$$

where $c$ is some parameter. Operators $\rho_n^\pm$ in (1.74) satisfy two constraints. First, they depend on the combination $\psi_n^\dagger\psi_n$ only, that is $\rho_n^\pm = \rho_n^\pm(\psi_n^\dagger\psi_n)$. Second,

$$\rho_n^+\rho_n^- = c + \frac{c^2\Delta^2}{4}\psi_n^\dagger\psi_n. \tag{1.75}$$

There are no other restrictions for the operators $\rho_n^\pm$. For instance, we can set one of them equal to the identity operator. Then the second operator is equal to the rhs of (1.75). We can also set them equal to the square root of the rhs of (1.75) and so on. Independently of the choice of $\rho_n^\pm$, the $L$-operator (1.74) satisfies the $RTT$-relation, if $c = i\eta$ where $\eta$ is the constant in the $R$-matrix (1.38). We will discuss this issue in more detail in Section 1.8.

Using the $L$-operator (1.74), one can build the monodromy matrix (1.50). Then, with an appropriate choice of the decomposition point (see, e.g., [A. G. Izergin and V. E. Korepin (1981a)]), the transfer matrix gives

[1] Again, following the tradition, we use the notation $\psi$, although usually it is reserved for fermions.

us a set of commuting operators.The corresponding model is known as a *lattice quantum nonlinear Schrödinger equation* (lattice QNLSE model). In the continuous limit, we obtain the following Hamiltonian:

$$H = \int_0^L \left(\partial_x \Psi^\dagger(x)\partial_x\Psi(x) + c\Psi^\dagger(x)\Psi^\dagger(x)\Psi(x)\Psi(x)\right) \mathrm{d}x. \tag{1.76}$$

Here $c$ is a coupling constant. Canonical Bose fields $\Psi(x)$ and $\Psi^\dagger(x)$ are periodic on the interval $[0, L]$ and obey commutation relations (1.71). The time evolution of $\Psi(x)$ and $\Psi^\dagger(x)$ is defined in a standard way:

$$\Psi(x,t) = e^{iHt}\Psi(x)e^{-iHt}, \qquad \Psi^\dagger(x,t) = e^{iHt}\Psi^\dagger(x)e^{-iHt}. \tag{1.77}$$

Then the Hamiltonian equation of motion is

$$i\frac{\partial\Psi}{\partial t} = [H,\Psi] = -\frac{\partial^2\Psi}{\partial x^2} + 2c\Psi^\dagger\Psi\Psi. \tag{1.78}$$

The form of this equation coincides with the well-known nonlinear Schrödinger equation. This is where the name of the model comes from.

Unfortunately, the method to obtain the Hamiltonian (1.76) from the transfer matrix is much more complicated than the derivation of the $XXX$ chain Hamiltonian. In particular, calculating the continuous limit, although being straightforward, is too cumbersome. Therefore, we do not give this derivation here. However, the reader can find some details in [N. M. Bogolyubov et al (1985); N. M. Bogolyubov and V. E. Korepin (1986); K. M. Case (1984); B. Davies (1988, 1990); B. Davies and T. D. Kieu (1986); B. Davies and V. E. Korepin (2011); A. G. Izergin et al (1981); V. E. Korepin et al (1993); E. K. Sklyanin (1990); H. B. Thacker (1978, 1981)]. We restrict ourselves only to giving the expansion of the transfer matrix, which generates the integrals of motion:

$$\begin{aligned}\frac{1}{ic}\log\left(e^{iuL/2}\mathcal{T}(u)\right) \xrightarrow{u\to+i\infty} \ &\frac{Q}{u} + \frac{1}{u^2}\left(P - \tfrac{ic}{2}Q\right)\\ &+ \frac{1}{u^3}\left(H - icP - \tfrac{c^2}{3}Q\right) + O(u^{-4}).\end{aligned} \tag{1.79}$$

Here

$$Q = \int_0^L \Psi^\dagger(x)\Psi(x)\,\mathrm{d}x, \tag{1.80}$$

$$P = -\frac{i}{2}\int_0^L \left(\Psi^\dagger(x)\partial_x\Psi(x) - \left(\partial_x\Psi^\dagger(x)\right)\Psi(x)\right)\mathrm{d}x, \tag{1.81}$$

and $H$ is given by (1.76). The operators $Q$ and $P$ are respectively called the *particles number operator* and the *momentum operator*. They commute with the Hamiltonian $H$ and have a common set of eigenstates. Within the framework of the coordinate Bethe ansatz, these eigenstates have the following form:

$$|\Omega_n\rangle = \frac{1}{n!}\int_0^L \chi_n(x_1,\dots,x_n)\left(\prod_{k=1}^{n}\Psi^\dagger(x_k)\,\mathrm{d}x_k\right)|0\rangle, \qquad n=0,1,\dots. \tag{1.82}$$

Here $\chi_n(x_1,\dots,x_n)$ is a function that is symmetric and periodic in all arguments $\chi_n(x_1,\dots,x_k+L,\dots,x_n) = \chi_n(x_1,\dots,x_k,\dots,x_n)$. Solving the eigenvalue problem

$$H|\Omega_n\rangle = E|\Omega_n\rangle \tag{1.83}$$

and substituting here $|\Omega_n\rangle$ in the form (1.82), we obtain a differential equation for the function $\chi_n$:

$$-\sum_{k=1}^{n}\frac{\partial^2\chi_n}{\partial x_k^2} + 2c\sum_{j>k}^{n}\delta(x_j-x_k)\chi_n = E\chi_n. \tag{1.84}$$

The form of equation (1.84) coincides with the stationary Schrödinger equation, which describes a system of $n$ particles on the segment $[0,L]$. The potential of this system defines a pairwise point interaction. For this reason, this model is also often referred to as a *one-dimensional Bose gas with pointwise interaction* [E. H. Lieb and W. Liniger (1963); E. H. Lieb (1963); E. Lieb and D. Mattis (1961); F. A. Berezin et al (1964); J. B. McGuire (1964, 1965, 1966); C. N. Yang and C. P. Yang (1969); M. Gaudin (1983); V. E. Korepin et al (1993)]. For $c>0$, the particles repel; for $c<0$, they are attracted. The model of one-dimensional bosons is also known as the *Lieb–Liniger model* [E. H. Lieb and W. Liniger (1963)].

It is clear that in any region in which $x_j \neq x_k$, the solution to equation (1.84) is a plane wave. At the boundary of these regions, we should continuously glue these plane waves. At the same time, we should provide a discontinuity of the first derivative. Then the second derivative of $\chi_n$ is just the $\delta$-function on the hyperplanes. It is not difficult to construct a function $\chi_n$ that has such properties. However, we will not do this yet. In Section 4.3.2, we will obtain the function $\chi_n$ in the algebraic Bethe ansatz framework.

### 1.6.3 *Trigonometric R-matrix*

Let us introduce two functions

$$f(u,v) = \frac{u-v+\eta}{u-v} \qquad \text{and} \qquad g(u,v) = \frac{\eta}{u-v}, \tag{1.85}$$

and consider a matrix

$$R(u,v) = \begin{pmatrix} f(u,v) & 0 & 0 & 0 \\ 0 & 1 & g(u,v) & 0 \\ 0 & g(u,v) & 1 & 0 \\ 0 & 0 & 0 & f(u,v) \end{pmatrix}. \tag{1.86}$$

Obviously, the matrix (1.86) differs from (1.39) by the factor $(u - v)$ only. As already explained, multiplication of the $R$-matrix by an arbitrary scalar function is always possible. This does not violate either the Yang–Baxter equation or the $RTT$-relation. Thus, we may well call the matrix (1.86) the $R$-matrix. Further on, we will use both matrices (1.86) and (1.39). Depending on the specific task, it is convenient to choose either one or another normalization.

**Remark 1.3.** Following the tradition we do not stress that the functions $f(u,v)$, $g(u,v)$, and the $R$-matrix (1.86) actually depend on the difference $u-v$. This will allow us to introduce a shorthand notation for the products of these functions.

Let us now consider a matrix defined by (1.86), however, this time we set

$$f(u,v) = \frac{\sinh(u - v + \eta)}{\sinh(u - v)} \qquad \text{and} \qquad g(u,v) = \frac{\sinh \eta}{\sinh(u - v)}, \tag{1.87}$$

where $\eta$ is some constant. This matrix also satisfies the Yang–Baxter equation. We suggest that the reader verify this for himself. The matrix (1.86) with the functions (1.87) is called *trigonometric $R$-matrix*,[2] while the matrix (1.86) with the functions (1.85) is called *rational $R$-matrix*. Obviously, in the limit $u = \epsilon u'$, $v = \epsilon v'$, $\eta = \epsilon \eta'$, $\epsilon \to 0$, the trigonometric $R$-matrix turns into the rational one.

Among the physical models described by the trigonometric $R$-matrix, the most famous are the $XXZ$ Heisenberg chain and the sine-Gordon model. In particular, the construction of the $XXZ$ chain Hamiltonian is completely analogous to the case of the $XXX$ chain discussed above. The $R$-matrix (1.86) with the functions (1.87) generates an $L$-operator

$$L_n(u) = \begin{pmatrix} \sinh(u + \frac{\eta}{2}\sigma_n^z) & \sigma_n^- \sinh \eta \\ \sigma_n^+ \sinh \eta & \sinh(u - \frac{\eta}{2}\sigma_n^z) \end{pmatrix}. \tag{1.88}$$

[2]This is a matter of taste to write trigonometric or hyperbolic sinus in the formulas (1.87).

Recall that any matrix-valued function is understood as the corresponding Taylor series, that leads to

$$\sinh(u \pm \tfrac{\eta}{2}\sigma_n^z) = \sinh u \cosh \tfrac{\eta}{2} \pm \sigma_n^z \cosh u \sinh \tfrac{\eta}{2}. \tag{1.89}$$

The monodromy matrix is equal to the product of the $L$-operators over all sites of the chain (see (1.50)), and the Hamiltonian can be obtained from the transfer matrix by

$$H = 2\sinh\eta \frac{\mathrm{d}\mathcal{T}(u)}{\mathrm{d}u}\, \mathcal{T}^{-1}(u)\Big|_{u=\frac{\eta}{2}} - N\Delta\,, \tag{1.90}$$

where $\Delta = \cosh\eta$. The reader can convince himself that the formula (1.90) leads to the Hamiltonian

$$H = \sum_{k=1}^{N} \Big(\sigma_k^x \sigma_{k+1}^x + \sigma_k^y \sigma_{k+1}^y + \Delta \sigma_k^z \sigma_{k+1}^z\Big). \tag{1.91}$$

Completely analogous to the $XXX$ chain, we can construct an inhomogeneous $XXZ$ chain that appears to be useful in intermediate calculations. A description of the quantum sine-Gordon model in the algebraic Bethe ansatz framework is given in [L. D. Faddeev et al (1979); A. G. Izergin and V. E. Korepin (1981c); N. M. Bogolyubov and A. G. Izergin (1984a,b); N. M. Bogolyubov et al (1985); V. E. Korepin et al (1993)].

## 1.7 Hilbert space

We have considered examples of $L$-operators and the corresponding monodromy matrices. Recall that the monodromy matrix elements act in some Hilbert space $\mathcal{H}$, which we have not yet discussed. In the framework of the algebraic Bethe ansatz, minimal requirements are imposed on this space. Namely, there must exist a vacuum vector[3] $|0\rangle$ such that

$$A(u)|0\rangle = a(u)|0\rangle, \qquad D(u)|0\rangle = d(u)|0\rangle, \qquad C(u)|0\rangle = 0. \tag{1.92}$$

Here $a(u)$ and $d(u)$ are some functions of $u$. Their explicit form depends on the specific model. In other words, the vector $|0\rangle$ is the eigenvector for the operators $A(u)$ and $D(u)$, while the operator $C(u)$ annihilates it. The action of the operator $B(u)$ on the vacuum is free. It is assumed that multiple action of this operator generates all the space $\mathcal{H}$ (taking into account the closure, if the space is infinite-dimensional).

In the examples considered above, the vacuum vector exists. In the lattice version of the QNLSE model, it coincides with the Fock vacuum (1.73).

[3] One can also meet with a terminology pseudovacuum or reference state.

In the case of the $XXX$ or $XXZ$ chains, this is a state with all spins up (ferromagnetic state):

$$|0\rangle = \left(\begin{smallmatrix}1\\0\end{smallmatrix}\right)_1 \otimes \cdots \otimes \left(\begin{smallmatrix}1\\0\end{smallmatrix}\right)_N. \tag{1.93}$$

Let us check this. To this end, we prove the following statement.

**Proposition 1.2.** *Let a monodromy matrix $T(u)$ be given by (1.50), and let every $L$-operator have the form*

$$L_n(u) = \begin{pmatrix}\alpha_n(u) & \beta_n(u)\\ \gamma_n(u) & \delta_n(u)\end{pmatrix}, \qquad n = 1, \ldots, N. \tag{1.94}$$

*Assume that there exists a vector $|0\rangle$ such that*

$$\begin{aligned}&\alpha_n(u)|0\rangle = \hat{\alpha}_n(u)|0\rangle,\\ &\delta_n(u)|0\rangle = \hat{\delta}_n(u)|0\rangle, \qquad n = 1, \ldots, N,\\ &\gamma_n(u)|0\rangle = 0,\end{aligned} \tag{1.95}$$

*where $\hat{\alpha}_n(u)$ and $\hat{\delta}_n(u)$ are some functions of $u$. Then the action of the monodromy matrix entries on $|0\rangle$ is given by formulas* (1.92), *and*

$$a(u) = \prod_{n=1}^{N} \hat{\alpha}_n(u), \qquad d(u) = \prod_{n=1}^{N} \hat{\delta}_n(u). \tag{1.96}$$

**Proof.** We use induction over $N$. Proposition 1.2 is obvious for $N = 1$. Let it be true for some $N-1$. Then a matrix

$$\tilde{T}(u) = L_{N-1}(u) \ldots L_1(u) = \begin{pmatrix}\tilde{A}(u) & \tilde{B}(u)\\ \tilde{C}(u) & \tilde{D}(u)\end{pmatrix}, \tag{1.97}$$

has the following properties:

$$\tilde{A}(u)|0\rangle = \tilde{a}(u)|0\rangle, \qquad \tilde{D}(u)|0\rangle = \tilde{d}(u)|0\rangle, \qquad \tilde{C}(u)|0\rangle = 0, \tag{1.98}$$

where

$$\tilde{a}(u) = \prod_{n=1}^{N-1} \hat{\alpha}_n(u), \qquad \tilde{d}(u) = \prod_{n=1}^{N-1} \hat{\delta}_n(u). \tag{1.99}$$

On the other hand, the matrix (1.50) is presented in the form

$$T(u) = L_N(u)\tilde{T}(u) = \begin{pmatrix}\alpha_N(u) & \beta_N(u)\\ \gamma_N(u) & \delta_N(u)\end{pmatrix}\begin{pmatrix}\tilde{A}(u) & \tilde{B}(u)\\ \tilde{C}(u) & \tilde{D}(u)\end{pmatrix}. \tag{1.100}$$

Hence,

$$\begin{aligned}T(u) &= \begin{pmatrix}A(u) & B(u)\\ C(u) & D(u)\end{pmatrix}\\ &= \begin{pmatrix}\alpha_N(u)\tilde{A}(u) + \beta_N(u)\tilde{C}(u) & \alpha_N(u)\tilde{B}(u) + \beta_N(u)\tilde{D}(u)\\ \gamma_N(u)\tilde{A}(u) + \delta_N(u)\tilde{C}(u) & \gamma_N(u)\tilde{B}(u) + \delta_N(u)\tilde{D}(u)\end{pmatrix}.\end{aligned} \tag{1.101}$$

From this representation and (1.98), we immediately obtain

$$C(u)|0\rangle = \big(\gamma_N(u)\tilde{A}(u) + \delta_N(u)\tilde{C}(u)\big)|0\rangle = 0, \tag{1.102}$$

because $\gamma_N(u)|0\rangle = 0$. Similarly

$$\begin{aligned} A(u)|0\rangle &= \big(\alpha_N(u)\tilde{A}(u) + \beta_N(u)\tilde{C}(u)\big)|0\rangle = \left(\prod_{n=1}^{N} \hat{\alpha}_n(u)\right)|0\rangle, \\ D(u)|0\rangle &= \big(\gamma_N(u)\tilde{B}(u) + \delta_N(u)\tilde{D}(u)\big)|0\rangle = \left(\prod_{n=1}^{N} \hat{\delta}_n(u)\right)|0\rangle, \end{aligned} \tag{1.103}$$

where we used[4] $[\gamma_N(u), \tilde{B}(u)] = 0$. Proposition 1.2 is proved. □

Note that the method described above also allows us to calculate the action of the operator $B(u)$ on the vector $|0\rangle$ in terms of actions of the local operators $\beta_n(u)$:

$$B(u)|0\rangle = \sum_{k=1}^{N}\left(\prod_{n=k+1}^{N} \hat{\alpha}_n(u)\right)\left(\prod_{n=1}^{k-1} \hat{\delta}_n(u)\right)\beta_k(u)|0\rangle. \tag{1.104}$$

This formula is easily proved by induction over $N$ using representation (1.101). However, within the algebraic Bethe ansatz, it is used rarely since the states of the form $B(u)|0\rangle$ correspond to single-particle excitations and are of little interest.

It is easy to see that in the case of the inhomogeneous $XXX$ chain, the functions $\hat{\alpha}_n(u)$ and $\hat{\delta}_n(u)$ have the form

$$\hat{\alpha}_n(u) = u - \xi_n + \eta, \qquad \hat{\delta}_n(u) = u - \xi_n. \tag{1.105}$$

Therefore,

$$a(u) = \prod_{n=1}^{N}(u - \xi_n + \eta), \qquad d(u) = \prod_{n=1}^{N}(u - \xi_n), \tag{1.106}$$

and the action of the $B(u)$ operator on $|0\rangle$ is given by

$$B(u)|0\rangle = \eta\sum_{k=1}^{N}\left(\prod_{n=k+1}^{N}(u - \xi_n + \eta)\right)\left(\prod_{n=1}^{k-1}(u - \xi_n)\right)\sigma_k^-|0\rangle. \tag{1.107}$$

For the homogeneous chain one should simply set all $\xi_n = \eta/2$.

In the case of the inhomogeneous $XXZ$ chain, we have

$$a(u) = \prod_{n=1}^{N}\sinh(u - \xi_n + \eta), \qquad d(u) = \prod_{n=1}^{N}\sinh(u - \xi_n), \tag{1.108}$$

[4]Recall that $[(L_N)_{ij}(u), \tilde{T}_{kl}(v)] = 0$ for any $u$ and $v$ and for any $i, j, k, l$.

and

$$B(u)|0\rangle = \sinh\eta \sum_{k=1}^{N} \left( \prod_{n=k+1}^{N} \sinh(u - \xi_n + \eta) \right) \left( \prod_{n=1}^{k-1} \sinh(u - \xi_n) \right) \sigma_k^- |0\rangle. \tag{1.109}$$

In the lattice QNLSE model, we have

$$\hat{\alpha}_n(u) = 1 - \frac{iu\Delta}{2}, \qquad \hat{\delta}_n(u) = 1 + \frac{iu\Delta}{2}. \tag{1.110}$$

Then

$$a(u) = \left(1 - \frac{iu\Delta}{2}\right)^N, \qquad d(u) = \left(1 + \frac{iu\Delta}{2}\right)^N, \tag{1.111}$$

and the action of the $B(u)$ operator on $|0\rangle$ takes the form

$$B(u)|0\rangle = -i\Delta\sqrt{c} \sum_{k=1}^{N} \left(1 - \frac{iu\Delta}{2}\right)^{N-k} \left(1 + \frac{iu\Delta}{2}\right)^{k-1} \psi_k^\dagger |0\rangle. \tag{1.112}$$

**Remark 1.4.** When deriving this formula, we assumed that $\rho_n^+ = \rho_n^-$ and $c > 0$. However, we could put, for example, $\rho_n^+ = 1$ and not impose restrictions on the coupling constant $c$. The freedom to choose $\rho_n^\pm$ leads only to an insignificant constant factor in the vector $B(u)|0\rangle$.

It is easy to verify that in the continuous limit $\Delta \to 0$, $N \to \infty$, $N\Delta = L = const$, we obtain

$$a(u) = e^{-iuL/2}, \qquad d(u) = e^{iuL/2}, \tag{1.113}$$

and

$$B(u)|0\rangle = -i\sqrt{c}\, e^{-iuL/2} \int_0^L e^{iux} \Psi^\dagger(x)\, \mathrm{d}x |0\rangle. \tag{1.114}$$

## 1.8 Commutation relations

In the case of the rational $R$-matrix (1.39), one can easily derive commutation relations of the matrix elements from the matrix equation (1.7). We have

$$R(u,v) = \sum_{j,k} (E_1^{jj} E_2^{kk} + g(u,v) E_1^{jk} E_2^{kj}), \tag{1.115}$$

where $g(u,v)$ is given by (1.85) and $j,k$ take the values $1,2$. Substituting this expression in (1.14) we immediately obtain

$$[T_{ij}(u), T_{kl}(v)] = g(u,v)\big(T_{kj}(v)T_{il}(u) - T_{kj}(u)T_{il}(v)\big). \tag{1.116}$$

Replacing in this formula $i \leftrightarrow k$, $j \leftrightarrow l$, and $u \leftrightarrow v$ we obtain one more commutation relation

$$[T_{ij}(u), T_{kl}(v)] = g(u,v)\big(T_{il}(u)T_{kj}(v) - T_{il}(v)T_{kj}(u)\big). \tag{1.117}$$

Let us give all commutation relations in detail:

$$[T_{jk}(u), T_{jk}(v)] = 0, \qquad j,k = 1,2, \tag{1.118}$$

$$[A(u), D(v)] = g(u,v)\big(C(v)B(u) - C(u)B(v)\big), \tag{1.119}$$
$$[D(u), A(v)] = g(u,v)\big(B(v)C(u) - B(u)C(v)\big), \tag{1.120}$$
$$[C(u), B(v)] = g(u,v)\big(A(v)D(u) - A(u)D(v)\big), \tag{1.121}$$
$$[B(u), C(v)] = g(u,v)\big(D(v)A(u) - D(u)A(v)\big), \tag{1.122}$$

$$A(v)B(u) = f(u,v)B(u)A(v) + g(v,u)B(v)A(u), \tag{1.123}$$
$$B(v)A(u) = f(u,v)A(u)B(v) + g(v,u)A(v)B(u), \tag{1.124}$$
$$D(v)C(u) = f(u,v)C(u)D(v) + g(v,u)C(v)D(u), \tag{1.125}$$
$$C(v)D(u) = f(u,v)D(u)C(v) + g(v,u)D(v)C(u), \tag{1.126}$$
$$A(u)C(v) = f(u,v)C(v)A(u) + g(v,u)C(u)A(v), \tag{1.127}$$
$$C(u)A(v) = f(u,v)A(v)C(u) + g(v,u)A(u)C(v), \tag{1.128}$$
$$B(u)D(v) = f(u,v)D(v)B(u) + g(v,u)D(u)B(v), \tag{1.129}$$
$$D(u)B(v) = f(u,v)B(v)D(u) + g(v,u)B(u)D(v). \tag{1.130}$$

Some of these relationships repeat each other. For instance, (1.120) follows from (1.119) and (1.121). One can also verify the consistency of certain relations. For instance, applying (1.124) to the rhs of (1.123), we reproduce the lhs of (1.123).

Note also that due to (1.118), any matrix element commutes with itself for any argument values. This means that any $T_{ij}(u)$ can be chosen to build some integrable system.

Remarkably, the commutation relations (1.118)–(1.130) are also preserved in the case of the trigonometric $R$-matrix if the functions $f$ and $g$ are defined by (1.87). This leads to important consequences. In particular, the methods for constructing eigenvectors of Hamiltonians in models with rational and trigonometric $R$-matrices are identical. The vast majority of formulas for scalar products written in terms of the functions $f$ and $g$ are also applicable in rational and trigonometric cases. Therefore, from now on, we will mainly use the notation $f(u,v)$ and $g(u,v)$ without specifying their explicit form. This way may not seem very convenient at first, but we hope that the reader will appreciate its advantage over time.

### 1.8.1 *Lattice QNLSE model*

In Section 1.6.2, we introduced the $L$-operator (1.74) of the lattice QNLSE model. We stated that this $L$-operator satisfies the $RTT$-relation with the rational $R$-matrix. Let us check this. For this, we should check the commutation relations (1.118)–(1.130). Matrix elements of the $L$-operator have the form

$$\begin{aligned} A(u) &= 1 - \frac{iu\Delta}{2} + \frac{c\Delta^2}{2}\psi_n^\dagger\psi_n, \\ D(u) &= 1 + \frac{iu\Delta}{2} + \frac{c\Delta^2}{2}\psi_n^\dagger\psi_n, \\ B(u) &= -i\Delta\psi_n^\dagger\rho_n^+, \\ C(u) &= i\Delta\rho_n^-\psi_n. \end{aligned} \tag{1.131}$$

Recall that $c = i\eta$. The operators $\rho_n^\pm$ depend only on $\psi_n^\dagger\psi_n$ and satisfy condition (1.75).

First of all, commutation relations (1.118) are trivially satisfied. It is also easy to verify the validity of (1.119) and (1.120) because in this case, the operators $B(u)$ and $C(u)$ do not depend on $u$.

Consider commutation relation (1.123). Since $B(u)$ does not depend on $u$, setting $B(u) = B$, we obtain

$$[A(v), B] = g(u,v)B\big(A(v) - A(u)\big). \tag{1.132}$$

Substituting here $A(u)$ and $B$ from (1.74) we arrive at

$$\left[\tfrac{c\Delta^2}{2}\psi_n^\dagger\psi_n, \psi_n^\dagger\rho_n^+\right] = \tfrac{c\Delta}{2}\psi_n^\dagger\rho_n^+. \tag{1.133}$$

If $\rho_n^+$ depends only on $\psi_n^\dagger\psi_n$, then (1.133) obviously holds. In this case, the specific dependence of $\rho_n^+$ on $\psi_n^\dagger\psi_n$ does not play a role. All that matters is that $[\rho_n^+, \psi_n^\dagger\psi_n] = 0$.

Commutation relations (1.124)–(1.130) are checked in the same way. They are fulfilled for any $\rho_n^\pm$ such that $[\rho_n^\pm, \psi_n^\dagger\psi_n] = 0$.

The only exceptions are the commutations of operators $C$ and $B$. Consider, for example, (1.121). We have in the rhs

$$g(u,v)\big(A(v)D(u) - A(u)D(v)\big) = c\Delta\left(1 + \tfrac{c\Delta^2}{2}\psi_n^\dagger\psi_n\right). \tag{1.134}$$

On the other hand,

$$\begin{aligned} [C(u), B(v)] &= \Delta^2(\rho_n^-\psi_n\psi_n^\dagger\rho_n^+ - \psi_n^\dagger\rho_n^+\rho_n^-\psi_n) \\ &= \Delta^2(\rho_n^-\rho_n^+\psi_n\psi_n^\dagger - \psi_n^\dagger\rho_n^+\rho_n^-\psi_n), \end{aligned} \tag{1.135}$$

where we used $[\psi_n\psi_n^\dagger, \rho_n^+] = 0$. Now we should use (1.75). Then

$$[C(u), B(v)] = c\Delta^2 \left\{\left(1 + \tfrac{c\Delta^2}{4}\psi_n^\dagger\psi_n\right)\psi_n\psi_n^\dagger - \psi_n^\dagger\left(1 + \tfrac{c\Delta^2}{4}\psi_n^\dagger\psi_n\right)\psi_n\right\}. \tag{1.136}$$

Now it is easy to check that (1.136) coincides with (1.134). Thus, the commutation relations (1.118)–(1.130) impose restrictions only on the product $\rho_n^-\rho_n^+$, but not on the operators $\rho_n^-$ and $\rho_n^+$ themselves.

### 1.8.2 *Quantum determinant*

The $RTT$-relation defines an algebra of the $T_{ij}$ operators with commutation relations (1.118)–(1.130). This algebra has a central element called the *quantum determinant* [A. G. Izergin and V. E. Korepin (1981a); P. P. Kulish and E. K. Sklyanin (1982a); V. E. Korepin (1982a); E. K. Sklyanin (1982b, 1983); V. G. Drinfeld (1983); A. Molev et al (1996); A. Molev (2007)]. It is defined as follows:

$$\operatorname{qdet} T(u) = A(u)D(u-\eta) - B(u)C(u-\eta). \tag{1.137}$$

Then

$$[T_{ij}(v), \operatorname{qdet} T(u)] = 0, \qquad i,j = 1,2, \tag{1.138}$$

for any complex $u$ and $v$. The interested reader can check (1.138) using commutation relations (1.118)–(1.130). However, this task is far from trivial. There is an easier way to verify the validity of (1.138). Namely, we can use that

$$\operatorname{qdet} T(u) \begin{pmatrix} 1 & 0 \\ 0 & 1 \end{pmatrix} = T(u)\sigma^y T^T(u-\eta)\sigma^y, \tag{1.139}$$

where $T^T$ means the transposed matrix. On the one hand, it is easy to verify that definition (1.139) coincides with (1.137) due to commutation relations (1.118)–(1.130). On the other hand, it follows immediately from (1.139) that

$$\operatorname{qdet}\left(T^{(1)}(u)T^{(2)}(u)\right) = \operatorname{qdet} T^{(1)}(u)\operatorname{qdet} T^{(2)}(u), \tag{1.140}$$

provided $T^{(1)}(u)$ and $T^{(2)}(u)$ satisfy the $RTT$-relation and their matrix elements commute:

$$[T_{ij}^{(1)}(u), T_{kl}^{(2)}(u)] = 0, \qquad \forall i,j,k,l. \tag{1.141}$$

It follows that the quantum determinant of the monodromy matrix is equal to the product of the quantum determinants of the $L$-operators

$$\operatorname{qdet} T(u) = \prod_{n=1}^{N} \operatorname{qdet} L_n(u). \tag{1.142}$$

It is no longer difficult to calculate the quantum determinant of any $L$-operator using its explicit representation in terms of local operators.

## Chapter 2

# Algebraic Bethe ansatz

In this chapter, we proceed to a direct study of the algebraic Bethe ansatz method. We will show how, within the framework of this method, the eigenvectors of the transfer matrix can be found. We also obtain a system of Bethe equations that describes these vectors. One of the main advantages of the algebraic Bethe ansatz is its universality: it is based only on the $RTT$-relation, and without any changes, it can be applied to a wide class of models that may have a different physical interpretation. From the viewpoint of this approach, one can say that different physical models are different representations of the same algebra. We will also return to the Yang–Baxter equation and discuss its role in the algebraic Bethe ansatz.

We have seen that the space $\mathcal{H}$, in which the elements of the monodromy matrix act, can be either a finite-dimensional (spin chains) or an infinite-dimensional (one-dimensional Bose gas). In the framework of the algebraic Bethe ansatz, this space is not specified. Also, the functions $a(u)$ and $d(u)$ remain free functional parameters. Nevertheless, these data are already sufficient to formulate certain conditions on the eigenvectors of the transfer matrix $\mathcal{T}(u) = A(u) + D(u)$. We recall that the transfer matrix $\mathcal{T}(u)$ is the generating function of the integrals of motion of the model. Thus, by construction, the eigenvectors (eigenstates) of the operator $\mathcal{T}(u)$ are simultaneously eigenvectors for all integrals of motion, including the Hamiltonian. Therefore, instead of looking for the eigenvectors of the Hamiltonian of a certain concrete model, we look for the eigenvectors of the transfer matrix.

We now proceed to the description of the content of this chapter. In Section 2.1, we derive the action of the transfer matrix on Bethe vectors. Here we formulate the basic rules that should be followed when using the $RTT$-relation. We obtain a system of Bethe equations, which determine the spectrum of the transfer matrix. In the same section, we introduce a shorthand notation for the products used throughout the book. Section 2.2 introduces the concepts of a twisted transfer matrix and its eigenvectors. They will also play a very important role in what follows. Sections 2.3 and 2.4 are devoted to the Bethe equations. Here we are considering the case of coincident roots. We also give an example of such solutions of the Bethe equations that do not generate the eigenvectors of the transfer matrix. In Section 2.5, we introduce dual Bethe vectors. Finally, in Section 2.6, we construct a generalized $L$-operator satisfying the $RTT$-relation with the rational $R$-matrix.

## 2.1 Actions of the monodromy matrix entries on vectors

### 2.1.1 *Shorthand notation*

First of all, let us agree on a new notation. We denote sets of variables by a bar:

$$\bar{u} = \{u_1, \dots, u_n\}, \qquad \bar{v} = \{v_1, \dots, v_m\}, \tag{2.1}$$

and so on. If necessary, the cardinalities of the sets are given in special comments.

We also use a shorthand notation for the products of the operators $T_{ij}(u)$, the functions $a(u)$ and $d(u)$, and the functions $f(u,v)$ and $g(u,v)$ (1.85), (1.87). Namely, if this operator or function depends on a set of variables, then this means that one should take a product of these operators or functions over the corresponding set. For example,

$$B(\bar{u}) = \prod_{u_j \in \bar{u}} B(u_j), \qquad C(\bar{u}) = \prod_{u_j \in \bar{u}} C(u_j), \qquad a(\bar{v}) = \prod_{v_j \in \bar{v}} a(v_j), \tag{2.2}$$

and so on. Recall that due to (1.21), we have $[B(u), B(v)] = 0$, $[C(u), C(v)] = 0$. Hence, the operator products in (2.2) are well defined.

Similarly, we have for the functions $g(u,v)$ and $f(u,v)$,

$$g(u, \bar{v}) = \prod_{v_i \in \bar{v}} g(u, v_i), \qquad f(\bar{u}, \bar{v}) = \prod_{u_j \in \bar{u}} \prod_{v_i \in \bar{v}} f(u_j, v_i). \tag{2.3}$$

In the second case, we deal with the double product.

Below we will often deal with sets of variables in which one of the elements is missing, for example, $\{u_1, \dots, u_n\} \setminus u_k$. We denote such sets by $\bar{u}_k = \{u_1, \dots, u_n\} \setminus u_k$. Respectively, we have

$$B(\bar{u}_k) = \prod_{\substack{u_j \in \bar{u} \\ j \neq k}} B(u_j), \qquad f(u_k, \bar{u}_k) = \prod_{\substack{u_i \in \bar{u} \\ i \neq k}} f(u_k, u_i), \tag{2.4}$$

and so on. Finally, we agree that a product over the empty set is equal to 1 by definition. A double product is equal to 1 if at least one of the sets is empty.

### 2.1.2 *Action of operators A and D on Bethe vectors*

The main idea of the algebraic Bethe ansatz is that we look for the eigenvectors of the transfer matrix in the form $B(\bar{u})|0\rangle$. In general, the set $\bar{u} = \{u_1, \ldots, u_n\}$ can contain an arbitrary number of elements, and the parameters $u_k$ can take arbitrary complex values.

Recall that the space $\mathcal{H}$ consists of the vectors $B(\bar{u})|0\rangle$ and their linear combinations. The ansatz is that we exclude linear combinations from our consideration. In other words, we assume that eigenvectors of the transfer matrix can be represented as monomials in operators $B$ applied to the vacuum. We will call vectors of this kind *Bethe vectors*. The parameters $\bar{u} = \{u_1, \ldots, u_n\}$ are called *Bethe parameters*. First, we need to check whether the Bethe vectors can indeed be eigenvectors of the operator $\mathcal{T}(v)$. Secondly, under what conditions is it possible. Third, whether the resulting vectors form a basis in the Hilbert space $\mathcal{H}$.

Strictly speaking, the last question goes beyond the framework of the algebraic Bethe ansatz. To answer it, we need to know much more about the space $\mathcal{H}$. Meanwhile, we do not even know whether it is finite-dimensional or not. Therefore, this issue should be resolved separately for each specific model. However, it is interesting to mention that even for the finite-dimensional spaces $\mathcal{H}$, the question of the completeness of the obtained system of eigenvectors may be non-trivial. In particular, this problem occurs for the $XXX$ chain. There was no exhaustive proof of completeness until very recently in the general case of chains of arbitrary length. It was given eventually in [E. Mukhin et al (2009, 2013)]. On the other hand, in the model of the one-dimensional Bose gas, in which the Hilbert space is infinite-dimensional, the completeness of the eigenvector system is relatively easy to prove, at least for the positive coupling constant [M. Gaudin (1983); T. C. Dorlas (1993); N. A. Slavnov (2011)].

Let us compute the action of the operator $A(v)$ on a vector $B(\bar{u})|0\rangle$

$$A(v)B(\bar{u})|0\rangle, \tag{2.5}$$

assuming that the parameters $v, u_1, \ldots, u_n$ are generic complex numbers (that is, they are pairwise distinct, they do not coincide with singular points of the function $a(u)$, and so on.). Formally, we have all the necessary tools for this. Using the commutation relations (1.123), we can move the operator

$A$ to the right until it arrives at the extreme right position. After this, it remains to act on the vacuum vector via (1.92). Thus, the $A(v)$ action on the Bethe vector reduces to an $n$-fold application of the formula (1.123). However, it is clear that we will obtain two terms at each step so that we will have $2^n$ terms in the end. For large $n$, such a result is difficult not only to obtain but even to write down. To solve this problem, we formulate the basic rules of the algebraic Bethe ansatz, which must be taken into account when permuting many operators.

- One should try to understand the general structure of the result. Namely, how the result depends on the operators $T_{ij}(u)$ (or their eigenvalues $a(u)$ and $d(u)$). At this stage, we ignore the functions $f(u, v)$ and $g(u, v)$, which arise when we permute the operators.
- It is necessary to fix a term in the final result and obtain just this term, neglecting all other contributions. In order to do this, it might be useful to reorder the operators in the original expression.

Let us see how these rules work in practice. Applying the first rule to the formula (2.5), we see that the number of linearly independent terms in the final result anyway does not exceed $n + 1$:

$$A(v)B(\bar{u})|0\rangle = a(v)\Lambda(v|\bar{u})B(\bar{u})|0\rangle + \sum_{k=1}^{n} a(u_k)\Lambda_k(v|\bar{u})B(v)B(\bar{u}_k)|0\rangle. \quad (2.6)$$

Here $\Lambda(v|\bar{u})$ and $\Lambda_k(v|\bar{u})$ are some unknown coefficients to be determined.

Indeed, it follows from (1.123) that when we permute the operators $A$ and $B$, they either preserve the original arguments or exchange them. However, the 'law of conservation of arguments' anyway takes place: if the operators in the lhs have arguments $u$ and $v$, then in the rhs, they have the same arguments. New arguments do not arise, and the original arguments do not disappear. In our case, in the lhs of the formula (2.6), the arguments of the operators are the parameters $v, u_1, \dots, u_n$. Therefore, the same parameters should appear in the rhs. It is easy to see that (2.6) is the most general expression of this kind. It contains a term in which, after moving $A(v)$ to the right, all operators retained their original arguments. Then the operator $A(v)$, acting on the vacuum, produces the $a(v)$ function. Besides, (2.6) contains terms in which, moving to the right, the operator $A(v)$ exchanged its argument one or more times with the operators $B(u_i)$. As a result, having reached the rightmost position, the operator $A$ receives some argument $u_k$ from the set $\bar{u}$. Thus, acting on the vacuum, we obtain the function $a(u_k)$. At the same time, the $B(u_k)$ operator disappears in the resulting product of the $B$ operators, and $B(v)$ appears instead. When operators are permuted, the functions $f(u,v)$ and $g(u,v)$ arise, but we ignore them, simply denoting the coefficients in the final result by $\Lambda(v|\bar{u})$ and $\Lambda_k(v|\bar{u})$. Later, we have to find them.

This concludes the first stage. We have determined the general structure of the final expression. Note that there are only $n+1$ terms instead of the expected $2^n$ terms.

**Remark 2.1.** Actually, the number of terms can be even less in specific models. The matter is that we considered Bethe vectors with different sets of the arguments as linearly independent. Generally speaking, this is not the case. For example, in the $XXX$ model, the number of linearly independent vectors can not exceed the dimension of the Hilbert space. The latter is $2^N$, where $N$ is the number of sites of the chain. On the other hand, we can take a vector $B(\bar{u})|0\rangle$, fix the parameters $u_2, \dots, u_n$, and move the variable $u_1$ within the whole complex plane. If with each new value $u_1$ we get a linearly independent vector of all the previous ones, then the number of such vectors is obviously infinite. Nevertheless, this observation does not invalidate formula (2.6). It still holds. However, if there are linearly dependent vectors in the rhs, then the number of terms in this formula can be reduced.

Let us now proceed to the second stage. We shall call the first term in the rhs of (1.123) *the first commutation scheme* (when the operators keep their arguments). The term in which the operators exchange their arguments will be called *the second commutation scheme.*

Let us compute the coefficient $\Lambda(v|\bar{u})$, that is, the coefficient of the vector $a(v)B(\bar{u})|0\rangle$. This coefficient arises in (2.6) if and only if we use the first commutation scheme when permuting the operators $A(v)$ and $B(u_j)$. Indeed, let the operator $A(v)$ exchange its argument with some operator $B(u_j)$. Then we obtain a vector

$$B(u_1)\dots B(u_{j-1})B(v)A(u_j)B(u_{j+1})\dots B(u_n)|0\rangle. \tag{2.7}$$

It is clear that when moving further to the right, the operator $A(u_j)$ can not accept the argument $v$ because it is to the left from $A$. Therefore, the function $a(v)$ cannot arise in the final result. Furthermore, the result will contain the operator $B(v)$, while this operator should not be in the term we are interested in.

Thus, to obtain the term $a(v)B(\bar{u})|0\rangle$, we can only use the first commutation scheme. Then, permuting $A(v)$ and $B(u_j)$ we obtain a coefficient

$f(u_j, v)$. After all permutations, we obtain a product of $f(u_j, v)$ over $u_j$, while the operator $A(v)$, approaching the extreme right position and acting on the vacuum, gives the function $a(v)$. As a result, we have

$$\Lambda(v|\bar{u}) = f(\bar{u}, v). \tag{2.8}$$

Recall that according to the convention on the shorthand notation introduced in Section 2.1.1, the rhs of (2.8) is the product of functions $f(u_j, v)$ over all parameters from the set $\bar{u}$.

Let us now find $\Lambda_k(v|\bar{u})$ that is, the coefficients of the vectors $a(u_k)B(v)B(\bar{u}_k)|0\rangle$, $k = 1, \dots, n$. For this, we fix some $k$. Using the fact that all the $B$ operators commute with each other, we reorder the product $B(\bar{u})$ so that the operator $B(u_k)$ stands in the extreme left position:

$$A(v)B(\bar{u})|0\rangle = A(v)B(u_k)B(\bar{u}_k)|0\rangle. \tag{2.9}$$

Now, in order to obtain a contribution proportional to $a(u_k)$, it is necessary to use the second commutation scheme when permuting $A(v)$ and $B(u_k)$. Overwise, we obtain a vector

$$B(u_k)A(v)B(\bar{u}_k)|0\rangle, \tag{2.10}$$

and we see that when moving further to the right, the operator $A(v)$ can not accept the argument $u_k$ because it is to the left from $A$. Besides, the result will contain the operator $B(u_k)$, while this operator is missing in the term of our interest. Thus, we find

$$A(v)B(u_k)B(\bar{u}_k)|0\rangle = g(v, u_k)B(v)A(u_k)B(\bar{u}_k)|0\rangle + \mathfrak{Z}, \tag{2.11}$$

where $\mathfrak{Z}$ means terms that do not contribute in any way to the coefficient under consideration. After this, when moving further to the right, $A(u_k)$ must keep its argument. Therefore, we should use the first commutation scheme. Indeed, if there is an exchange at least once, then the argument $u_k$ remains on the left from $A$, and we can not get the function $a(u_k)$ in the final result. After every permutation with the operators $B(u_j)$, we will get the factors $f(u_j, u_k)$. Thus, we obtain

$$\Lambda_k(v|\bar{u}) = g(v, u_k)f(\bar{u}_k, u_k), \qquad k = 1, \dots, n. \tag{2.12}$$

We have found the action of the operator $A(v)$ on the Bethe vector $B(\bar{u})|0\rangle$. It is given by equation (2.6), where the coefficients $\Lambda(v|\bar{u})$ and $\Lambda_k(v|\bar{u})$ respectively are given by (2.8) and (2.12).

The operator $D(v)$ action on the Bethe vector $B(\bar{u})|0\rangle$ can be found in the same way. One can, however, do even easier. Observe that the commutation relations of the operators $D$ and $B$ (1.130) can be obtained from those of $A$ and $B$ (1.123) under the formal replacement of the constant $\eta$ with $-\eta$. Indeed, it is easy to see that

$$f(u,v)\Big|_{\eta\to-\eta} = f(v,u), \qquad g(u,v)|_{\eta\to-\eta} = g(v,u). \tag{2.13}$$

Thus, the action of $D(v)$ on the vector $B(\bar{u})|0\rangle$ is given by (2.6), where one should replace $\eta$ with $-\eta$ and the function $a$ with the function $d$:

$$D(v)B(\bar{u})|0\rangle = d(v)\tilde{\Lambda}(v|\bar{u})B(\bar{u})|0\rangle + \sum_{k=1}^{n} d(u_k)\tilde{\Lambda}_k(v|\bar{u})B(v)B(\bar{u}_k)|0\rangle, \tag{2.14}$$

where

$$\tilde{\Lambda}(v|\bar{u}) = f(v,\bar{u}), \qquad \tilde{\Lambda}_k(v|\bar{u}) = g(u_k,v)f(u_k,\bar{u}_k). \tag{2.15}$$

The operator $C(v)$ action on the Bethe vector $B(\bar{u})|0\rangle$ will be calculated in Section 5.1. However, we suggest that the reader try to do it independently since all the necessary tools for this calculation are already explained.

### 2.1.3 *Eigenvectors and Bethe equations*

Now we know the action of the transfer matrix on the vector $B(\bar{u})|0\rangle$

$$\begin{aligned}\mathcal{T}(v)B(\bar{u})|0\rangle &= \Big(a(v)\Lambda(v|\bar{u}) + d(v)\tilde{\Lambda}(v|\bar{u})\Big)B(\bar{u})|0\rangle \\ &\quad + \sum_{k=1}^{n}\Big(a(u_k)\Lambda_k(v|\bar{u}) + d(u_k)\tilde{\Lambda}_k(v|\bar{u})\Big)B(v)B(\bar{u}_k)|0\rangle.\end{aligned} \tag{2.16}$$

In the first line of the rhs of this equation, we have the same vector $B(\bar{u})|0\rangle$ as in the lhs. In the second line, we have a sum of vectors in which one of the operators $B$ depends on the variable $v$. There are no such vectors in the lhs. However, if we impose conditions

$$a(u_k)\Lambda_k(v|\bar{u}) + d(u_k)\tilde{\Lambda}_k(v|\bar{u}) = 0, \qquad k = 1,\dots,n, \tag{2.17}$$

for any complex $v$, then the sum in the second line of (2.16) vanishes. Then the vector $B(\bar{u})|0\rangle$ becomes an eigenvector of the operator $\mathcal{T}(v)$ with an eigenvalue

$$\tau(v|\bar{u}) = a(v)\Lambda(v|\bar{u}) + d(v)\tilde{\Lambda}(v|\bar{u}) = a(v)f(\bar{u},v) + d(v)f(v,\bar{u}). \tag{2.18}$$

In its turn, the condition (2.17) is usually written as follows:

$$\frac{a(u_k)}{d(u_k)} = \frac{f(u_k,\bar{u}_k)}{f(\bar{u}_k,u_k)}, \qquad k = 1,\dots,n. \tag{2.19}$$

The system (2.19) is called the system of *Bethe equations*. The vector $B(\bar{u})|0\rangle$, in which the parameters $\bar{u}$ satisfy the system (2.19), is often called the *on-shell Bethe vector*. Thus, the on-shell Bethe vector is synonymous with the eigenvector of the transfer matrix. If no restrictions are imposed on the parameters $\bar{u}$, this vector is often called the *off-shell Bethe vector*.

**Remark 2.2.** To date, there is no commonly accepted interpretation of the concepts of on-shell and off-shell Bethe vectors. Different authors put different meanings into these concepts. It should also be noted that the fulfillment of the Bethe equations does not yet guarantee that the corresponding vector $B(\bar{u})|0\rangle$ is an eigenvector of the transfer matrix $\mathcal{T}(v)$. There are so-called spurious (inadmissible, non-physical) solutions. The vectors $B(\bar{u})|0\rangle$ corresponding to such solutions are not eigenvectors of the transfer matrix. We will discuss this issue in more detail in Sections 2.3.1–2.4.1. In what follows, we will call an on-shell Bethe vector such a vector $B(\bar{u})|0\rangle$ that is an eigenvector of $\mathcal{T}(v)$ and in which the parameters $\bar{u}$ are the roots of the Bethe equations. We will use the term off-shell Bethe vector if it is necessary to emphasize that the set $\bar{u}$, generally speaking, is not a solution of the Bethe equations. Finally, we will often refer to the vector $B(\bar{u})|0\rangle$ simply as the Bethe vector, if it does not matter whether the parameters $\bar{u}$ satisfy the Bethe equations or not.

It is worth mentioning that the obtained on-shell Bethe vectors are eigenvectors of the operator $\mathcal{T}(v)$ for any $v$:

$$\mathcal{T}(v)B(\bar{u})|0\rangle = \tau(v|\bar{u})B(\bar{u})|0\rangle, \qquad \forall v \in \mathbb{C}. \tag{2.20}$$

Therefore, if the integrals of motion $I_k$ of the model under consideration arise in the expansion

$$\mathcal{T}(v) = \sum_k (v - v_0)^k I_k, \tag{2.21}$$

then their eigenvalues $\hat{I}_k(\bar{u})$ arise in the similar expansion of the transfer matrix eigenvalue

$$\tau(v|\bar{u}) = \sum_k (v - v_0)^k \hat{I}_k(\bar{u}). \tag{2.22}$$

Differentiating equation (2.20) $k$ times over $v$ at $v = v_0$, we obtain

$$I_k B(\bar{u})|0\rangle = \hat{I}_k(\bar{u})\, B(\bar{u})|0\rangle. \tag{2.23}$$

Thus, if we solve the system of Bethe equations, we thereby find the eigenvalues of all the integrals of motion of the model under consideration and the corresponding eigenvectors.

Let us find, for example, the energy of the eigenstate $B(\bar{u})|0\rangle$ in the $XXX$ chain. For this, we should use equation (1.57). Replacing there the transfer matrix with its eigenvalue $\tau(v|\bar{u})$ we find

$$E_n(\bar{u}) = 2\eta \frac{\mathrm{d}\log\tau(v|\bar{u})}{\mathrm{d}v}\Big|_{v=\frac{\eta}{2}} - N. \tag{2.24}$$

Substituting here (2.18) with $a(v) = (v+\eta/2)^N$ and $d(v) = (v-\eta/2)^N$ we obtain

$$E_n(\bar{u}) = N + \sum_{k=1}^{n} \frac{2\eta^2}{u_k^2 - \eta^2/4}, \tag{2.25}$$

where $n = \#\bar{u}$. An analog of this formula for the $XXZ$ chain is

$$E_n(\bar{u}) = N\Delta + \sum_{k=1}^{n} \frac{2\sinh^2\eta}{\sinh(u_k - \frac{\eta}{2})\sinh(u_k + \frac{\eta}{2})}. \tag{2.26}$$

The eigenvalues of the Hamiltonian of the one-dimensional Bose gas are found from expansion (1.79)

$$E_n(\bar{u}) = \sum_{k=1}^{n} u_k^2. \tag{2.27}$$

Let us give one more useful interpretation of the Bethe equations. Note that the eigenvalue of the transfer matrix $\tau(v|\bar{u})$ (2.18) has poles at the points $v = u_k$. Consider the residues at these points

$$\frac{1}{\eta}\operatorname{Res}\tau(v|\bar{u})\Big|_{v=u_k} = d(u_k)f(u_k, \bar{u}_k) - a(u_k)f(\bar{u}_k, u_k). \tag{2.28}$$

We see that if the Bethe equations are fulfilled, then the residue (2.28) vanishes. Thus, we can interpret the Bethe equations as the absence of singularities of the transfer matrix eigenvalue in the points $v = u_k$.

This property has a simple explanation. Consider, for example, the transfer matrix $\mathcal{T}(v)$ of the $XXX$ chain. This operator is a polynomial in $v$ of degree $N$. At the same time, the eigenvector of the transfer matrix does not depend on $v$. Then the eigenvalue $\tau(v|\bar{u})$ must be a polynomial in $v$ of degree $N$, and it certainly cannot have poles at $v = u_k$. The Bethe equations guarantee the absence of such poles.

Note that the derivation of Bethe equations is universal in the sense that it is model-independent. We used only the commutation relations between the elements of the monodromy matrix (in fact, the $RTT$-relation) and the formulas of the operators' action on the vacuum. If we consider a specific model, we should only substitute the concrete values of the functions $a(u)$

and $d(u)$ into the system (2.19). Thus, for the $XXX$ chain, the system of Bethe equations has the form

$$\left(\frac{u_k+\frac{\eta}{2}}{u_k-\frac{\eta}{2}}\right)^N=\frac{f(u_k,\bar{u}_k)}{f(\bar{u}_k,u_k)}=\prod_{\substack{j=1\\j\neq k}}^{n}\frac{u_k-u_j+\eta}{u_k-u_j-\eta},\qquad k=1,\dots,n. \tag{2.29}$$

Pay attention that we can always change variables $u_k\to\eta u_k$, which leads to the complete disappearance of the parameter $\eta$ from the Bethe equations. This fact again says that this parameter does not play any role in the $XXX$ model.

In the continuous limit of the lattice QNLSE model, we have

$$e^{-iu_kL}=\frac{f(u_k,\bar{u}_k)}{f(\bar{u}_k,u_k)}=\prod_{\substack{j=1\\j\neq k}}^{n}\frac{u_k-u_j+\eta}{u_k-u_j-\eta},\qquad k=1,\dots,n. \tag{2.30}$$

Recall that now the parameter $\eta$ is associated with the interaction constant $c$ of the Hamiltonian by the condition $\eta=-ic$. Now we cannot get rid of this parameter by simply rescaling the roots of the Bethe system.

One can prove that all solutions to the system (2.30) are real if $c>0$ (see Section 10.1). If $c<0$, then along with real solutions, equations (2.30) also have complex conjugate roots. Such roots are commonly called *strings.* String solutions also exist in the Bethe equations for the $XXX$ and $XXZ$ chains. However, the properties of solutions to the Bethe equations usually become significant only when studying the thermodynamic limit in a particular model.

Having solutions of systems (2.29) or (2.30), we can construct on-shell vectors $B(\bar{u})|0\rangle$. The question of whether the obtained vectors form a basis, of course, remains open.

### 2.1.4 *Example of on-shell Bethe vector*

Consider a homogeneous $XXX$ chain and a Bethe vector $B(u)|0\rangle$, that depends on one variable $u$ only. We have an explicit representation for this vector (see (1.107))

$$B(u)|0\rangle=\eta\sum_{k=1}^{N}(u+\tfrac{\eta}{2})^{N-k}(u-\tfrac{\eta}{2})^{k-1}\,\sigma_k^-|0\rangle. \tag{2.31}$$

The operator $\sigma_k^-$, when acting on the state $|0\rangle$, flips the spin in the $k$th site:

$$\sigma_k^-|0\rangle=\left(\begin{smallmatrix}1\\0\end{smallmatrix}\right)_1\otimes\cdots\otimes\left(\begin{smallmatrix}0\\1\end{smallmatrix}\right)_k\otimes\cdots\otimes\left(\begin{smallmatrix}1\\0\end{smallmatrix}\right)_N. \tag{2.32}$$

Thus, the vector $B(u)|0\rangle$ is a linear combination of the states with one spin down and $N-1$ spins up. Obviously, vectors with one spin down and $N-1$ spins up form an $N$-dimensional subspace in the space of states.

**Remark 2.3.** In Section 4.3.1, we will obtain an explicit form of Bethe vectors in the $XXX$ and $XXZ$ chains by expressing them in terms of the $\sigma_k^-$ operators and the vacuum vector $|0\rangle$ (see (4.37), (4.38)). It follows from this explicit form that the Bethe vector $B(\bar{u})|0\rangle$ with $\#\bar{u} = n$ belongs to the subspace with $n$ spins down, and $N-n$ spins up. However, this fact can be proved right now without using the explicit form of Bethe vectors. We invite the reader to think about this question.

Let the Bethe vector (2.31) be an on-shell. Then the parameter $u$ satisfies the system of Bethe equations (2.29) for $n = 1$:

$$\left(\frac{u+\frac{\eta}{2}}{u-\frac{\eta}{2}}\right)^N = 1. \tag{2.33}$$

This equation admits an explicit solution

$$\frac{u_\ell+\frac{\eta}{2}}{u_\ell-\frac{\eta}{2}} = e^{2\pi i\ell/N}, \qquad \ell = 0, 1, \dots, N-1. \tag{2.34}$$

Note that for $\ell = 0$, we obtain $u_0 \to \infty$. However, we should also take this infinite root into account, since otherwise, we will not be able to construct a basis.

Let

$$|\phi_\ell\rangle = \frac{B(u_\ell)|0\rangle}{\eta(u_\ell-\frac{\eta}{2})^{N-1}\sqrt{N}}. \tag{2.35}$$

Then we have

$$|\phi_\ell\rangle = \frac{1}{\sqrt{N}}\sum_{k=1}^{N}\left(\frac{u_\ell+\frac{\eta}{2}}{u_\ell-\frac{\eta}{2}}\right)^{N-k}\sigma_k^-|0\rangle = \frac{1}{\sqrt{N}}\sum_{k=1}^{N} e^{-2\pi ik\ell/N}\sigma_k^-|0\rangle. \tag{2.36}$$

If we introduce $\langle\phi_\ell|$ by $\langle\phi_\ell| = |\phi_\ell\rangle^\dagger$, then we immediately obtain

$$\langle\phi_{\ell'}|\phi_\ell\rangle = \delta_{\ell,\ell'}. \tag{2.37}$$

Thus, we see that the system of on-shell Bethe vectors constructed from the solutions of equation (2.29) forms an orthogonal basis in the subspace with one spin down. Thanks to the simple explicit form of on-shell vectors, we also made this basis normalized. Note also that the infinite root $u_0$ does not stand out from the rest. This is not surprising since the spectral parameter $u$ has no physical meaning and plays the role of some parametrization of

Bethe vectors. We could have chosen a different parametrization from the beginning, such as $u' = (u+\eta/2)/(u-\eta/2)$, and then we would not have infinite roots in the Bethe equations.

Note also that the presence of infinite roots of the Bethe equations is a feature of the $XXX$ chain. The reader will easily see that the analog of equation (2.33) for the $XXZ$ chain no longer has an infinite root.

In this example, it is easy to verify by direct calculation that the vectors (2.36) are indeed eigenvectors of the $XXX$ Hamiltonian. One can also calculate the corresponding eigenvalues and compare them with formula (2.25). We suggest that the reader do this exercise.

### 2.1.5 *Compatibility of commutation relations*

Let us write down the action formula of $A(v)$ on the Bethe vector $B(\bar{u})|0\rangle$ (2.6) in more detail

$$A(v)B(\bar{u})|0\rangle = a(v)f(\bar{u},v)B(\bar{u})|0\rangle + \sum_{k=1}^{n} a(u_k)g(v,u_k)f(\bar{u}_k,u_k)B(v)B(\bar{u}_k)|0\rangle. \tag{2.38}$$

This formula is a consequence of the commutation relation (1.123). There are two terms in the rhs of this commutation relation: one is proportional to the function $f$, another is proportional to $g$. Acting with $A(v)$ on $B(\bar{u})|0\rangle$ we applied (1.123) several times. Therefore, products of the functions $f$ appear in (2.38). Surprisingly, however, this does not happen with the $g$ function. This function is included in the action (2.38) only linearly. What is the reason for such inequality between the functions $f$ and $g$?

To answer this question, let us consider commutation relations of one operator $A(v)$ with a product of two operators $B(u_1)$ and $B(u_2)$. Using (1.123) we obtain

$$A(v)B(u_1)B(u_2) = \Big(f(u_1,v)B(u_1)A(v) + g(v,u_1)B(v)A(u_1)\Big)B(u_2). \tag{2.39}$$

Applying (1.123) once more we arrive at

$$\begin{aligned} A(v)B(u_1)B(u_2) = {} & f(u_1,v)f(u_2,v)B(u_1)B(u_2)A(v) \\ & + g(v,u_1)f(u_2,u_1)B(v)B(u_2)A(u_1) \\ & + \Big(g(v,u_2)f(u_1,v) + g(v,u_1)g(u_1,u_2)\Big) \\ & B(v)B(u_1)A(u_2). \end{aligned} \tag{2.40}$$

On the other hand, the operators $B(u_1)$ and $B(u_2)$ commute with each other. Hence,

$$A(v)B(u_1)B(u_2) = A(v)B(u_2)B(u_1). \tag{2.41}$$

Thus, expression (2.40) does not change, if we replace there $u_1$ with $u_2$:

$$\begin{aligned} A(v)B(u_1)B(u_2) = & \, f(u_1,v)f(u_2,v)B(u_1)B(u_2)A(v) \\ & + g(v,u_2)f(u_1,u_2)B(v)B(u_1)A(u_2) \\ & + \Big(g(v,u_1)f(u_2,v) + g(v,u_2)g(u_2,u_1)\Big) \\ & B(v)B(u_2)A(u_1). \end{aligned} \tag{2.42}$$

Comparing (2.40) and (2.42), we see that, generally speaking, these expressions are different. The first terms in these formulas coincide, while the second and the third terms have a different structure. In order to have the coincidence we should require the following identity:

$$g(v,u_2)f(u_1,v) + g(v,u_1)g(u_1,u_2) = g(v,u_2)f(u_1,u_2). \tag{2.43}$$

It is easy to check that if we substitute here $f(u,v)$ and $g(u,v)$ from either (1.85) or (1.87), then (2.43) does become an identity. This equation shows the mechanism by which the expressions quadratic in $g$ (that is, the term $g(v,u_1)g(u_1,u_2)$) reduces to linear expressions in $g$.

This example clearly demonstrates the role of the Yang–Baxter equation in permutations of the operators $T_{ij}$. We recall that this equation was obtained as a compatibility condition for different permutations of three monodromy matrices. In particular, the Yang–Baxter equation guarantees that the result of the permutation of $A(v)$ and $B(u_1)B(u_2)$ coincides with the result of the permutation of $A(v)$ and $B(u_2)B(u_1)$. In fact, identity (2.43) is one of the scalar relations that arise if we write down the matrix Yang–Baxter equation in the components. Similarly, we can take any other three operators and rearrange them in various ways. The result will be the same regardless of the sequence of permutations.

Thus, we used the Yang–Baxter equation to derive the action formula (2.38), although this fact did not manifest itself explicitly.

## 2.2 Twisted monodromy matrix

Let a $2 \times 2$ matrix $\hat{\kappa}$ be such that its tensor square commutes with the $R$-matrix

$$[\hat{\kappa}_1\hat{\kappa}_2, R_{12}(u,v)] = 0. \tag{2.44}$$

If a monodromy matrix $T(u)$ satisfies the $RTT$-relation, then the matrix $\hat{\kappa}T(u)$ also satisfies this relation

$$R_{12}(u,v)\big(\hat{\kappa}T(u)\big)_1\big(\hat{\kappa}T(v)\big)_2 = \big(\hat{\kappa}T(v)\big)_2\big(\hat{\kappa}T(u)\big)_1 R_{12}(u,v). \tag{2.45}$$

In fact, equation (2.45) follows from proposition 1.1. Indeed, we can treat the matrix $\hat{\kappa}$ as a monodromy matrix whose matrix elements are numbers independent of the spectral parameter. Then the relation (2.44) is nothing else than the $RTT$-relation for such a monodromy matrix. Equation (2.45) holds due to the $RTT$-relation for the original matrix $T(u)$ and Proposition 1.1.

However, it is not difficult to prove (2.45) directly by simply multiplying the original $RTT$-relation by $\hat{\kappa}_1\hat{\kappa}_2$ from the left. Then we obtain in the lhs of the $RTT$-relation

$$\begin{aligned} \hat{\kappa}_1\hat{\kappa}_2 R_{12}(u,v)T_1(u)T_2(v) &= R_{12}(u,v)\hat{\kappa}_1\hat{\kappa}_2 T_1(u)T_2(v) \\ &= R_{12}(u,v)\big(\hat{\kappa}T(u)\big)_1\big(\hat{\kappa}T(v)\big)_2. \end{aligned} \tag{2.46}$$

We have used the condition (2.44) and the fact that the entries of the matrix $\hat{\kappa}$ commute with all matrix elements of the monodromy matrix, which leads to $\hat{\kappa}_2 T_1(u) = T_1(u)\hat{\kappa}_2$. In the rhs of the $RTT$-relation, we have

$$\hat{\kappa}_1\hat{\kappa}_2 T_2(v)T_1(u)R_{12}(u,v) = \big(\hat{\kappa}T(v)\big)_2\big(\hat{\kappa}T(u)\big)_1 R_{12}(u,v), \tag{2.47}$$

and we arrive at (2.45). It is easy to check that the initial monodromy matrix can also be multiplied by $\hat{\kappa}$ from the right, and the new matrix $T(u)\hat{\kappa}$ also satisfies the $RTT$-relation.

It is obvious that the $R$-matrix $R(u,v) = I + g(u,v)P$ commutes with $\hat{\kappa}_1\hat{\kappa}_2$ for an arbitrary $2\times 2$ matrix $\hat{\kappa}$. For this reason, this $R$-matrix is called $GL(2)$-invariant. However, as a result of multiplying $T(u)$ by $\hat{\kappa}$, we can break the properties of the vacuum vector. Indeed, if

$$\hat{\kappa} = \begin{pmatrix} \kappa_{11} & \kappa_{12} \\ \kappa_{21} & \kappa_{22} \end{pmatrix}, \tag{2.48}$$

then

$$\hat{\kappa}T(u) = \begin{pmatrix} A_\kappa(u) & B_\kappa(u) \\ C_\kappa(u) & D_\kappa(u) \end{pmatrix} = \begin{pmatrix} \kappa_{11}A(u) + \kappa_{12}C(u) & \kappa_{11}B(u) + \kappa_{12}D(u) \\ \kappa_{21}A(u) + \kappa_{22}C(u) & \kappa_{21}B(u) + \kappa_{22}D(u) \end{pmatrix}. \tag{2.49}$$

We see that the vacuum vector $|0\rangle$ is no longer annihilated by the operator $C_\kappa(u)$ and is not an eigenvector for the operator $D_\kappa(u)$.

**Remark 2.4.** However, note that the violation of these properties of the vector $|0\rangle$ is not very crucial. The algebraic Bethe ansatz can also be used in this case, although some techniques vary slightly (see, e.g., [S. Belliard and N. Crampé (2013); S. Belliard (2014); N. Crampé (2015); S. Belliard and R. Pimenta (2015a,b, 2016); J. Avan et al (2015); S. Belliard et al (2018a,b); J. Cao et al (2013a,b); Y. Wang et al (2015)]).

If the matrix $\hat{\kappa}$ is diagonal, say, $\hat{\kappa} = \mathrm{diag}(\kappa_1, \kappa_2)$, then the properties of the vacuum are not violated. In fact, without loss of generality, we can take the matrix $\hat{\kappa}$ in the form $\hat{\kappa} = \mathrm{diag}(1, \kappa)$ because multiplication of the monodromy matrix by an arbitrary constant is certainly not essential.

If the matrix $\hat{\kappa}$ is diagonal, then its tensor square commutes with the trigonometric $R$-matrix. Therefore, in this case we can also consider similar deformation of the original monodromy matrix.

The new matrix $\hat{\kappa}T(u)$ obtained in this way is called a *twisted monodromy matrix*. Its trace in the auxiliary space $A(u) + \kappa D(u)$ is called a *twisted transfer matrix*. The parameter $\kappa$ is called the *twist parameter*. The properties of the twisted monodromy matrix are entirely analogous to the properties of the original monodromy matrix. Only the value of the function $d(u)$ changes. Indeed, it is easy to see that

$$D_\kappa(u)|0\rangle = \kappa d(u)|0\rangle. \tag{2.50}$$

Therefore, in the action of the $D_\kappa(v)$ operator on the Bethe vectors $B(\bar{u})|0\rangle$, the parameter $\kappa$ arises. In particular, the action of the twisted transfer matrix $\mathcal{T}_\kappa(v) = A(v) + \kappa D(v)$ on Bethe vectors has the form

$$\begin{aligned} \mathcal{T}_\kappa(v)B(\bar{u})|0\rangle &= \Big(a(v)\Lambda(v|\bar{u}) + \kappa d(v)\tilde{\Lambda}(v|\bar{u})\Big)B(\bar{u})|0\rangle \\ &\quad + \sum_{k=1}^{n}\Big(a(u_k)\Lambda_k(v|\bar{u}) + \kappa d(u_k)\tilde{\Lambda}_k(v|\bar{u})\Big)B(v)B(\bar{u}_k)|0\rangle. \end{aligned} \tag{2.51}$$

It is easy to see that the only difference between this formula and (2.16) is that the function $d$ is multiplied by $\kappa$. This leads to the appearance of this parameter in the *twisted Bethe equations*

$$\frac{a(u_k)}{d(u_k)} = \kappa\frac{f(u_k, \bar{u}_k)}{f(\bar{u}_k, u_k)}, \qquad k = 1, \dots, n. \tag{2.52}$$

Eigenvectors of the twisted transfer matrix are called *twisted on-shell Bethe vectors*. They still have the form $B(\bar{u})|0\rangle$. However, the parameters $\bar{u}$ should satisfy twisted Bethe equations (2.52).

For possible physical applications, it makes sense to consider $\kappa = e^{i\varphi}$, where $\varphi$ is a real number. Then we can study models not only with periodic but also with quasi-periodic boundary conditions. However, the role of parameter $\kappa$ is much more significant. In what follows, we will have the opportunity to see this many times.

## 2.3 Bethe equations

The roots of the Bethe equations determine the spectrum of the transfer matrix and, thus, the spectrum of the Hamiltonian of the model under consideration. Of course, in the overwhelming majority of cases, it is impossible to find these roots analytically. This is not surprising. Indeed, suppose that we have found solutions to the Bethe equations for the $XXX$ model. It would mean that using such a method, we were able to solve an algebraic equation (1.5) of degree $2^N$ explicitly. It would be naive to hope to be so lucky.

Nevertheless, even without an explicit analytical solution, the Bethe equations give us much useful information about the quantum system that we are studying. In particular, it is possible to find the ground state of the system and to describe excitations. In cases where the system consists of many particles, one can describe its thermodynamic properties. Besides, the system of Bethe equations can be solved numerically, which ultimately makes it possible to compare the results of theoretical calculations with experimental data obtained in real physical systems.

Let us now discuss the question of spurious solutions of the Bethe equations.

### 2.3.1 *Coinciding roots*

Formally, the system of Bethe equations

$$\frac{a(u_k)}{d(u_k)} = \frac{f(u_k,\bar{u}_k)}{f(\bar{u}_k,u_k)}, \qquad k=1,\dots,n, \tag{2.53}$$

might have coinciding roots. For instance, we can look for a solution of (2.53), in which all the roots are equal to each other $u_j = u$, $j=1,\dots,n$. Then the system turns into one equation

$$\frac{a(u)}{d(u)} = (-1)^{n-1}, \tag{2.54}$$

since

$$\lim_{u\to v}\frac{f(u,v)}{f(v,u)} = -1. \tag{2.55}$$

For the models considered above, we obtain

$$\begin{aligned}
&\left(\frac{u+\frac{\eta}{2}}{u-\frac{\eta}{2}}\right)^N = (-1)^{n-1}, && XXX \text{ chain},\\
&\left(\frac{\sinh(u+\frac{\eta}{2})}{\sinh(u-\frac{\eta}{2})}\right)^N = (-1)^{n-1}, && XXZ \text{ chain},\\
&e^{-iuL} = (-1)^{n-1}, && \text{one-dimensional Bose gas.}
\end{aligned} \tag{2.56}$$

All these equations have solutions. However, the vector $B(\bar{u})|0\rangle$ with $u_j = u$, $j = 1, \dots, n$, generally speaking, is not an eigenvector even if $u$ solves equations (2.56). The point is that when deriving the Bethe equations, we assumed that all the parameters $u_j$ were distinct.

Let us turn back to the action formula (2.16)

$$\mathcal{T}(v)B(\bar{u})|0\rangle = \tau(v|\bar{u})B(\bar{u})|0\rangle + \sum_{k=1}^{n} g(v,u_k)\Big(a(u_k)f(\bar{u}_k,u_k) - d(u_k)f(u_k,\bar{u}_k)\Big) \\ B(v)B(\bar{u}_k)|0\rangle. \tag{2.57}$$

For simplicity, consider the case $n = 2$ and let the $R$-matrix be rational. Then (2.57) takes the form

$$\begin{aligned}\mathcal{T}(v)&B(\bar{u})|0\rangle \\ &= \tau(v|\bar{u})B(\bar{u})|0\rangle + \eta\,\frac{a(u_1)(u_2-u_1+\eta) - d(u_1)(u_2-u_1-\eta)}{(v-u_1)(u_2-u_1)}B(v)B(u_2)|0\rangle \\ &-\eta\,\frac{a(u_2)(u_1-u_2+\eta) - d(u_2)(u_1-u_2-\eta)}{(v-u_2)(u_2-u_1)}B(v)B(u_1)|0\rangle. \end{aligned}\tag{2.58}$$

Obviously, a singularity arises at $u_1 = u_2$. Taking the limit $u_1 \to u_2 = u$ we find

$$\begin{aligned}\mathcal{T}(v)B^2(u)|0\rangle &= \tau(v|\bar{u})B^2(u)|0\rangle + \frac{\eta^2}{v-u}\Big(a(u)+d(u)\Big)B(v)B'(u)|0\rangle \\ &+ \frac{\eta^2}{v-u}\Big(\tfrac{2}{\eta}\big(a(u)-d(u)\big) - a'(u) - d'(u) - \frac{a(u)+d(u)}{v-u}\Big) \\ &B(v)B(u)|0\rangle. \end{aligned}\tag{2.59}$$

We see that besides the original vector $B^2(u)|0\rangle$, we obtain two new vectors. First, this is the vector $B(v)B(u)|0\rangle$. Second, this is the vector $B(v)B'(u)|0\rangle$, which contains a derivative of the operator $B(u)$. One should set the coefficients of both vectors equal to zero. Otherwise, the original vector is not on-shell. Hence, we obtain two conditions

$$\begin{aligned}&a(u)+d(u) = 0, \\ &2\big(a(u)-d(u)\big) - \eta\big(a'(u)+d'(u)\big) = 0.\end{aligned}\tag{2.60}$$

Thus, although the number of variables has decreased (instead of $u_1$ and $u_2$, we have only one variable $u$), we still have two equations. The first of these equations formally coincides with the Bethe equations for $u_1 = u_2 = u$. However, we also have an additional equation, and the system is overdetermined. The solvability of such an overdetermined system depends on the specific values of the parameters of the model under consideration. For example, in the one-dimensional Bose gas model, the system (2.60) has a solution under the condition $cL = -4$. Obviously, there are no solutions for a positive coupling constant. However, even with a negative coupling constant, the solution exists only in an exceptional case. Any deviation of the length of the interval $L$ from a given value leads to an unsolvable system (2.60).

It is now clear what happens in the general case when the Bethe vector $B(\bar{u})|0\rangle$ depends on $n$ variables $u_1, \dots, u_n$. If there are coinciding ones among them, then singular terms appear in formula (2.57) since the function $f(u, v)$ has a pole at $u = v$. After solving these singularities, we get vectors containing derivatives of the $B$ operator. It is easy to verify that we will always have $n$ vectors of different types in the end. Their coefficients should be set equal to zero. Therefore, we always obtain a system of $n$ equations. At the same time, the number of variables decreases. Therefore, the resulting system will be overdetermined. Generally, such an overdetermined system is unsolvable.

Let us stress once more that the system (2.53) does have coinciding roots. However, as we have seen, a Bethe vector corresponding to this solution generically is not on-shell.

## 2.4 Spurious solutions in spin chains

Writing Bethe equations in the form of (2.53) is traditional but not entirely correct. Indeed, it follows from (2.57) that the conditions of the vanishing of the coefficients for unwanted vectors have the form

$$a(u_k) f(\bar{u}_k, u_k) = d(u_k) f(u_k, \bar{u}_k), \qquad k = 1, \dots, n. \tag{2.61}$$

The system (2.61) is equivalent to (2.53) if both sides of (2.61) do not vanish for all $k = 1, \dots, n$. However, in the $XXX$ model, one can find such solutions of (2.61), that both sides are equal to zero. Let us consider the case $n = 2$ again. Then the system (2.61) takes the form

$$\begin{aligned} (u_1 + \tfrac{\eta}{2})^N (u_1 - u_2 - \eta) &= (u_1 - \tfrac{\eta}{2})^N (u_1 - u_2 + \eta), \\ (u_2 + \tfrac{\eta}{2})^N (u_2 - u_1 - \eta) &= (u_2 - \tfrac{\eta}{2})^N (u_2 - u_1 + \eta). \end{aligned} \tag{2.62}$$

It is easy to see that the values $u_1 = \frac{\eta}{2}$ and $u_2 = -\frac{\eta}{2}$ are the roots of the system (2.62). At first glance, these roots are no worse or better than all other possible roots. However, these roots show a very curious feature of the system (2.62) (see, e.g., [R. J. Baxter (2002); V. Tarasov and A. Varchenko

(1995a); R. I. Nepomechie and C. Wang (2014)]). The matter is that the Bethe vector corresponding to this solution turns out to be a zero vector. A complete proof of this fact will be given later when we learn how to calculate scalar products of Bethe vectors. In the meantime, we will consider a specific example of a chain consisting of four sites. This example ($N = 4$, $n = 2$) is remarkable in that, on the one hand, it is non-trivial, and on the other hand, Bethe equations are explicitly solvable.

So, we want to build a vector $B(-\frac{\eta}{2})B(\frac{\eta}{2})|0\rangle$. First, let us find a vector $|\phi\rangle = B(\frac{\eta}{2})|0\rangle$. Setting $N = 4$ in (2.31) we obtain

$$B(u)|0\rangle = \eta \sum_{k=1}^{4} (u + \tfrac{\eta}{2})^{4-k}(u - \tfrac{\eta}{2})^{k-1} \cdot \sigma_k^-|0\rangle. \tag{2.63}$$

Setting here $u = \frac{\eta}{2}$ we see that only the term $k = 1$ survives in the sum. Therefore,

$$|\phi\rangle = \eta^4 \sigma_1^-|0\rangle, \tag{2.64}$$

or equivalently

$$|\phi\rangle = \eta^4 \left(\begin{smallmatrix} 0 \\ 1 \end{smallmatrix}\right)_1 \otimes \left(\begin{smallmatrix} 1 \\ 0 \end{smallmatrix}\right)_2 \otimes \left(\begin{smallmatrix} 1 \\ 0 \end{smallmatrix}\right)_3 \otimes \left(\begin{smallmatrix} 1 \\ 0 \end{smallmatrix}\right)_4. \tag{2.65}$$

Let us find now the operator $B(-\frac{\eta}{2})$. The $L$-operator at $u = -\frac{\eta}{2}$ has the form

$$L_n(-\tfrac{\eta}{2}) = \eta \begin{pmatrix} -E_n^{22} & E_n^{21} \\ E_n^{12} & -E_n^{11} \end{pmatrix}. \tag{2.66}$$

Multiplying the $L$-operators over the four sites of the chain we find an explicit expression for $B(-\frac{\eta}{2})$:

$$\begin{aligned} B(-\tfrac{\eta}{2}) = &-\eta^4(E_4^{22}E_3^{22} + E_4^{21}E_3^{12})(E_2^{22}E_1^{21} + E_2^{21}E_1^{11}) \\ &- \eta^4(E_4^{22}E_3^{21} + E_4^{21}E_3^{11})(E_2^{12}E_1^{21} + E_2^{11}E_1^{11}). \end{aligned} \tag{2.67}$$

It remains to act with this operator on the vector $|\phi\rangle$ (2.65). Taking into account obvious formulas

$$\begin{aligned} &E_k^{22}|\phi\rangle = 0, \qquad E_k^{12}|\phi\rangle = 0, \qquad k > 2, \\ &E_1^{11}|\phi\rangle = 0, \qquad E_1^{21}|\phi\rangle = 0, \end{aligned} \tag{2.68}$$

we easily convince ourselves that $B(-\frac{\eta}{2})|\phi\rangle = 0$.

Thus, the Bethe vector corresponding to the solution $u_1 = \frac{\eta}{2}$, $u_2 = -\frac{\eta}{2}$ is the zero vector. It seems that it is not difficult to fix the situation. Indeed, we can take the vector $B(u_1)B(u_2)|0\rangle$ and compute its norm $\mathcal{N}(u_1, u_2)$ in the case when the parameters $u_1$ and $u_2$ are generic complex numbers. Then the norm of the vector

$$\frac{1}{\mathcal{N}(u_1,u_2)}\,B(u_1)B(u_2)|0\rangle, \tag{2.69}$$

is equal to 1 for any $u_1$ and $u_2$. Hence, we can never obtain the zero vector. It remains only to take the limit $u_1 \to \frac{\eta}{2}$, $u_2 \to -\frac{\eta}{2}$ in (2.69). The problem, however, is that this limit depends on how $u_1$ and $u_2$ tend to their limit values (see Section 4.4 for more details). In our case, we do not have any additional prescriptions on how $u_1$ and $u_2$ go to $\frac{\eta}{2}$ and $-\frac{\eta}{2}$. One of the possible ways of solving the problem is introducing an additional parameter in the system of Bethe equations.

### 2.4.1 *Spurious solution of twisted Bethe equations*

To illustrate the role of the twist parameter $\kappa$ in the twisted Bethe equations, we continue to consider the example of a chain consisting of four sites: $N = 4$. For the sake of generality, let it now be the $XXZ$ chain. Let us construct the Bethe vector $B(u_1)B(u_2)|0\rangle$, dependent on two parameters $u_1$ and $u_2$. This vector belongs to the subspace with two spins up and two spins down. The dimension of this subspace is $\binom{4}{2} = 6$.

We require the parameters $u_1$ and $u_2$ to satisfy the twisted Bethe equations, which can be written in the form similar to (2.62):

$$\begin{aligned}
\sinh^4(u_1+\tfrac{\eta}{2})\sinh(u_1-u_2-\eta) &= \kappa\,\sinh^4(u_1-\tfrac{\eta}{2})\sinh(u_1-u_2+\eta),\\
\sinh^4(u_2+\tfrac{\eta}{2})\sinh(u_2-u_1-\eta) &= \kappa\,\sinh^4(u_2-\tfrac{\eta}{2})\sinh(u_2-u_1+\eta).
\end{aligned} \tag{2.70}$$

According to the results of [V. Tarasov and A. Varchenko (1995a)], eigenvectors of the twisted transfer matrix correspond to such solutions of the system (2.70), for which $u_1 \ne u_2$ and both sides of the equations do not vanish. Thus, the solution $u_1 = \eta/2$ and $u_2 = -\eta/2$ is forbidden in the sense that the corresponding vector is not a twisted on-shell vector. We have seen that a vector constructed from such the solution in the $XXX$ chain is the zero vector. It is easy to verify that the same is true for the $XXZ$ chain. Note also that this solution does not depend on $\kappa$.

If both sides of (2.70) do not vanish, then we can rewrite these equations in the form

$$\begin{aligned}
\left(\frac{\sinh(u_1+\frac{\eta}{2})}{\sinh(u_1-\frac{\eta}{2})}\right)^4\frac{\sinh(u_1-u_2-\eta)}{\sinh(u_1-u_2+\eta)} &= \kappa,\\
\left(\frac{\sinh(u_2+\frac{\eta}{2})}{\sinh(u_2-\frac{\eta}{2})}\right)^4\frac{\sinh(u_2-u_1-\eta)}{\sinh(u_2-u_1+\eta)} &= \kappa.
\end{aligned} \tag{2.71}$$

Setting here $\kappa = \theta^{-2}$ and

$$\frac{\sinh(u_j - \frac{\eta}{2})}{\sinh(u_j + \frac{\eta}{2})} = w_j, \qquad j = 1, 2, \tag{2.72}$$

we obtain a new system for $w_1$ and $w_2$:

$$\begin{aligned} w_1^4 \cdot \frac{1 + w_1 w_2 - 2\Delta w_2}{1 + w_1 w_2 - 2\Delta w_1} &= -\theta^2, \\ w_2^4 \cdot \frac{1 + w_1 w_2 - 2\Delta w_1}{1 + w_1 w_2 - 2\Delta w_2} &= -\theta^2, \end{aligned} \tag{2.73}$$

where $\Delta = \cosh \eta$. Observe that in such the parametrization, the system of Bethe equations is equally suitable for both $XXZ$ and $XXX$ chains. In the latter case, we simply put $\Delta = 1$. Note also that the roots $u_1 = \eta/2$ and $u_2 = -\eta/2$ correspond to the values $w_1 = 0$ and $w_2 = \infty$ in the new parametrization.

Taking the product of both equations we obtain $(w_1 w_2)^4 = \theta^4$. Hence, there are four possibilities: $w_1 w_2 = \pm i\theta$ and $w_1 w_2 = \pm\theta$.

Let $w_2 = \pm i\theta / w_1$. Substituting this into the first equation (2.73) we obtain

$$[w_1^2 \mp i\theta][(1 \pm i\theta)(w_1^2 \pm i\theta) \mp 2i\Delta\theta w_1] = 0. \tag{2.74}$$

If $w_1^2 = \pm i\theta$, then $w_1 = w_2$, and this solution does not correspond to an eigenvector of the twisted transfer matrix. Thus, it remains to solve the quadratic equation

$$(1 + i\theta)(w_1^2 + i\theta) - 2i\Delta\theta w_1 = 0, \quad \text{for} \quad w_1 w_2 = i\theta, \tag{2.75}$$

$$(1 - i\theta)(w_1^2 - i\theta) + 2i\Delta\theta w_1 = 0, \quad \text{for} \quad w_1 w_2 = -i\theta. \tag{2.76}$$

If $w_1 w_2 = i\theta$, then we obtain a solution

$$w_1 = \frac{i\Delta\theta}{1 + i\theta} + \sqrt{-i\theta - \frac{\Delta^2\theta^2}{(1 + i\theta)^2}}, \qquad w_2 = \frac{i\Delta\theta}{1 + i\theta} - \sqrt{-i\theta - \frac{\Delta^2\theta^2}{(1 + i\theta)^2}}. \tag{2.77}$$

**Remark 2.5.** Equation (2.75) has two roots $w_1^+$ and $w_1^-$. They are different by the sign of the square root. We have set $w_1 = w_1^+$ in the solution (2.77). It is easy to see that if we put $w_1 = w_1^-$, then it would be equivalent to the replacement $w_1$ with $w_2$ in the solution of the system (2.73). However, we do not distinguish between the Bethe equations solutions, which differ only by permutation, since they correspond to the same eigenvector.

Similarly, for the case $w_1 w_2 = -i\theta$, we find

$$w_1 = \frac{-i\Delta\theta}{1-i\theta} + \sqrt{i\theta - \frac{\Delta^2\theta^2}{(1-i\theta)^2}}, \qquad w_2 = \frac{-i\Delta\theta}{1-i\theta} - \sqrt{i\theta - \frac{\Delta^2\theta^2}{(1-i\theta)^2}}. \tag{2.78}$$

Let now $w_1 w_2 = \pm\theta$. Then we obtain the following equation for the parameter $w_1$:

$$(1 \pm \theta)(w_1^4 + \theta^2) \mp 2\Delta\theta w_1 (w_1^2 \pm \theta) = 0. \tag{2.79}$$

Dividing this equation by $w_1^2$ and setting

$$x = w_1 \pm \theta / w_1, \tag{2.80}$$

we obtain a quadratic equation for $x$:

$$(1 \pm \theta)(x^2 \mp 2\theta) \mp 2\Delta\theta x = 0. \tag{2.81}$$

Thus, in the case $w_1 w_2 = \theta$, we obtain two solutions

$$w_1 = \frac{x}{2} + \sqrt{\frac{x^2}{4} - \theta}, \qquad w_2 = \frac{x}{2} - \sqrt{\frac{x^2}{4} - \theta}, \tag{2.82}$$

where $x$ can take two values

$$x = \frac{\Delta\theta}{1+\theta} \pm \sqrt{\frac{\Delta^2\theta^2}{(1+\theta)^2} + 2\theta}. \tag{2.83}$$

Similarly, in the case $w_1 w_2 = -\theta$, we have

$$w_1 = \frac{x}{2} + \sqrt{\frac{x^2}{4} + \theta}, \qquad w_2 = \frac{x}{2} - \sqrt{\frac{x^2}{4} + \theta}, \tag{2.84}$$

where

$$x = -\frac{\Delta\theta}{1-\theta} \pm \sqrt{\frac{\Delta^2\theta^2}{(1-\theta)^2} - 2\theta}. \tag{2.85}$$

Thus, we have obtained six solutions to the twisted Bethe equations, for which we can construct six eigenvectors of the twisted transfer matrix. Of course, it is also necessary to verify the linear independence of these vectors. However, at least the number of eigenvectors obtained coincides with the dimension of the subspace with two spins up and two spins down. Therefore, there is a chance that this system of vectors forms a basis.

Let us now see what happens if, from the very beginning, we consider the usual system of Bethe equations and not the twisted one. To do this, we simply put $\theta = 1$ in the equations (2.73). Then for $w_1 w_2 = -1$ equation (2.79) takes the form

$$w_1(w_1^2 - 1) = 0, \tag{2.86}$$

leading to only one solution of Bethe equations: $w_1 = 1$ and $w_2 = -1$ (or vice versa). The solution $w_1 = 0$ does not seem to suit us, as in this case we can not satisfy the condition $w_1w_2 = -1$. However, looking at the formula (2.85) for $\theta \neq 1$, we immediately see that this solution can not be discarded. Indeed, for $w_1w_2 = -1$ and $\theta \to 1$, we obtain either $x = 0$ or $x = \infty$. In the first case, we reproduce the solution $w_1 = 1$ and $w_2 = -1$, and in the second, we obtain $w_1 = 0$ and $w_2 = \infty$, and we need to provide that the product of zero and infinity is $-1$. Note that it is this solution that corresponds to the values $u_1 = \eta/2$ and $u_2 = -\eta/2$ in the original parametrization. Clearly, we could not have found such a 'strange solution' if we had worked with the usual Bethe equations from the very beginning. As a result, we would get only five solutions, and accordingly, five on-shell vectors. The latter obviously can not form a basis in the 6-dimensional space.

Thus, we are dealing with a very curious situation. We have six admissible solutions of twisted Bethe equations for $\kappa \neq 1$. However, in the limit $\kappa \to 1$, one of these admissible solutions tends to the spurious solution $u_1 = \eta/2$, $u_2 = -\eta/2$. Recall once more that the latter exists for any $\kappa$. If we considered the usual Bethe equations (with $\kappa = 1$) from the beginning, we could not distinguish between the admissible and the spurious solutions.

We have already mentioned that if we normalize the original vector $B(u_1)B(u_2)|0\rangle$ and then consider the limit $u_1 \to \eta/2$, $u_2 \to -\eta/2$, then the result depends on how $u_1$ and $u_2$ tend to their limit values. The example considered above gives a hint to solve the problem. We should consider the twisted Bethe equations and take such a solution $u_1(\kappa)$ and $u_2(\kappa)$, that goes to the spurious solution in the limit $\kappa \to 1$. Then the normalized Bethe vector limit must be uniquely determined, since this time, we have fixed how $u_1$ and $u_2$ go to the limit. We will consider this issue in Section 4.4.

### 2.4.2 *On the completeness of the twisted on-shell Bethe vectors*

As one more example of using twisted Bethe equations, we consider the question of the completeness of the system of twisted on-shell Bethe vectors. As we have already mentioned, the proof of the completeness of the system of ordinary on-shell Bethe vectors in spin chains is very non-trivial. However, introducing an additional parameter to the Bethe equations makes it easy to prove completeness for at least some values of this parameter.

Consider an inhomogeneous $XXZ$ chain, in which all the inhomogeneities are pairwise distinct and satisfy a condition $\xi_j - \xi_k \pm \eta \neq 0$. The system of twisted Bethe equations reads

$$\left(\prod_{k=1}^{N} \sinh(u_j - \xi_k + \eta)\right) \left(\prod_{\ell=1}^{n} \sinh(u_j - u_\ell - \eta)\right)$$
$$= -\kappa \left(\prod_{k=1}^{N} \sinh(u_j - \xi_k)\right) \left(\prod_{\ell=1}^{n} \sinh(u_j - u_\ell + \eta)\right), \quad j = 1, \dots, n. \tag{2.87}$$

Recall that the eigenvectors of the twisted transfer matrix correspond to such solutions of the system (2.87), for which all the parameters $u_j$ are pairwise distinct, and both sides of the equations do not vanish for all $j = 1, \dots, n$ (admissible solutions). Spurious solutions, therefore, are those for which both sides of the equations (2.87) vanish at least for one $j$. In the case of the inhomogeneous $XXZ$ chain, such spurious solutions are, for example, $u_j = \xi_s$ and $u_k = \xi_s - \eta$ for some $j, k \in \{1, \dots, n\}$, and some $s \in \{1, \dots, N\}$. These are analogs of the solutions discussed above.

Another type of spurious solution arises with special values of the parameter $\eta = i\pi p/q$, where $p$ and $q$ are positive integers, with $p < q$ and $q \leq n$. It is easy to check that in this case, a subset of the parameters $u_1, \dots, u_q$, such that $u_j = w + \eta(j-1)$, satisfies the equations with $j = 1, \dots, q$, and again, both sides of these equations vanish. Remarkably, the constant $w$ is not fixed, and thus, these solutions are not isolated from each other.

All the spurious solutions can be characterized by the property that they do not depend on $\kappa$. However, if both sides of the Bethe equations do not vanish, then the roots of the system (2.87) depend on the parameter $\kappa$, that is, $u_j = u_j(\kappa)$. Therefore, we can treat the twisted Bethe equations as the specification of implicit functions $u_j(\kappa)$. For $\kappa = 0$, the system (2.87) turns into

$$\left(\prod_{k=1}^{N} \sinh(u_j - \xi_k + \eta)\right) \left(\prod_{\ell=1}^{n} \sinh(u_j - u_\ell - \eta)\right) = 0, \qquad j = 1, \dots, n, \tag{2.88}$$

and has obvious admissible solutions $u_j = \xi_{s_j} - \eta$, where $\xi_{s_j}$ is any from the set $\bar{\xi}$. In total, there are $\binom{N}{n}$ sets of roots $\bar{u} = \{u_1, \dots, u_n\}$, consisting of pairwise distinct numbers. Thus, the number of solutions to the twisted Bethe equations for $\kappa = 0$ coincides with the dimension of the subspace with $n$ spins down and $N - n$ spins up.

The matrix of the first derivatives of the system (2.88) in the point $u_j = \xi_{s_j} - \eta$, $(j = 1, \ldots, n)$, has the form

$$
\begin{aligned}
&\frac{\partial}{\partial u_m}\left(\prod_{k=1}^{N}\sinh(u_j-\xi_k+\eta)\right)\left(\prod_{\ell=1}^{n}\sinh(u_j-u_\ell-\eta)\right)\Big|_{u_j=\xi_{s_j}-\eta}\\
&\quad=\delta_{jm}\left(\prod_{\substack{k=1\\k\neq s_j}}^{N}\sinh(\xi_{s_j}-\xi_k)\right)\left(\prod_{\ell=1}^{n}\sinh(\xi_{s_j}-\xi_{s_\ell}-\eta)\right). \qquad (2.89)
\end{aligned}
$$

The Jacobian of the system is non-degenerate. It was for this purpose that we considered the inhomogeneous model and imposed the condition $\xi_j - \xi_k \pm \eta \neq 0$. Then all the conditions of the implicit function theorem are satisfied, and we can formulate

**Proposition 2.1.** *There exists $\kappa_0 > 0$ such that for $|\kappa| < \kappa_0$, the system (2.87) has solutions $u_j = u_j(\kappa)$, and $u_j(\kappa)$ are analytic functions of $\kappa$, such that $u_j(0) = \xi_{s_j} - \eta$.*

Thus, for $|\kappa|$ small enough, the system (2.87) has exactly $\binom{N}{n}$ admissible solutions. In Section 6.4.1, we will prove the linear independence of the corresponding twisted on-shell Bethe vectors. Hence, these vectors form a basis in the subspace of states with $n$ spins down and $N - n$ spins up.

In the case of the homogeneous chain $\xi_k = \eta/2$, the Jacobian of the system of twisted Bethe equations becomes degenerate for $\kappa = 0$, and the implicit function theorem becomes inapplicable in its classical formulation. However, this theorem has many generalizations, which allow us to consider this case [S. G. Krantz and H. R. Parks (2013)]. Then the assertion about the existence of the necessary number of admissible solutions is made for a punctured vicinity $0 < |\kappa| < \kappa_0$. Therefore, the condition of inhomogeneity of the chain is not fundamental but only facilitates the proof of the completeness of the system of twisted on-shell vectors. More important is the presence of the twist parameter $\kappa$.

The implicit function theorem says nothing about the size of the neighborhood $|\kappa| < \kappa_0$, and we can not say that the roots $u_j(\kappa)$ can be analytically continued to the point $\kappa = 1$. Consequently, the completeness of the admissible solutions of the twisted Bethe equations does not yet imply the completeness of the admissible solutions of the ordinary Bethe equations. Nevertheless, the completeness of the system of twisted on-shell vectors, at least for sufficiently small $|\kappa|$, is an important fact. We will use it in

Section 11.5 when calculating the correlation functions of the $XXZ$ chain. Recall, however, that so far, we have only proved that the number of admissible solutions of the twisted Bethe equations coincides with the dimension of the subspace in which the corresponding twisted on-shell Bethe vectors lie. Completeness will follow after the proof of their linear independence.

## 2.5 Dual Bethe vectors

To study scalar products of Bethe vectors, we need to introduce the concept of *dual Bethe vectors*.

The monodromy matrix elements act in a Hilbert space $\mathcal{H}$ with the vacuum vector $|0\rangle$. Assume that they also act in a dual space $\mathcal{H}^*$ with a dual vacuum vector $\langle 0|$, which has properties

$$\langle 0|A(u) = a(u)\langle 0|, \qquad \langle 0|D(u) = d(u)\langle 0|, \qquad \langle 0|B(u) = 0. \tag{2.90}$$

Here functions $a(u)$ and $d(u)$ are the same as in the case of the vacuum vector $|0\rangle$. We also assume that

$$\langle 0| = |0\rangle^\dagger, \qquad \langle 0|0\rangle = 1. \tag{2.91}$$

In the dual space, the operator $C$ plays the role of the creation operator. Using this operator, one can construct the dual Bethe vectors $\langle 0|C(\bar{u})$. Properties of these vectors are entirely analogous to the properties of the ordinary Bethe vectors. In particular, it is easy to show that the dual vector becomes an eigenvector of the transfer matrix (a dual on-shell Bethe vector) if the parameters $\bar{u}$ satisfy the system of Bethe equations.

The dual on-shell vectors are not necessarily Hermitian conjugate to the ordinary on-shell Bethe vectors. For example, for the $XXX$ chain with $\eta = i$, we have $D(u^*) = A^\dagger(u)$ and $C(u^*) = -B^\dagger(u)$. These properties can easily be proved by induction on the number of sites in the chain. For one $L$-operator

$$L_n(u) = \begin{pmatrix} u + \frac{i}{2}\,\sigma_n^z & i\,\sigma_n^- \\ i\,\sigma_n^+ & u - \frac{i}{2}\,\sigma_n^z \end{pmatrix}, \tag{2.92}$$

they are correct. Next, we assume that they are valid for $T(u)$, which is the product of $(N-1)$ $L$-operators, and multiply this monodromy matrix with the $L$-operator in the $N$th site. By a direct calculation, we see that the relations between the elements of the monodromy matrices are still valid.

Hence, if $\bar{u}^* = \bar{u}$ (that is, the set $\bar{u} = \{u_1, \dots, u_n\}$ consists of real or complex conjugated numbers), then

$$\langle 0|C(\bar{u}) = (-1)^n \big(B(\bar{u})|0\rangle\big)^\dagger. \tag{2.93}$$

The solutions of the Bethe equations do have the property $\bar{u}^* = \bar{u}$ (see, e.g., [A. A. Vladimirov (1986)]). Therefore, in the case of on-shell vectors, the property (2.93) is satisfied.[1] However, if we consider the inhomogeneous $XXX$ chain or deal with twisted Bethe equations, all of the above properties of the operators and vectors can be violated. Generally speaking, for such 'non-physical' models, the basis of the on-shell vectors $B(\bar{u})|0\rangle$ is not orthogonal. Therefore, in this case, we should find a dual basis in the dual space. This dual basis does not have to be Hermitian conjugated to the original one. As we will see later, the dual Bethe vectors $\langle 0|C(\bar{u})$ do form the desired dual basis.

## 2.6 Generalized $L$-operator

It is clear that multiplying the monodromy matrix by an arbitrary scalar function $T(u) \to \phi(u)T(u)$ leads only to a change of the functions $a(u) \to \phi(u)a(u)$ and $d(u) \to \phi(u)d(u)$. Thus, essentially different monodromy matrices are parameterized by different ratios $r(u) = a(u)/d(u)$. It is not accidental that just this ratio is included in the Bethe equations. In this section, we consider models with the rational $R$-matrix and show that we can construct a monodromy matrix for which $r(u)$ is the ratio of two arbitrary polynomials of the same degree. This observation underlies the concept of the generalized model, which we will discuss in Section 5.1.2.

Recall the $L$-operator (1.74) of the lattice QNLSE model

$$L_n(u) = \begin{pmatrix} 1 - \frac{iu\Delta}{2} + \frac{c\Delta^2}{2}\psi_n^\dagger\psi_n & -i\Delta\psi_n^\dagger\rho_n^+ \\ i\Delta\rho_n^-\psi_n & 1 + \frac{iu\Delta}{2} + \frac{c\Delta^2}{2}\psi_n^\dagger\psi_n \end{pmatrix}. \tag{2.94}$$

Here $\rho_n^\pm = \rho_n^\pm(\psi_n^\dagger\psi_n)$ and

$$\rho_n^+\rho_n^- = c + \frac{c^2\Delta^2}{4}\psi_n^\dagger\psi_n. \tag{2.95}$$

Recall that $\Delta$ is an arbitrary complex parameter. We can add three more arbitrary constants to the $L$-operator (2.94). First, one can shift the spectral parameter $u$ by some number $\xi_n$ because the $R$-matrix depends on the difference of the arguments. Secondly, we can multiply the $L$-operator by a

[1]The factor $(-1)^n$ does not play any role because eigenvectors are determined up to a phase factor.

twist matrix $\hat{\kappa}_n = \operatorname{diag}(\kappa_{n1}, \kappa_{n2})$. As a result, we obtain

$$\tilde{L}_n(u) = \begin{pmatrix} \kappa_{n1}\Big(1 - \frac{i(u-\xi_n)\Delta}{2} \\ \qquad + \frac{c\Delta^2}{2}\psi_n^\dagger\psi_n\Big) & -i\kappa_{n1}\Delta\psi_n^\dagger\rho_n^+ \\ i\kappa_{n2}\Delta\rho_n^-\psi_n & \kappa_{n2}\Big(1 + \frac{i(u-\xi_n)\Delta}{2} \\ & \qquad + \frac{c\Delta^2}{2}\psi_n^\dagger\psi_n\Big) \end{pmatrix}. \tag{2.96}$$

**Remark 2.6.** We can also consider the case where one of the parameters $\kappa_{nj}$ is zero, but then we will get a too poor model. Therefore, we assume that the parameters $\kappa_{n1}$ and $\kappa_{n2}$ are non-vanishing.

The action of the diagonal matrix elements on the vacuum $|0\rangle$ is now given by

$$\big(\tilde{L}_n\big)_{11}(u)|0\rangle = (a_n^{(1)}u + a_n^{(0)})|0\rangle, \qquad \big(\tilde{L}_n\big)_{22}(u)|0\rangle = (d_n^{(1)}u + d_n^{(0)})|0\rangle, \tag{2.97}$$

where

$$\begin{aligned} a_n^{(1)} &= -\frac{i\kappa_{n1}\Delta}{2}, & a_n^{(0)} &= \kappa_{n1} + \frac{i\kappa_{n1}\xi_n\Delta}{2}, \\ d_n^{(1)} &= \frac{i\kappa_{n2}\Delta}{2}, & d_n^{(0)} &= \kappa_{n2} - \frac{i\kappa_{n2}\xi_n\Delta}{2}. \end{aligned} \tag{2.98}$$

Since $\Delta$, $\kappa_{n1}$, $\kappa_{n2}$, and $\xi_n$, are arbitrary complex, the values of $a_n^{(0)}$ and $d_n^{(0)}$ also are arbitrary, while $a_n^{(1)}$, $d_n^{(1)}$ are arbitrary non-vanishing (see the remark above).

Now we can build a monodromy matrix via the usual method

$$\tilde{T}(u) = \tilde{L}_N(u)\cdots\tilde{L}_1(u). \tag{2.99}$$

Then the ratio of the vacuum eigenvalues of the operators $A(u)$ and $D(u)$ is

$$r(u) = \frac{a(u)}{d(u)} = \prod_{n=1}^{N} \frac{a_n^{(1)}u + a_n^{(0)}}{d_n^{(1)}u + d_n^{(0)}}, \tag{2.100}$$

where $N$ is an arbitrary natural number. Thus, we have constructed a monodromy matrix for which the ratio $a(u)/d(u)$ is an arbitrary rational function bounded at infinity by a nonzero constant. Furthermore, in the limit $N \to \infty$, we can obtain functions that have essential singularities, cuts, and so on. In particular, in the continuous limit of the lattice QNLSE model, the functions $a(u)$ and $d(u)$ become exponents (1.113). Therefore, we can say that the matrix (2.99) is the monodromy matrix of the most general form, which satisfies the $RTT$-relation with the rational $R$-matrix and has a vacuum vector [A. G. Izergin and V. E. Korepin (1985)].

In particular, we should be able to reproduce the monodromy matrix of the $XXX$ chain. Indeed, this is possible. Let

$$\tilde{L}_n(u) = \frac{2i}{\Delta}\,\sigma^z L_n(u) = \begin{pmatrix} u + \frac{2i}{\Delta} + ic\Delta\psi_n^\dagger\psi_n & 2\psi_n^\dagger\rho_n^+ \\ 2\rho_n^-\psi_n & u - \frac{2i}{\Delta} - ic\Delta\psi_n^\dagger\psi_n \end{pmatrix}. \tag{2.101}$$

It is easy to see that the $L$-operator (2.101) can be presented in the form

$$\tilde{L}_n(u) = u - ic \sum_{\alpha=x,y,z} \sigma^\alpha t_n^\alpha, \tag{2.102}$$

where

$$\begin{aligned} t_n^x &= \frac{i}{c}(\psi_n^\dagger\rho_n^+ + \rho_n^-\psi_n), \\ t_n^y &= \frac{1}{c}(\rho_n^-\psi_n - \psi_n^\dagger\rho_n^+), \\ t_n^z &= -\frac{2}{c\Delta} - \Delta\psi_n^\dagger\psi_n. \end{aligned} \tag{2.103}$$

A direct calculation shows that the operators $t_n^\alpha$ obey commutation relations

$$[t_n^\alpha, t_n^\beta] = i\varepsilon_{\alpha\beta\gamma} t_n^\gamma. \tag{2.104}$$

Here $\varepsilon_{\alpha\beta\gamma}$ is the antisymmetric tensor. It is also easy to check that

$$(t_n)^2 \equiv (t_n^x)^2 + (t_n^y)^2 + (t_n^z)^2 = s(s+1), \qquad \text{where} \qquad s = -\frac{2}{c\Delta}, \tag{2.105}$$

and hence, we obtained a representation of the $su(2)$ algebra of spin $s$. In fact, we have reproduced a well-known Holstein–Primakov transformation from the creation and annihilation operators to the spin operators [T. Holstein, H. Primakoff (1940)]. Generally speaking, the constructed representation is infinite-dimensional. However, in the particular case $s = 1/2$, we can put $t_n^\alpha = \frac{\sigma_n^\alpha}{2}$, $\alpha = x, y, z$. Then, taking into account that $-ic = \eta$, we do reproduce the $L$-operator of the $XXX$ chain.

# Chapter 3

# Quantum inverse problem

We construct a monodromy matrix as a product of $L$-operators over all lattice sites (1.50). Therefore, operators $A$, $B$, $C$, and $D$ are global in the sense that they depend on local operators in all sites. We had already seen an example of such a global operator when we constructed the $B$ operator for the $XXX$ spin chain of four sites (see (2.67)). It depends on the standard basis matrices $E_k^{ab}$ acting in all local quantum spaces. In this example, we could obtain an explicit formula for the $B$ operator because the chain was relatively short. However, for long chains, one can hardly hope to obtain a compact and easy-to-use expression for the monodromy matrix elements in terms of local operators. The idea of expressing local operators in terms of global $A$, $B$, $C$, and $D$ seems even more hopeless and absurd.

However, the example of $XXX$ and $XXZ$ chains disproves all of these 'obvious' arguments. In these models, one can obtain simple and very nice formulas for the local operators in terms of the monodromy matrix entries [N. Kitanine et al (1999)]. These formulas are called the *solution of the quantum inverse problem*, which is considered in this chapter. We consider an inhomogeneous $XXZ$ chain for definiteness, although all our considerations remain valid in a more general case of the $XYZ$ chain. In Section 3.1, we prove that the transfer matrix of the $XXZ$ chain turns into a cyclic permutation operator if its argument coincides with an inhomogeneity. Using this property, we obtain an explicit solution to the quantum inverse problem in Section 3.2.

## 3.1 Cyclic permutations in the monodromy matrix

We consider an inhomogeneous $XXZ$ chain, in which the monodromy matrix depends on inhomogeneities $\bar{\xi} = \{\xi_1, \ldots, \xi_N\}$

$$
\begin{aligned}
T_0(u) &= T_{0;1\ldots N}(u|\{\xi\}) = R_{0N}(u, \xi_N) \cdots R_{01}(u, \xi_1) \\
&= \begin{pmatrix} A_{1\ldots N}(u|\{\xi\}) & B_{1\ldots N}(u|\{\xi\}) \\ C_{1\ldots N}(u|\{\xi\}) & D_{1\ldots N}(u|\{\xi\}) \end{pmatrix}_0 .
\end{aligned} \tag{3.1}
$$

This matrix acts in the tensor product of spaces $V_0 \otimes V_1 \otimes \cdots \otimes V_N$, where each space $V_k \sim \mathbb{C}^2$. The space $V_0$ is auxiliary. In this space, the monodromy matrix can be written as a $2\times 2$ matrix, whose matrix elements are operators acting in the space $V_1 \otimes \cdots \otimes V_N$. The notation in (3.1) stresses all this. The short notation $T_0(u)$ means that being a $2 \times 2$ matrix, the monodromy matrix acts in the space $V_0$. The more detailed notation $T_{0;1\ldots N}(u|\{\xi\})$ shows that the matrix elements of $T_0(u)$ act in the space $V_1 \otimes \cdots \otimes V_N$, and each space $V_k$ is associated with the inhomogeneity $\xi_k$. The subscript 0 is separated from the rest of the indices by a semicolon to emphasize that $V_0$ is the auxiliary space. The same meaning is given by the subscript of the $2 \times 2$ matrix in the second line of (3.1).

Let the $R$-matrix be normalized so as it turns into the permutation matrix if its arguments coincide: $R(u, u) = P$. Consider the transfer matrix $\mathcal{T}(u)$ at $u = \xi_k$, $k = 1, \ldots, N$:

$$
\mathcal{T}(\xi_k) = \mathrm{tr}_0 T_0(\xi_k) = A_{1\ldots N}(\xi_k|\{\xi\}) + D_{1\ldots N}(\xi_k|\{\xi\}). \tag{3.2}
$$

The trace is taken over the space $V_0$, which is stressed by the subscript 0 of the symbol tr. Substituting the product of the $R$-matrices from (3.1) into (3.2) and taking into account the cyclicity of the trace, we find

$$
\begin{aligned}
\mathcal{T}(\xi_k) &= \mathrm{tr}_0 T_{0;1\ldots N}(\xi_k|\{\xi\}) = \mathrm{tr}_0\Big(R_{0N}(\xi_k, \xi_N) \cdots R_{01}(\xi_k, \xi_1)\Big) \\
&= \mathrm{tr}_0\Big(P_{0k} R_{0k-1}(\xi_k, \xi_{k-1}) \cdots R_{01}(\xi_k, \xi_1) R_{0N}(\xi_k, \xi_N) \cdots \\
&\qquad \times R_{0k+1}(\xi_k, \xi_{k+1})\Big).
\end{aligned} \tag{3.3}
$$

Here we also used $R_{0k}(\xi_k, \xi_k) = P_{0k}$. Let us add the identity operator in the form $\mathbf{1} = P_{0k}^2$ on the right of the product of all the matrices:

$$
\begin{aligned}
\mathcal{T}(\xi_k) = \mathrm{tr}_0\Big(&P_{0k} R_{0k-1}(\xi_k, \xi_{k-1}) \cdots R_{01}(\xi_k, \xi_1) \\
&\times R_{0N}(\xi_k, \xi_N) \cdots R_{0k+1}(\xi_k, \xi_{k+1}) P_{0k} P_{0k}\Big).
\end{aligned} \tag{3.4}
$$

Due to the properties of the permutation matrix, we should replace the space $V_0$ by $V_k$ in the product $R_{0k-1}(\xi_k,\xi_{k-1})\cdots R_{0k+1}(\xi_k,\xi_{k+1})$ between two $P_{0k}$. We obtain

$$\begin{aligned}\mathcal{T}(\xi_k) = \mathrm{tr}_0\Big(&R_{kk-1}(\xi_k,\xi_{k-1})\cdots R_{k1}(\xi_k,\xi_1)\\ &\times R_{kN}(\xi_k,\xi_N)\cdots R_{kk+1}(\xi_k,\xi_{k+1})P_{0k}\Big).\end{aligned} \tag{3.5}$$

Now only $P_{0k}$ acts in the space $V_0$. All other $R$-matrices can be taken out from the trace symbol $\mathrm{tr}_0$. Since $\mathrm{tr}_0 P_{0k} = \mathbf{1}$, we find

$$\mathcal{T}(\xi_k) = R_{kk-1}(\xi_k,\xi_{k-1})\cdots R_{k1}(\xi_k,\xi_1)R_{kN}(\xi_k,\xi_N)\cdots R_{kk+1}(\xi_k,\xi_{k+1}). \tag{3.6}$$

It turns out that the products of transfer matrices $\mathcal{T}(\xi_k)$ are the operators of cyclic permutations. More precisely,

$$\begin{aligned}&\prod_{k=1}^{m}\mathcal{T}(\xi_k)\cdot T_{0;1\ldots N}(u|\{\xi\})\cdot\prod_{k=1}^{m}\mathcal{T}^{-1}(\xi_k)\\ &\quad= R_{0m}(u,\xi_m)\cdots R_{01}(u,\xi_1)R_{0N}(u,\xi_N)\cdots R_{0m+1}(u,\xi_{m+1}).\end{aligned} \tag{3.7}$$

In other words, the operators $\prod_{k=1}^{m}\mathcal{T}(\xi_k)$ make cyclic permutations of the $R$-matrices when acting on the monodromy matrix $T_{0;1\ldots N}(u|\{\xi\})$. Let us check this.

We begin with the case $m=1$. First of all, we recall the Yang–Baxter equation in the tensor product of the spaces $V_a\otimes V_b\otimes V_c$:

$$R_{ab}(\mu_a,\mu_b)R_{ac}(\mu_a,\mu_c)R_{bc}(\mu_b,\mu_c) = R_{bc}(\mu_b,\mu_c)R_{ac}(\mu_a,\mu_c)R_{ab}(\mu_a,\mu_b), \tag{3.8}$$

for arbitrary complex $\mu_a$, $\mu_b$, and $\mu_c$.

Let us consider the product $\mathcal{T}(\xi_1)T_0(u)$:

$$\mathcal{T}(\xi_1)T_{0;1\ldots N}(u|\{\xi\}) = \mathcal{T}(\xi_1)R_{0N}(u,\xi_N)\cdots R_{02}(u,\xi_2)R_{01}(u,\xi_1). \tag{3.9}$$

Substituting here $\mathcal{T}(\xi_1)$ from (3.6), we find

$$\begin{aligned}\mathcal{T}(\xi_1)T_{0;1\ldots N}(u|\{\xi\}) = R_{1N}(\xi_1,\xi_N)\cdots &\underline{R_{12}(\xi_1,\xi_2)}\\ \times R_{0N}(u,\xi_N)\cdots &\underline{R_{02}(u,\xi_2)R_{01}(u,\xi_1)}.\end{aligned} \tag{3.10}$$

Observe that the combination of the underlined terms coincides with the rhs of the equation (3.8), if we put in the latter $a=0$, $b=1$, $c=2$ and $\mu_a=u$, $\mu_b=\xi_1$, $\mu_c=\xi_2$. In addition, the $R$-matrix $R_{12}(\xi_1,\xi_2)$ acts trivially in the space $V_0\otimes V_3\otimes\cdots\otimes V_N$. Therefore, it commutes with all $R$-matrices $R_{0k}(u,\xi_k)$ for $k=3,\ldots,N$. Thus, we can carry $R_{12}(\xi_1,\xi_2)$ to the right through the product $R_{0N}(u,\xi_N)\cdots R_{03}(u,\xi_3)$. After this, we get precisely the rhs of the equation (3.8). Replacing it with the lhs, we arrive at the following expression:

$$\begin{aligned}\mathcal{T}(\xi_1)T_{0;1\dots N}(u|\{\xi\}) = R_{1N}(\xi_1,\xi_N)\cdots \underline{R_{13}(\xi_1,\xi_3)}R_{0N}(u,\xi_N)\cdots \\ \times \underline{R_{03}(u,\xi_3)R_{01}(u,\xi_1)}R_{02}(u,\xi_2)R_{12}(\xi_1,\xi_2). \quad (3.11)\end{aligned}$$

We see that we again can apply the Yang–Baxter equation to the new underlined terms in (3.11), which gives us

$$\begin{aligned}\mathcal{T}(\xi_1)T_{0;1\dots N}(u|\{\xi\}) = R_{1N}(\xi_1,\xi_N)\cdots \underline{R_{14}(\xi_1,\xi_4)}R_{0N}(u,\xi_N)\cdots \\ \times \underline{R_{04}(u,\xi_4)R_{01}(u,\xi_1)}R_{03}(u,\xi_3) \\ \times R_{02}(u,\xi_2)R_{13}(\xi_1,\xi_3)R_{12}(\xi_1,\xi_2), \qquad (3.12)\end{aligned}$$

and so on. Continuing this process we move all the $R$-matrices $R_{1k}(\xi_1,\xi_k)$ to the right and eventually obtain

$$\begin{aligned}&\mathcal{T}(\xi_1)T_{0;1\dots N}(u|\{\xi\}) \\ &\quad= R_{01}(u,\xi_1)R_{0N}(u,\xi_N)\cdots R_{02}(u,\xi_2)\times R_{1N}(\xi_1,\xi_N)\cdots R_{12}(\xi_1,\xi_2) \\ &\quad= R_{01}(u,\xi_1)R_{0N}(u,\xi_N)\cdots R_{02}(u,\xi_2)\mathcal{T}(\xi_1). \qquad (3.13)\end{aligned}$$

Thus, equation (3.7) is proved for $m=1$.

To prove (3.7) for $m=2$ we should check that

$$\begin{aligned}&\mathcal{T}(\xi_2)R_{01}(u,\xi_1)R_{0N}(u,\xi_N)\cdots R_{02}(u,\xi_2)\mathcal{T}^{-1}(\xi_2) \\ &\quad= R_{02}(u,\xi_2)R_{01}(u,\xi_1)R_{0N}(u,\xi_N)\cdots R_{03}(u,\xi_3). \qquad (3.14)\end{aligned}$$

We could repeat all the above arguments related to the permutation of $R$-matrices via the Yang–Baxter equation. However, this is not necessary since a simple change of variables reduces our problem to the case $m=1$. Indeed, let us make a cyclic change of variables $\xi_k$ to $\xi'_k$ and the spaces $V_k$ to $V'_k$:

$$\begin{array}{cccc} \xi_1 & \xi_2 \;\dots & \xi_{N-1} & \xi_N \\ \downarrow & \downarrow & \downarrow & \downarrow \\ \xi'_N & \xi'_1 \;\dots & \xi'_{N-2} & \xi'_{N-1} \end{array} \qquad \begin{array}{cccc} V_1 & V_2 \;\dots & V_{N-1} & V_N \\ \downarrow & \downarrow & \downarrow & \downarrow \\ V'_N & V'_1 \;\dots & V'_{N-2} & V'_{N-1} \end{array} \qquad (3.15)$$

Then, using the new notation we recast (3.14) in the form

$$\begin{aligned}&\mathcal{T}(\xi'_1)R_{0N}(u,\xi'_N)R_{0N-1}(u,\xi'_{N-1})\cdots R_{01}(u,\xi'_1)\mathcal{T}^{-1}(\xi'_1) \\ &\quad= R_{01}(u,\xi'_1)R_{0N}(u,\xi'_N)R_{0N-1}(u,\xi'_{N-1})\cdots R_{02}(u,\xi'_2). \qquad (3.16)\end{aligned}$$

This form precisely coincides with already proved (3.13). It is clear that the process can be continued for $m=3,4,\dots$, and we thus prove (3.7) for an arbitrary $m$.

## 3.2 Quantum inverse problem

The formulas given above are preparatory for our main goal: to express an arbitrary matrix $E_m$ that acts non-trivially in the space $V_m$ in terms of the matrix $E_0$ acting in the auxiliary space $V_0$. For this, we consider the following expression:

$$\mathrm{tr}_0\Big(T_{0;1\dots N}(\xi_m|\{\xi\})E_0\Big). \tag{3.17}$$

Using equation (3.7) we can rewrite it as

$$\begin{aligned}
&\mathrm{tr}_0\Big(T_{0;1\dots N}(\xi_m|\{\xi\})E_0\Big)\\
&\quad = \mathrm{tr}_0\left[\left(\prod_{k=1}^{m}\mathcal{T}^{-1}(\xi_k)\right)P_{0m}R_{0m-1}(\xi_m,\xi_{m-1})\cdots R_{01}(\xi_m,\xi_1)\right.\\
&\qquad\quad \left.\times\ R_{0N}(\xi_m,\xi_N)\cdots R_{0m+1}(\xi_m,\xi_{m+1})\left(\prod_{k=1}^{m}\mathcal{T}(\xi_k)\right)E_0\right],
\end{aligned} \tag{3.18}$$

where we took into account that $R_{0m}(\xi_m,\xi_m) = P_{0m}$. Since the transfer matrices $\mathcal{T}(\xi_k)$ trivially act in $V_0$, we can take them out of the trace $\mathrm{tr}_0$

$$\begin{aligned}
&\mathrm{tr}_0\Big(T_{0;1\dots N}(\xi_m|\{\xi\})E_0\Big)\\
&\quad = \left(\prod_{k=1}^{m}\mathcal{T}^{-1}(\xi_k)\right)\mathrm{tr}_0\Big[P_{0m}R_{0m-1}(\xi_m,\xi_{m-1})\cdots R_{01}(\xi_m,\xi_1)\\
&\qquad\quad \times R_{0N}(\xi_m,\xi_N)\cdots R_{0m+1}(\xi_m,\xi_{m+1})E_0P_{0m}P_{0m}\Big]\left(\prod_{k=1}^{m}\mathcal{T}(\xi_k)\right).
\end{aligned} \tag{3.19}$$

Here, similarly to (3.4), we have inserted the identity operator in the form $\mathbf{1} = P_{0m}^2$. Using the basic property of the permutation matrix we obtain

$$\begin{aligned}
&\mathrm{tr}_0\Big(T_{0;1\dots N}(\xi_m|\{\xi\})E_0\Big)\\
&\quad = \left(\prod_{k=1}^{m}\mathcal{T}^{-1}(\xi_k)\right)\mathrm{tr}_0\Big[R_{mm-1}(\xi_m,\xi_{m-1})\cdots R_{m1}(\xi_m,\xi_1)\\
&\qquad\quad \times R_{mN}(\xi_m,\xi_N)\cdots R_{mm+1}(\xi_m,\xi_{m+1})E_mP_{0m}\Big]\left(\prod_{k=1}^{m}\mathcal{T}(\xi_k)\right).
\end{aligned} \tag{3.20}$$

Now we can take the trace over the space $V_0$ as we already did above. This gives us

$$\begin{aligned}\mathrm{tr}_0\Big(T_{0;1\dots N}(\xi_m|\{\xi\})E_0\Big) &= \left(\prod_{k=1}^{m}\mathcal{T}^{-1}(\xi_k)\right) R_{mm-1}(\xi_m,\xi_{m-1})\cdots\\ &\quad\times R_{m1}(\xi_m,\xi_1)R_{mN}(\xi_m,\xi_N)\cdots\\ &\quad\times R_{mm+1}(\xi_m,\xi_{m+1})E_m\left(\prod_{k=1}^{m}\mathcal{T}(\xi_k)\right), \qquad (3.21)\end{aligned}$$

and due to (3.6) we arrive at

$$\begin{aligned}\mathrm{tr}_0\Big(T_{0;1\dots N}(\xi_m|\{\xi\})E_0\Big) &= \left(\prod_{k=1}^{m}\mathcal{T}^{-1}(\xi_k)\right)\mathcal{T}(\xi_m)E_m\left(\prod_{k=1}^{m}\mathcal{T}(\xi_k)\right)\\ &= \left(\prod_{k=1}^{m-1}\mathcal{T}^{-1}(\xi_k)\right)E_m\left(\prod_{k=1}^{m}\mathcal{T}(\xi_k)\right). \qquad (3.22)\end{aligned}$$

Hence, we finally obtain

$$E_m = \left(\prod_{k=1}^{m-1}\mathcal{T}(\xi_k)\right)\mathrm{tr}_0\Big(T_0(\xi_m)E_0\Big)\left(\prod_{k=1}^{m}\mathcal{T}^{-1}(\xi_k)\right). \qquad (3.23)$$

Formula (3.23) is the one we are looking for. In this formula, the operator $E_m$ acting in the local space $V_m$ is expressed in terms of the operator $E_0$ acting in the auxiliary space $V_0$ and the transfer matrix.

**Remark 3.1.** The derivation was done for the $XXZ$ chain. However, we did not use the explicit form of the $R$-matrix. We essentially used two conditions:

- the monodromy matrix is the product of the $R$-matrices of the form (3.1);
- the $R$-matrix becomes the permutation matrix at some values of the arguments.

If these two conditions are satisfied, then the inverse problem is solvable, and its solution is still given by the formula (3.23). In particular, all of the above is true for the $XXX$ chain and even for the $XYZ$ chain. Furthermore, we have never used the fact that the monodromy matrix was a $2\times 2$ matrix. Therefore, the solution of the inverse problem can be obtained in the case of $n\times n$ matrices with $n > 2$ *without any changes*. The possible generalizations of the inverse problem can be found in [J. M. Maillet and V. Terras (2000); F. Göhmann and V.E. Korepin (2000)].

Let us write down explicit representations for the standard basis matrices $E_m^{ab}$ acting in the local quantum space $V_m$:

$$\begin{aligned} E_m^{11} &= \left(\prod_{l=1}^{m-1} \mathcal{T}(\xi_l)\right) A(\xi_m) \left(\prod_{l=1}^{m} \mathcal{T}^{-1}(\xi_l)\right), \\ E_m^{22} &= \left(\prod_{l=1}^{m-1} \mathcal{T}(\xi_l)\right) D(\xi_m) \left(\prod_{l=1}^{m} \mathcal{T}^{-1}(\xi_l)\right), \\ E_m^{12} &= \left(\prod_{l=1}^{m-1} \mathcal{T}(\xi_l)\right) C(\xi_m) \left(\prod_{l=1}^{m} \mathcal{T}^{-1}(\xi_l)\right), \\ E_m^{21} &= \left(\prod_{l=1}^{m-1} \mathcal{T}(\xi_l)\right) B(\xi_m) \left(\prod_{l=1}^{m} \mathcal{T}^{-1}(\xi_l)\right). \end{aligned} \tag{3.24}$$

Note that if we take the sum of the first two formulas, we get an obvious identity since $A(\xi_m) + D(\xi_m) = \mathcal{T}(\xi_m)$. Note also that in the homogeneous limit, we can simply put all $\xi_m = \eta/2$, $m = 1, \dots, N$. In fact, we could solve the quantum inverse problem directly for the homogeneous chain. However, the use of inhomogeneities $\xi_m$ is convenient since each $\xi_m$ is a kind of label of the $m$th site.

Finally, we recall that we solved the quantum inverse problem with a special normalization of the $R$-matrix: $R(u,u) = P$. However, the final result does not depend on the normalization. Indeed, it is easy to see that formulas (3.24) are invariant under the multiplication of the monodromy matrix by an arbitrary scalar function.

We conclude this chapter with a few words about possible generalizations of the quantum inverse problem. For Heisenberg spin-1/2 chains, the quantum space of the $L$-operator is isomorphic to the auxiliary space. This fact was essentially used in deriving formula (3.23). However, a solution to the quantum inverse problem also exists for more general models, such as higher spin $XXX$ and $XXZ$ chains [J. M. Maillet and V. Terras (2000)]. Explicit expressions for local operators in terms of the monodromy matrix entries in the lattice QNLSE and sine-Gordon models were obtained in [T. Oota (2004)].

# Chapter 4

# Composite model

We will see in Section 9.1 that the formulas of the quantum inverse problem make it easy to calculate matrix elements of local operators in spin chains. However, other methods can be used for this purpose. One such way is based on a composite model [A. G. Izergin and V. E. Korepin (1984)]. Within the framework of this model, we select an intermediate point of an interval on which the initial model is determined (for example, some point $x$ on the interval $[0, L]$ in a one-dimensional Bose gas). Thus, we can describe the operators associated with this intermediate point. The composite model can also be used to compute scalar products (see Section 5.1). Finally, it is easy to establish a connection between the algebraic and coordinate Bethe ansatz using the composite model.

Section 4.1 is auxiliary. Here we define the sums over partitions of a set of parameters into subsets. We introduce a special notation for such sums and consider a specific example. Section 4.2 defines a composite model. Here we prove a formula for the Bethe vector in the composite model. In Section 4.3, we discuss a multi-composite model. It is the link between the algebraic and coordinate Bethe ansatz. We obtain formulas for Bethe vectors in terms of local operators in the $XXZ$ chain and the one-dimensional Bose gas. In Section 4.4, we consider a twisted on-shell Bethe vector, which corresponds to a special solution of the Bethe equations found in Section 2.4.1. We show that, with correct normalization, this vector becomes an eigenvector of the $XXZ$ Hamiltonian in the limit $\kappa \to 1$.

## 4.1 Summation over partitions

We will often deal with sums over partitions of sets of parameters into disjoint subsets from now on. To denote subsets, we mainly use Roman numerals (all other notations will be specified separately). For example, notation $\bar{v} \mapsto \{\bar{v}_{\rm I}, \bar{v}_{\rm II}\}$ means that the set $\bar{v}$ is divided into subsets $\bar{v}_{\rm I}$ and $\bar{v}_{\rm II}$, such that $\bar{v}_{\rm I} \cap \bar{v}_{\rm II} = \emptyset$ and $\bar{v}_{\rm I} \cup \bar{v}_{\rm II} = \bar{v}$. Summation over partitions of some set of parameters means taking a sum over all possible subsets of this set (including the empty set). Let us give an example. Let $\bar{v} = \{v_1, v_2, v_3\}$. Then

$$\begin{aligned}\sum_{\bar{v}\mapsto\{\bar{v}_{\rm I},\bar{v}_{\rm II}\}} & f(\bar{v}_{\rm I}, \bar{v}_{\rm II}) \\ &= 1 + \big[f(v_1,v_2)f(v_1,v_3) + f(v_2,v_1)f(v_2,v_3) + f(v_3,v_1)f(v_3,v_2)\big] \\ &\quad + \big[f(v_2,v_1)f(v_3,v_1) + f(v_1,v_2)f(v_3,v_2) + f(v_1,v_3)f(v_2,v_3)\big] + 1. \quad (4.1)\end{aligned}$$

Recall that $f(\bar{v}_{\rm I}, \bar{v}_{\rm II})$ is the shorthand notation for the product

$$f(\bar{v}_{\rm I}, \bar{v}_{\rm II}) = \prod_{v_i\in\bar{v}_{\rm I}} \prod_{v_k\in\bar{v}_{\rm II}} f(v_i, v_k), \tag{4.2}$$

and if at least one of the subsets is empty, then the product is equal to 1. Then the first term in (4.1) corresponds to the partition $\bar{v}_{\rm I} = \emptyset$, $\bar{v}_{\rm II} = \{v_1, v_2, v_3\}$. Then a group of three terms follows that corresponds to three different partitions in which $\#\bar{v}_{\rm I} = 1$ and $\#\bar{v}_{\rm II} = 2$. The following three terms correspond to three partitions with $\#\bar{v}_{\rm I} = 2$ and $\#\bar{v}_{\rm II} = 1$. Finally, the last term arises from the partition $\bar{v}_{\rm I} = \{v_1, v_2, v_3\}$, $\bar{v}_{\rm II} = \emptyset$. Pay attention to the fact that the order of the elements in each subset is not essential. So, we consider the subsets $\{v_1, v_2\}$ and $\{v_2, v_1\}$ as equivalent. We do not take them into account twice when summing up.

Sometimes we will deal with sums over partitions with restrictions on the cardinalities of the subsets. For example, if we impose a restriction on the sum (4.1) $\#\bar{v}_{\rm I} = 1$ and $\#\bar{v}_{\rm II} = 2$, then only three terms in the square brackets survive in the second line of the sum (4.1). To facilitate the notation, we will often indicate restrictions on subsets only in the comments to the formulas.

If there are no restrictions on cardinalities, one should always consider the empty set among possible subsets when summing over partitions.

## 4.2 Definition of composite model

Suppose that we have some lattice model. Then its monodromy matrix is a product of $L$-operators (1.50). If we deal with a continuous model, then we can take the definition (1.50) as a lattice approximation of the monodromy matrix of this model.

Let us fix a site $m$ and introduce *partial monodromy matrices* $T^{(1)}(u)$ and $T^{(2)}(u)$ as follows:

$$T^{(2)}(u) = L_N(u)\dots L_{m+1}(u), \qquad T^{(1)}(u) = L_m(u)\dots L_1(u). \tag{4.3}$$

Then the *total monodromy matrix* is equal to the product of the partial monodromy matrices

$$T(u) = T^{(2)}(u)T^{(1)}(u). \tag{4.4}$$

We call this model a *composite model.*[1]

On the one hand, introducing the intermediate point $m$ allows us to access local operators in this site. On the other hand, we can still work with monodromy matrices and the $R$-matrix within the standard algebraic Bethe ansatz framework.

Let the matrix elements of the matrices $T^{(1)}(u)$ and $T^{(2)}(u)$ be equipped with the superscript

$$T^{(j)}(u) = \begin{pmatrix} A^{(j)}(u) & B^{(j)}(u) \\ C^{(j)}(u) & D^{(j)}(u) \end{pmatrix}, \qquad j = 1, 2. \tag{4.5}$$

Each of the partial monodromy matrices satisfies the $RTT$-relation with the same $R$-matrix as the total monodromy matrix. This property follows from the fact that they are equal to the product of $L$-operators, where every $L$-operator satisfies the $RTT$-relation. The matrix elements of the different partial matrices commute with each other. Each of the matrices $T^{(j)}(u)$ possesses a vector $|0\rangle^{(j)}$ and a dual vector $\langle 0|^{(j)}$, where $|0\rangle = |0\rangle^{(2)} \otimes |0\rangle^{(1)}$. The action formulas of the operators on these vectors are similar to the action formulas for the total monodromy matrix

$$\begin{aligned} A^{(j)}(u)|0\rangle^{(j)} &= a^{(j)}(u)|0\rangle^{(j)}, & \langle 0|^{(j)}A^{(j)}(u) &= a^{(j)}(u)\langle 0|^{(j)}, \\ D^{(j)}(u)|0\rangle^{(j)} &= d^{(j)}(u)|0\rangle^{(j)}, & \langle 0|^{(j)}D^{(j)}(u) &= d^{(j)}(u)\langle 0|^{(j)}, \\ C^{(j)}(u)|0\rangle^{(j)} &= 0, & \langle 0|^{(j)}B^{(j)}(u) &= 0, \end{aligned} \tag{4.6}$$

where $a^{(j)}(u)$ and $d^{(j)}(u)$ are some functions. We do not specify their explicit form. It is easy to see that $a(u) = a^{(1)}(u)a^{(2)}(u)$ and $d(u) = d^{(1)}(u)d^{(2)}(u)$.

[1]This model was introduced in [A. G. Izergin and V. E. Korepin (1984)]. It was called there a generalized two-site model. Later, in [N. A. Slavnov (2007)], the term two-component model was used. Unfortunately, both these terms can lead to misunderstanding: the first—when considering spin chains, the second—when considering multi-component gases.

To simplify further formulas, we extend the convention on the shorthand notation for products of partial operators and partial vacuum eigenvalues. For example,

$$B^{(j)}(\bar{v}) = \prod_{v_k\in\bar{v}} B^{(j)}(v_k), \qquad a^{(j)}(\bar{u}_{\rm II}) = \prod_{u_l\in\bar{u}_{\rm II}} a^{(j)}(u_l), \tag{4.7}$$

and so on.

We define partial Bethe vectors and their dual in the standard way

$$B^{(j)}(\bar{u})|0\rangle^{(j)}, \qquad \langle 0|^{(j)}C^{(j)}(\bar{u}). \tag{4.8}$$

The actions of the partial operators $A^{(j)}$, $B^{(j)}$, $C^{(j)}$, and $D^{(j)}$ on the corresponding Bethe vectors are the same as the actions of the total operators on the total Bethe vectors. One should only replace everywhere the functions $a(u)$ and $d(u)$ with $a^{(j)}(u)$ and $d^{(j)}(u)$. It is in this way, the dependence on a selected site arises in the formulas for the scalar products and form factors of local operators.[2] If, for instance, we deal with a homogeneous model and the action of the diagonal matrix elements of the $L$-operators on the vacuum is given by

$$\big(L_n\big)_{11}(u)|0\rangle = \alpha(u)|0\rangle, \qquad \big(L_n\big)_{22}(u)|0\rangle = \delta(u)|0\rangle, \tag{4.9}$$

then

$$a(u) = \alpha^N(u), \qquad d(u) = \delta^N(u). \tag{4.10}$$

At the same time,

$$\begin{aligned} a^{(1)}(u) &= \alpha^m(u), & d^{(1)}(u) &= \delta^m(u),\\ a^{(2)}(u) &= \alpha^{N-m}(u), & d^{(2)}(u) &= \delta^{N-m}(u). \end{aligned} \tag{4.11}$$

As we shall see further, the formulas for form factors of local operators somehow include the functions $a^{(j)}(u)$ and $d^{(j)}(u)$. In this way, they depend on the intermediate $m$th site.

In order to obtain formulas for the action of partial operators on the total Bethe vector $B(\bar{u})|0\rangle$, it is necessary to express the total vector in terms of the partial ones (4.8).

**Proposition 4.1.** *A generic total Bethe vector can be expressed in terms of the partial vectors as follows [A. G. Izergin and V. E. Korepin (1984)]:*

$$B(\bar{u})|0\rangle = \sum_{\bar{u}\mapsto\{\bar{u}_{\rm I},\bar{u}_{\rm II}\}} a^{(2)}(\bar{u}_{\rm I})d^{(1)}(\bar{u}_{\rm II})f(\bar{u}_{\rm II},\bar{u}_{\rm I}) \times B^{(1)}(\bar{u}_{\rm I})|0\rangle^{(1)} \otimes B^{(2)}(\bar{u}_{\rm II})|0\rangle^{(2)}. \tag{4.12}$$

*Here the sum is taken over all possible partitions of the set $\bar{u}$ into two subsets $\bar{u}_{\rm I}$ and $\bar{u}_{\rm II}$.*

---

[2] Recall that we call the form factor of a local operator the matrix element of this operator in the basis of physical states.

**Proof.** For proving (4.12), we use induction over $n = \#\bar{u}$. Obviously, equation (4.12) is valid for $n = 1$. Indeed, it follows from (4.4) that

$$B(u) = A^{(2)}(u)B^{(1)}(u) + B^{(2)}(u)D^{(1)}(u), \tag{4.13}$$

which implies (4.12) for $n = 1$. Let (4.12) be valid for $n - 1$:

$$\begin{aligned} B(\bar{u}_n)|0\rangle = &\sum_{\bar{u}_n \mapsto \{\bar{u}_{\rm I}, \bar{u}_{\rm II}\}} a^{(2)}(\bar{u}_{\rm I})d^{(1)}(\bar{u}_{\rm II})f(\bar{u}_{\rm II}, \bar{u}_{\rm I}) \\ &\times B^{(1)}(\bar{u}_{\rm I})|0\rangle^{(1)} \otimes B^{(2)}(\bar{u}_{\rm II})|0\rangle^{(2)}, \end{aligned} \tag{4.14}$$

where the sum is taken over partitions of the set $\bar{u}_n = \bar{u} \setminus u_n$. Taking $B(u_n)$ in the form (4.13) we obtain

$$B(\bar{u})|0\rangle = B(u_n)B(\bar{u}_n)|0\rangle = J_A + J_D, \tag{4.15}$$

where

$$\begin{aligned} J_A &= A^{(2)}(u_n)B^{(1)}(u_n)B(\bar{u}_n)|0\rangle, \\ J_D &= B^{(2)}(u_n)D^{(1)}(u_n)B(\bar{u}_n)|0\rangle. \end{aligned} \tag{4.16}$$

Consider the contribution $J_A$. Due to (4.14) we have

$$\begin{aligned} J_A = &\sum_{\bar{u}_n \mapsto \{\bar{u}_{\rm I}, \bar{u}_{\rm II}\}} a^{(2)}(\bar{u}_{\rm I})d^{(1)}(\bar{u}_{\rm II})f(\bar{u}_{\rm II}, \bar{u}_{\rm I})B^{(1)}(u_n)B^{(1)}(\bar{u}_{\rm I})|0\rangle^{(1)} \\ &\otimes A^{(2)}(u_n)B^{(2)}(\bar{u}_{\rm II})|0\rangle^{(2)}. \end{aligned} \tag{4.17}$$

Now we should act with the operator $A^{(2)}(u_n)$ on the vector $B^{(2)}(\bar{u}_{\rm II})|0\rangle^{(2)}$. For this, we use (2.38) replacing everywhere $a$, $B$, and $|0\rangle$ respectively with $a^{(2)}$, $B^{(2)}$, and $|0\rangle^{(2)}$. It is convenient to write down the action formula in the following form:

$$\begin{aligned} A^{(2)}(v)B^{(2)}(\bar{w})|0\rangle^{(2)} = f(\bar{w}, v)a^{(2)}(v)B^{(2)}(\bar{w})|0\rangle^{(2)} + &\sum_{\bar{w} \mapsto \{\bar{w}_{\rm i}, \bar{w}_{\rm ii}\}} a^{(2)}(\bar{w}_{\rm i}) \\ \times g(v, \bar{w}_{\rm i})f(\bar{w}_{\rm ii}, \bar{w}_{\rm i})B^{(2)}(v)B^{(2)}(\bar{w}_{\rm ii})|0\rangle^{(2)}. \end{aligned} \tag{4.18}$$

In the second line of this formula, the sum is taken over partitions of the set $\bar{w}$ into subsets $\bar{w}_{\rm i}$ and $\bar{w}_{\rm ii}$, such that the subset $\bar{w}_{\rm i}$ consists of one element.

Then $J_A$, in turn, splits into two contributions $J_A = J_A^{(1)} + J_A^{(2)}$. The contribution $J_A^{(1)}$ corresponds to the first line of (4.18):

$$\begin{aligned} J_A^{(1)} = &\sum_{\bar{u}_n \mapsto \{\bar{u}_{\rm I}, \bar{u}_{\rm II}\}} a^{(2)}(\bar{u}_{\rm I})d^{(1)}(\bar{u}_{\rm II})f(\bar{u}_{\rm II}, \bar{u}_{\rm I})B^{(1)}(u_n)B^{(1)}(\bar{u}_{\rm I})|0\rangle^{(1)} \\ &\otimes a^{(2)}(u_n)f(\bar{u}_{\rm II}, u_n)B^{(2)}(\bar{u}_{\rm II})|0\rangle^{(2)}. \end{aligned} \tag{4.19}$$

The contribution $J_A^{(2)}$ corresponds to the second line of (4.18):

$$J_A^{(2)} = \sum_{\bar{u}_n \mapsto \{\bar{u}_{\rm I}, \bar{u}_{\rm i}, \bar{u}_{\rm ii}\}} g(u_n, \bar{u}_{\rm i}) a^{(2)}(\bar{u}_{\rm i}) a^{(2)}(\bar{u}_{\rm I}) d^{(1)}(\bar{u}_{\rm i}) d^{(1)}(\bar{u}_{\rm ii}) \\ \times f(\bar{u}_{\rm i}, \bar{u}_{\rm I}) f(\bar{u}_{\rm ii}, \bar{u}_{\rm I}) f(\bar{u}_{\rm ii}, \bar{u}_{\rm i}) \\ \times B^{(1)}(u_n) B^{(1)}(\bar{u}_{\rm I}) |0\rangle^{(1)} \otimes B^{(2)}(u_n) B^{(2)}(\bar{u}_{\rm ii}) |0\rangle^{(2)}. \tag{4.20}$$

Let us comment on this formula. The set $\bar{u}_n$ was already divided into subsets $\bar{u}_{\rm I}$ and $\bar{u}_{\rm II}$ in equation (4.17). After the action of the operator $A^{(2)}(u_n)$ on the vector $B^{(2)}(\bar{u}_{\rm II})|0\rangle$ according to (4.18), we obtain an additional partitions of the set $\bar{u}_{\rm II}$ into subsets $\bar{u}_{\rm i}$ and $\bar{u}_{\rm ii}$. As a result, the set $\bar{u}_n$ is divided into three subsets $\bar{u}_n \mapsto \{\bar{u}_{\rm I}, \bar{u}_{\rm i}, \bar{u}_{\rm ii}\}$, where $\bar{u}_{\rm i}$ consists of one element. It is also obvious that $d^{(1)}(\bar{u}_{\rm II}) = d^{(1)}(\bar{u}_{\rm i}) d^{(1)}(\bar{u}_{\rm ii})$ and $f(\bar{u}_{\rm II}, \bar{u}_{\rm I}) = f(\bar{u}_{\rm i}, \bar{u}_{\rm I}) f(\bar{u}_{\rm ii}, \bar{u}_{\rm I})$.

Now we do the same with the contribution $J_D$

$$J_D = \sum_{\bar{u}_n \mapsto \{\bar{u}_{\rm I}, \bar{u}_{\rm II}\}} a^{(2)}(\bar{u}_{\rm I}) d^{(1)}(\bar{u}_{\rm II}) f(\bar{u}_{\rm II}, \bar{u}_{\rm I}) D^{(1)}(u_n) B^{(1)}(\bar{u}_{\rm I}) |0\rangle^{(1)} \\ \otimes B^{(2)}(u_n) B^{(2)}(\bar{u}_{\rm II}) |0\rangle^{(2)}. \tag{4.21}$$

Calculating the action of $D^{(1)}(u_n)$ on $B^{(1)}(\bar{u}_{\rm I})|0\rangle^{(1)}$ we obtain $J_D = J_D^{(1)} + J_D^{(2)}$, where

$$J_D^{(1)} = \sum_{\bar{u}_n \mapsto \{\bar{u}_{\rm I}, \bar{u}_{\rm II}\}} a^{(2)}(\bar{u}_{\rm I}) d^{(1)}(u_n) d^{(1)}(\bar{u}_{\rm II}) f(u_n, \bar{u}_{\rm I}) f(\bar{u}_{\rm II}, \bar{u}_{\rm I}) \\ \times B^{(1)}(\bar{u}_{\rm I}) |0\rangle^{(1)} \otimes B^{(2)}(u_n) B^{(2)}(\bar{u}_{\rm II}) |0\rangle^{(2)}, \tag{4.22}$$

and

$$J_D^{(2)} = \sum_{\bar{u}_n \mapsto \{\bar{u}_{\rm ii}, \bar{u}_{\rm i}, \bar{u}_{\rm II}\}} g(\bar{u}_{\rm i}, u_n) a^{(2)}(\bar{u}_{\rm i}) a^{(2)}(\bar{u}_{\rm ii}) d^{(1)}(\bar{u}_{\rm i}) d^{(1)}(\bar{u}_{\rm II}) \\ \times f(\bar{u}_{\rm i}, \bar{u}_{\rm ii}) f(\bar{u}_{\rm II}, \bar{u}_{\rm ii}) f(\bar{u}_{\rm II}, \bar{u}_{\rm i}) \\ \times B^{(1)}(u_n) B^{(1)}(\bar{u}_{\rm ii}) |0\rangle^{(1)} \otimes B^{(2)}(u_n) B^{(2)}(\bar{u}_{\rm II}) |0\rangle^{(2)}. \tag{4.23}$$

Here, similarly to (4.20), the sum is taken over partitions $\bar{u}_n \mapsto \{\bar{u}_{\rm I}, \bar{u}_{\rm i}, \bar{u}_{\rm ii}\}$ such that $\bar{u}_{\rm i}$ consists of one element.

At first sight, it seems that equations (4.20) and (4.23) do not have anything in common with each other. In reality, they differ from each other in sign only. In order to see this, it is enough to change the subsets labels in one of these formulas so that the products of the operators $B^{(1)}$ and $B^{(2)}$

in both formulas would depend on the same subsets. For example, we can first replace $\bar{u}_{\rm ii}$ with $\bar{u}_{\rm II}$ in the formula (4.20), and then replace $\bar{u}_{\rm I}$ with $\bar{u}_{\rm ii}$. Then the contribution (4.20) takes the form

$$J_A^{(2)} = \sum_{\bar{u}_n \mapsto \{\bar{u}_{\rm ii}, \bar{u}_{\rm i}, \bar{u}_{\rm II}\}} g(u_n, \bar{u}_{\rm i}) a^{(2)}(\bar{u}_{\rm i}) a^{(2)}(\bar{u}_{\rm ii}) d^{(1)}(\bar{u}_{\rm i}) d^{(1)}(\bar{u}_{\rm II}) \times f(\bar{u}_{\rm i}, \bar{u}_{\rm ii}) f(\bar{u}_{\rm II}, \bar{u}_{\rm ii}) f(\bar{u}_{\rm II}, \bar{u}_{\rm i}) \times B^{(1)}(u_n) B^{(1)}(\bar{u}_{\rm ii}) |0\rangle^{(1)} \otimes B^{(2)}(u_n) B^{(2)}(\bar{u}_{\rm II}) |0\rangle^{(2)}. \tag{4.24}$$

We see now that due to $g(u_n, \bar{u}_{\rm i}) = -g(\bar{u}_{\rm i}, u_n)$, expression (4.24) differs from (4.23) in the sign only. Thus, $J_A^{(2)} + J_D^{(2)} = 0$, and we arrive at

$$B(\bar{u})|0\rangle = J_A^{(1)} + J_D^{(1)}. \tag{4.25}$$

Substituting here $J_A^{(1)}$ and $J_D^{(1)}$ respectively from (4.19) and (4.22) we find

$$B(\bar{u})|0\rangle = \sum_{\bar{u}_n \mapsto \{\bar{u}_{\rm I}, \bar{u}_{\rm II}\}} a^{(2)}(u_n) a^{(2)}(\bar{u}_{\rm I}) d^{(1)}(\bar{u}_{\rm II}) f(\bar{u}_{\rm II}, u_n) f(\bar{u}_{\rm II}, \bar{u}_{\rm I}) \times B^{(1)}(u_n) B^{(1)}(\bar{u}_{\rm I}) |0\rangle^{(1)} \otimes B^{(2)}(\bar{u}_{\rm II}) |0\rangle^{(2)} + \sum_{\bar{u}_n \mapsto \{\bar{u}_{\rm I}, \bar{u}_{\rm II}\}} a^{(2)}(\bar{u}_{\rm I}) d^{(1)}(u_n) d^{(1)}(\bar{u}_{\rm II}) f(u_n, \bar{u}_{\rm I}) f(\bar{u}_{\rm II}, \bar{u}_{\rm I}) \times B^{(1)}(\bar{u}_{\rm I}) |0\rangle^{(1)} \otimes B^{(2)}(u_n) B^{(2)}(\bar{u}_{\rm II}) |0\rangle^{(2)}. \tag{4.26}$$

It is easy to see that this is exactly the formula (4.12). Indeed, if we have partitions of the set $\bar{u}$ into two subsets, then the parameter $u_n$ can belong to either $\bar{u}_{\rm I}$ or $\bar{u}_{\rm II}$. Each of the two terms in (4.26) corresponds to one of the two possibilities mentioned above. □

A similar formula exists for the dual vectors

$$\langle 0|C(\bar{u}) = \sum_{\bar{u} \mapsto \{\bar{u}_{\rm I}, \bar{u}_{\rm II}\}} a^{(1)}(\bar{u}_{\rm II}) d^{(2)}(\bar{u}_{\rm I}) f(\bar{u}_{\rm I}, \bar{u}_{\rm II}) \times \langle 0|^{(1)} C^{(1)}(\bar{u}_{\rm I}) \otimes \langle 0|^{(2)} C^{(2)}(\bar{u}_{\rm II}). \tag{4.27}$$

The proof of this formula is completely analogous to the one given above.

## 4.3 Multi-composite model

To calculate the form factors of local operators, the composite model described in the previous section is already sufficient. We will get acquainted with the details of this calculation in Section 9.3. In the meantime, we note that we can go further and split the total monodromy matrix not by two but by a larger number of partial matrices $T^{(j)}(u)$ [A. G. Izergin et al (1987); V. E. Korepin et al (1993)]

$$T(u) = T^{(M)}(u) \dots T^{(1)}(u). \tag{4.28}$$

This model is called a *multi-composite model*. We can define partial Bethe vectors $B^{(j)}|0\rangle^{(j)}$ for each monodromy matrix $T^{(j)}(u)$. Then the total Bethe vector expressed in terms of the partial ones as follows:

$$B(\bar{u})|0\rangle = \sum_{\bar{u}\mapsto\{\bar{u}^1,\dots,\bar{u}^M\}} \prod_{1\le k<j\le M} \left\{a^{(j)}(\bar{u}^k) d^{(k)}(\bar{u}^j) f(\bar{u}^j, \bar{u}^k)\right\}$$
$$\times \prod_{j=1}^{M} \left\{B^{(j)}(\bar{u}^j)\right\} |0\rangle. \tag{4.29}$$

Because of a large number of subsets, we do not use Roman numerals in this formula but denote subsets with superscripts. We also used the shorthand notation

$$\prod_{j=1}^{M} \left\{B^{(j)}(\bar{u}^j)\right\} |0\rangle = B^{(1)}(\bar{u}^1)\, |0\rangle^{(1)} \otimes \dots \otimes B^{(M)}(\bar{u}^M)\, |0\rangle^{(M)}. \tag{4.30}$$

The sum is taken over all partitions of the original set $\bar{u}$ into $M$ subsets $\bar{u}^1, \dots, \bar{u}^M$. The functions $a^{(j)}(u)$ and $d^{(j)}(u)$ respectively are the vacuum eigenvalues of the operators $A^{(j)}(u)$ and $D^{(j)}(u)$. It is important to stress that the number of partial monodromy matrices $M$ is not related to the number of Bethe parameters $n = \#\bar{u}$. In particular, we can consider the case $M > n$. Then some subsets in (4.29) necessarily are empty. However, we should not forget that for $M \le n$, we also can have empty subsets in this formula.

Formula (4.29) can easily be proved by induction on $M$. Assuming that it is valid for $M-1$ partial monodromy matrices, we apply the equality (4.12) to the partial vector $B^{(M-1)}(\bar{u}^{M-1})|0\rangle^{(M-1)}$. This immediately gives us (4.29) for $M$ partial monodromy matrices.

### 4.3.1 *Spin chains*

Introducing the multi-composite model is a convenient way to express Bethe vectors in terms of local operators. Consider, for example, an inhomogeneous $XXX$ chain consisting of $N$ sites. The monodromy matrix is defined as the product of local $L$-operators

$$T(u) = L_N(u - \xi_N) \dots L_1(u - \xi_1), \tag{4.31}$$

where $\xi_k$ are inhomogeneities, and

$$L_n(u) = \frac{1}{u}\begin{pmatrix} u+\eta\, E_n^{11} & \eta\, E_n^{21} \\ \eta\, E_n^{12} & u+\eta\, E_n^{22} \end{pmatrix}. \tag{4.32}$$

The normalizing factor $1/u$ is chosen so as to have $d(u) = 1$. Recall that a vacuum vector is a state with all spins up

$$|0\rangle = \left(\begin{smallmatrix}1\\0\end{smallmatrix}\right)_1 \otimes \cdots \otimes \left(\begin{smallmatrix}1\\0\end{smallmatrix}\right)_N\,. \tag{4.33}$$

Consider a multi-composite model consisting of $N$ partial monodromy matrices $T^{(j)}$. This means that each matrix $T^{(j)}$ coincides with the $L$-operator $L_j(u-\xi_j)$, and the partial vacuum vector has the form $|0\rangle^{(j)} = \left(\begin{smallmatrix}1\\0\end{smallmatrix}\right)_j$. Then every partial Bethe vector in formula (4.29) corresponds to the $j$th site of the chain. Therefore, by virtue of (4.32), we get

$$B^{(j)}(\bar u^j)|0\rangle^{(j)} = g(\bar u^j,\xi_j)\left(\sigma_j^-\right)^{n_j}\left(\begin{smallmatrix}1\\0\end{smallmatrix}\right)_j, \tag{4.34}$$

where $n_j$ is the cardinality of the subset $\bar u^j$. Obviously, this vector vanishes for $n_j > 1$ because $\left(\sigma_j^-\right)^2 = 0$. Thus, we conclude that all $n_j \le 1$ and the subsets $\bar u^j$ are either empty or consist of one element. Let the total number of variables in the set $\bar u$ be equal to $n$, and let the subsets $\bar u^{j_k}$ ($k = 1,\dots,n$), corresponding to lattice sites with numbers $j_1,\dots,j_n$ contain one element $u_k$. The remaining subsets are empty. Then the sum over partitions of the set $\bar u$ turns into the sum over all possible lattice sites $j_1,\dots,j_n$ corresponding to non-empty subsets, followed by the sum over permutations of the parameters $\bar u$.

It is easy to see that

$$\begin{aligned} &\frac{u-\xi_j+\eta E_j^{11}}{u-\xi_j}\left(\begin{smallmatrix}1\\0\end{smallmatrix}\right)_j = f(u,\xi_j)\left(\begin{smallmatrix}1\\0\end{smallmatrix}\right)_j,\\ &\frac{u-\xi_j+\eta E_j^{22}}{u-\xi_j}\left(\begin{smallmatrix}1\\0\end{smallmatrix}\right)_j = \left(\begin{smallmatrix}1\\0\end{smallmatrix}\right)_j, \end{aligned} \tag{4.35}$$

and hence,

$$a^{(j)}(u) = f(u,\xi_j), \qquad d^{(j)}(u) = 1. \tag{4.36}$$

Then equation (4.29) takes the form

$$\begin{aligned} B(\bar u)|0\rangle = \operatorname*{Sym}_{\bar u} \prod_{1\le k<j\le n} f(u_j,u_k) \sum_{N\ge j_n>\cdots>j_1\ge 1}\prod_{k=1}^{n} \\ \times\left[\Bigl(\prod_{m=j_k+1}^{N} f(u_k,\xi_m)\Bigr) g(u_k,\xi_{j_k})\sigma_{j_k}^-\right]|0\rangle, \end{aligned} \tag{4.37}$$

where symbol Sym means symmetrization (that is, the sum over permutations) over the set indicated in the subscript. In formula (4.37), the symmetrization acts on the entire expression that depends on $\bar{u}$. In the homogeneous limit $\xi_k = \eta/2$, this expression coincides with the formula for the eigenvector of the $XXX$ Hamiltonian found by H. Bethe in 1931 (see [H. Bethe (1931); M. Gaudin (1983)]).

All the arguments above are completely applicable to the $XXZ$ chain, and the formula for the Bethe vector (4.37) remains the same. We present it in the detailed form for the homogeneous model

$$B(\bar{u})|0\rangle = \prod_{j=1}^{n} \frac{\sinh\eta}{\sinh(u_j - \frac{\eta}{2})} \operatorname*{Sym}_{\bar{u}} \prod_{1\le k<j\le n} \frac{\sinh(u_j - u_k + \eta)}{\sinh(u_j - u_k)} \times \sum_{N\ge j_n>\cdots>j_1\ge 1} \prod_{k=1}^{n} \left(\frac{\sinh(u_k + \frac{\eta}{2})}{\sinh(u_k - \frac{\eta}{2})}\right)^{N-j_k} \sigma^-_{j_k}|0\rangle. \tag{4.38}$$

If we replace all hyperbolic functions with rational ones (that is, replace $u_k \to \epsilon u_k$, $\eta \to \epsilon\eta$ and consider the limit $\epsilon \to 0$) in this formula, we obtain the Bethe vector of the $XXX$ chain.

The advantage of (4.38) is that the Bethe vector is expressed in terms of local operators. Therefore, we can easily calculate the action of the local spin operators $\sigma^\alpha_k$, $\alpha = x, y, z$, on this state. However, if we compare the rhs of (4.38) with the lhs, it is clear that the latter looks somewhat simpler.

### 4.3.2 *One-dimensional Bose gas*

We now construct a Bethe vector in the continuous limit of the lattice QNLSE model [T. C. Dorlas (1993)]. It is convenient to set $\rho^+_n = 1$ in the $L$-operator (1.74):

$$L_n(u) = \begin{pmatrix} 1 - \frac{iu\Delta}{2} + \frac{c\Delta^2}{2}\psi^\dagger_n\psi_n & -i\Delta\psi^\dagger_n \\ i\Delta\rho^-_n\psi_n & 1 + \frac{iu\Delta}{2} + \frac{c\Delta^2}{2}\psi^\dagger_n\psi_n \end{pmatrix}. \tag{4.39}$$

We again consider the case when the number of partial monodromy matrices $M$ coincides with the number of lattice sites $N$. Then

$$B^{(j)}(\bar{u}^j)|0\rangle = \left(-i\Delta\psi^\dagger_j\right)^{n_j}|0\rangle, \tag{4.40}$$

where $n_j$ is the cardinality of the subset $\bar{u}^j$. The main difference from the previous case is that now the cardinalities $n_j$ can be greater than 1. However, as we will see later, contributions with $n_j > 1$ vanish in the continuous limit.

Taking into account (4.40) and

$$a^{(j)}(u) \equiv a_0(u) = 1 - \frac{iu\Delta}{2}, \qquad j = 1, \ldots, N, \tag{4.41}$$
$$d^{(j)}(u) \equiv d_0(u) = 1 + \frac{iu\Delta}{2},$$

we recast (4.29) as follows:

$$B(\bar{u})|0\rangle = (-i\Delta)^n \sum_{\bar{u} \mapsto \{\bar{u}^1, \ldots, \bar{u}^N\}} \prod_{1 \le k < j \le N} f(\bar{u}^j, \bar{u}^k) \times \left( \prod_{j=1}^{N} \big(a_0(\bar{u}^j)\big)^{N-j} \big(d_0(\bar{u}^j)\big)^{j-1} \big(\psi_j^\dagger\big)^{n_j} \right) |0\rangle, \tag{4.42}$$

where $n = \#\bar{u}$.

Let us now consider such partitions for which the condition $n_j \le 1$ is satisfied. We denote the corresponding contribution to the Bethe vector by the symbol $B(\bar{u})|0\rangle_{\rm restr}$:

$$B(\bar{u})|0\rangle_{\rm restr} = (-i\Delta)^n \sum_{\substack{\bar{u} \mapsto \{\bar{u}^1, \ldots, \bar{u}^N\} \\ \#\bar{u}^j \le 1}} \prod_{1 \le k < j \le N} f(\bar{u}^j, \bar{u}^k) \times \left( \prod_{j=1}^{N} \big(a_0(\bar{u}^j)\big)^{N-j} \big(d_0(\bar{u}^j)\big)^{j-1} \big(\psi_j^\dagger\big)^{n_j} \right) |0\rangle. \tag{4.43}$$

As in the case of the $XXX$ chain, we introduce the subsets $\bar{u}^{j_k}$ ($k = 1, \ldots, n$), which correspond to lattice sites with numbers $j_1, \ldots, j_n$ and contain one element $u_k$. All other subsets are empty. Then the sum over partitions of the set $\bar{u}$ turns into the sum over all possible lattice sites $j_1, \ldots, j_n$, followed by the sum over permutations of the parameters $\bar{u}$:

$$B(\bar{u})|0\rangle_{\rm restr} = (-i\Delta)^n \operatorname*{Sym}_{\bar{u}} \prod_{1 \le k < j \le n} f(u_j, u_k) \times \sum_{N \ge j_n > \cdots > j_1 \ge 1} \left( \prod_{k=1}^{n} a_0^{N-j_k}(u_k)\, d_0^{j_k - 1}(u_k)\, \psi_{j_k}^\dagger \right) |0\rangle. \tag{4.44}$$

Now let us proceed to the continuous limit in (4.44). To do this, we introduce a coordinate $x_k = k\Delta$ of the $k$th site. Obviously,

$$\lim_{\Delta \to 0} d_0^{j_k - 1}(u) = \lim_{\Delta \to 0} \left(1 + \frac{iu\Delta}{2}\right)^{(x_{j_k}/\Delta) - 1} = e^{iux_{j_k}/2},$$
$$\lim_{\Delta \to 0} a_0^{N-j_k}(u) = \lim_{\Delta \to 0} \left(1 - \frac{iu\Delta}{2}\right)^{(L - x_{j_k})/\Delta} = e^{-iu(L - x_{j_k})/2}, \tag{4.45}$$

where $L = N\Delta$. Substituting (4.45) into (4.44), we obtain

$$B(\bar{u})|0\rangle_{\text{restr}} = (-i)^n \operatorname*{Sym}_{\bar{u}} \prod_{1\le k<j\le n} f(u_j, u_k) \sum_{N\ge j_n>\cdots>j_1\ge 1} \times \left( \prod_{k=1}^{n} e^{iu_k(x_{j_k}-L/2)}(x_{j_k+1}-x_{j_k})\psi^{\dagger}_{j_k} \right) |0\rangle, \quad (4.46)$$

where we presented $\Delta$ as $\Delta = x_{j_k+1} - x_{j_k}$ for all $k = 1, \dots, n$. The $n$-fold sum over the lattice sites turns into the $n$-fold integral in the continuous limit $\Delta \to 0$:

$$\lim_{\Delta\to 0} B(\bar{u})|0\rangle_{\text{restr}} = (-i)^n \operatorname*{Sym}_{\bar{u}} \prod_{1\le k<j\le n} f(u_j, u_k) \times \int_{\mathcal{D}} \left( \prod_{k=1}^{n} e^{iu_k(x_k-L/2)}\Psi^{\dagger}(x_k)\,\mathrm{d}x_k \right) |0\rangle, \quad (4.47)$$

where $\mathcal{D} = 0 \le x_1 \le \cdots \le x_n \le L$.

Let us now check that all other terms in the sum over partitions do not survive in the continuous limit. To do this, note that the $n$-fold integral in formula (4.47) appeared as the limit of the $n$-fold sum in (4.46). In its turn, the $n$-fold sum occurs because we have precisely $n$ non-empty subsets in the partition. If we consider such partitions in which at least one subset contains more than one element, then the number of non-empty subsets decreases. Therefore, we obtain an $m$-fold sum over lattice sites with $m < n$. At the same time, we still have the factor $\Delta^n$ in equation (4.44). Therefore, in the continuous limit, this contribution vanishes.

For example, consider the case $n = 2$. Consider such partitions when one subset contains two elements, and all the others are empty. The sum over such partitions is just the sum over lattice sites $j$ corresponding to the non-empty subset:

$$B(\bar{u})|0\rangle - B(\bar{u})|0\rangle_{\text{restr}} = (-i\Delta)^2 \sum_{j=1}^{N} \left[ \prod_{l=1}^{2} a_0^{N-j}(u_l) d_0^{j-1}(u_l) \right] (\psi_j^{\dagger})^2 |0\rangle. \quad (4.48)$$

In the continuous limit, we obtain

$$\lim_{\Delta\to 0} \left( B(\bar{u})|0\rangle - B(\bar{u})|0\rangle_{\text{restr}} \right) = -\lim_{\Delta\to 0} \Delta \int_0^L e^{i(u_1+u_2)(x-L/2)} \left(\Psi^{\dagger}(x)\right)^2 \mathrm{d}x\, |0\rangle = 0. \quad (4.49)$$

Thus, in the continuous limit of the sum over partitions, only such terms remain in which each subset contains at most one element. We arrive at the following representation for the Bethe vector in the one-dimensional Bose gas model:

$$B(\bar{u})|0\rangle = \operatorname*{Sym}_{\bar{u}} \prod_{1\le k<j\le n} f(u_j,u_k) \int_{\mathcal{D}} \left(\prod_{k=1}^{n} e^{iu_kx_k}\Psi^\dagger(x_k)\,\mathrm{d}x_k\right)|0\rangle, \quad (4.50)$$

where we omitted the insignificant phase factors $(-i)^n$ and $e^{-iu_kL/2}$. Traditionally, Bethe vectors are written slightly differently

$$B(\bar{u})|0\rangle = \int_0^L \chi_n(\bar{x}|\bar{u}) \left(\prod_{k=1}^{n} \Psi^\dagger(x_k)\,\mathrm{d}x_k\right)|0\rangle, \quad (4.51)$$

where

$$\chi_n(\bar{x}|\bar{u}) = \sum_p (-1)^{[p]} \prod_{j>k}^{n} \bigl(u_{p(j)} - u_{p(k)} - ic\,\mathrm{sign}(x_j - x_k)\bigr) \prod_{j=1}^{n} e^{ix_ju_{p(j)}}. \quad (4.52)$$

Here the sum is taken over permutations in the set $\bar{u}$, the symbol $[p]$ denotes the parity of the permutation, $\mathrm{sign}(x)$ is the sign-function. It is easy to check that the function $\chi_n(\bar{x}|\bar{u})$ is symmetric over $\bar{x}$. Therefore, the integral in (4.51) can be restricted to the domain $\mathcal{D}$. Then representation (4.51) coincides with (4.50) up to normalization.

On the other hand, it is also easy to see that the function $\chi_n(\bar{x}|\bar{u})$ is continuous, despite the presence of the sign-functions. Only the first derivatives on hyperplanes $x_j = x_k$, $j,k = 1,\dots,n$, are discontinuous:

$$\left.\frac{\partial\chi_n(\bar{x}|\bar{u})}{\partial x_j}\right|_{x_k-0}^{x_k+0} = c\chi_n(\bar{x}|\bar{u})\Big|_{x_j=x_k}. \quad (4.53)$$

Therefore, the function $\chi_n(\bar{x}|\bar{u})$ satisfies equation (1.84) with the eigenvalue of the energy

$$E = \sum_{j=1}^{n} u_j^2. \quad (4.54)$$

However, we emphasize that vector (4.51) is still an off-shell vector, even though $\chi_n$ satisfies equation (1.84). This vector becomes on-shell if we require that the function $\chi_n$ be periodic for each $x_k$, $k = 1,\dots,n$. We suggest that the reader check that the periodicity condition for $\chi_n$ yields the Bethe equations for the parameters $\bar{u}$.

## 4.4 Special on-shell Bethe vector

In Section 2.4.1, we found solutions to a system of two twisted Bethe equations for the $XXZ$ chain of four sites. One of these solutions has the property $u_1 \to \eta/2$ and $u_2 \to -\eta/2$ at $\kappa \to 1$. Now we construct the corresponding on-shell Bethe vector.

Let us write down representation (4.38) for $N = 4$ and $n = 2$. For further convenience, we multiply this vector by a coefficient

$$\frac{\sinh(u_1 - u_2)\sinh(u_1 + \frac{\eta}{2})\sinh(u_2 - \frac{\eta}{2})}{\sinh^2\eta\,\sinh(u_1 - u_2 + \eta)}. \tag{4.55}$$

Then

$$\begin{aligned} B(\bar{u})|0\rangle = \frac{\sinh(u_1 + \frac{\eta}{2})}{\sinh(u_1 - \frac{\eta}{2})} \sum_{j_2 > j_1}^{4} \Bigg\{ & \left[\frac{\sinh(u_2 - \frac{\eta}{2})}{\sinh(u_2 + \frac{\eta}{2})}\right]^{j_1 - 4} \left[\frac{\sinh(u_1 - \frac{\eta}{2})}{\sinh(u_1 + \frac{\eta}{2})}\right]^{j_2 - 4} \\ & + \frac{\sinh(u_2 - u_1 + \eta)}{\sinh(u_2 - u_1 - \eta)} \left[\frac{\sinh(u_1 - \frac{\eta}{2})}{\sinh(u_1 + \frac{\eta}{2})}\right]^{j_1 - 4} \\ & \times \left[\frac{\sinh(u_2 - \frac{\eta}{2})}{\sinh(u_2 + \frac{\eta}{2})}\right]^{j_2 - 4} \Bigg\} \sigma^-_{j_1}\sigma^-_{j_2}|0\rangle. \end{aligned} \tag{4.56}$$

It is already clear from this formula which problems arise in the limit $u_1 \to \eta/2$ and $u_2 \to -\eta/2$. Indeed, if we put $u_1 = \epsilon_1 + \eta/2$ and $u_2 = \epsilon_2 - \eta/2$, then we should fix how the ratio $\epsilon_1/\epsilon_2$ behaves when $\epsilon_k \to 0$, $k = 1, 2$. Moreover, we also should say how the difference $\epsilon_1 - \epsilon_2$ behaves in this limit because the formula (4.56) has terms proportional to $u_2 - u_1 + \eta = \epsilon_2 - \epsilon_1$. Clearly, we can get different limits depending on how we fix $\epsilon_1/\epsilon_2$ and $\epsilon_1 - \epsilon_2$. However, we do not have any prescription for how to do this.[3] Let us show how to solve this problem using twisted Bethe equations.

In Section 2.4.1, we parameterized the twist parameter as $\kappa = \theta^{-2}$ and did a change of variables

$$\frac{\sinh(u_k - \frac{\eta}{2})}{\sinh(u_k + \frac{\eta}{2})} = w_k, \qquad k = 1, 2. \tag{4.57}$$

[3]Actually, we can substitute $u_1 = \epsilon_1 + \eta/2$ and $u_2 = \epsilon_2 - \eta/2$ into Bethe equations (2.71) (with $\kappa = 1$) and require that there exists a limit of these equations for $\epsilon_k \to 0$. This will give additional restrictions on the parameters $\epsilon_k$.

It also follows from the twisted Bethe equations that

$$\frac{\sinh(u_2 - u_1 + \eta)}{\sinh(u_2 - u_1 - \eta)} = \theta^{-2}\left(\frac{\sinh(u_1 - \frac{\eta}{2})}{\sinh(u_1 + \frac{\eta}{2})}\right)^4 = \theta^{-2} w_1^4. \tag{4.58}$$

We are interested in a solution in which $w_1 \to 0$ and $w_2 \to \infty$ so that $w_1 w_2 = -\theta$. At the same time, the parameter $\theta$ itself goes to 1.

Let us require that Bethe vector (4.56) corresponds to this solution of the twisted Bethe equations. Then writing this vector in terms of $w_1$ and $w_2$, we have

$$B(\bar{u})|0\rangle = \sum_{4 \geq j_2 > j_1 \geq 1} \left(w_2^{j_1-4} w_1^{j_2-5} + \theta^{-2} w_1^{j_1-1} w_2^{j_2-4}\right)\sigma_{j_1}^{-}\sigma_{j_2}^{-}|0\rangle, \tag{4.59}$$

where we used (4.58). Let us now use $w_1 w_2 = -\theta$ to get rid of the parameter $w_2$. Then

$$B(\bar{u})|0\rangle = \sum_{4 \geq j_2 > j_1 \geq 1} \left((-\theta)^{j_1-4} w_1^{j_2-j_1-1} + (-\theta)^{j_2-6} w_1^{j_1-j_2+3}\right)\sigma_{j_1}^{-}\sigma_{j_2}^{-}|0\rangle. \tag{4.60}$$

Now the limit $w_1 \to 0$ and $\theta \to 1$ can be easily computed. In the first term, we obtain the terms with $j_2 = j_1 + 1$, while in the second term, only the term with $j_1 = 1$ and $j_2 = 4$ survives. We obtain

$$B(\bar{u})|0\rangle = \sum_{j=1}^{4}(-1)^j \sigma_j^{-}\sigma_{j+1}^{-}|0\rangle, \qquad u_1 \to \tfrac{\eta}{2}, \quad u_2 \to -\tfrac{\eta}{2}, \tag{4.61}$$

where we used the periodicity condition $\sigma_5^{-} \equiv \sigma_1^{-}$. We suggest that the reader check that the constructed vector is indeed the eigenvector of the $XXZ$ chain Hamiltonian. This vector must also be the $XXX$ Hamiltonian eigenvector since it does not depend on $\Delta$.

So, we have shown that in some cases, the procedure for constructing the eigenvectors of spin chain Hamiltonians is more sophisticated than usual. This situation occurs when such a vector corresponds to an admissible solution of the Bethe equations, which coincides with a spurious solution. In this case, the eigenvector can be constructed as the limit of the twisted on-shell vector. Then the twist parameter $\kappa$ plays the role of regularization.

## Chapter 5

# Scalar products of off-shell Bethe vectors

We proceed to the study of scalar products of Bethe vectors. It is to these objects that the calculation of all correlation functions eventually reduces. We will see later that scalar products are a kind of elementary blocks from which correlation functions are constructed. That is why it is crucial to get simple and compact representations for these elementary blocks that we can work with further.

This chapter is devoted to the most general case of scalar products of Bethe vectors. As a rule, to calculate correlation functions, we need to know a special case of scalar products when at least one of the vectors is on-shell. However, the general case also turns out to be useful. By calculating the scalar product of arbitrary Bethe vectors, we can then consider any particular cases.

Scalar products can be treated as an example of multiple commutation relations in which many operators need to be reordered. In fact, we had already encountered similar multiple commutation relations when we rearranged operator $A(v)$ with the product of several operators $B(\bar{u})$. In this simplest example, we saw that the straightforward application of commutation relation (1.123) leads to the appearance of too many terms in the final result. However, thanks to the Yang–Baxter equation, this huge number of terms can be significantly reduced. In Section 2.1.2, we developed special rules for applying commutation relations (1.123). These rules allow us to get the desired result in the most compact form.

Now we have a more difficult task ahead of us. We need to rearrange the product of several operators $C(\bar{v})$ with the product of operators $B(\bar{u})$. As in the case of the permutation of $A(v)$ and $B(\bar{u})$, the straightforward application of the commutation relations (1.118)–(1.130) leads to a highly cumbersome result. Using the algebraic Bethe ansatz rules allows us to write down the result in a relatively 'compact' form. However, we did not accidentally use quotations for the word compact. In general, the scalar product of the vector $B(\bar{u})|0\rangle$ and the dual vector $\langle 0|C(\bar{v})$ is the sum over the partitions of the parameters $\bar{v}$ and $\bar{u}$. If $\#\bar{u} = \#\bar{v} = n$, then this sum contains $\binom{2n}{n}$ terms. Using this formula is too inconvenient. It can only be considered as an intermediate result. In the particular case where one of the vectors is an on-shell Bethe vector, this formula is greatly simplified. However, we will consider this issue in the next chapter.

The study of scalar products in the algebraic Bethe ansatz framework was started in work [V. E. Korepin (1982b)], where the recurrent properties of scalar products were established, and the Gaudin hypothesis [M. Gaudin (1972); M. Gaudin et al (1981); M. Gaudin (1983)] on the norm of on-shell vectors was proved. In this work, the concept of the highest coefficient was also introduced. An explicit formula for the highest coefficient in the form of a determinant was obtained in work [A. G. Izergin (1987)] (see also [A. G. Izergin et al (1992)]). We also note that in work [V. E. Korepin (1987)], auxiliary quantum operators (dual fields) were introduced. Using these operators, one can present the scalar products as the vacuum average of a determinant in the auxiliary quantum space.

The content of this chapter is as follows. In Section 5.1, we define the scalar product and calculate the action of the $C$ operator on an arbitrary off-shell Bethe vector. Here we introduce the reader to the generalized model, which plays an essential role in calculating scalar products. We also introduce the concept of the highest coefficient and derive a representation for the scalar product of off-shell Bethe vectors in the form of a bilinear combination of the highest coefficients. Section 5.2 is devoted to the highest coefficient. Here we establish its properties and obtain an explicit determinant representation for it. In Section 5.3, we present several formulas for the scalar product of off-shell Bethe vectors and consider an example of using these formulas.

For definiteness, in this chapter, we consider models with the rational $R$-matrix. However, all the presentations can be easily generalized to the case of the trigonometric $R$-matrix. In particular, for this reason, most of the formulas are given in the universal form (that is, in terms of the functions $f$, $g$, and their combinations). We recommend the reader to check independently (if necessary) the validity of different statements for the case of the trigonometric $R$-matrix.

## 5.1 Scalar product of off-shell Bethe vectors

Traditionally, the scalar product of Bethe vectors $\langle 0|C(\bar{v})$ and $B(\bar{u})|0\rangle$ is called the vacuum expectation value

$$S_n(\bar{v}|\bar{u}) = \langle 0|C(\bar{v})B(\bar{u})|0\rangle. \tag{5.1}$$

Here $\bar{v} = \{v_1, \dots, v_n\}$, $\bar{u} = \{u_1, \dots, u_n\}$, $n = 0, 1, \dots$. If $\#\bar{v} \neq \#\bar{u}$, then the scalar product vanishes (see below). The parameters $\bar{v}$ and $\bar{u}$ are arbitrary complex numbers. In other words, Bethe vectors $\langle 0|C(\bar{v})$ and $B(\bar{u})|0\rangle$ are off-shell. To calculate the scalar product, we should normally order the operators in the expression (5.1). For this, using the commutation relations of the $RTT$ algebra, we should move all the $C$ operators to the rightmost position and the $B$ operators to the extreme left.

Often we will also deal with scalar products normalized to the product of the $d$-functions:

$$\boldsymbol{S}_n(\bar{v}|\bar{u}) = \frac{\langle 0|C(\bar{v})B(\bar{u})|0\rangle}{d(\bar{v})d(\bar{u})}. \tag{5.2}$$

Normalization to the product of the $d(z)$ functions does not play a fundamental role. However, it is convenient because it allows us to express the scalar product only in terms of the ratio $r(z) = a(z)/d(z)$. The exceptions are cases when among the parameters $\bar{v}$ and $\bar{u}$, there are such that the function $d(z)$ vanishes at these points. For example, in inhomogeneous spin chains, such points are inhomogeneities. In such cases, we will use the original definition (5.1).

Using commutation relation (1.121) or (1.122), we easily find for $n = 1$,

$$S_1(v|u) = \langle 0|C(v)B(u)|0\rangle = g(v, u)\big(d(v)a(u) - d(u)a(v)\big), \tag{5.3}$$

which implies

$$\boldsymbol{S}_1(v|u) = g(v, u)\big(r(u) - r(v)\big), \qquad r(u) = \frac{a(u)}{d(u)}. \tag{5.4}$$

However, even for $n = 2$, calculating the scalar product becomes a nontrivial task. To date, there are several methods for calculating scalar products. The traditional method was proposed in [V. E. Korepin (1982b)] and described in detail in the book [V. E. Korepin et al (1993)]. It is based on the standard reasoning of the algebraic Bethe ansatz formulated in Section 2.1 on page 34. Another method uses a formula for the action of the product of operators $C(v_i)$ on the vector $B(\bar{u})|0\rangle$. We will give this formula later.

We describe here the third method, which is based on the composite model. This approach is quite close to the traditional one, but it has one essential advantage. Namely, it is applied almost unchanged to the nested algebraic Bethe ansatz solvable models in which the size of the monodromy matrix is greater than $2 \times 2$ [A. Hutsalyuk et al (2017a)].

### 5.1.1 *Action of $C(v)$ on Bethe vectors*

First of all, we consider the action of one operator $C(v)$ on an off-shell vector $B(\bar{u})|0\rangle$. To do this, we use the standard scheme of the algebraic Bethe ansatz. First, we find a general form of the final result. After that, we fix one of the vectors in this result and calculate the corresponding coefficient.

Permuting the operator $C(v)$ with the product $B(\bar{u})$, we should obtain at some stage the commutator $[C(v), B(u_\ell)]$, because $C(v)|0\rangle = 0$:

$$C(v)B(\bar{u})|0\rangle = \sum_{\ell=1}^{n} B(u_1)\cdots B(u_{\ell-1})[C(v), B(u_\ell)]B(u_{\ell+1})\cdots B(u_n)|0\rangle. \tag{5.5}$$

Due to relation (1.122), we obtain

$$\begin{aligned} C(v)B(\bar{u})|0\rangle = \sum_{\ell=1}^{n} & g(v, u_\ell)B(u_1)\cdots B(u_{\ell-1}) \\ & \times\big(D(v)A(u_\ell) - D(u_\ell)A(v)\big)B(u_{\ell+1})\cdots B(u_n)|0\rangle. \end{aligned} \tag{5.6}$$

Then we should move the operators $A$ and $D$ to the right until they act on $|0\rangle$ and respectively give the functions $a$ and $d$.

Note that there exist three possibilities after the operators $A$ and $D$ have approached the extreme right position: the operator $A$ depends on $v$, while the operator $D$ depends on some $u_k$; the operator $A$ depends on some $u_k$, while the operator $D$ depends on $v$; the operator $A$ depends on some $u_j$, while the operator $D$ depends on some $u_k$. As a result of the action of these operators on the vacuum vector $|0\rangle$ we obtain contributions proportional to either $a(v)d(u_k)$, or $d(v)a(u_k)$, or $a(u_j)d(u_k)$. It is also easy to see that we lose the operator $B(u_k)$ in the product of the $B$ operators in the first and second cases. In the third case, we lose the operators $B(u_k)$ and $B(u_j)$, but instead, we obtain a new operator $B(v)$. Therefore, the result can be written in the following form:

$$C(v)B(\bar{u})|0\rangle = \sum_{k=1}^{n}\big(d(v)a(u_k)X_k + a(v)d(u_k)\tilde{X}_k\big)B(\bar{u}_k)|0\rangle + \sum_{j=1}^{n}\sum_{\substack{k=1\\k\neq j}}^{n} a(u_j)d(u_k)Y_{jk}B(v)B(\bar{u}_{jk})|0\rangle. \tag{5.7}$$

Here $X_k$, $\tilde{X}_k$, and $Y_{jk}$ are some numerical coefficients to be determined, and the symbol $\bar{u}_{jk}$ means $\bar{u}_{jk} = \bar{u} \setminus \{u_j, u_k\}$. For our goal, it is enough to find $X_k$ only.

Before calculating $X_k$, we note that the action of one operator $C$ on the vector $B(\bar{u})|0\rangle$ gives a linear combination of off-shell Bethe vectors in which the number of operators $B$ is equal to $n-1$. Then it is clear that if in the scalar product $\#\bar{v} = m$, $\#\bar{u} = n$, and $m \neq n$, it is equal to zero. Indeed, if $m > n$, then after the action of $m$ operators $C$ on $B(\bar{u})|0\rangle$, we will have extra operators $C$ that act on the vacuum $|0\rangle$. Similarly, if $m < n$, we will have extra operators $B$ acting on the dual vacuum $\langle 0|$.

Let us now fix some $k \in \{1, \dots, n\}$ and compute the coefficient $X_k$. All other contributions should be neglected. Using commutativity of the $B$ operators, we can move $B(u_k)$ to the extreme left position with respect to all other $B$ operators:

$$C(v)B(\bar{u})|0\rangle = C(v)B(u_k)B(\bar{u}_k)|0\rangle. \tag{5.8}$$

Now it becomes obvious that at the first step, the commutator $[C(v), B(u_k)]$ should appear. Otherwise, we obtain $B(u_k)C(v)B(\bar{u}_k)|0\rangle$, and hence, the operator $B(u_k)$ is to the left from the operator $C(v)$. Therefore, when moving $C(v)$ further to the right, we do not touch $B(u_k)$, and this operator remains in the final result. At the same time, the coefficient $d(v)a(u_k)X_k$ is multiplied by a vector that does not depend on $B(u_k)$. Hence, we obtain

$$C(v)B(\bar{u})|0\rangle = g(v,u_k)\big(D(v)A(u_k) - D(u_k)A(v)\big)B(\bar{u}_k)|0\rangle + \mathfrak{Z}\,, \tag{5.9}$$

where we have denoted by $\mathfrak{Z}$ the terms that do not contribute to the coefficient $X_k$.

It is easy to see that $D(u_k)A(v)$ also does not contribute to the desired result. Indeed, when the operator $A(v)$ moves to the right, it can keep its argument $v$ or exchange it with an argument of one of the $B$ operators. In the first case, after the action on the vacuum, it gives $a(v)$; in the second case, it gives $a(u_j)$ with $u_j \neq u_k$. No one of these functions can appear with the coefficient $X_k$. Thus,

$$C(v)B(\bar{u})|0\rangle = g(v,u_k)D(v)A(u_k)B(\bar{u}_k)|0\rangle + \mathfrak{Z}. \tag{5.10}$$

Now, when moving $A(u_k)$ to the right, we should keep its argument $u_k$. Otherwise, we cannot obtain the function $a(u_k)$ in the final result. This means that

$$A(u_k)B(u_j) = f(u_j, u_k)B(u_j)A(u_k) + \mathfrak{Z}\,, \tag{5.11}$$

where $\mathfrak{Z}$ still means the terms that do not contribute to the result. Then

$$C(v)B(\bar{u})|0\rangle = g(v, u_k)a(u_k)f(\bar{u}_k, u_k)\; D(v)B(\bar{u}_k)|0\rangle + \mathfrak{Z}. \tag{5.12}$$

Similarly, we conclude that the operator $D(v)$ should keep its argument when moving to the right. Then we can write the commutation relation for the operators $D$ and $B$ in the form

$$D(v)B(u_j) = f(v, u_j)B(u_j)D(v) + \mathfrak{Z}. \tag{5.13}$$

Thus, we obtain

$$C(v)B(\bar{u})|0\rangle = g(v, u_k)a(u_k)d(v)f(\bar{u}_k, u_k)f(v, \bar{u}_k)\; B(\bar{u}_k)|0\rangle + \mathfrak{Z}, \tag{5.14}$$

which implies

$$X_k = g(v, u_k)f(\bar{u}_k, u_k)f(v, \bar{u}_k). \tag{5.15}$$

The coefficients $\tilde{X}_k$ and $Y_{jk}$ in formula (5.7) can be found using similar reasoning. We do not need them now, but we give explicit expressions for them for the sake of completeness:

$$\begin{aligned} \tilde{X}_k &= g(u_k, v)f(u_k, \bar{u}_k)f(\bar{u}_k, v),\\ Y_{jk} &= g(u_k, v)g(v, u_j)f(u_k, u_j)f(u_k, \bar{u}_{jk})f(\bar{u}_{jk}, u_j), \end{aligned} \tag{5.16}$$

where $\bar{u}_{jk} = \bar{u}\setminus\{u_j, u_k\}$. In Section 5.3, we will give a more general formula for the action of the product of several $C$ operators on the vector $B(\bar{u})|0\rangle$ (see (5.74)). From this formula, explicit representations can be obtained for all the coefficients in formula (5.7).

Now it is easy to see that

$$\langle 0|C(\bar{v})B(\bar{u})|0\rangle = \sum_{\substack{\bar{w}\mapsto\{\bar{w}_{\rm I},\bar{w}_{\rm II}\}\\ \#\bar{w}_{\rm I}=n}} a(\bar{w}_{\rm I})d(\bar{w}_{\rm II})W_{\rm part}(\bar{w}_{\rm I}, \bar{w}_{\rm II}), \tag{5.17}$$

where $\bar{w} = \{\bar{v}, \bar{u}\}$ and $W_{\rm part}(\bar{w}_{\rm I}, \bar{w}_{\rm II})$ are rational coefficients dependent on the partitions. The sum is taken over partitions $\bar{w} \mapsto \{\bar{w}_{\rm I}, \bar{w}_{\rm II}\}$ such that $\#\bar{w}_{\rm I} = n$.

In fact, formula (5.17) arises as a result of the $n$-fold application of (5.7). Indeed, after each operator $C(v_i)$ action on a vector, we obtain a pair of functions $a$ and $d$. They always occur in pairs and depend on the parameters from the union $\{\bar{v}, \bar{u}\}$. As a result, we always get in the final formula the product of $n$ functions $a$ and the product of $n$ functions $d$. Each of these functions may depend on any parameter from the union $\bar{w} = \{\bar{v}, \bar{u}\}$. Besides, for all the numerous permutations of the operators, the functions $f$ and $g$ appear in the commutation relations (see (1.119)–(1.130)). All these functions are collected in coefficients $W_{\text{part}}$. The main problem is to find them. It is important to emphasize that these coefficients are completely determined by the $R$-matrix.

Let $\bar{w}_{\text{I}} = \{\bar{v}_{\text{i}}, \bar{u}_{\text{ii}}\}$ and $\bar{w}_{\text{II}} = \{\bar{v}_{\text{ii}}, \bar{u}_{\text{i}}\}$. Due to $\#\bar{v} = \#\bar{u} = n$ we have $\#\bar{v}_{\text{i}} = \#\bar{u}_{\text{i}}$ and $\#\bar{v}_{\text{ii}} = \#\bar{u}_{\text{ii}}$. Equation (5.17) then turns into

$$S_n(\bar{v}|\bar{u}) = \sum_{\substack{\bar{u}\mapsto\{\bar{u}_{\text{i}},\bar{u}_{\text{ii}}\} \\ \bar{v}\mapsto\{\bar{v}_{\text{i}},\bar{v}_{\text{ii}}\} \\ \#\bar{u}_{\text{i}}=\#\bar{v}_{\text{i}}}} a(\bar{v}_{\text{i}})d(\bar{u}_{\text{i}})a(\bar{u}_{\text{ii}})d(\bar{v}_{\text{ii}})W_{\text{part}}(\bar{u}_{\text{i}}, \bar{u}_{\text{ii}}, \bar{v}_{\text{i}}, \bar{v}_{\text{ii}}). \tag{5.18}$$

The sum is taken over partitions $\bar{v} \mapsto \{\bar{v}_{\text{i}}, \bar{v}_{\text{ii}}\}$ and $\bar{u} \mapsto \{\bar{u}_{\text{i}}, \bar{u}_{\text{ii}}\}$ such that $\#\bar{v}_{\text{i}} = \#\bar{u}_{\text{i}} = k$, $k = 0, 1, \dots, n$. Dividing (5.18) by the product $d(\bar{v})d(\bar{u})$ we find

$$\boldsymbol{S}_n(\bar{v}|\bar{u}) = \sum_{\substack{\bar{u}\mapsto\{\bar{u}_{\text{i}},\bar{u}_{\text{ii}}\} \\ \bar{v}\mapsto\{\bar{v}_{\text{i}},\bar{v}_{\text{ii}}\} \\ \#\bar{u}_{\text{i}}=\#\bar{v}_{\text{i}}}} r(\bar{v}_{\text{i}})r(\bar{u}_{\text{ii}})W_{\text{part}}(\bar{u}_{\text{i}}, \bar{u}_{\text{ii}}, \bar{v}_{\text{i}}, \bar{v}_{\text{ii}}), \tag{5.19}$$

where we used the shorthand notation for the products of the functions $r$ over subsets $\bar{v}_{\text{i}}$ and $\bar{u}_{\text{ii}}$.

Before moving on, we introduce the concept of a generalized model.

### 5.1.2 *Generalized model and scalar products*

A generalized model was introduced in [V. E. Korepin (1982a)]. In the early stages of developing the algebraic Bethe ansatz, it was actively used [V. E. Korepin (1982b); A. G. Izergin and V. E. Korepin (1984); V. E. Korepin (1984); N. A. Slavnov (1989)] but then was undeservedly forgotten.

In fact, the generalized model is not one model but a class of models. The monodromy matrix of each representative of this class satisfies the $RTT$-relation with a fixed $R$-matrix and has a vacuum vector (and also a dual vector). Different representatives of this class are parameterized by different functions $r(u) = a(u)/d(u)$. We can say that a representative of the generalized model is a concrete representation of the operator algebra given by the $RTT$-relation.

We have already seen how one can build a model in which the function $r(u)$ is an arbitrary rational function with a finite limit at infinity[1] (2.100). We can also consider the limits of this expression when an arbitrary natural $N$ tends to infinity. Thus, we have a vast reserve of functions $r(u)$. We can say that if the variable $u$ is fixed, then running through different representatives of the generalized model, we can always find such of them that $r(u)$ takes any prescribed value. In this sense, we can consider the variables $u$ and $r(u)$ as independent variables. Then the scalar product of two generic Bethe vectors becomes a function of variables of two types: on the one hand, these are parameters $u_k$ and $v_k$; on the other hand, these are parameters $r(u_k)$ and $r(v_k)$. In the generalized model, both types of variables take arbitrary values independently of each other.

If one or both vectors are on-shell, then the situation changes. Then a constraint arises for the two types of variables, which is specified by Bethe equations. However, it is essential to emphasize that within the framework of the generalized model, the Bethe equations are no longer equations. Instead, they become conditions that connect two types of variables. If a representative of the generalized model is fixed (that is, a concrete function $r(u)$ is specified), then the Bethe equations again become equations: one should find possible values of the parameters $u_k$ for a given function $r(u)$. Within the generalized model framework, we can look at these constraints from a different point of view. Namely, we can consider the variables $u_k$ as arbitrary complex numbers and determine the function $r(u)$ values in the points $u_k$.

Let us explain this with an example. Consider the Bethe equations for $n$ variables $\bar{u} = \{u_1, \dots, u_n\}$ in the inhomogeneous $XXZ$ (or $XXX$) chain

$$r(u_j) = \frac{f(u_j, \bar{u}_j)}{f(\bar{u}_j, u_j)}, \qquad j = 1, \dots, n, \quad \text{where} \quad r(u) = \prod_{l=1}^{N} f(u, \xi_\ell). \quad (5.20)$$

We arbitrarily fix the parameters $\bar{u}$ and consider (5.20) as a system of equations for inhomogeneities $\xi_l$. Since the length of the chain $N$ is not limited in any way, we can always choose $N$ so that the system will have a solution (in fact, many solutions). In other words, we can always find such a representative of the generalized model that the Bethe equations will be satisfied for any set of complex variables $\bar{u}$. Note that in this example, we have restricted ourselves only to inhomogeneous spin chains of spin 1/2. In fact, the generalized model includes many more representatives.

[1]In Section 2.6, we explicitly constructed such a model for the case of the rational $R$-matrix, but a similar result can be obtained for the trigonometric $R$-matrix [V. Tarasov (1984)].

### 5.1.3 *Highest coefficient of the scalar products*

Let us turn back to equation (5.19). We introduce the notion of the *highest coefficient*.

**Definition 5.1.** The coefficient $W_{\rm part}(\emptyset, \bar u, \emptyset, \bar v)$ corresponding to the partition $\bar v_{\rm i} = \bar u_{\rm i} = \emptyset$ is called the highest coefficient. We denote it by $\mathbb{K}_n(\bar v|\bar u)$:

$$\mathbb{K}_n(\bar v|\bar u) = W_{\rm part}(\emptyset, \bar u, \emptyset, \bar v). \tag{5.21}$$

In other words, $\mathbb{K}_n(\bar v|\bar u)$ is the coefficient of the product $r(\bar u)$ in (5.19).

The coefficient $W_{\rm part}(\bar u, \emptyset, \bar v, \emptyset)$ corresponding to the partition $\bar v_{\rm ii} = \bar u_{\rm ii} = \emptyset$ is called a *conjugated highest coefficient*. We denote it by $\overline{\mathbb{K}}_n(\bar v|\bar u)$:

$$\overline{\mathbb{K}}_n(\bar v|\bar u) = W_{\rm part}(\bar u, \emptyset, \bar v, \emptyset). \tag{5.22}$$

In other words, $\overline{\mathbb{K}}_n(\bar v|\bar u)$ is the coefficient of the product $r(\bar v)$ in (5.19).

It is not difficult to show that $\overline{\mathbb{K}}_n(\bar v|\bar u) = \mathbb{K}_n(\bar u|\bar v)$. However, we will not use this property yet.

**Proposition 5.1.** *An arbitrary coefficient* $W_{\rm part}(\bar u_{\rm i}, \bar u_{\rm ii}, \bar v_{\rm i}, \bar v_{\rm ii})$ *in formulas* (5.18), (5.19) *is expressed in terms of* $\mathbb{K}$ *and* $\overline{\mathbb{K}}$ *as follows:*

$$W_{\rm part}(\bar u_{\rm i}, \bar u_{\rm ii}, \bar v_{\rm i}, \bar v_{\rm ii}) = f(\bar u_{\rm i}, \bar u_{\rm ii}) f(\bar v_{\rm ii}, \bar v_{\rm i}) \overline{\mathbb{K}}_k(\bar v_{\rm i}|\bar u_{\rm i}) \mathbb{K}_{n-k}(\bar v_{\rm ii}|\bar u_{\rm ii}). \tag{5.23}$$

*Here* $k = \#\bar v_{\rm i} = \#\bar u_{\rm i}$.

**Proof.** We prove this proposition in the framework of the generalized model. We have mentioned already that the coefficients $W_{\rm part}$ are determined by the $R$-matrix only. Therefore, they do not depend on the concrete representative of the generalized model. If so, we can choose such a representative to simplify calculating the coefficient $W_{\rm part}$ as much as possible. If we find this coefficient for some concrete model, then we find it in the general case.

Note that the coefficients $W_{\rm part}$ are rational functions of the parameters[2] $\bar v$ and $\bar u$. Hence, equality (5.23) is a rational identity. Any rational function is entirely determined by its values at a finite number of points. Therefore, to prove any rational identity, it is sufficient to check that it holds at a finite number of points. In our case, it is sufficient to consider a finite number of fixed sets of parameters $\bar v$ and $\bar u$ and check that (5.23) holds for each fixed pair of the sets $\bar v$ and $\bar u$. Let us show how to do this for a fixed pair.

[2]In the case of the trigonometric $R$-matrix, they are reduced to rational functions after replacement $x_j = e^{2v_j}$ and $y_j = e^{2u_j}$.

Let $\bar{v}$ and $\bar{u}$ be fixed complex numbers. Let us fix some partition $\bar{v} \mapsto \{\bar{v}_{\rm I}, \bar{v}_{\rm II}\}$ and $\bar{u} \mapsto \{\bar{u}_{\rm I}, \bar{u}_{\rm II}\}$ such that $\#\bar{v}_{\rm I} = \#\bar{u}_{\rm I} = k$, where $k = 0, 1, \dots, n$. Consider a concrete composite model (see Chapter 4)

$$T(z) = T^{(2)}(z)T^{(1)}(z), \tag{5.24}$$

where the functions $a^{(j)}(z)$ and $d^{(j)}(z)$ are such that the following conditions hold:

$$\begin{aligned} &a^{(1)}(z) = 0, \quad \text{if} \quad z \in \bar{v}_{\rm II}; \qquad && d^{(1)}(z) = 0, \quad \text{if} \quad z \in \bar{u}_{\rm II}; \\ &a^{(2)}(z) = 0, \quad \text{if} \quad z \in \bar{u}_{\rm I}; \qquad && d^{(2)}(z) = 0, \quad \text{if} \quad z \in \bar{v}_{\rm I}. \end{aligned} \tag{5.25}$$

For example, we can take an inhomogeneous $XXX$ chain and choose the inhomogeneities to satisfy conditions (5.25) for given $\bar{v}$ and $\bar{u}$ and fixed partition.

Now we calculate the scalar product (5.1), substituting there representations for the total Bethe vectors in terms of the partial ones (see (4.12), (4.27)). Let us recall them

$$\begin{aligned} B(\bar{u})|0\rangle = &\sum_{\bar{u}\mapsto\{\bar{u}_{\rm i},\bar{u}_{\rm ii}\}} a^{(2)}(\bar{u}_{\rm i})d^{(1)}(\bar{u}_{\rm ii})f(\bar{u}_{\rm ii}, \bar{u}_{\rm i}) \\ &\times B^{(1)}(\bar{u}_{\rm i})|0\rangle^{(1)} \otimes B^{(2)}(\bar{u}_{\rm ii})|0\rangle^{(2)}, \end{aligned} \tag{5.26}$$

and

$$\begin{aligned} \langle 0|C(\bar{v}) = &\sum_{\bar{v}\mapsto\{\bar{v}_{\rm i},\bar{v}_{\rm ii}\}} a^{(1)}(\bar{v}_{\rm ii})d^{(2)}(\bar{v}_{\rm i})f(\bar{v}_{\rm i}, \bar{v}_{\rm ii}) \\ &\times \langle 0|^{(1)}C^{(1)}(\bar{v}_{\rm i}) \otimes \langle 0|^{(2)}C^{(2)}(\bar{v}_{\rm ii}). \end{aligned} \tag{5.27}$$

Then we have

$$\begin{aligned} \langle 0|C(\bar{v})B(\bar{u})|0\rangle = &\sum_{\substack{\bar{v}\mapsto\{\bar{v}_{\rm i},\bar{v}_{\rm ii}\} \\ \bar{u}\mapsto\{\bar{u}_{\rm i},\bar{u}_{\rm ii}\} \\ \#\bar{v}_{\rm i}=\#\bar{u}_{\rm i}}} a^{(1)}(\bar{v}_{\rm ii})d^{(2)}(\bar{v}_{\rm i})a^{(2)}(\bar{u}_{\rm i})d^{(1)}(\bar{u}_{\rm ii})f(\bar{u}_{\rm ii}, \bar{u}_{\rm i}) \\ &\times f(\bar{v}_{\rm i}, \bar{v}_{\rm ii})\langle 0|^{(1)}C^{(1)}(\bar{v}_{\rm i})B^{(1)}(\bar{u}_{\rm i})|0\rangle^{(1)} \\ &\cdot\langle 0|^{(2)}C^{(2)}(\bar{v}_{\rm ii})B^{(2)}(\bar{u}_{\rm ii})|0\rangle^{(2)}. \end{aligned} \tag{5.28}$$

The condition $\#\bar{v}_{\rm i} = \#\bar{u}_{\rm i}$ (and thus, $\#\bar{v}_{\rm ii} = \#\bar{u}_{\rm ii}$) arises because otherwise, the scalar products in (5.28) vanish.

Let $\#\bar{v}_{\rm i} = \#\bar{u}_{\rm i} = k'$. Then $\#\bar{v}_{\rm ii} = \#\bar{u}_{\rm ii} = n - k'$. Due to the presence of the products $a^{(1)}(\bar{v}_{\rm ii})$ and $a^{(2)}(\bar{u}_{\rm i})$, we conclude that $\bar{v}_{\rm ii} \subset \bar{v}_{\rm I}$ and $\bar{u}_{\rm i} \subset \bar{u}_{\rm II}$, otherwise, we obtain vanishing contribution. Then $n-k' \le k$ and $k' \le n-k$, otherwise we cannot provide $\bar{v}_{\rm ii} \subset \bar{v}_{\rm I}$ and $\bar{u}_{\rm i} \subset \bar{u}_{\rm II}$. But the system $n-k' \le k$ and $k' \le n-k$ has only one solution $k' = n-k$. This is possible if and only if $\bar{u}_{\rm i} = \bar{u}_{\rm II}$, $\bar{v}_{\rm i} = \bar{v}_{\rm II}$, $\bar{u}_{\rm ii} = \bar{u}_{\rm I}$, and $\bar{v}_{\rm ii} = \bar{v}_{\rm I}$. Thus, only one term in the sum over partitions survives, and we arrive at

$$\begin{aligned}\langle 0|C(\bar{v})B(\bar{u})|0\rangle &= a^{(1)}(\bar{v}_{\rm I})d^{(2)}(\bar{v}_{\rm II})a^{(2)}(\bar{u}_{\rm II})d^{(1)}(\bar{u}_{\rm I})f(\bar{u}_{\rm I},\bar{u}_{\rm II})f(\bar{v}_{\rm II},\bar{v}_{\rm I})\\ &\quad\times\langle 0|^{(1)}C^{(1)}(\bar{v}_{\rm II})B^{(1)}(\bar{u}_{\rm II})|0\rangle^{(1)}\\ &\quad\cdot\langle 0|^{(2)}C^{(2)}(\bar{v}_{\rm I})B^{(2)}(\bar{u}_{\rm I})|0\rangle^{(2)}.\end{aligned} \tag{5.29}$$

To calculate scalar products in (5.29), we can use formula (5.18). We should only replace $a \to a^{(1)}$ and $d \to d^{(1)}$ for the first scalar product and $a \to a^{(2)}$ and $d \to d^{(2)}$ for the second scalar product. Due to conditions (5.25), we obtain

$$\begin{aligned}\langle 0|^{(1)}C^{(1)}(\bar{v}_{\rm II})B^{(1)}(\bar{u}_{\rm II})|0\rangle^{(1)} &= d^{(1)}(\bar{v}_{\rm II})a^{(1)}(\bar{u}_{\rm II})\mathbb{K}_{n-k}(\bar{v}_{\rm II}|\bar{u}_{\rm II}),\\ \langle 0|^{(2)}C^{(2)}(\bar{v}_{\rm I})B^{(2)}(\bar{u}_{\rm I})|0\rangle^{(2)} &= a^{(2)}(\bar{v}_{\rm I})d^{(2)}(\bar{u}_{\rm I})\overline{\mathbb{K}}_{k}(\bar{v}_{\rm I}|\bar{u}_{\rm I}).\end{aligned} \tag{5.30}$$

Thus, using $a^{(1)}(z)a^{(2)}(z) = a(z)$ and $d^{(1)}(z)d^{(2)}(z) = d(z)$, we finally arrive at

$$\begin{aligned}\langle 0|C(\bar{v})B(\bar{u})|0\rangle &= a(\bar{v}_{\rm I})d(\bar{v}_{\rm II})a(\bar{u}_{\rm II})d(\bar{u}_{\rm I})f(\bar{u}_{\rm I},\bar{u}_{\rm II})f(\bar{v}_{\rm II},\bar{v}_{\rm I})\\ &\quad\times\mathbb{K}_{n-k}(\bar{v}_{\rm II}|\bar{u}_{\rm II})\overline{\mathbb{K}}_{k}(\bar{v}_{\rm I}|\bar{u}_{\rm I}).\end{aligned} \tag{5.31}$$

On the other hand, it follows from conditions (5.25) that

$$\begin{aligned} a(z) &= 0, \quad \text{if} \quad z \in \{\bar{v}_{\rm II}, \bar{u}_{\rm I}\},\\ d(z) &= 0, \quad \text{if} \quad z \in \{\bar{v}_{\rm I}, \bar{u}_{\rm II}\}.\end{aligned} \tag{5.32}$$

Then equation (5.18) implies

$$\langle 0|C(\bar{v})B(\bar{u})|0\rangle = a(\bar{v}_{\rm I})d(\bar{v}_{\rm II})a(\bar{u}_{\rm II})d(\bar{u}_{\rm I})W_{\rm part}(\bar{u}_{\rm I},\bar{u}_{\rm II},\bar{v}_{\rm I},\bar{v}_{\rm II}). \tag{5.33}$$

Comparing (5.33) and (5.31), we find

$$W_{\rm part}(\bar{u}_{\rm I},\bar{u}_{\rm II},\bar{v}_{\rm I},\bar{v}_{\rm II}) = f(\bar{u}_{\rm I},\bar{u}_{\rm II})f(\bar{v}_{\rm II},\bar{v}_{\rm I})\mathbb{K}_{n-k}(\bar{v}_{\rm II}|\bar{u}_{\rm II})\overline{\mathbb{K}}_{k}(\bar{v}_{\rm I}|\bar{u}_{\rm I}). \tag{5.34}$$

Relabeling the subsets, we arrive at (5.23).

Fixing the sets $\bar{v}$ and $\bar{u}$ in some other way, we can repeat the same reasoning and check (5.23) for a new pair of sets, etc. Since the number of representatives of the generalized model is unlimited, we can check equality (5.23) for any finite number of fixed pairs $\bar{v}$ and $\bar{u}$. It follows that (5.23) is an identity. □

Thus, we have expressed an arbitrary coefficient $W_{\rm part}$ in terms of $\mathbb{K}$ and $\overline{\mathbb{K}}$. Thanks to the special choice of the representative of the generalized model, all the sums over partitions were reduced to a single term. This calculation method is especially well suited for models in which the monodromy matrix size exceeds $2 \times 2$. In these models, one can find the actions of the annihilation operators (lower-triangular elements of the monodromy matrix) on Bethe vectors; however, applying these formulas to the calculation of scalar products becomes technically extremely difficult. At the same time, the method described above still works without any significant complications.

To complete the solution, we should find the highest coefficients $\mathbb{K}$ and $\overline{\mathbb{K}}$. This is not an easy task. However, it is not difficult to obtain a recurrence relation for these coefficients. Indeed, let $\bar{u}$ be a fixed set of complex numbers. Let us take a representative of the generalized model in which $d(u_j) = 0$ for all $u_j \in \bar{u}$. Then it follows from (5.18) that only one term survives in the sum over partitions

$$\langle 0|C(\bar{v})B(\bar{u})|0\rangle = d(\bar{v})a(\bar{u})\mathbb{K}_n(\bar{v}|\bar{u}), \qquad d(u_j) = 0. \tag{5.35}$$

On the other hand, we can use (5.7) for the operator $C(v_n)$ action on the vector $B(\bar{u})|0\rangle$. If $d(u_j) = 0$, then only terms proportional to $d(v)a(u_k)X_k$ survive in it. Hence,

$$\begin{aligned}\langle 0|C(\bar{v})B(\bar{u})|0\rangle &= \langle 0|C(\bar{v}_n)C(v_n)B(\bar{u})|0\rangle \\ &= \sum_{k=1}^{n} d(v_n)a(u_k)X_k\langle 0|C(\bar{v}_n)B(\bar{u}_k)|0\rangle, \quad d(u_j) = 0. \end{aligned} \tag{5.36}$$

Using

$$\langle 0|C(\bar{v}_n)B(\bar{u}_k)|0\rangle = d(\bar{v}_n)a(\bar{u}_k)\mathbb{K}_{n-1}(\bar{v}_n|\bar{u}_k), \qquad d(u_j) = 0, \tag{5.37}$$

we immediately arrive at a recursion for the highest coefficient

$$\mathbb{K}_n(\bar{v}|\bar{u}) = \sum_{k=1}^{n} g(v_n, u_k)f(\bar{u}_k, u_k)f(v_n, \bar{u}_k)\mathbb{K}_{n-1}(\bar{v}_n|\bar{u}_k), \tag{5.38}$$

where we used (5.15). The initial condition for this recursion follows from (5.4): $\mathbb{K}_1(v|u) = g(v, u)$. We will find a solution to this recursion in the next section.

**Remark 5.1.** A recursion for the conjugated highest coefficient is derived similarly:

$$\overline{\mathbb{K}}_n(\bar{v}|\bar{u}) = \sum_{k=1}^{n} g(u_k, v_n)f(u_k, \bar{u}_k)f(\bar{u}_k, v_n)\overline{\mathbb{K}}_{n-1}(\bar{v}_n|\bar{u}_k). \tag{5.39}$$

We invite the reader to do this exercise.

Comparing (5.38) and (5.39), we find that

$$\overline{\mathbb{K}}_n(\bar{v}|\bar{u}) = \mathbb{K}_n(\bar{v}|\bar{u})\Big|_{\eta\to-\eta}. \tag{5.40}$$

## 5.2 Determinant representation for the highest coefficient

**Proposition 5.2.** *[A. G. Izergin (1987)] The highest coefficient $\mathbb{K}_n(\bar{v}|\bar{u})$ has the following determinant representation:*

$$\mathbb{K}_n(\bar{v}|\bar{u}) = \left(\prod_{n\geq \ell>m\geq 1} g(v_\ell, v_m)g(u_m, u_\ell)\right) \frac{f(\bar{v},\bar{u})}{g(\bar{v},\bar{u})} \det_n \left(\frac{g^2(v_j,u_k)}{f(v_j,u_k)}\right). \tag{5.41}$$

Let us also give (5.41) in a more detailed form for the trigonometric $R$-matrix

$$\mathbb{K}_n(\bar{v}|\bar{u}) = \frac{\prod_{\ell,m=1}^{n}\sinh(v_\ell - u_m + \eta)}{\prod_{n\geq\ell>m\geq 1}\sinh(v_\ell - v_m)\sinh(u_m - u_\ell)} \times \det_n \left(\frac{\sinh\eta}{\sinh(v_j - u_k)\sinh(v_j - u_k + \eta)}\right). \tag{5.42}$$

In the case of the rational $R$-matrix, it is enough to make the replacement $v_j \to \epsilon v_j$, $u_j \to \epsilon u_j$, $\eta \to \epsilon\eta$ and take the limit $\epsilon \to 0$:

$$\mathbb{K}_n(\bar{v}|\bar{u}) = \frac{\prod_{\ell,m=1}^{n}(v_\ell - u_m + \eta)}{\prod_{n\geq\ell>m\geq 1}(v_\ell - v_m)(u_m - u_\ell)} \times \det_n \left(\frac{\eta}{(v_j - u_k)(v_j - u_k + \eta)}\right). \tag{5.43}$$

**Remark 5.2.** It follows from (5.40) that an explicit representation for the conjugated highest coefficient is obtained from (5.41) when replacing $\eta$ with $-\eta$. It is easy to see that this is equivalent to replacing $\bar{u} \leftrightarrow \bar{v}$:

$$\overline{\mathbb{K}}_n(\bar{v}|\bar{u}) = \mathbb{K}_n(\bar{u}|\bar{v}). \tag{5.44}$$

It follows from (5.41) that $\mathbb{K}_1(v|u) = g(v,u)$. Thus, the initial condition is satisfied. Therefore, to prove the determinant representation (5.41), we only need to check that this function satisfies recursion (5.38). However, before doing this, we need to establish several important properties of this function.

### 5.2.1 *Properties of the determinant representation*

Obviously, $\mathbb{K}_n(\bar{v}|\bar{u})$ (5.41) is symmetric in $\bar{v}$ and is symmetric in $\bar{u}$. In the case of the rational $R$-matrix, this is a rational function of $\bar{v}$ and $\bar{u}$ (see (5.43)). It decreases if one of the arguments goes to infinity at other arguments fixed. Indeed, if $v_n \to \infty$, then the prefactor of the determinant behaves as $v_n$. At the same time, the determinant behaves as $v_n^{-2}$ because all the elements in the last row of the corresponding matrix behave as $v_n^{-2}$. As a result, $\mathbb{K}_n(\bar{v}|\bar{u})$ (5.43) behaves as $v_n^{-1}$.

In the case of the trigonometric $R$-matrix, we can make a change of variables $e^{2v_j} = x_j$, $e^{2u_j} = y_j$, $(j = 1, \dots, n)$, and $e^{\eta} = q$. Then (5.42) takes the form

$$\mathbb{K}_n(\bar{v}|\bar{u}) = \left(\prod_{\ell=1}^{n} \sqrt{x_\ell\, y_\ell}\right) \widetilde{\mathbb{K}}_n(\bar{x}|\bar{y}), \tag{5.45}$$

where $\widetilde{\mathbb{K}}_n(\bar{x}|\bar{y})$ is a rational function of $\bar{x}$ and $\bar{y}$. Similarly to the rational case, this function also decreases if one of the arguments goes to infinity at other arguments fixed. Other properties of the function $\mathbb{K}_n(\bar{v}|\bar{u})$ can be formulated in a general form, which is valid for both the rational and trigonometric cases.

Function (5.41) has simple poles in $v_j = u_k$, $j, k = 1, \dots, n$. The residues of $\mathbb{K}_n$ can be expressed in terms of $\mathbb{K}_{n-1}$ as follows:

$$\mathbb{K}_n(\bar{v}|\bar{u})\Big|_{v_n \to u_n} = g(v_n, u_n) f(\bar{v}_n, v_n) f(u_n, \bar{u}_n)\, \mathbb{K}_{n-1}(\bar{v}_n|\bar{u}_n) + \text{Reg}, \tag{5.46}$$

where Reg means regular terms. The residues in all other poles can be restored by symmetry.

The proof of this property is straightforward. Let $v_n \to u_n$. Then the matrix element $g^2(v_n, u_n)/f(v_n, u_n)$ becomes singular, and the determinant reduces to the product of this matrix element by the corresponding minor

$$\det_n \left(\frac{g^2(v_j, u_k)}{f(v_j, u_k)}\right)\Big|_{v_n \to u_n} = g(v_n, u_n) \det_{n-1} \left(\frac{g^2(v_j, u_k)}{f(v_j, u_k)}\right) + \text{Reg}. \tag{5.47}$$

Setting $v_n = u_n$ in the prefactor we after elementary algebra arrive at (5.46).

One more recursion has the following form:

$$\mathbb{K}_{n+1}(\{\bar{v}, z - \eta\}|\{\bar{u}, z\}) = \mathbb{K}_{n+1}(\{\bar{v}, z\}|\{\bar{u}, z + \eta\}) = -\mathbb{K}_n(\bar{v}|\bar{u}). \tag{5.48}$$

The proof of this formula is similar to the previous one. Indeed, if we replace $\bar{v}$ with $\{\bar{v}, z - \eta\}$ and $\bar{u}$ with $\{\bar{u}, z\}$ in (5.41), then the double product in this formula vanishes because $f(z - \eta, z) = 0$ for any $z$. On the other hand, one of the matrix elements in the determinant becomes singular. Therefore, the determinant reduces to the product of this element by the corresponding minor, which ultimately leads to (5.48).

We will also need the properties of the function (5.41) under the replacement of the sets $\bar{v} \leftrightarrow \bar{u}$. We have mentioned already that this replacement is equivalent to the change $\eta$ by $-\eta$ (see (5.40) and (5.44)). The replacement of the sets $\bar{v}$ and $\bar{u}$ also can be achieved due to

$$\mathbb{K}_n(\bar{v}-\eta|\bar{u}) = \mathbb{K}_n(\bar{v}|\bar{u}+\eta) = (-1)^n \frac{\mathbb{K}_n(\bar{u}|\bar{v})}{f(\bar{u},\bar{v})}. \tag{5.49}$$

This formula can be easily derived from the explicit representation (5.41) and the fact that

$$f(u-\eta, v) = f(u, v+\eta) = \frac{1}{f(v,u)}. \tag{5.50}$$

### 5.2.2 *Summation identity*

Now we will establish a more sophisticated identity for the function (5.41). For definiteness and simplicity, we consider models described by the rational $R$-matrix. However, all of the following entirely applies to the models with trigonometric $R$-matrix (unless otherwise specified).

**Proposition 5.3.** *Let $\bar{v}$, $\bar{u}$, and $\bar{w}$ be sets of complex numbers such that $\#\bar{v}=m$, $\#\bar{u}=n$, and $\#\bar{w}=m+n$. Then*

$$\sum_{\bar{w}\mapsto\{\bar{w}_{\rm I},\bar{w}_{\rm II}\}} \mathbb{K}_m(\bar{w}_{\rm I}|\bar{v})\mathbb{K}_n(\bar{u}|\bar{w}_{\rm II})f(\bar{w}_{\rm II},\bar{w}_{\rm I}) = (-1)^m f(\bar{w},\bar{v})\mathbb{K}_{m+n}(\{\bar{v}-\eta,\bar{u}\}|\bar{w}). \tag{5.51}$$

*The sum is taken over partitions of the set $\bar{w}$ into two subsets $\bar{w}_{\rm I}$ and $\bar{w}_{\rm II}$, such that $\#\bar{w}_{\rm I}=m$ and $\#\bar{w}_{\rm II}=n$.*

**Remark 5.3.** We have explicitly indicated that we are dealing with summation over partitions with restrictions on the cardinalities of the subsets. However, these restrictions automatically follow from the form of the formula. Indeed, by definition, the function (5.41) has an equal number of arguments to the left and the right from the vertical line. Then $\#\bar{v}=m$ yields $\#\bar{w}_{\rm I}=m$, and $\#\bar{u}=n$ yields $\#\bar{w}_{\rm II}=n$. Formula (5.51) makes sense only in this case.

Before proving identity (5.51), we explain its meaning. To do this, consider the case $m=1$ when the set $\bar{v}$ consists of a single element $v$. Then, we can set $\bar{w}_{\rm I}=w_j$ and $\bar{w}_{\rm II}=\bar{w}_j$, $j=1,\dots,n+1$. Using the fact that $\mathbb{K}_1(u|v)=g(u,v)$, we then recast (5.51) for $m=1$ as follows:

$$\sum_{j=1}^{n+1} g(w_j,v)\mathbb{K}_n(\bar{u}|\bar{w}_j)f(\bar{w}_j,w_j) = -f(\bar{w},v)\mathbb{K}_{n+1}(\{v-\eta,\bar{u}\}|\bar{w}). \tag{5.52}$$

It is easy to verify that the sum in the lhs is nothing but the partial fraction decomposition of the rhs. Namely, let $\chi(z)$ be a rational function of $z$ that decreases at $z \to \infty$ and has simple poles at the points $z_1, \dots, z_N$. Then

$$\chi(z) = \sum_{k=1}^{N} \frac{1}{z - z_k} \operatorname{Res} \chi(z)\Big|_{z=z_k}. \tag{5.53}$$

We are dealing with the same sum in (5.52). Indeed, the function in the rhs of (5.52) decreases at $v \to \infty$. It has simple poles at $v = w_j$, $j = 1, \dots, n+1$, due to the product $f(\bar{w}, v)$. The poles of $\mathbb{K}_{n+1}(\{v - \eta, \bar{u}\}|\bar{w})$ at $v - \eta = w_j$ are compensated by zeros of the same product $f(\bar{w}, v)$. Therefore, we can decompose the rhs via (5.53)

$$f(\bar{w}, v)\mathbb{K}_{n+1}(\{v - \eta, \bar{u}\}|\bar{w}) = \sum_{j=1}^{n+1} g(w_j, v) f(\bar{w}_j, w_j)\mathbb{K}_{n+1}(\{w_j - \eta, \bar{u}\}|\bar{w}). \tag{5.54}$$

It remains to use (5.48), and we arrive at (5.52).

If the set $\bar{v}$ consists of several parameters, we can apply this procedure successively, decomposing the rhs of (5.51) first by the poles at $v_1 = w_j$, then by the poles at $v_2 = w_j$, and so on. As a result, the sum over the partitions in (5.51) turns out to be the expansion of the rhs with respect to the poles at the points $v_k = w_j$, $k = 1, \dots, m$, $j = 1, \dots, n + m$.

**Proof of Proposition 5.3.** First of all, let us find the properties of both sides of (5.51) in the poles. Consider, for example, the lhs as a function of $v_1$ at all other variables fixed. Let

$$F^{(l)}_{m,n}(\bar{v}|\bar{w}) = \sum_{\bar{w} \mapsto \{\bar{w}_{\rm I}, \bar{w}_{\rm II}\}} \mathbb{K}_m(\bar{w}_{\rm I}|\bar{v})\mathbb{K}_n(\bar{u}|\bar{w}_{\rm II}) f(\bar{w}_{\rm II}, \bar{w}_{\rm I}). \tag{5.55}$$

The function $F^{(l)}_{m,n}(\bar{v}|\bar{w})$ is the analytic function of $v_1$ in the whole complex plane except the points $w_k$, where it has simple poles. Here $w_k$ is an arbitrary element of the set $\bar{w}$. Indeed, the parameter $v_1$ is included in the function $\mathbb{K}_m(\bar{w}_{\rm I}|\bar{v})$ only. The latter has simple poles at the points where $v_1 = w_k$. Let us fix $w_k$. Then the singularity in the rhs of (5.55) arises if and only if $w_k \in \bar{w}_{\rm I}$. Let us put $\bar{w}_{\rm I'} = \bar{w}_{\rm I} \setminus w_k$, $\bar{w}_k = \bar{w} \setminus w_k$, and $\bar{v}_1 = \bar{v} \setminus v_1$. Then, sending $v_1 \to w_k$ and using (5.46), we obtain

$$\begin{aligned} F^{(l)}_{m,n}(\bar{v}|\bar{w})\Big|_{v_1 \to w_k} = \sum_{\bar{w}_k \mapsto \{\bar{w}_{\rm I'}, \bar{w}_{\rm II}\}} & g(w_k, v_1) f(v_1, \bar{v}_1) f(\bar{w}_{\rm I'}, w_k)\mathbb{K}_{m-1} \\ & \times (\bar{w}_{\rm I'}|\bar{v}_1)\mathbb{K}_n(\bar{u}|\bar{w}_{\rm II}) f(\bar{w}_{\rm II}, \bar{w}_{\rm I'}) f(\bar{w}_{\rm II}, w_k) + \text{Reg}, \end{aligned} \tag{5.56}$$

and we recall that Reg means the terms that remain regular at $v_1 = w_k$. We also have used an equality

$$f(\bar{w}_{\mathrm{II}}, \bar{w}_{\mathrm{I}}) = f(\bar{w}_{\mathrm{II}}, \bar{w}_{\mathrm{I}'})f(\bar{w}_{\mathrm{II}}, w_k). \tag{5.57}$$

Note that now the sum is taken over the partitions of the set $\bar{w}_k$ into subsets $\bar{w}_{\mathrm{I}'}$ and $\bar{w}_{\mathrm{II}}$. Then, the factors $g(w_k, v_1)f(v_1, \bar{v}_1)$ do not depend on the partitions. Hence, they can be taken outside the sum. Similarly, it is easy to see that the product $f(\bar{w}_{\mathrm{I}'}, w_k)f(\bar{w}_{\mathrm{II}}, w_k) = f(\bar{w}_k, w_k)$ does not depend on the partitions, and it thus also can be taken out the sum. We obtain

$$F^{(l)}_{m,n}(\bar{v}|\bar{w})\Big|_{v_1\to w_k} = g(w_k, v_1)f(v_1, \bar{v}_1)f(\bar{w}_k, w_k) \sum_{\bar{w}_k\mapsto\{\bar{w}_{\mathrm{I}'},\bar{w}_{\mathrm{II}}\}} \times \mathbb{K}_{m-1}(\bar{w}_{\mathrm{I}'}|\bar{v}_1)\mathbb{K}_n(\bar{u}|\bar{w}_{\mathrm{II}})f(\bar{w}_{\mathrm{II}}, \bar{w}_{\mathrm{I}'}) + \mathrm{Reg}. \tag{5.58}$$

Obviously, the remaining sum over partitions is nothing but $F^{(l)}_{m-1,n}(\bar{v}_1|\bar{w}_k)$. Hence,

$$F^{(l)}_{m,n}(\bar{v}|\bar{w})\Big|_{v_1\to w_k} = g(w_k, v_1)\ f(v_1, \bar{v}_1)f(\bar{w}_k, w_k)F^{(l)}_{m-1,n}(\bar{v}_1|\bar{w}_k) + \mathrm{Reg}. \tag{5.59}$$

Thus, the residue of the function $F^{(l)}_{m,n}(\bar{v}|\bar{w})$ at $v_1 \to w_k$ reduces to the function $F^{(l)}_{m-1,n}(\bar{v}_1|\bar{w}_k)$.

Consider now the behavior of the rhs of (5.51) at $v_1 \to w_k$. Let

$$F^{(r)}_{m,n}(\bar{v}|\bar{w}) = (-1)^m f(\bar{w}, \bar{v})\mathbb{K}_{m+n}(\{\bar{v} - \eta, \bar{u}\}|\bar{w}). \tag{5.60}$$

As in the case considered above, the function $F^{(r)}_{m,n}(\bar{v}|\bar{w})$ is an analytic function of the variable $v_1$ in the whole complex plane except the points $w_k$, where it has simple poles. Function $\mathbb{K}_{m+n}(\{\bar{v} - \eta, \bar{u}\}|\bar{w})$ has poles at $v_1 - \eta = w_k$, but the product $f(\bar{w}, \bar{v})$ vanishes at these points. Therefore, the function $F^{(r)}_{m,n}(\bar{v}|\bar{w})$ turns out to be holomorphic at $v_1 - \eta = w_k$. The only singularities arise in the product $f(\bar{w}, \bar{v})$ at $v_1 = w_k$. It is obvious that

$$f(\bar{w}, \bar{v})\Big|_{v_1\to w_k} = g(w_k, v_1)f(v_1, \bar{v}_1)f(\bar{w}_k, w_k)f(\bar{w}_k, \bar{v}_1) + \mathrm{Reg}. \tag{5.61}$$

Due to (5.48), we also have

$$\mathbb{K}_{m+n}(\{\bar{v} - \eta, \bar{u}\}|\bar{w})\Big|_{v_1=w_k} = -\mathbb{K}_{m+n-1}(\{\bar{v}_1 - \eta, \bar{u}\}|\bar{w}_k). \tag{5.62}$$

Substituting this into (5.60), we find

$$F^{(r)}_{m,n}(\bar{v}|\bar{w})\Big|_{v_1\to w_k} = g(w_k, v_1)\ f(v_1, \bar{v}_1)f(\bar{w}_k, w_k)F^{(r)}_{m-1,n}(\bar{v}_1|\bar{w}_k) + \mathrm{Reg}. \tag{5.63}$$

Hence, taking into account (5.59), we obtain

$$\Big(F^{(l)}_{m,n}(\bar v|\bar w) - F^{(r)}_{m,n}(\bar v|\bar w)\Big)\Big|_{v_1\to w_k}$$
$$= g(w_k,v_1)\,f(v_1,\bar v_1)f(\bar w_k,w_k)\Big(F^{(l)}_{m-1,n}(\bar v_1|\bar w_k) - F^{(r)}_{m-1,n}(\bar v_1|\bar w_k)\Big) + \text{Reg.} \tag{5.64}$$

Equation (5.64) allows us to prove proposition 5.3 by induction. Let $m = 0$. Then $\bar v = \emptyset$, and we have only one term in the sum over partitions: $\bar w_{\rm I} = \emptyset$, $\bar w_{\rm II} = \bar w$. The statement of the proposition becomes trivial. Suppose that the proposition holds for some $m-1$ and consider the difference $F^{(l)}_{m,n}(\bar v|\bar w) - F^{(r)}_{m,n}(\bar v|\bar w)$ as a function of $v_1$ for all other variables fixed. Then due to induction assumption and (5.64), the function $F^{(l)}_{m,n}(\bar v|\bar w) - F^{(r)}_{m,n}(\bar v|\bar w)$ has no poles at the points $v_1 = w_k$. In other words, the function $F^{(l)}_{m,n}(\bar v|\bar w) - F^{(r)}_{m,n}(\bar v|\bar w)$ turns out to be holomorphic in the whole complex plane. Hence, due to Liouville's theorem, it is identically constant. Since $\mathbb{K}_m$ and $\mathbb{K}_{m+n}$ tend to zero as $v_1 \to \infty$, this constant is zero. Thus, $F^{(l)}_{m,n}(\bar v|\bar w) - F^{(r)}_{m,n}(\bar v|\bar w) = 0$. □

**Corollary 5.1.** *Under the conditions of proposition 5.3,*

$$\sum_{\bar w\mapsto\{\bar w_{\rm I},\bar w_{\rm II}\}} \mathbb{K}_m(\bar w_{\rm I}|\bar v)\mathbb{K}_n(\bar u|\bar w_{\rm II})f(\bar w_{\rm II},\bar w_{\rm I}) = (-1)^n f(\bar u,\bar w)\mathbb{K}_{m+n}(\bar w|\{\bar v,\bar u+\eta\}). \tag{5.65}$$

Identity (5.65) follows from (5.51) due to (5.49).

Now we can easily prove recursion (5.38). It is a special case of identity (5.51). To see this, we set $m=1$ and replace $n\to n-1$ in (5.51). We also set $\bar v = v_n+\eta$, $\bar u = \bar v_n$, and $\bar w = \bar u$. Then (5.51) takes the form

$$\sum_{\bar u\mapsto\{\bar u_{\rm I},\bar u_{\rm II}\}} \mathbb{K}_1(\bar u_{\rm I}|v_n+\eta)\mathbb{K}_{n-1}(\bar v_n|\bar u_{\rm II})f(\bar u_{\rm II},\bar u_{\rm I}) = -f(\bar u,v_n+\eta)\mathbb{K}_n(\bar v|\bar u). \tag{5.66}$$

The sum is taken over partitions of the set $\bar u$ into two subsets $\bar u_{\rm I}$ and $\bar u_{\rm II}$, such that $\#\bar u_{\rm I} = 1$ and $\#\bar u_{\rm II} = n-1$. Setting $\bar u_{\rm I} = u_k$ and $\bar u_{\rm II} = \bar u_k$, and using (5.50) we obtain

$$f(v_n,\bar u)\sum_{k=1}^{n}\mathbb{K}_1(u_k|v_n+\eta)\mathbb{K}_{n-1}(\bar v_n|\bar u_k)f(\bar u_k,u_k) = -\mathbb{K}_n(\bar v|\bar u). \tag{5.67}$$

It remains to be noted that

$$f(v_n,\bar u)\mathbb{K}_1(u_k|v_n+\eta) = -g(v_n,u_k)f(v_n,\bar u_k), \tag{5.68}$$

and we arrive at (5.38).

Thus, we have proved that the function $\mathbb{K}_n(\bar{v}|\bar{u})$ given by equation (5.41) satisfies the initial condition $\mathbb{K}_1(v|u) = g(v,u)$ and recursion (5.38). Therefore, it is indeed the highest coefficient of the scalar product.

The concept of the highest coefficient was introduced in work [V. E. Korepin (1982b)], where it was shown that it coincides with the partition function of the six-vertex model with so-called domain wall boundary conditions. For this reason, the function $\mathbb{K}_n(\bar{v}|\bar{u})$ is often called the partition function of the six-vertex model with domain wall boundary conditions. However, this name is too long, so that we will use the term highest coefficient. In paper [V. E. Korepin (1982b)], recurrent properties of the highest coefficient (residues at the poles (5.46)) were also found, which uniquely fixed it. Explicit representation (5.41) for the highest coefficient was obtained in work [A. G. Izergin (1987)]. We also note that the determinant of the matrix included in (5.43) was already found in early works by M. Gaudin on scalar products of Bethe vectors in the one-dimensional Bose gas (see [M. Gaudin (1983)]).

## 5.3 Different representations for the scalar product

Using (5.18), (5.23), and (5.44) we find

$$\begin{aligned}\langle 0|C(\bar{v})B(\bar{u})|0\rangle = \sum_{\substack{\bar{v}\mapsto\{\bar{v}_{\rm I},\bar{v}_{\rm II}\}\\ \bar{u}\mapsto\{\bar{u}_{\rm I},\bar{u}_{\rm II}\}}} & d(\bar{v}_{\rm I})a(\bar{u}_{\rm I})d(\bar{u}_{\rm II})a(\bar{v}_{\rm II})\\ &\times \mathbb{K}_k(\bar{v}_{\rm I}|\bar{u}_{\rm I})\mathbb{K}_{n-k}(\bar{u}_{\rm II}|\bar{v}_{\rm II})\, f(\bar{v}_{\rm I},\bar{v}_{\rm II})f(\bar{u}_{\rm II},\bar{u}_{\rm I}). \end{aligned} \tag{5.69}$$

Hence,

$$\boldsymbol{S}_n(\bar{v}|\bar{u}) = \sum_{\substack{\bar{v}\mapsto\{\bar{v}_{\rm I},\bar{v}_{\rm II}\}\\ \bar{u}\mapsto\{\bar{u}_{\rm I},\bar{u}_{\rm II}\}}} r(\bar{u}_{\rm I})r(\bar{v}_{\rm II})\,\mathbb{K}_k(\bar{v}_{\rm I}|\bar{u}_{\rm I})\mathbb{K}_{n-k}(\bar{u}_{\rm II}|\bar{v}_{\rm II})\, f(\bar{v}_{\rm I},\bar{v}_{\rm II})f(\bar{u}_{\rm II},\bar{u}_{\rm I}). \tag{5.70}$$

The sums in (5.69) and (5.70) are taken over the partitions $\bar{v}\mapsto\{\bar{v}_{\rm I},\bar{v}_{\rm II}\}$ and $\bar{u}\mapsto\{\bar{u}_{\rm I},\bar{u}_{\rm II}\}$ so that $\#\bar{v}_{\rm I} = \#\bar{u}_{\rm I} = k$, $k = 0,1,\dots,n$. Note again that the restriction $\#\bar{v}_{\rm I} = \#\bar{u}_{\rm I}$ follows from the form of formulas (5.69) and (5.70) since the highest coefficients exist only if this condition is fulfilled.

Representation (5.70) yields

$$\boldsymbol{S}_n(\bar{v}|\bar{u}) = \boldsymbol{S}_n(\bar{u}|\bar{v}). \tag{5.71}$$

Indeed, the replacement $\bar{u}\leftrightarrow\bar{v}$ in (5.70) leads only to the change of the summation indices I $\leftrightarrow$ II.

If we introduce a union $\bar{w} = \{\bar{v}, \bar{u}\}$, we can write down equation (5.70) in the form

$$\boldsymbol{S}_n(\bar{v}|\bar{u}) = \sum_{\bar{w} \mapsto \{\bar{w}_{\rm I}, \bar{w}_{\rm II}\}} r(\bar{w}_{\rm II}) \mathbb{K}_n(\bar{v}|\bar{w}_{\rm I} + \eta) \mathbb{K}_n(\bar{w}_{\rm II}|\bar{v} + \eta) f(\bar{w}_{\rm I}, \bar{w}_{\rm II}). \quad (5.72)$$

The sum is taken over the partitions $\bar{w} \mapsto \{\bar{w}_{\rm I}, \bar{w}_{\rm II}\}$ such that $\#\bar{w}_{\rm I} = \#\bar{w}_{\rm II} = n$.

To make sure that formulas (5.72) and (5.70) are equivalent, we should set

$$\begin{aligned} &\bar{w}_{\rm I} = \{\bar{v}_{\rm I}, \bar{u}_{\rm II}\}, \qquad && \#\bar{v}_{\rm I} = \#\bar{u}_{\rm I} = k, \\ &\bar{w}_{\rm II} = \{\bar{v}_{\rm II}, \bar{u}_{\rm I}\}, \qquad && \#\bar{v}_{\rm II} = \#\bar{u}_{\rm II} = n - k, \end{aligned} \qquad k = 0, 1, \dots, n. \quad (5.73)$$

Substituting these subsets in (5.72) and using the properties (5.48) and (5.49) of the highest coefficient, we arrive at (5.70). We invite the reader to check this statement.

Formula (5.72), in turn, is a particular case of a more general formula that gives the action of the product of the $C$ operators on a Bethe vector $B(\bar{u})|0\rangle$. Let $\bar{v} = \{v_1, \dots, v_m\}$, $\bar{u} = \{u_1, \dots, u_n\}$, and $\bar{w} = \{\bar{v}, \bar{u}\}$. Then for $m \le n$,

$$\begin{aligned} C(\bar{v})B(\bar{u})|0\rangle = &\sum_{\bar{w} \mapsto \{\bar{w}_{\rm I}, \bar{w}_{\rm II}, \bar{w}_{\rm III}\}} d(\bar{w}_{\rm I}) a(\bar{w}_{\rm II}) \mathbb{K}_m(\bar{v}|\bar{w}_{\rm I} + \eta) \mathbb{K}_m(\bar{w}_{\rm II}|\bar{v} + \eta) \\ &\times f(\bar{w}_{\rm I}, \bar{w}_{\rm II}) f(\bar{w}_{\rm I}, \bar{w}_{\rm III}) f(\bar{w}_{\rm III}, \bar{w}_{\rm II})\, B(\bar{w}_{\rm III})|0\rangle. \end{aligned} \quad (5.74)$$

Here the sum is taken over partitions of the set $\{\bar{v}, \bar{u}\}$ into three subsets $\bar{w}_{\rm I}$, $\bar{w}_{\rm II}$, and $\bar{w}_{\rm III}$, such that $\#\bar{w}_{\rm I} = \#\bar{w}_{\rm II} = m$. Formula (5.74) can be proved using double induction in $n$ and $m$ and identity (5.51). For $m = 1$, it gives the action of a single operator $C(v)$ on the Bethe vector $B(\bar{u})|0\rangle$. The reader can compare it with the formula (5.7) and reproduce formulas (5.15), (5.16). We do not give proof of (5.74) here since it is rather cumbersome. However, the reader can find detailed proof in works [N. A. Slavnov (2017); S. Belliard et al (2013a)].

Note also that formula (5.74) is very typical for multiple commutation relations, when the product of several operators $T_{ij}(\bar{v})$ must be permuted with the product of operators $T_{kl}(\bar{u})$ (see [N. A. Slavnov (2017); S. Belliard et al (2013a)]). In all such formulas, the highest coefficient appears. The scalar product is an example of such multiple commutation relations.[3] If we

[3] Other examples of multiple commutation relations are given in Section 7.1.3 (see (7.62) and (7.63))

put $m = n$ in (5.74), then $\bar{w}_{\rm III} = \emptyset$. The Bethe vector $B(\bar{w}_{\rm III})|0\rangle$ turns into the vacuum vector, and after multiplying by the dual vector $\langle 0|$, formula (5.74) gives the scalar product (5.72).

Let us consider an example of applying the scalar product formulas.

**Proposition 5.4.** *Let* $\bar{u} = \{u_1, \dots, u_n\}$ *and* $n \geq 2$. *Let for some* $j$ *and* $k$ *from the set* $\{1, \dots, n\}$, *parameters* $u_j$ *and* $u_k$ *be such that* $a(u_j) = 0$, $d(u_k) = 0$, *and* $u_k = u_j + \eta$. *Then a vector* $B(\bar{u})|0\rangle$ *is the zero vector.*

**Proof.** Consider equation (5.69), in which $\langle 0|C(\bar{v})$ is an arbitrary dual vector, while the vector $B(\bar{u})|0\rangle$ obeys the condition of the proposition. It follows from the conditions $a(u_j) = 0$ and $d(u_k) = 0$ that (5.69) contains only those partitions of the set $\bar{u}$, for which $u_j \in \bar{u}_{\rm II}$ and $u_k \in \bar{u}_{\rm I}$. Then the product $f(\bar{u}_{\rm II}, \bar{u}_{\rm I})$ contains the function $f(u_j, u_k)$, which vanishes due to the condition $u_k = u_j + \eta$. Thus, the scalar product of the vector $B(\bar{u})|0\rangle$ with any dual vector vanishes. Hence, $B(\bar{u})|0\rangle$ is the zero vector. □

**Remark 5.4.** The above proof uses the standard assumption of the algebraic Bethe ansatz that multiple actions of the operators $C(v)$ generate the dual Hilbert space. The validity of this assumption should be checked for each specific model. Another proof of proposition 5.4, based on the quantum inverse problem, can be found in [N. Kitanine et al (2007)].

The situation described in proposition 5.4 has already been considered earlier for the particular case of the $XXX$ chain consisting of four sites (see Section 2.4). In an inhomogeneous $XXX$ chain, $d(\xi_l) = 0$ and $a(\xi_l - \eta) = 0$, where $\xi_l$ is any of the inhomogeneities. Therefore, any Bethe vector containing the operators $B(\xi_l)$ and $B(\xi_l - \eta)$ turns to the zero vector. However, as we saw in Section 4.4, with a suitable normalization and a proper limiting procedure, such a vector can become an on-shell Bethe vector.

# Chapter 6

# Scalar products with on-shell Bethe vectors

In this chapter, we consider scalar products in which one of the vectors is a (twisted) on-shell vector. Then the sum over the partitions of the Bethe parameters in formula (5.70) can be explicitly calculated as a single matrix determinant (a determinant representation).

The general formula (5.70) is very cumbersome. It is inconvenient not only because it contains a large number of terms. For example, it is challenging to calculate the norm of the Bethe vector using this formula. Indeed, if we put $\bar{v} = \bar{u}$ in (5.70), we obtain many singular terms since the highest coefficients have poles for coinciding arguments. Of course, all these singularities eventually cancel each other since the commutation relations in the $RTT$ algebra are non-singular. And so it happens. However, it is not easy to see this cancellation explicitly. In contrast, in the determinant representation, the limit $\bar{v} = \bar{u}$ is easily calculated by standard methods. Thus, we can obtain an explicit representation of the norm of an on-shell Bethe vector. The latter is necessary for calculating correlation functions.

Determinant representations are quite convenient not only for analytical but also for numerical calculations. The fact is that as the size of the matrix increases, the time required to calculate its determinant increases polynomially. Therefore, a modern computer can easily cope with calculating the scalar product, even with a large number of Bethe parameters.

M. Gaudin was the first who formulated the hypothesis of the existence of determinant representations for the norms of the eigenvectors of integrable quantum Hamiltonians [M. Gaudin (1972); M. Gaudin et al (1981); M. Gaudin (1983)]. This hypothesis was subsequently proved in [V. E. Korepin (1982b)] using the algebraic Bethe ansatz. A determinant representation for the scalar product of an on-shell and a twisted on-shell vector was obtained in [A. N. Kirillov and F. A. Smirnov (1989)]. The case of the scalar product of an on-shell vector and an arbitrary Bethe vector was considered in [N. A. Slavnov (1989)] (see also [N. A. Slavnov (2000)]). In work [N. Kitanine et al (1999)], it was found that the determinant representation for the scalar product is closely related to the Jacobian of the transfer matrix eigenvalue (see also [S. Belliard and N.A. Slavnov (2019b)]).

Let us now give a brief description of the contents of this chapter. In Section 6.1, we consider sums over partitions containing determinants of matrices of a general form. Here we prove some identities, which are then used in Section 6.2. In this section, we derive a determinant representation for a scalar product in which one of the vectors is a (twisted) on-shell Bethe vector. Using this determinant representation, we prove the orthogonality of the twisted on-shell vectors in Section 6.3. In Section 6.4, we compute their norm. Here we also prove that the twisted on-shell vectors form a basis in the $XXZ$ chain for a sufficiently small twist parameter. Finally, in Section 6.5, we consider some particular cases of scalar products used later.

Throughout this chapter, we are dealing with the twisted on-shell vectors instead of the ordinary ones. We do it solely for the sake of generality. In all formulas (except Subsection 6.4.1), we can set the twist parameter equal to 1. We also consider models with the rational $R$-matrix for simplicity. However, most of the formulas are given in the universal form, which is also suitable for models with the trigonometric $R$-matrix.

## 6.1 Identities for determinants

In this section, we obtain some identities for determinants. These identities will be used in the next section to derive a determinant representation for the scalar product.

So far, we have dealt with the sums over partitions of functions that were symmetric in their arguments. Therefore, the order of the elements in each subset was not essential. When we are dealing with determinants, we have functions that are antisymmetric in their arguments. Therefore, we must agree upon the order of the elements in each subset.

To avoid double indices, in this section, we will denote subsets by Roman superscripts, for instance, $\bar{u} \mapsto \{\bar{u}^{\mathrm{I}}, \bar{u}^{\mathrm{II}}\}$. Then notation $u^{\mathrm{I}}_k$ (resp. $u^{\mathrm{II}}_k$) means the $k$th element of the subset $u^{\mathrm{I}}$ (resp. $u^{\mathrm{II}}$). We assume that if a set $\bar{u}$ is divided into two (or more) subsets $\bar{u}^{\mathrm{I}}$ and $\bar{u}^{\mathrm{II}}$, the elements are ordered in increasing order of indices in each of them. For example: $\bar{u} = \{u_1, \dots, u_5\}$, $\bar{u}^{\mathrm{I}} = \{u_1, u_4\}$ and $\bar{u}^{\mathrm{II}} = \{u_2, u_3, u_5\}$. Such an ordering will be called the *natural order*.

We also introduce the notion of *parity of partition*. By the parity of the partition $P(\bar{u}^{\text{I}},\bar{u}^{\text{II}})$, we mean the parity of the permutation that maps the union of two subsets $\{\bar{u}^{\text{I}},\bar{u}^{\text{II}}\}$ (in each of them, the elements should be in the natural order) to the naturally ordered set $\bar{u}$. Thus, in the above example, the union of two subsets has the form $\{\bar{u}^{\text{I}},\bar{u}^{\text{II}}\} = \{u_1,u_4,u_2,u_3,u_5\}$. It is easy to see that the permutation that maps this sequence to the naturally ordered set $\bar{u} = \{u_1,u_2,u_3,u_4,u_5\}$ is even. Therefore, this partition also is even.

Finally, let us introduce a new notation for special products

$$\Delta(\bar{u}) = \prod_{1\le j<k\le n} g(u_k,u_j), \qquad \Delta'(\bar{u}) = \prod_{1\le j<k\le n} g(u_j,u_k), \tag{6.1}$$

where $n = \#\bar{u} \ge 2$. It is clear that $\Delta'(\bar{u}) = (-1)^{n(n-1)/2}\Delta(\bar{u})$. For $n = 0,1$, we set $\Delta(\bar{u}) = \Delta'(\bar{u}) = 1$ by definition.

**Proposition 6.1.** *For any partition of a set $\bar{u}$ into two subsets $\bar{u}^{\text{I}}$ and $\bar{u}^{\text{II}}$, such that $\#\bar{u}^{\text{I}} = m$, $\#\bar{u}^{\text{II}} = n-m$, where $m$ is any integer from the set $0,1,\dots,n$, the following identity holds:*

$$\Delta(\bar{u}) = (-1)^{P(\bar{u}^{\text{I}},\bar{u}^{\text{II}})}\Delta(\bar{u}^{\text{I}})\Delta(\bar{u}^{\text{II}})\, g(\bar{u}^{\text{II}},\bar{u}^{\text{I}}). \tag{6.2}$$

**Proof.** The product $\Delta(\bar{u})^{-1}$ is the Vandermonde determinant of the parameters $\bar{u}$ up to a constant factor:

$$\Delta(\bar{u})^{-1} = \eta^{-n(n-1)/2} \det_n \left(u_k^{j-1}\right). \tag{6.3}$$

For any fixed partition of the set $\bar{u}$ into two subsets $\bar{u}^{\text{I}}$ and $\bar{u}^{\text{II}}$, we reorder the columns of the determinant so that all $u_k \in \bar{u}^{\text{I}}$ would be on the left, and all $u_k \in \bar{u}^{\text{II}}$ would be on the right. In each subset $\bar{u}^{\text{I}}$ and $\bar{u}^{\text{II}}$, the columns are ordered in the natural order. Then

$$\Delta(\bar{u})^{-1} = (-1)^{P(\bar{u}^{\text{I}},\bar{u}^{\text{II}})}\eta^{-n(n-1)/2} \det_n\Big((u_k^{\text{I}})^{j-1} \,\Big|\, (u_k^{\text{II}})^{j-1}\Big), \tag{6.4}$$

and we recall that $u_k^{\text{I}}$ (resp. $u_k^{\text{II}}$) means the $k$th element of the subset $u^{\text{I}}$ (resp. $u^{\text{II}}$). The determinant in the rhs of (6.4) is equal to

$$\begin{aligned}\det_n\Big((u_k^{\text{I}})^{j-1} \,\Big|\, (u_k^{\text{II}})^{j-1}\Big) = &\prod_{\substack{j<k\\ u_j,u_k\in\bar{u}^{\text{I}}}} (u_k-u_j) \prod_{\substack{j<k\\ u_j,u_k\in\bar{u}^{\text{II}}}} (u_k-u_j)\\ &\times \prod_{u_j\in\bar{u}^{\text{I}}}\prod_{u_k\in\bar{u}^{\text{II}}} (u_k-u_j),\end{aligned} \tag{6.5}$$

whence follows the statement of proposition 6.1. □

We now turn to the expansion of determinants into sums over partitions. Suppose that we now have two sets: $\bar{v} = \{v_1, \dots, v_n\}$ and $\bar{u} = \{u_1, \dots, u_n\}$. Let us define on these sets two functions of two variables $\Phi_\ell(v_j, u_k)$, $\ell = 1, 2$, and form a matrix

$$M_{jk} = \Phi_1(v_j, u_k) + \Phi_2(v_j, u_k). \tag{6.6}$$

**Proposition 6.2.**

$$\det_n(M_{jk}) = \sum_{\bar{u} \mapsto \{\bar{u}^{\rm I}, \bar{u}^{\rm II}\}} (-1)^{P(\bar{u}^{\rm I}, \bar{u}^{\rm II})} \det_n \Big[ \Phi_1(v_j, u_k^{\rm I}) \; \Big| \; \Phi_2(v_j, u_k^{\rm II}) \Big]. \tag{6.7}$$

*The sum is taken over partitions of the set $\bar{u}$ into two naturally ordered subsets $\bar{u}^{\rm I}$ and $\bar{u}^{\rm II}$, such that $\#\bar{u}^{\rm I} = m$, $\#\bar{u}^{\rm II} = n - m$, and $m$ runs through the set $\{0, 1, \dots, n\}$. The matrix in the rhs of (6.7) consists of two parts. The first $m$ columns correspond to the subset $\bar{u}^{\rm I}$. In these columns, the matrix elements are equal to the function $\Phi_1(v_j, u_k^{\rm I})$. The remaining $n - m$ columns correspond to the subset $\bar{u}^{\rm II}$. In these columns, the matrix elements are equal to the function $\Phi_2(v_j, u_k^{\rm II})$.*

First of all, we illustrate the idea of the proof by a simple example. Let $n = 2$. We use the property of linearity of any determinant with respect to the elements of the first column. Then

$$\det_2 \begin{pmatrix} M_{11} & M_{12} \\ M_{21} & M_{22} \end{pmatrix} = \det_2 \begin{pmatrix} \Phi_1(v_1, u_1) & M_{12} \\ \Phi_1(v_2, u_1) & M_{22} \end{pmatrix} + \det_2 \begin{pmatrix} \Phi_2(v_1, u_1) & M_{12} \\ \Phi_2(v_2, u_1) & M_{22} \end{pmatrix}. \tag{6.8}$$

Using the linearity of the obtained determinants with respect to the elements of the second columns, we find

$$\begin{aligned} \det_2 \begin{pmatrix} M_{11} & M_{12} \\ M_{21} & M_{22} \end{pmatrix} &= \det_2 \begin{pmatrix} \Phi_1(v_1, u_1) & \Phi_1(v_1, u_2) \\ \Phi_1(v_2, u_1) & \Phi_1(v_2, u_2) \end{pmatrix} + \det_2 \begin{pmatrix} \Phi_1(v_1, u_1) & \Phi_2(v_1, u_2) \\ \Phi_1(v_2, u_1) & \Phi_2(v_2, u_2) \end{pmatrix} \\ &+ \det_2 \begin{pmatrix} \Phi_2(v_1, u_1) & \Phi_1(v_1, u_2) \\ \Phi_2(v_2, u_1) & \Phi_1(v_2, u_2) \end{pmatrix} + \det_2 \begin{pmatrix} \Phi_2(v_1, u_1) & \Phi_2(v_1, u_2) \\ \Phi_2(v_2, u_1) & \Phi_2(v_2, u_2) \end{pmatrix}. \end{aligned} \tag{6.9}$$

Now we reorder the columns of the obtained determinants so that all the columns containing $\Phi_1$ would be on the left, and all the columns containing $\Phi_2$ would be on the right. In fact, only the columns of the third determinant in the rhs of (6.9) should be permuted because, in all other determinants, the columns are already ordered in the required way. We obtain

$$\det_2 \begin{pmatrix} M_{11} & M_{12} \\ M_{21} & M_{22} \end{pmatrix}$$
$$= \det_2 \begin{pmatrix} \Phi_1(v_1,u_1) & \Phi_1(v_1,u_2) \\ \Phi_1(v_2,u_1) & \Phi_1(v_2,u_2) \end{pmatrix} + \det_2 \begin{pmatrix} \Phi_1(v_1,u_1) & \Phi_2(v_1,u_2) \\ \Phi_1(v_2,u_1) & \Phi_2(v_2,u_2) \end{pmatrix}$$
$$- \det_2 \begin{pmatrix} \Phi_1(v_1,u_2) & \Phi_2(v_1,u_1) \\ \Phi_1(v_2,u_2) & \Phi_2(v_2,u_1) \end{pmatrix} + \det_2 \begin{pmatrix} \Phi_2(v_1,u_1) & \Phi_2(v_1,u_2) \\ \Phi_2(v_2,u_1) & \Phi_2(v_2,u_2) \end{pmatrix}. \quad (6.10)$$

It is easy to see that the four terms in the rhs of (6.10) correspond to four possible partitions of the set $\bar{u}$ into two subsets:

$$\begin{aligned} &\bar{u}^{\mathrm{I}} = \{u_1, u_2\}, \qquad &&\bar{u}^{\mathrm{II}} = \emptyset; \\ &\bar{u}^{\mathrm{I}} = u_1, &&\bar{u}^{\mathrm{II}} = u_2; \\ &\bar{u}^{\mathrm{I}} = u_2, &&\bar{u}^{\mathrm{II}} = u_1; \\ &\bar{u}^{\mathrm{I}} = \emptyset, &&\bar{u}^{\mathrm{II}} = \{u_1, u_2\}. \end{aligned} \quad (6.11)$$

It is also clear that the parity of the third partition is odd. Therefore, the third determinant in the rhs of (6.10) has a minus sign. The parities of all other partitions are even, and thus, the corresponding determinants enter (6.10) with a plus sign.

**Proof of Proposition 6.2.** The proof of Proposition 6.2 in the general case now becomes evident. Using the linearity of the determinant with respect to the elements of any column successively, we arrive at

$$\det_n(M_{jk}) = \sum_{\bar{u}\mapsto\{\bar{u}^{\mathrm{I}},\bar{u}^{\mathrm{II}}\}} \det_n\big(\Phi(v_j,u_k)\big), \quad (6.12)$$

where

$$\Phi(v_j,u_k) = \begin{cases} \Phi_1(v_j,u_k), & \text{if} \quad u_k \in \bar{u}^{\mathrm{I}}, \\ \Phi_2(v_j,u_k), & \text{if} \quad u_k \in \bar{u}^{\mathrm{II}}. \end{cases} \quad (6.13)$$

Moving all the columns with $\Phi_1$ to the left, we obtain the sign $(-1)^{P(\bar{u}^{\mathrm{I}},\bar{u}^{\mathrm{II}})}$ and reproduce (6.7). □

**Corollary 6.1.** *Let $M$ be the matrix* (6.6). *Then*

$$\Delta(\bar{u}) \det_n(M_{jk}) = \sum_{\bar{u}\mapsto\{\bar{u}^{\mathrm{I}},\bar{u}^{\mathrm{II}}\}} \Delta(\bar{u}^{\mathrm{I}})\Delta(\bar{u}^{\mathrm{II}}) g(\bar{u}^{\mathrm{II}},\bar{u}^{\mathrm{I}})$$
$$\times \det_n\Big[\Phi_1(v_j,u^{\mathrm{I}}_k)\ \Big|\ \Phi_2(v_j,u^{\mathrm{II}}_k)\Big]. \quad (6.14)$$

*Here the sum over partitions and the notation is the same as in proposition 6.2*

**Proof.** This formula immediately follows from propositions 6.1 and 6.2. □

**Proposition 6.3.** *Let $\Omega(\bar{u}|\bar{v})$ be a function of two sets of variables $\bar{u} = \{u_1, \dots, u_n\}$ and $\bar{v} = \{v_1, \dots, v_n\}$. Let this function have the following representation:*

$$\Omega(\bar{u}|\bar{v}) = \Delta(\bar{u}) \det_n \bigl(\Phi(v_j, u_k)\bigr), \tag{6.15}$$

*where $\Phi(v_j, u_k)$ is a function of two variables. Then*

$$\sum_{\bar{u} \mapsto \{\bar{u}^{\mathrm{I}}, \bar{u}^{\mathrm{II}}\}} f(\bar{u}^{\mathrm{II}}, \bar{u}^{\mathrm{I}}) \Omega(\{\bar{u}^{\mathrm{I}} - \eta, \bar{u}^{\mathrm{II}}\}|\bar{v}) = \Delta(\bar{u}) \det_n \Bigl(\Phi(v_j, u_k - \eta) + \Phi(v_j, u_k)\Bigr). \tag{6.16}$$

*Here the sum is taken over all partitions of the set $\bar{u}$ into two subsets $\bar{u}^{\mathrm{I}}$ and $\bar{u}^{\mathrm{II}}$, such that $\#\bar{u}^{\mathrm{I}} = m$, $\#\bar{u}^{\mathrm{II}} = n - m$, and $m$ runs through the set $\{0, 1, \dots, n\}$.*

**Proof.** Let us fix a partition $\bar{u} \mapsto \{\bar{u}^{\mathrm{I}}, \bar{u}^{\mathrm{II}}\}$. Since $\Omega(\bar{u}|\bar{v})$ is symmetric over $\bar{u}$, we can reorder the columns of $\det_n \bigl(\Phi(v_j, u_k)\bigr)$ so that all the columns depending on $u_k \in \bar{u}^{\mathrm{I}}$ would be on the left in the natural order. Of course, the same reordering of the parameters $\bar{u}$ should be done in the product $\Delta(\bar{u})$. Then we obtain

$$\begin{aligned} \Omega(\{\bar{u}^{\mathrm{I}} - \eta, \bar{u}^{\mathrm{II}}\}|\bar{v}) &= \Delta(\{\bar{u}^{\mathrm{I}} - \eta, \bar{u}^{\mathrm{II}}\}) \det_n \Bigl[\Phi(v_j, u^{\mathrm{I}}_k - \eta) \Bigm| \Phi(v_j, u^{\mathrm{II}}_k)\Bigr] \\ &= \Delta(\bar{u}^{\mathrm{I}}) \Delta(\bar{u}^{\mathrm{II}}) g(\bar{u}^{\mathrm{II}}, \bar{u}^{\mathrm{I}} - \eta) \det_n \Bigl[\Phi(v_j, u^{\mathrm{I}}_k - \eta) \Bigm| \Phi(v_j, u^{\mathrm{II}}_k)\Bigr]. \end{aligned} \tag{6.17}$$

Using

$$g(x, y - \eta) = \frac{\eta}{x - y + \eta} = \frac{g(x, y)}{f(x, y)}, \tag{6.18}$$

we find

$$\begin{aligned} \Omega(\{\bar{u}^{\mathrm{I}} - \eta, \bar{u}^{\mathrm{II}}\}|\bar{v}) f(\bar{u}^{\mathrm{II}}, \bar{u}^{\mathrm{I}}) &= \Delta(\bar{u}^{\mathrm{I}}) \Delta(\bar{u}^{\mathrm{II}}) g(\bar{u}^{\mathrm{II}}, \bar{u}^{\mathrm{I}}) \\ &\quad \times \det_n \Bigl[\Phi(v_j, u^{\mathrm{I}}_k - \eta) \Bigm| \Phi(v_j, u^{\mathrm{II}}_k)\Bigr]. \end{aligned} \tag{6.19}$$

Applying corollary 6.1 to the sum (6.16), we complete the proof. □

**Corollary 6.2.** *Let $\alpha_1(z)$ and $\alpha_2(z)$ be two functions of complex variable $z$. Then under the conditions of proposition 6.3*

$$\begin{aligned} &\sum_{\bar{u} \mapsto \{\bar{u}^{\mathrm{I}}, \bar{u}^{\mathrm{II}}\}} \alpha_1(\bar{u}^{\mathrm{I}}) \alpha_2(\bar{u}^{\mathrm{II}}) f(\bar{u}^{\mathrm{II}}, \bar{u}^{\mathrm{I}}) \Omega(\{\bar{u}^{\mathrm{I}} - \eta, \bar{u}^{\mathrm{II}}\}|\bar{v}) \\ &\qquad = \Delta(\bar{u}) \det_n \Bigl(\alpha_1(u_k) \Phi(v_j, u_k - \eta) + \alpha_2(u_k) \Phi(v_j, u_k)\Bigr). \end{aligned} \tag{6.20}$$

*Here we have extended the convention on the shorthand notation to the products of the functions $\alpha_\ell(z)$ ($\ell = 1, 2$).*

**Proof.** Multiplying (6.19) by $\alpha_1(\bar{u}^{\text{I}})\alpha_2(\bar{u}^{\text{II}})$ we obtain

$$\begin{aligned}
&\alpha_1(\bar{u}^{\text{I}})\alpha_2(\bar{u}^{\text{II}})f(\bar{u}^{\text{II}}, \bar{u}^{\text{I}})\Omega(\{\bar{u}^{\text{I}} - \eta, \bar{u}^{\text{II}}\}|\bar{v}) \\
&\quad = \alpha_1(\bar{u}^{\text{I}})\alpha_2(\bar{u}^{\text{II}})\Delta(\bar{u}^{\text{I}})\Delta(\bar{u}^{\text{II}})g(\bar{u}^{\text{II}}, \bar{u}^{\text{I}}) \det_n \Big[\Phi(v_j, u_k^{\text{I}} - \eta) \;\Big|\; \Phi(v_j, u_k^{\text{II}})\Big] \\
&\quad = \Delta(\bar{u}^{\text{I}})\Delta(\bar{u}^{\text{II}})g(\bar{u}^{\text{II}}, \bar{u}^{\text{I}}) \det_n \Big[\alpha_1(\bar{u}_k^{\text{I}})\Phi(v_j, u_k^{\text{I}} - \eta) \;\Big|\; \alpha_2(\bar{u}_k^{\text{II}})\Phi(v_j, u_k^{\text{II}})\Big]. \qquad (6.21)
\end{aligned}$$

Then the use of corollary 6.1 completes the proof. □

## 6.2 Determinant formula for the scalar product

Before moving on, we introduce two new functions

$$h(u, v) = \frac{f(u, v)}{g(u, v)}, \qquad t(u, v) = \frac{g(u, v)}{h(u, v)}. \qquad (6.22)$$

In what follows, we will often deal with these combinations. Therefore, it makes sense to introduce a special notation for them. We will also extend to these functions the convention on the shorthand notation for the products. We have

$$h(u, v) = \frac{u - v + \eta}{\eta}, \qquad t(u, v) = \frac{\eta^2}{(u - v)(u - v + \eta)} \qquad (6.23)$$

in the case of the rational $R$-matrix and

$$h(u, v) = \frac{\sinh(u - v + \eta)}{\sinh \eta}, \qquad t(u, v) = \frac{\sinh^2 \eta}{\sinh(u - v)\sinh(u - v + \eta)} \qquad (6.24)$$

in the case of the trigonometric $R$-matrix.

Using these functions, we can write down the explicit expression for the highest coefficient in the form

$$\mathbb{K}_n(\bar{u}|\bar{v}) = \Delta'(\bar{v})\Delta(\bar{u}) \det_n \Big(t(u_k, v_j)h(u_k, \bar{v})\Big). \qquad (6.25)$$

Let us return to equation (5.70), which gives the scalar product of two arbitrary Bethe vectors. In this formula, $\bar{v}$ and $\bar{u}$ are sets of arbitrary complex numbers. No conditions have been imposed on them yet. Assume now that one of the vectors is an eigenvector of the twisted transfer matrix.[1]

[1]Recall that we consider twisted on-shell vectors solely for the sake of generality. In particular, the twist parameter $\kappa$ can be equal to 1.

Let it be, for instance, $\langle 0|C(\bar{v})$. This means that the parameters $\bar{v}$ satisfy twisted Bethe equations

$$r(v_j) = \kappa \frac{f(v_j, \bar{v}_j)}{f(\bar{v}_j, v_j)}, \qquad j = 1, \dots, n. \tag{6.26}$$

It follows from this that for any partition $\bar{v} \mapsto \{\bar{v}_{\rm I}, \bar{v}_{\rm II}\}$,

$$r(\bar{v}_{\rm II}) f(\bar{v}_{\rm I}, \bar{v}_{\rm II}) = \kappa^{n-k} f(\bar{v}_{\rm II}, \bar{v}_{\rm I}), \tag{6.27}$$

where $n - k = \#\bar{v}_{\rm II}$. Substituting (6.27) into (5.70) we obtain

$$\boldsymbol{S}_n(\bar{v}|\bar{u}) = \sum_{\substack{\bar{v} \mapsto \{\bar{v}_{\rm I}, \bar{v}_{\rm II}\} \\ \bar{u} \mapsto \{\bar{u}_{\rm I}, \bar{u}_{\rm II}\}}} \kappa^{n-k} r(\bar{u}_{\rm I}) \mathbb{K}_k(\bar{v}_{\rm I}|\bar{u}_{\rm I}) \mathbb{K}_{n-k}(\bar{u}_{\rm II}|\bar{v}_{\rm II}) f(\bar{v}_{\rm II}, \bar{v}_{\rm I}) f(\bar{u}_{\rm II}, \bar{u}_{\rm I}). \tag{6.28}$$

We see that for any fixed $k$, the sum over partitions of the set $\bar{v}$ can be computed via proposition 5.3:

$$\sum_{\bar{v} \mapsto \{\bar{v}_{\rm I}, \bar{v}_{\rm II}\}} \mathbb{K}_k(\bar{v}_{\rm I}|\bar{u}_{\rm I}) \mathbb{K}_{n-k}(\bar{u}_{\rm II}|\bar{v}_{\rm II}) f(\bar{v}_{\rm II}, \bar{v}_{\rm I}) = (-1)^k f(\bar{v}, \bar{u}_{\rm I}) \mathbb{K}_n(\{\bar{u}_{\rm I} - \eta, \bar{u}_{\rm II}\}|\bar{v}). \tag{6.29}$$

Thus, we arrive at

$$\boldsymbol{S}_n(\bar{v}|\bar{u}) = \sum_{\bar{u} \mapsto \{\bar{u}_{\rm I}, \bar{u}_{\rm II}\}} (-1)^k \kappa^{n-k} r(\bar{u}_{\rm I}) \mathbb{K}_n(\{\bar{u}_{\rm I} - \eta, \bar{u}_{\rm II}\}|\bar{v}) f(\bar{u}_{\rm II}, \bar{u}_{\rm I}) f(\bar{v}, \bar{u}_{\rm I}). \tag{6.30}$$

Summation in (6.30) is carried out over all partitions $\bar{u} \mapsto \{\bar{u}_{\rm I}, \bar{u}_{\rm II}\}$ such that $\#\bar{u}_{\rm I} = k$ and $k = 0, 1, \dots, n$. This sum can be calculated via corollary 6.2. Indeed, if we set in (6.20)

$$\Phi(v_j, u_k) = t(u_k, v_j) h(u_k, \bar{v}), \qquad \alpha_1(z) = -r(z) f(\bar{v}, z), \qquad \alpha_2(z) = \kappa, \tag{6.31}$$

then equation (6.20) turns into[2] (6.30). Thus, due to corollary 6.2, we find

$$\boldsymbol{S}_n(\bar{v}|\bar{u}) = \Delta'(\bar{v}) \Delta(\bar{u}) \det_n \Big( \kappa t(u_k, v_j) h(u_k, \bar{v}) - r(u_k) f(\bar{v}, u_k) t(u_k - \eta, v_j) h(u_k - \eta, \bar{v}) \Big). \tag{6.32}$$

Using obvious identities

$$t(u - \eta, v) = t(v, u), \qquad f(v, u) h(u - \eta, v) = -h(v, u), \tag{6.33}$$

[2]The fact that $\Phi(v_j, u_k)$ and $\alpha_1(z)$ have additional dependence on the set $\bar{v}$ does not play any role.

we arrive at

$$\begin{aligned}\boldsymbol{S}_n(\bar{v}|\bar{u}) &= \Delta'(\bar{v})\Delta(\bar{u})\det_n \\ &\times\left[\kappa\, t(u_k,v_j)h(u_k,\bar{v}) + (-1)^{n-1}r(u_k)t(v_j,u_k)h(\bar{v},u_k)\right]. \qquad (6.34)\end{aligned}$$

After simple algebra, this formula also can be written in the form

$$\boldsymbol{S}_n(\bar{v}|\bar{u}) = \Delta'(\bar{v})\Delta(\bar{u})\; h(\bar{u},\bar{v})\;\det_n \mathcal{M}, \qquad (6.35)$$

where

$$\mathcal{M}_{jk} = t(u_k,v_j)\left(\kappa - r(u_k)\frac{f(\bar{v}_j,u_k)}{f(u_k,\bar{v}_j)}\right). \qquad (6.36)$$

**Remark 6.1.** We have considered the scalar product in which the dual vector $\langle 0|C(\bar{v})$ is the twisted on-shell vector, and the vector $B(\bar{u})|0\rangle$ is arbitrary. One can consider the case when the vector $B(\bar{u})|0\rangle$ is an eigenvector of the twisted transfer matrix, while the dual vector $\langle 0|C(\bar{v})$ is arbitrary. It is enough to use the equation (5.71), and the problem reduces to the case considered above. The result is still given by the formulas (6.35), (6.36), where one needs to make the replacement $\bar{v} \leftrightarrow \bar{u}$. The reader can easily verify this independently.

**Remark 6.2.** All the considerations above remain valid for the models with trigonometric $R$-matrix.

To conclude this section, we note that one can rewrite the formula for the scalar product (6.34) in terms of the Jacobian of the twisted transfer matrix eigenvalue [N. Kitanine et al (1999)]

$$\tau_\kappa(u|\bar{v}) = a(u)f(\bar{v},u) + \kappa d(u)f(u,\bar{v}). \qquad (6.37)$$

Indeed, it is not difficult to check that representation (6.34) is equivalent to the formula[3]

$$\boldsymbol{S}_n(\bar{v}|\bar{u}) = \Delta'(\bar{v})\Delta(\bar{u})\;\det_n\left(\frac{\eta}{d(u_k)g(u_k,\bar{v})}\frac{\partial\tau_\kappa(u_k|\bar{v})}{\partial v_j}\right). \qquad (6.38)$$

In this equation, we first assume that $\bar{v}$ are arbitrary complex numbers. After taking derivatives, we send this set to a solution of the system of twisted Bethe equations.

[3]In the trigonometric case, the coefficient $\eta$ should be replaced with $\sinh\eta$.

## 6.3 Orthogonality of the eigenvectors

Formula (6.35) allows us to consider a particular case of the scalar product when both vectors are twisted on-shell Bethe vectors. We will see that if the two vectors are different, then their scalar product vanishes in complete agreement with the general theory.

We start with the case when $v_j \neq u_k$ for all $j, k = 1, \dots, n$. If the vector $B(\bar{u})|0\rangle$ is an eigenvector of the twisted transfer matrix, then the parameters $\bar{u}$ satisfy twisted Bethe equations

$$r(u_k) = \kappa \frac{f(u_k, \bar{u}_k)}{f(\bar{u}_k, u_k)}. \tag{6.39}$$

Substituting the function $r(u_k)$ (6.39) into (6.36) we obtain after simple algebra

$$\begin{aligned} \mathcal{M}_{jk} &= \kappa t(u_k, v_j) + \kappa t(v_j, u_k)\, Y_k^{(n)} \\ &= \frac{\kappa\, \eta^2}{(u_k - v_j)(u_k - v_j + \eta)} + \frac{\kappa\, \eta^2 Y_k^{(n)}}{(u_k - v_j)(u_k - v_j - \eta)}, \end{aligned} \tag{6.40}$$

where

$$Y_k^{(n)} = \frac{h(u_k, \bar{u})h(\bar{v}, u_k)}{h(\bar{u}, u_k)h(u_k, \bar{v})} = \prod_{m=1}^{n} \frac{(u_k - v_m - \eta)}{(u_k - v_m + \eta)} \frac{(u_k - u_m + \eta)}{(u_k - u_m - \eta)}. \tag{6.41}$$

In these formulas, we give expressions for $\mathcal{M}_{jk}$ and $Y_k^{(n)}$ in two forms: first, in terms of the functions $h$ and $t$; then explicitly. The first form is universal because it is valid for the models with rational and trigonometric $R$-matrices.

Let us prove that $\det_n \mathcal{M}_{jk} = 0$. For this, we prove that the rows of the matrix $\mathcal{M}_{jk}$ are linearly dependent.

**Proposition 6.4.** *Let*

$$\nu_j^{(n)} = \frac{g(v_j, \bar{v}_j)}{g(v_j, \bar{u})} = \frac{\prod_{m=1}^{n}(v_j - u_m)}{\eta \prod_{m=1;\, m\neq j}^{n}(v_j - v_m)}. \tag{6.42}$$

*Then*

$$\frac{1}{\kappa} \sum_{j=1}^{m} \nu_j^{(n)} \mathcal{M}_{jk} = 0. \tag{6.43}$$

**Proof.** The linear combination (6.43) can be presented in the form

$$\frac{1}{\kappa} \sum_{j=1}^{n} \nu_j^{(n)} \mathcal{M}_{jk} = U_k^{+} + U_k^{-}\, Y_k^{(n)}, \tag{6.44}$$

where

$$U_k^{\pm} = \sum_{j=1}^{n} \frac{\nu_j^{(n)} \eta^2}{(u_k - v_j)(u_k - v_j \pm \eta)}. \tag{6.45}$$

To compute $U_k^{\pm}$, we consider an auxiliary contour integral over a circle of a big radius

$$I_{\pm} = \frac{1}{2\pi i} \oint_{|z|=R\to\infty} \frac{\eta\, \mathrm{d}z}{(u_k - z)(u_k - z \pm \eta)} \prod_{m=1}^{n} \frac{z - u_m}{z - v_m}. \tag{6.46}$$

Since the integrand behaves as $z^{-2}$ at $z \to \infty$, we conclude that $I_{\pm} = 0$. On the other hand, the integral (6.46) is equal to the sum of the residues in the poles within the integration contour. First, these are poles at $z = v_j$, $j = 1, \dots, n$. It is easy to see that the sum of the residues in these poles gives exactly $U_k^{\pm}$. The pole in the point $z = u_k$ is spurious because the factor $(z - u_k)$ compensates it. There is one more pole in the point $z = u_k \pm \eta$. Obviously,

$$\operatorname{Res} \frac{\eta}{(u_k - z)(u_k - z \pm \eta)} \prod_{m=1}^{n} \frac{z - u_m}{z - v_m}\bigg|_{z=u_k\pm\eta} = \pm \prod_{m=1}^{n} \frac{u_k - u_m \pm \eta}{u_k - v_m \pm \eta}. \tag{6.47}$$

We arrive at

$$I_{\pm} = 0 = U_k^{\pm} \pm \prod_{m=1}^{n} \frac{u_k - u_m \pm \eta}{u_k - v_m \pm \eta}, \tag{6.48}$$

which implies

$$U_k^{\pm} = \mp \prod_{m=1}^{n} \frac{u_k - u_m \pm \eta}{u_k - v_m \pm \eta}. \tag{6.49}$$

Substituting these expressions for $U_k^{\pm}$ into (6.44), we immediately arrive at (6.43).

We suggest that the reader check how this method works in the trigonometric case. We only note that the linear combination coefficients still are given by (6.42) if we use the trigonometric analog of the function $g$.

The proof of orthogonality slightly changes if some variables from the set $\bar{v}$ coincide with the variables from the set $\bar{u}$. In this case, some matrix elements of the matrix (6.36) have indeterminate form. Without loss of generality, we can assume that $u_j = v_j$ for $j = \ell + 1, \dots, n$, where $\ell < n$, while $u_j \neq v_k$ for $j, k = 1, \dots, \ell$. Consider the diagonal elements of the matrix $\mathcal{M}_{jk}$ (6.36) at $j = \ell + 1, \dots, n$

$$\mathcal{M}_{jj} = \frac{\eta^2}{(u_j - v_j)(u_j - v_j + \eta)} \left( \kappa - r(u_j) \prod_{\substack{m=1 \\ m\neq j}}^{n} \frac{u_j - v_m - \eta}{u_j - v_m + \eta} \right). \tag{6.50}$$

We see that if we set $u_j = v_j$, then we obtain zero in the denominator, but at the same time, the expression in the parenthesis vanishes due to the twisted Bethe equations

$$\kappa - r(u_j) \prod_{\substack{m=1\\ m\neq j}}^{n} \frac{u_j - v_m - \eta}{u_j - v_m + \eta}\bigg|_{u_j=v_j} = \kappa - r(v_j) \prod_{\substack{m=1\\ m\neq j}}^{n} \frac{v_j - v_m - \eta}{v_j - v_m + \eta} = 0. \quad (6.51)$$

Therefore, we should first solve this indeterminateness, and only then express the function $r(u_j)$ in terms of solutions of the twisted Bethe equations (6.39). We will consider this issue in more detail in the following section, but for now, we will study the matrix elements of $\mathcal{M}_{jk}$ for $j = 1, \dots, \ell$. There are no singularities in these matrix elements. Therefore, as before, we can express the function $r(u_k)$ in terms of solutions of the twisted Bethe equations and obtain

$$\mathcal{M}_{jk} = \frac{\kappa \eta^2}{(u_k - v_j)(u_k - v_j + \eta)} + \frac{\kappa \eta^2 Y_k^{(\ell)}}{(u_k - v_j)(u_k - v_j - \eta)}, \quad (6.52)$$

where

$$Y_k^{(\ell)} = Y_k^{(n)}\bigg|_{\substack{u_s=v_s\\ s>\ell}} = \prod_{m=1}^{\ell} \frac{(u_k - v_m - \eta)}{(u_k - v_m + \eta)} \frac{(u_k - u_m + \eta)}{(u_k - u_m - \eta)}. \quad (6.53)$$

Note that the function $Y_k^{(\ell)}$ differs from the previous function $Y_k^{(n)}$ (6.41) only in that the upper limit in the product over $m$ is now bounded by $\ell$. We can thus say that the matrix elements $\mathcal{M}_{jk}$ with $j = 1, \dots, \ell$ depend on variables $u_1, \dots, u_\ell$ and $v_1, \dots, v_\ell$ only.

A similar effect occurs with the coefficients $\nu_j^{(n)}$ $j = 1, \dots, \ell$:

$$\nu_j^{(n)}\bigg|_{\substack{u_s=v_s\\ s>\ell}} = \nu_j^{(\ell)} = \frac{\prod_{m=1}^{\ell}(v_j - u_m)}{\eta \prod_{m=1;\, m\neq j}^{\ell}(v_j - v_m)}, \qquad j = 1, \dots, \ell. \quad (6.54)$$

Therefore, replacing $n$ by $\ell$ in the previous proof, we immediately obtain

$$\sum_{j=1}^{\ell} \nu_j^{(\ell)} \mathcal{M}_{jk} = 0. \quad (6.55)$$

The remaining part of the linear combination vanishes since $\nu_j^{(n)} = 0$ for $j = \ell + 1, \dots, n$. Therefore, regardless of the explicit form of the matrix elements in the rows with $j = \ell+1, \dots, n$, this part of the linear combination does not contribute. Hence, equation (6.43) is still valid, and the two vectors are still orthogonal. □

## 6.4 Norm of the twisted on-shell vector

The only case when the orthogonality proof does not work is the case $\bar{v} = \bar{u}$. Then we are dealing with a scalar product of an eigenvector with itself, that is, the square of the norm.[4] It is natural to expect that the square of the eigenvector's norm is not equal to zero. Indeed, if $\bar{v} = \bar{u}$, then all the coefficients $\nu_j^{(n)}$ vanish. Therefore, we cannot prove the linear dependence of the rows of the matrix $\mathcal{M}_{jk}$.

As we already noted, in the case when the parameters $\bar{v}$ and $\bar{u}$ coincide, the diagonal elements of the matrix $\mathcal{M}_{jj}$ have the indeterminant form (see (6.50)) that should be resolved. It is important to emphasize that it would be wrong first to express the functions $r(u_j)$ in terms of solutions of the Bethe equations in (6.50), and then to take the limit $v_j \to u_j$. Indeed, one cannot send two different solutions to any system of equations to each other. Let us illustrate this with a simple example.

Consider the scalar product $S_1(v|u)$, in which the vectors $\langle 0|C(v)$ and $B(u)|0\rangle$ depend on one parameter each. This scalar product is elementarily calculated using commutation relations (1.121) or formula (5.70):

$$\boldsymbol{S}_1(v|u) = -\eta\,\frac{r(v) - r(u)}{v - u}. \tag{6.56}$$

This formula holds for any complex $v$ and $u$. In particular, for $v = u$, we obtain

$$\boldsymbol{S}_1(u|u) = -\eta\, r'(u). \tag{6.57}$$

On the other hand, if both vectors are eigenvectors of the twisted transfer matrix, then $v$ and $u$ satisfy the system of twisted Bethe equations. The latter, in this case, has the form

$$r(v) = \kappa, \qquad r(u) = \kappa. \tag{6.58}$$

If we substitute this into (6.56), we immediately obtain $\boldsymbol{S}_1(v|u) = 0$. Taking the limit $v = u$, we still have zero. Of course, the described calculation is incorrect. Indeed, consider, for example, the one-dimensional Bose gas, in which (6.58) takes the form

$$e^{-ivL} = \kappa, \qquad e^{-iuL} = \kappa. \tag{6.59}$$

Hence,

$$v = \frac{i\log\kappa}{L} + \frac{2\pi s}{L}, \qquad u = \frac{i\log\kappa}{L} + \frac{2\pi r}{L}, \tag{6.60}$$

---

[4] Such a scalar product is usually called the norm's square, even if the dual vector is not Hermitian conjugated to the initial vector.

where $s$ and $r$ are integers. If we assume from the very beginning that $v \neq u$, then we obtain that $s \neq r$, and we cannot consider the limit in which two different integers are going to each other. If, on the other hand, we say that $v = u$, then we must return to the formula (6.56), calculate the limit (6.57), and only then substitute an explicit expression for $u$ (6.60) into the obtained result (if necessary).

One can also require that one of the variables, say $u$, is a root of the twisted Bethe equation, while $v$ is generic. Then the formula (6.56) turns into

$$\boldsymbol{S}_1(v|u) = -\eta \, \frac{r(v) - \kappa}{v - u}. \tag{6.61}$$

Here the limiting procedure $v \to u$ is possible since $v$ remains an arbitrary complex number, and we can consider the limit when this complex number goes to the root of the twisted Bethe equation. The only wrong method is to take the limit in which two different roots go to each other.

Let us turn back to calculating the norm of the twisted on-shell Bethe vector in the general case. This can be done in fairly general settings, starting from representation (6.38). As usual, we will consider the rational $R$-matrix for simplicity, allowing the reader to do similar calculations in the trigonometric case.

Let us select the pole part in the twisted transfer matrix eigenvalue

$$\tau_\kappa(z|\bar{v}) = g(z, \bar{v})\mathcal{Y}_\kappa(z|\bar{v}), \tag{6.62}$$

Here $\mathcal{Y}_\kappa(z|\bar{v})$ is a smooth function of complex parameters $z$ and $\bar{v}$, such that it has no singularities at $z = v_k$, $k = 1, \dots, n$. In Section 2.1.3, we established that if the parameters $\bar{v}$ satisfy Bethe equations, then the eigenvalue of the transfer matrix has no poles at the points $z = v_k$. The same property, of course, holds for the twisted transfer matrix. Therefore, if $\bar{v}$ satisfy twisted Bethe equations, then

$$\mathcal{Y}_\kappa(v_k|\bar{v}) = 0, \qquad k = 1, \dots, n. \tag{6.63}$$

Let $\partial_0$ denote the partial derivative of $\mathcal{Y}_\kappa(z|\bar{v})$ over $z$, and $\partial_j$ denote the partial derivative of $\mathcal{Y}_\kappa(z|\bar{v})$ over the $j$th element of the set $\bar{v}$. Then we have[5]

$$\frac{\partial \mathcal{Y}_\kappa(v_k|\bar{v})}{\partial v_j} = \delta_{jk}\partial_0\mathcal{Y}_\kappa(v_k|\bar{v}) + \partial_j\mathcal{Y}_\kappa(v_k|\bar{v}). \tag{6.64}$$

[5]As before, we first assume that $\bar{v}$ are arbitrary complex numbers, and after taking derivatives, we send this set to the roots of the twisted Bethe equations.

On the other hand,

$$\frac{\eta}{g(u_k,\bar v)}\frac{\partial\tau_\kappa(u_k|\bar v)}{\partial v_j}=g(u_k,v_j)\mathcal{Y}_\kappa(u_k|\bar v)+\eta\partial_j\mathcal{Y}_\kappa(u_k|\bar v). \tag{6.65}$$

Let the system (6.63) be fulfilled. Then it follows from (6.65) that

$$\begin{aligned}\frac{\eta}{g(u_k,\bar v)}&\frac{\partial\tau_\kappa(u_k|\bar v)}{\partial v_j}\Big|_{u_k=v_k}\\&=g(u_k,v_j)\big(\mathcal{Y}_\kappa(u_k|\bar v)-\mathcal{Y}_\kappa(v_k|\bar v)\big)\Big|_{u_k=v_k}+\eta\partial_j\mathcal{Y}_\kappa(v_k|\bar v)\\&=\eta\Big(\delta_{jk}\partial_0\mathcal{Y}_\kappa(v_k|\bar v)+\partial_j\mathcal{Y}_\kappa(v_k|\bar v)\Big).\end{aligned} \tag{6.66}$$

Comparing (6.66) and (6.64), we arrive at

$$\frac{\eta}{g(u_k,\bar v)}\frac{\partial\tau_\kappa(u_k|\bar v)}{\partial v_j}\Big|_{u_k=v_k}=\eta\frac{\partial\mathcal{Y}_\kappa(v_k|\bar v)}{\partial v_j}. \tag{6.67}$$

Hence,

$$\lim_{\bar u\to\bar v}\frac{\eta}{d(u_k)g(u_k,\bar v)}\frac{\partial\tau_\kappa(u_k|\bar v)}{\partial v_j}=\frac{\eta}{d(v_k)}\frac{\partial\mathcal{Y}_\kappa(v_k|\bar v)}{\partial v_j}. \tag{6.68}$$

Substituting this in (6.38), we obtain the following determinant representation for the square of the norm of the twisted on-shell Bethe vector:

$$\boldsymbol{S}_n(\bar v|\bar v)=\Delta'(\bar v)\Delta(\bar v)\det_n\left(\frac{\eta}{d(v_k)}\frac{\partial\mathcal{Y}_\kappa(v_k|\bar v)}{\partial v_j}\right). \tag{6.69}$$

Let us rewrite this representation in a more explicit form. Consider for a change the case of the trigonometric $R$-matrix. Then

$$\boldsymbol{S}_n(\bar v|\bar v)=\Delta'(\bar v)\Delta(\bar v)\det_n\left(\frac{\sinh\eta}{d(v_k)}\frac{\partial\mathcal{Y}_\kappa(v_k|\bar v)}{\partial v_j}\right), \tag{6.70}$$

and

$$\begin{aligned}\mathcal{Y}_\kappa(v_k|\bar v)=&\frac{\kappa d(v_k)}{\sinh^n\eta}\prod_{m=1}^n\sinh(v_k-v_m+\eta)\\&+\frac{a(v_k)}{\sinh^n\eta}\prod_{m=1}^n\sinh(v_k-v_m-\eta).\end{aligned} \tag{6.71}$$

Recall once more that we need to take derivatives and only after this impose twisted Bethe equations $\mathcal{Y}_\kappa(v_k|\bar v)=0$. For $j\ne k$, we obtain

$$\frac{\partial\mathcal{Y}_\kappa(v_k|\bar v)}{\partial v_j}=\frac{\kappa d(v_k)}{\sinh^n\eta}\left(\prod_{m=1}^n\sinh(v_k-v_m+\eta)\right)\mathcal{K}(v_k-v_j),\quad j\ne k, \tag{6.72}$$

where

$$\mathcal{K}(u) = \frac{\sinh 2\eta}{\sinh(u+\eta)\sinh(u-\eta)}. \tag{6.73}$$

For diagonal elements, we have

$$\frac{\partial \mathcal{Y}_\kappa(v_k|\bar{v})}{\partial v_k} = -\frac{\kappa d(v_k)}{\sinh^n \eta}\left(\prod_{m=1}^{n} \sinh(v_k - v_m + \eta)\right) \times \left(Z_k + \sum_{\substack{m=1\\ m\neq k}}^{n} \mathcal{K}(v_k - v_m)\right), \tag{6.74}$$

where

$$Z_k = \frac{\mathrm{d}}{\mathrm{d}z}\log r(z)\Big|_{z=v_k}. \tag{6.75}$$

Thus, representation (6.69) takes the form

$$\boldsymbol{S}_n(\bar{v}|\bar{v}) = (\kappa \sinh\eta)^n \left(\prod_{\substack{a,b=1\\ a\neq b}}^{n} \frac{\sinh(v_a - v_b + \eta)}{\sinh(v_a - v_b)}\right) \det_n \mathcal{N}, \tag{6.76}$$

where

$$\mathcal{N}_{jk} = -\delta_{jk}\Big(Z_k + \sum_{m=1}^{n} \mathcal{K}(v_k - v_m)\Big) + \mathcal{K}(v_k - v_j). \tag{6.77}$$

All calculations in models with the rational $R$-matrix are entirely analogous. The result is given by formulas (6.75)–(6.77), in which one should replace $v_j \to \epsilon v_j$, $\eta \to \epsilon\eta$, and $Z_k \to Z_k/\epsilon$, and then proceed to the limit $\epsilon \to 0$.

Matrix elements $\mathcal{N}_{jk}$ can also be associated with Bethe equations in logarithmic form. Let

$$\Psi_j \equiv \Psi_j(\bar{v}) = \log r(v_j) + \sum_{\substack{m=1\\ m\neq j}}^{n} \log\frac{\sinh(\eta + v_m - v_j)}{\sinh(\eta - v_m + v_j)}. \tag{6.78}$$

Then twisted Bethe equations for the parameters $\bar{v}$ take a very simple form

$$\Psi_j = 2\pi i I_j + \log\kappa, \tag{6.79}$$

where $I_j$ are integers or half-integers, depending on the parity of $n$. In fact, $\Psi_j(\bar{v})$ is the logarithm of the system of twisted Bethe equations. It is not difficult to check that

$$\mathcal{N}_{jk} = -\frac{\partial \Psi_j}{\partial v_k}. \tag{6.80}$$

### 6.4.1 *Basis of twisted on-shell Bethe vectors*

Consider an inhomogeneous $XXZ$ (or $XXX$) chain of the length $N$. Let the inhomogeneities $\bar{\xi}$ satisfy the following constraint:

$$\xi_j - \xi_k \neq 0, \qquad \xi_j - \xi_k \pm \eta \neq 0, \qquad j,k = 1,\dots,N. \tag{6.81}$$

Due to proposition 2.1, for $|\kappa|$ small enough, the system of twisted Bethe equations has $\binom{N}{n}$ admissible solutions, to which twisted on-shell Bethe vectors correspond. Let us denote these solutions by $\bar{v}^{(\ell)}$, $\ell = 1,\dots,\binom{N}{n}$. The roots $v_j^{(\ell)}(\kappa)$ of each such solution are analytic functions of $\kappa$ in some neighborhood $|\kappa| < \kappa_0$. To list these solutions, consider all possible partitions of inhomogeneities into two subsets $\bar{\xi} \mapsto \{\bar{\xi}^{\rm I}, \bar{\xi}^{\rm II}\}$, such that $\#\bar{\xi}^{\rm I} = n$. Then each solution $\bar{v}^{(\ell)}(\kappa)$ can be characterized by its value at $\kappa = 0$: $\bar{v}^{(\ell)}(0) = \bar{\xi}^{\rm I} - \eta$. Different solutions correspond to different subsets $\bar{\xi}^{\rm I}$.

**Proposition 6.5.** *The twisted on-shell Bethe vectors $B(\bar{v}^{(\ell)})|0\rangle$ corresponding to different solutions $\bar{v}^{(\ell)}$ are linearly independent.*

**Proof.** The proof goes along the standard line. Suppose that the vectors $B(\bar{v}^{(\ell)})|0\rangle$ are linearly dependent. Then there exists a linear combination of these vectors that vanishes:

$$\sum_{\ell=1}^{\binom{N}{n}} \alpha_\ell B(\bar{v}^{(\ell)})|0\rangle = 0. \tag{6.82}$$

Here at least one of the numeric coefficients $\alpha_\ell$ should be non-vanishing. Suppose without loss of generality that $\alpha_1 \neq 0$. Then multiplying (6.82) by $\langle 0|C(\bar{v}^{(1)})$ and using the orthogonality of the twisted on-shell Bethe vectors, we obtain

$$\alpha_1 S_n(\bar{v}^{(1)}|\bar{v}^{(1)}) = \alpha_1 \langle 0|C(\bar{v}^{(1)})B(\bar{v}^{(1)})|0\rangle = 0. \tag{6.83}$$

Since $\alpha_1 \neq 0$, we conclude that $S_n(\bar{v}^{(1)}|\bar{v}^{(1)}) = 0$. Let us prove that this is not the case.

**Remark 6.3.** We recall that for arbitrary complex inhomogeneities $\bar{\xi}$ and an arbitrary twist parameter $\kappa$, the twisted transfer matrix is not a self-adjoint operator. Therefore, generally speaking, $\langle 0|C(\bar{v}^{(1)}) \neq \big(B(\bar{v}^{(1)})|0\rangle\big)^\dagger$. Thus, we cannot use the fact that the norm of a nonzero vector is non-vanishing.

Multiplying (6.76) by the product $d^2(\bar{v}^{(1)})$, we obtain

$$S_n(\bar{v}^{(1)}|\bar{v}^{(1)}) = a(\bar{v}^{(1)})d(\bar{v}^{(1)})(\sinh\eta)^n \times\left(\prod_{\substack{a,b=1\\ a\neq b}}^{n} \frac{\sinh(v_a^{(1)} - v_b^{(1)} + \eta)}{\sinh(v_a^{(1)} - v_b^{(1)})}\right) \det_n \mathcal{N}, \tag{6.84}$$

where $\mathcal{N}$ is given by (6.77). Here we have used that

$$a(\bar{v}^{(1)}) = \kappa^n d(\bar{v}^{(1)}), \tag{6.85}$$

due to the twisted Bethe equations. Since the roots $v_k^{(1)}(\kappa)$ are analytic functions of $\kappa$ for $|\kappa| < \kappa_0$, the scalar product $S_n(\bar{v}^{(1)}|\bar{v}^{(1)})$ is also an analytic function of $\kappa$ in the same neighborhood. Therefore, it is sufficient to prove that $S_n(\bar{v}^{(1)}|\bar{v}^{(1)}) \neq 0$ at least for $\kappa = 0$. It follows that $S_n(\bar{v}^{(1)}|\bar{v}^{(1)}) \neq 0$ in some neighborhood of the point $\kappa = 0$.

Let $v_k^{(1)}(0) = \xi_k^{\mathrm{I}} - \eta$, $k = 1,\dots,n$, where $\bar{\xi}^{\mathrm{I}}$ is a fixed subset of the inhomogeneities of the cardinality $n$. Then $a\big(v_k^{(1)}(0)\big) = 0$, because

$$a(z) = \prod_{m=1}^{N} \sinh(z - \xi_m + \eta). \tag{6.86}$$

On the other hand, some entries of the matrix $\mathcal{N}$ become singular because

$$\begin{aligned}\frac{\mathrm{d}}{\mathrm{d}z}\log r(z) &= \sum_{m=1}^{N} \frac{\mathrm{d}}{\mathrm{d}z}\log\frac{\sinh(z-\xi_m+\eta)}{\sinh(z-\xi_m)}\\ &= -\sum_{m=1}^{N} \frac{\sinh\eta}{\sinh(z-\xi_m+\eta)\sinh(z-\xi_m)}.\end{aligned} \tag{6.87}$$

Multiplying the $k$th column of the matrix $\mathcal{N}_{jk}$ by $a\big(v_k^{(1)}\big)$, we find

$$\lim_{\kappa\to 0} a\big(v_k^{(1)}\big)\mathcal{N}_{jk} = -\delta_{jk} \prod_{\xi_m\neq\xi_k^{\mathrm{I}}} \sinh(\xi_k^{\mathrm{I}} - \xi_m). \tag{6.88}$$

Hence, we obtain a diagonal matrix, and

$$\lim_{\kappa\to 0}\det_n\Big(a\big(v_k^{(1)}\big)\mathcal{N}_{jk}\Big) = (-1)^n \prod_{k=1}^{n}\prod_{\xi_m\neq\xi_k^{\mathrm{I}}} \sinh(\xi_k^{\mathrm{I}} - \xi_m). \tag{6.89}$$

Substituting this into (6.84), we, after elementary algebra, arrive at

$$S_n(\bar{v}^{(1)}|\bar{v}^{(1)})\Big|_{\kappa=0} = \prod_{a,b=1}^{n} \sinh^2(\xi_a^{\mathrm{I}} - \xi_b^{\mathrm{I}} - \eta) \times \prod_{a=1}^{n}\prod_{b=1}^{N-n} \sinh(\xi_a^{\mathrm{I}} - \xi_b^{\mathrm{II}})\sinh(\xi_a^{\mathrm{I}} - \xi_b^{\mathrm{II}} - \eta). \tag{6.90}$$

We see that $S_n(\bar{v}^{(1)}|\bar{v}^{(1)}) \neq 0$ at $\kappa = 0$ due to constraint (6.81). We have come to a contradiction. Hence, our initial assumption about the linear dependence of the vectors $B(\bar{v}^{(\ell)})|0\rangle$ was incorrect. □

We can now formulate the following statement.

**Proposition 6.6.** *Consider an inhomogeneous $XXZ$ (or $XXX$) spin chain with $N$ sites and inhomogeneities satisfying condition* (6.81). *There exists $\kappa_0 > 0$ such that for $|\kappa| < \kappa_0$, the system of the twisted on-shell Bethe vectors $B(\bar{v}^{(\ell)})|0\rangle$ forms a basis in the subspace with $n$ spins down, and $N - n$ spins up.*

We add that this statement remains true in the case of a homogeneous chain if $\kappa$ belongs to a punctured neighborhood $0 < |\kappa| < \kappa_0$ (see [N. Kitanine et al (2005d)]).

## 6.5 Particular cases of scalar products

In this section, we obtain several identities that arise when computing scalar products in the generalized model. We will need these identities when calculating form factors and correlation functions.

For arbitrary complex $\bar{v} = \{v_1, \dots, v_n\}$, $\bar{u} = \{u_1, \dots, u_n\}$, and $\kappa$, we define a function

$$\boldsymbol{S}_n^{(\kappa)}(\bar{v}|\bar{u}) = \sum_{\substack{\bar{v} \mapsto \{\bar{v}_{\mathrm{I}}, \bar{v}_{\mathrm{II}}\} \\ \bar{u} \mapsto \{\bar{u}_{\mathrm{I}}, \bar{u}_{\mathrm{II}}\}}} \kappa^{n_{\mathrm{II}}} \mathbb{K}_{n_{\mathrm{I}}}(\bar{v}_{\mathrm{I}}|\bar{u}_{\mathrm{I}}) \mathbb{K}_{n_{\mathrm{II}}}(\bar{u}_{\mathrm{II}}|\bar{v}_{\mathrm{II}})\, f(\bar{v}_{\mathrm{II}}, \bar{v}_{\mathrm{I}}) f(\bar{u}_{\mathrm{I}}, \bar{u}_{\mathrm{II}}). \tag{6.91}$$

Here the sum is taken over partitions of the sets $\bar{u}$ and $\bar{v}$ into two subsets each. Both partitions are independent except for a restriction on the cardinalities of the subsets $\#\bar{u}_{\mathrm{I}} = \#\bar{v}_{\mathrm{I}} = n_{\mathrm{I}}$, $\#\bar{u}_{\mathrm{II}} = \#\bar{v}_{\mathrm{II}} = n_{\mathrm{II}} = n - n_{\mathrm{I}}$, $n_{\mathrm{I}} = 0, 1, \dots, n$.

**Proposition 6.7.** *The function $\boldsymbol{S}_n^{(\kappa)}(\bar{v}|\bar{u})$ has the following determinant representations:*

(i)

$$\boldsymbol{S}_n^{(\kappa)}(\bar{v}|\bar{u}) = \Delta'(\bar{v})\Delta(\bar{u})\, h(\bar{v}, \bar{u})\, \det_n \mathcal{M}^{(\kappa)}, \tag{6.92}$$

*where*

$$\mathcal{M}_{jk}^{(\kappa)} = t(v_j, u_k) + \kappa\, t(u_k, v_j) \frac{h(\bar{u}, u_k) h(u_k, \bar{v})}{h(u_k, \bar{u}) h(\bar{v}, u_k)}; \tag{6.93}$$

(ii)

$$\boldsymbol{S}_n^{(\kappa)}(\bar v|\bar u) = \Delta'(\bar u)\Delta(\bar v)\, h(\bar v,\bar u)\, \det_n \widetilde{\mathcal{M}}^{(\kappa)}, \tag{6.94}$$

*where*

$$\widetilde{\mathcal{M}}_{jk}^{(\kappa)} = t(v_k,u_j) + \kappa\, t(u_j,v_k)\frac{h(\bar u,v_k)h(v_k,\bar v)}{h(v_k,\bar u)h(\bar v,v_k)}. \tag{6.95}$$

**Proof.** First, we briefly describe the proof of proposition 6.7 by traditional methods. To prove representation (6.92), one should first calculate the sum over the partitions of the parameters $\bar v$ using proposition 5.3. After that, the sum over the partitions of the parameters $\bar u$ is calculated using corollary 6.2. To prove representation (6.94), one should do the opposite: first, calculate the sum over the partitions of the parameters $\bar u$ using proposition 5.3, and then calculate the sum over the partitions of the parameters $\bar v$ using corollary 6.2.

We will prove proposition 6.7 within the generalized model framework. Note that sum (6.91) has the meaning of the scalar product of twisted and ordinary on-shell vectors. Indeed, in the generic case, the scalar product is given by formula (5.70). Let us assume that the dual vector $\langle 0|C(\bar v)$ is a twisted on-shell vector, and the vector $B(\bar u)|0\rangle$ is an ordinary on-shell vector. Then, due to the (twisted) Bethe equations, we have

$$r(\bar u_{\rm I})f(\bar u_{\rm II},\bar u_{\rm I}) = f(\bar u_{\rm I},\bar u_{\rm II}), \qquad r(\bar v_{\rm II})f(\bar v_{\rm I},\bar v_{\rm II}) = \kappa^{n_{\rm II}} f(\bar v_{\rm II},\bar v_{\rm I}). \tag{6.96}$$

Substituting these expressions into (5.70), we reproduce the sum (6.91). Thus, we are dealing with a particular case of the scalar product considered in Section 6.2, where the vector $\langle 0|C(\bar v)$ was a twisted on-shell vector, and the vector $B(\bar u)|0\rangle$ was an arbitrary Bethe vector. In particular, it can also be an on-shell vector. The result for such a scalar product is given by formulas (6.35) and (6.36). However, now we can use Bethe equations and express functions $r(u_k)$ in terms of the $f$ functions products. Then, after simple transformations, we arrive at (6.92).

To obtain representation (6.94), we use (5.71)

$$\boldsymbol{S}_n(\bar v|\bar u) = \frac{\langle 0|C(\bar v)B(\bar u)|0\rangle}{d(\bar v)d(\bar u)} = \frac{\langle 0|C(\bar u)B(\bar v)|0\rangle}{d(\bar v)d(\bar u)}, \tag{6.97}$$

for arbitrary Bethe vectors $B(\bar u)|0\rangle$ and $\langle 0|C(\bar v)$. Now we can say that the vector $\langle 0|C(\bar u)$ is the twisted on-shell vector with $\kappa = 1$, while the vector $B(\bar v)|0\rangle$ is an arbitrary Bethe vector (which can be a twisted on-shell vector in the particular case). Again, the result for this scalar product is given by (6.35) and (6.36), where one should replace $\bar u \leftrightarrow \bar v$ and set $\kappa = 1$:

$$\boldsymbol{S}_n(\bar v|\bar u) = \Delta'(\bar v)\Delta(\bar u)\, h(\bar v,\bar u)\, \det_n \widetilde{\mathcal{M}}^{(\kappa)}, \tag{6.98}$$

where

$$\widetilde{\mathcal{M}}^{(\kappa)}_{jk} = t(v_k, u_j)\left(1 - r(v_k)\frac{f(\bar{u}_j, v_k)}{f(v_k, \bar{u}_j)}\right), \tag{6.99}$$

and we used $\Delta'(\bar{u})\Delta(\bar{v}) = \Delta'(\bar{v})\Delta(\bar{u})$. Now we should express the functions $r(v_k)$ in terms of the $f$ functions products via twisted Bethe equations. Then we arrive at (6.94).

Thus, within the framework of the generalized model, representations (6.92) and (6.94) are a consequence of the determinant formula for the scalar product of the (twisted) on-shell Bethe vector and an arbitrary Bethe vector. We only need to repeat the reasoning that we already carried out in the proof of Proposition 5.1.

Consider, for example, representation (6.92). Both sides of this formula are rational functions of the parameters $u_j$ and $v_j$ (or the parameters $e^{2u_j}$ and $e^{2v_j}$ in the trigonometric case). Therefore, to prove that (6.92) is an identity, it suffices to verify that it holds at a finite number of points. We fix the sets $\bar{u}$ and $\bar{v}$. After that, we find a representative of the generalized model, for which the vectors $\langle 0|C(\bar{v})$ and $B(\bar{u})|0\rangle$ are, respectively, the twisted on-shell Bethe vector and the ordinary on-shell Bethe vector. In other words, we impose constraints (Bethe equations) *on the parameters* $r(u_j)$ *and* $r(v_k)$ for fixed $\bar{u}$ and $\bar{v}$. Then equality (6.92) is satisfied. Repeating this reasoning for other choices of $\bar{u}$ and $\bar{v}$, we prove that (6.92) is fulfilled at the required number of points. This implies that (6.92) is an identity. Representation (6.94) is proved similarly. □

**Corollary 6.3.** *For arbitrary complex* $\bar{v}$ *and* $\bar{u}$*,*

$$\boldsymbol{S}^{(\kappa)}_n(\bar{v}|\bar{u})\Big|_{\kappa=1} = \delta_{n0}. \tag{6.100}$$

*Here* $\boldsymbol{S}^{(\kappa)}_0(\emptyset|\emptyset) = 1$ *by definition.*

**Proof.** We can use, for example, representation (6.92) and prove that the rows of the matrix $\mathcal{M}^{(\kappa)}$ become linearly dependent at $\kappa = 1$. This proof completely repeats the proof of the orthogonality of the twisted on-shell Bethe vectors from Section 6.3. The coefficients of the vanishing linear combination are given by formulas (6.42). Similarly, one can prove the linear dependence of the rows of the matrix $\widetilde{\mathcal{M}}^{(\kappa)}$.

From the point of view of the generalized model, $\boldsymbol{S}^{(\kappa)}_n(\bar{v}|\bar{u})$ is the scalar product of two on-shell vectors at $\kappa = 1$. If we fix the sets $\bar{u}$ and $\bar{v}$ so that $u_j \ne v_k$, $j, k = 1, \dots, n$, then such the scalar product must vanish. An exception is the case $n = 0$ when we have $\langle 0|0\rangle = 1$. Having proved (6.100) for some finite number of fixed sets $\bar{v}$ and $\bar{u}$, we conclude that (6.100) is an identity. It is important to emphasize that $\boldsymbol{S}_n^{(\kappa)}(\bar{v}|\bar{u})$ is not the norm of the on-shell Bethe vector for $\bar{u} = \bar{v}$ and $\kappa = 1$ (see the reasoning at the beginning of Section 6.4). □

# Chapter 7

# Alternative methods to compute scalar products

In this chapter, we continue to study scalar products of Bethe vectors and describe alternative methods for their calculation. All these methods have been developed relatively recently. They do not give any new formulas for scalar products but only reproduce already obtained ones. However, these methods have several advantages.

In Section 7.1, we calculate the action of the product of the twisted transfer matrices on off-shell Bethe vectors. We prove that one of the terms in this multiple action coincides with the scalar product of the twisted on-shell vector and the off-shell Bethe vector. Below, we will use this result to calculate the correlation functions. In Section 7.2, we describe another method for calculating scalar products, in which one of the vectors is on-shell. We show that such scalar products satisfy a system of linear equations. A distinctive feature of this approach is the minimal use of the $RTT$-relation. Besides, this method explains why the scalar product can be presented in the form of determinants.

For definiteness, we consider models with the rational $R$-matrix. However, all the derivations and the results are also true for models with the trigonometric $R$-matrix.

## 7.1 Scalar product and multiple action of transfer matrix

This section establishes a connection between the scalar product of a twisted on-shell vector and an off-shell vector with a special coefficient in the multiple action of the twisted transfer matrix on Bethe vectors.

### 7.1.1 *Scalar product and completely indirect action*

Let us turn back to the twisted transfer matrix action on an off-shell Bethe vector $B(\bar{u})|0\rangle$ (2.51). Recall that in Section 6.4, we introduced a function $\mathcal{Y}_\kappa(x|\bar{u})$ associated with the eigenvalue of the twisted transfer matrix (see (6.62)). If the latter is given by (6.37), then

$$\mathcal{Y}_\kappa(x|\bar{u}) = (-1)^n a(x)h(\bar{u},x) + \kappa d(x)h(x,\bar{u}), \tag{7.1}$$

where $\#\bar{u} = n$. Equation (2.51) then takes the form

$$\begin{aligned}\mathcal{T}_\kappa(x)B(\bar{u})|0\rangle &= \tau_\kappa(x|\bar{u})B(\bar{u})|0\rangle \\ &\quad + \sum_{k=1}^{n} g(u_k,x)g(u_k,\bar{u}_k)\mathcal{Y}_\kappa(u_k|\bar{u})B(x)B(\bar{u}_k)|0\rangle. \end{aligned}\tag{7.2}$$

Schematically it can be written as

$$\mathcal{T}_\kappa(x)B(\bar{u})|0\rangle = \sum_{\substack{\bar{w}\mapsto\{\bar{w}_{\rm I},\bar{w}_{\rm II}\}\\ \#\bar{w}_{\rm I}=1}} \mathcal{A}_1^{(n)}(\bar{w}_{\rm I}|\bar{w}_{\rm II})B(\bar{w}_{\rm II})|0\rangle, \tag{7.3}$$

where $\bar{w} = \{x,\bar{u}\}$. Here $\mathcal{A}_1^{(n)}$ denotes a numerical coefficient. We do not specify its explicit form in equation (7.3). However, it is easy to see that the first term in (7.2) corresponds to the partition $\bar{w}_{\rm I} = x$ and $\bar{w}_{\rm II} = \bar{u}$, while the sum over $k$ in (7.2) corresponds to the partitions $\bar{w}_{\rm I} = u_k$ and $\bar{w}_{\rm II} = \{x,\bar{u}_k\}$.

Schematic equation (7.3) almost does not change in a more general case when we act on the Bethe vector with several twisted transfer matrices. Indeed, let $\bar{x} = \{x_1,\dots,x_p\}$, where $p \le n$. Then the multiple action of the product of the twisted transfer matrices $\mathcal{T}_\kappa(\bar{x})$ is given by the expression

$$\mathcal{T}_\kappa(\bar{x})B(\bar{u})|0\rangle = \sum_{\substack{\bar{w}\mapsto\{\bar{w}_{\rm I},\bar{w}_{\rm II}\}\\ \#\bar{w}_{\rm I}=p}} \mathcal{A}_p^{(n)}(\bar{w}_{\rm I}|\bar{w}_{\rm II})B(\bar{w}_{\rm II})|0\rangle. \tag{7.4}$$

The only difference with respect to (7.3) is that now the cardinality of the subset $\bar{w}_{\rm I}$ is fixed by the condition $\#\bar{w}_{\rm I} = p = \#\bar{x}$. In this formula, $\bar{w} = \{\bar{x},\bar{u}\}$, and $\mathcal{A}_p^{(n)}(\bar{w}_{\rm I}|\bar{w}_{\rm II})$ are numerical coefficients depending on the partitions. Our goal is to find them.

We introduce the concept of direct and indirect actions of the twisted transfer matrix on the Bethe vector. We will call a *direct action* such contributions to the action in which the Bethe vector does not change. In other words, this is the first term in (7.2). All other terms in equation (7.2) will be called an *indirect action*. In these terms, the twisted transfer matrix argument becomes one of the Bethe vector arguments.

Similarly, one can introduce direct and indirect action for the product of twisted transfer matrices. With direct action, the Bethe vector does not change. In other words, this is a term corresponding to the partition $\bar{w}_{\rm I} = \bar{x}$ and $\bar{w}_{\rm II} = \bar{u}$ in equation (7.4). We denote the direct action by the symbol $\xrightarrow{\rm dir}$ and the corresponding coefficient by $\mathcal{D}_p^{(n)}$:

$$\mathcal{T}_\kappa(\bar{x})B(\bar{u})|0\rangle \xrightarrow{\rm dir} \mathcal{D}_p^{(n)}(\bar{x}|\bar{u})B(\bar{u})|0\rangle. \tag{7.5}$$

Obviously,

$$\mathcal{D}_p^{(n)}(\bar{x}|\bar{u}) = \prod_{j=1}^{p} \tau_\kappa(x_j|\bar{u}). \tag{7.6}$$

All other terms in (7.4) are called indirect action. Among them, there are those in which all arguments of the twisted transfer matrices $\mathcal{T}_\kappa(x_j)$ become arguments of the Bethe vectors. We will call such terms *completely indirect action*. We denote this action by the symbol $\xrightarrow{\rm c.-indir}$ and the corresponding coefficients by $\mathcal{I}_p^{(n)}$.

Let $\bar{w}_{\rm I} = \{\bar{u}_{\rm I}, \bar{x}_{\rm II}\}$ and $\bar{w}_{\rm II} = \{\bar{u}_{\rm II}, \bar{x}_{\rm I}\}$ in (7.4). Then the completely indirect action corresponds to the partitions such that $\bar{x}_{\rm I} = \bar{x}$ and $\bar{x}_{\rm II} = \emptyset$. The sum in (7.4) then is taken over partitions of the set $\bar{u}$ only:

$$\mathcal{T}_\kappa(\bar{x})B(\bar{u})|0\rangle \xrightarrow{\rm c.-indir} \sum_{\substack{\bar{u}\mapsto\{\bar{u}_{\rm I},\bar{u}_{\rm II}\} \\ \#\bar{u}_{\rm I}=p}} \mathcal{I}_p^{(n)}(\bar{u}_{\rm I}|\{\bar{x},\bar{u}_{\rm II}\})B(\bar{x})B(\bar{u}_{\rm II})|0\rangle. \tag{7.7}$$

In the particular case $p = n$, we have $\bar{u}_{\rm II} = \emptyset$, and the sum over partitions in (7.7) consists of only one term

$$\mathcal{T}_\kappa(\bar{x})B(\bar{u})|0\rangle \xrightarrow{\rm c.-indir} \mathcal{I}_n^{(n)}(\bar{u}|\bar{x})B(\bar{x})|0\rangle, \qquad \#\bar{x} = \#\bar{u} = n. \tag{7.8}$$

It turns out that in this particular case, the completely indirect action is closely related to the scalar product of a twisted on-shell vector and off-shell vector [N. Kitanine et al (2002a); A. Liashyk (2020)].

**Proposition 7.1.** *Let $\langle 0|C(\bar{v})$ and $B(\bar{u})|0\rangle$ respectively be a dual twisted on-shell Bethe vector and an off-shell Bethe vector. Then their scalar product is proportional to the completely indirect action of the product $\mathcal{T}_\kappa(\bar{v})$ on $B(\bar{u})|0\rangle$:*

$$\langle 0|C(\bar{v})B(\bar{u})|0\rangle = d(\bar{v})\mathcal{I}_n^{(n)}(\bar{u}|\bar{v}). \tag{7.9}$$

To prove this statement, we again use the generalized model. Recall that in the framework of the generalized model, we can consider Bethe parameters $u_j$ and functions $r(u_j) = a(u_j)/d(u_j)$ as independent variables.

If we are dealing with (twisted) on-shell Bethe vectors, then constraints between these variables arise. However, the derivatives of the functions $r(z)$ at $z = v_j$ are not fixed by Bethe equations. Therefore, these derivatives and $v_j$ variables can still be treated as independent variables.

Indeed, let[1]

$$Z_j = -\eta \frac{\mathrm{d}\log r(v)}{\mathrm{d}v}\Big|_{v=v_j}. \tag{7.10}$$

Suppose that the parameters $\bar{v} = \{v_1, \dots, v_n\}$ satisfy twisted Bethe equations. Then the values of the functions $r(v_j)$ are fixed at the points $v_j$. However, there are still infinitely many such functions $r(v)$ in the generalized model. They differ from each other, particularly by the values of the derivatives at the points $v_j$. Therefore, if we choose different representatives of the generalized model, the variables $Z_j$ take any prescribed values. In this sense, $Z_j$ and $v_j$ are independent variables, even if $r(v_j)$ and $v_j$ are related by the twisted Bethe equations.

The variables $Z_j$ arise when calculating scalar products $\langle 0|C(\bar{v})B(\bar{u})|0\rangle$ if some of the parameters from the sets $\bar{v}$ and $\bar{u}$ coincide. We have already seen this effect in Section 5 in the case of the norm of the twisted on-shell Bethe vectors (see (6.57), (6.74)). Let us explain the mechanism by which these variables arise.

Consider the limit $u_n \to v_n$ in a scalar product $\langle 0|C(\bar{v})B(\bar{u})|0\rangle$. In calculating the scalar product we obtain

$$S_n(\bar{v}|\bar{u}) = g(v_n, u_n)\langle 0|C(\bar{v}_n)\big(A(u_n)D(v_n) - A(v_n)D(u_n)\big)B(\bar{u}_n)|0\rangle + \mathfrak{Z}. \tag{7.11}$$

Here, as usual, the $\mathfrak{Z}$ symbol denotes terms that are not essential in our consideration. Moving the operators $A$ and $D$ to the right, we obtain, in particular, the following contribution:

$$\begin{aligned} S_n(\bar{v}|\bar{u}) = g(v_n, u_n)\Big(&a(u_n)d(v_n)f(\bar{u}_n, u_n)f(v_n, \bar{u}_n) \\ &- a(v_n)d(u_n)f(\bar{u}_n, v_n)f(u_n, \bar{u}_n)\Big)S_{n-1}(\bar{v}_n|\bar{u}_n) + \mathfrak{Z}. \end{aligned} \tag{7.12}$$

[1]We have previously introduced similar logarithmic derivatives of the function $r(v)$ (see (6.75)). This section makes it more convenient for us to include the constant $-\eta$ in the definition of the $Z_j$ parameters.

Taking the limit $u_n \to v_n$, we arrive at[2]

$$\lim_{u_n \to v_n} S_n(\bar{v}|\bar{u}) = Z_n a(v_n) d(v_n) f(\bar{u}_n, v_n) f(v_n, \bar{u}_n) S_{n-1}(\bar{v}_n|\bar{u}_n) + \mathfrak{Z}. \quad (7.13)$$

**Remark 7.1.** When the operators $A$ and $D$ move to the right, they can exchange arguments with the operators $B$. The resulting contributions can also be proportional to $Z_n$ in the limit $u_n \to v_n$. Therefore, equation (7.13) gives only one of the possible terms containing the variable $Z_n$. However, upon further consideration, all other contributions will be insignificant for us.

To prove proposition 7.1, we consider the scalar product $\langle 0|C(\bar{v})B(\bar{u})|0\rangle$ in the limit when all parameters $u_j$ go to $v_j$, that is, $\bar{u} \to \bar{v}$. The dual vector $\langle 0|C(\bar{v})$ can be either off-shell or (twisted) on-shell. The goal is to find a contribution that depends on all $Z_j$, $j = 1, \dots, n$:

$$\lim_{\bar{u} \to \bar{v}} S_n(\bar{v}|\bar{u}) = \chi_n(\bar{v}) \prod_{j=1}^{n} Z_j + \mathfrak{Z}. \quad (7.14)$$

Here $\chi_n(\bar{v})$ is a numerical coefficient to be determined. The term $\mathfrak{Z}$ contains all contributions which do not depend at least on one $Z_j$. Obviously,

$$\chi_n(\bar{v}) = \frac{\partial^n}{\partial Z_1 \dots \partial Z_n} \lim_{\bar{u} \to \bar{v}} S_n(\bar{v}|\bar{u}). \quad (7.15)$$

In fact, we have already derived all necessary formulas for finding $\chi_n(\bar{v})$. First, to obtain the parameter $Z_n$ in the final result, we should use equation (7.11). Indeed, this is the only way to obtain the pole $g(v_n, u_n)$ that ultimately leads to the appearance of $Z_n$. Further, as a result of the action of the operators $A$ and $D$ on the vector $B(\bar{u}_n)|0\rangle$, we must obtain the same vector $B(\bar{u}_n)|0\rangle$. Otherwise, in the remaining scalar product, not all arguments of the two vectors will coincide. Therefore, we will not get all the remaining $Z_j$ with $j = 1, \dots, n-1$. Thus, we conclude that we should use equation (7.12). Finally, taking the limit $u_n \to v_n$, we arrive at (7.13), in which one should also take the remaining limits $u_j \to v_j$, $j = 1, \dots, n-1$:

$$\begin{aligned} \lim_{\bar{u} \to \bar{v}} S_n(\bar{v}|\bar{u}) &= Z_n a(v_n) d(v_n) f(\bar{v}_n, v_n) f(v_n, \bar{v}_n) \\ &\quad \times \lim_{\bar{u}_n \to \bar{v}_n} S_{n-1}(\bar{v}_n|\bar{u}_n) + \mathfrak{Z}. \end{aligned} \quad (7.16)$$

In its turn, the obtained equation immediately implies a recursion for $\chi_n(\bar{v})$

$$\chi_n(\bar{v}) = a(v_n) d(v_n) f(\bar{v}_n, v_n) f(v_n, \bar{v}_n) \chi_{n-1}(\bar{v}_n), \quad (7.17)$$

[2] When calculating the limit, we are only interested in those contributions in which derivatives of the functions $a$ and $d$ arise.

with the initial condition $\chi_1(v) = a(v)d(v)$. Solving this simple recursion, we find

$$\chi_n(\bar v) = a(\bar v)d(\bar v)\prod_{\substack{j,k=1\\ j\ne k}}^{n} f(v_j, v_k). \tag{7.18}$$

Now we are ready to prove Proposition 7.1.

**Proof of Proposition 7.1.** Consider equation (7.4) for $\#\bar x = \#\bar u = n$. Let us multiply this equation from the left by a dual twisted on-shell Bethe vector $\langle 0|C(\bar v)$

$$\langle 0|C(\bar v)\mathcal{T}_\kappa(\bar x)B(\bar u)|0\rangle = \sum_{\substack{\bar w\mapsto\{\bar w_{\rm I},\bar w_{\rm II}\}\\ \#\bar w_{\rm I}=n}} \mathcal{A}_n^{(n)}(\bar w_{\rm I}|\bar w_{\rm II})\langle 0|C(\bar v)B(\bar w_{\rm II})|0\rangle. \tag{7.19}$$

Acting with $\mathcal{T}_\kappa(\bar x)$ on $\langle 0|C(\bar v)$ in the lhs of this equation, we obtain the product of the twisted transfer matrix eigenvalues

$$\left(\prod_{j=1}^{n}\tau_\kappa(x_j|\bar v)\right) S_n(\bar v|\bar u) = \sum_{\substack{\bar w\mapsto\{\bar w_{\rm I},\bar w_{\rm II}\}\\ \#\bar w_{\rm I}=n}} \mathcal{A}_n^{(n)}(\bar w_{\rm I}|\bar w_{\rm II})S_n(\bar v|\bar w_{\rm II}). \tag{7.20}$$

Recall that in this equation, $\bar x$ and $\bar u$ are arbitrary complex numbers.

Consider the limit $x_j \to v_j$, $j = 1, \dots, n$. In the lhs of (7.20), this limit should be taken in the twisted transfer matrix eigenvalues. It is easy to see that for any $x_j \in \bar x$,

$$\lim_{x_j\to v_j}\tau_\kappa(x_j|\bar v) = \kappa d(v_j)f(v_j,\bar v_j)Z_j + (Z_j\text{-independent terms}), \tag{7.21}$$

where we used the fact that the parameters $\bar v$ satisfy twisted Bethe equations. Hence, the lhs of (7.20) contains a term proportional to the product of all $Z_j$, and

$$\frac{\partial^n}{\partial Z_1\dots\partial Z_n}\lim_{\bar x\to\bar v}\left(\prod_{j=1}^{n}\tau_\kappa(x_j|\bar v)\right) S_n(\bar v|\bar u) = \kappa^n d(\bar v)S_n(\bar v|\bar u)\prod_{\substack{j,k=1\\ j\ne k}}^{n} f(v_j, v_k). \tag{7.22}$$

The same term, proportional to the product of all $Z_j$, should also appear in the rhs of (7.20) in the limit $\bar x \to \bar v$. However, the coefficients $\mathcal{A}_n^{(n)}(\bar w_{\rm I}|\bar w_{\rm II})$ are independent of the variables $\bar v$. Therefore, the appearance of $Z_j$ is possible only in the scalar product. In turn, the appearance of all $Z_j$ is possible if and only if $\bar{w}_{\rm II} = \bar{x}$. Thus, the product of all $Z_j$ is produced by the completely indirect action only

$$\frac{\partial^n}{\partial Z_1 \dots \partial Z_n} \lim_{\bar{x}\to\bar{v}} \sum_{\substack{\bar{w}\mapsto\{\bar{w}_{\rm I},\bar{w}_{\rm II}\}\\ \#\bar{w}_{\rm I}=n}} \mathcal{A}_n^{(n)}(\bar{w}_{\rm I}|\bar{w}_{\rm II}) S_n(\bar{v}|\bar{w}_{\rm II})$$
$$= \mathcal{I}_n^{(n)}(\bar{u}|\bar{v}) \frac{\partial^n}{\partial Z_1 \dots \partial Z_n} \lim_{\bar{x}\to\bar{v}} S_n(\bar{v}|\bar{x}). \tag{7.23}$$

The corresponding coefficient is exactly $\chi_n(\bar{v})$ (7.18). Thus,

$$\frac{\partial^n}{\partial Z_1 \dots \partial Z_n} \lim_{\bar{x}\to\bar{v}} \sum_{\substack{\bar{w}\mapsto\{\bar{w}_{\rm I},\bar{w}_{\rm II}\}\\ \#\bar{w}_{\rm I}=n}} \mathcal{A}_n^{(n)}(\bar{w}_{\rm I}|\bar{w}_{\rm II}) S_n(\bar{v}|\bar{w}_{\rm II})$$
$$= a(\bar{v})d(\bar{v})\mathcal{I}_n^{(n)}(\bar{u}|\bar{v}) \prod_{\substack{j,k=1\\ j\neq k}}^{n} f(v_j, v_k). \tag{7.24}$$

Comparing (7.24) and (7.22), we immediately arrive at

$$\langle 0|C(\bar{v})B(\bar{u})|0\rangle = \kappa^{-n} a(\bar{v})\mathcal{I}_n^{(n)}(\bar{u}|\bar{v}). \tag{7.25}$$

Finally, taking the product of all twisted Bethe equations, we obtain $\kappa^{-n}a(\bar{v}) = d(\bar{v})$ and thus prove (7.9). □

### 7.1.2 *Completely indirect action*

Proposition 7.1 opens a new way to calculate the scalar product of Bethe vectors if one of them is twisted on-shell. Namely, it is enough to calculate the completely indirect contribution to the multiple action of the twisted transfer matrices. We already know that the action of one twisted transfer matrix on the Bethe vector $B(\bar{u})|0\rangle$ is completely determined by the twisted transfer matrix eigenvalue and its residues at the points $u_j$. Therefore, the scalar product is also associated with the eigenvalue of the twisted transfer matrix.

In this section, we obtain an explicit representation of the completely indirect action. For this, we introduce an auxiliary function[3]

$$\mathcal{L}_p^{(n)}(\bar{x},\bar{y}|\bar{z}) = \Delta'(\bar{y})\Delta(\bar{x}) \det_p \left( \frac{\eta}{g(y_k,\bar{x})} \frac{\partial \tau_\kappa(y_k|\{\bar{x},\bar{z}\})}{\partial x_j} \right), \tag{7.26}$$

depending on three sets of complex variables: $\bar{x}$, $\bar{y}$, and $\bar{z}$. We assume that $\#\bar{x} = \#\bar{y} = p$, while the cardinality of the set $\bar{z}$ is $\#\bar{z} = n - p$, and $p \leq n$.

[3]In the trigonometric case, the coefficient $\eta$ should be replaced with $\sinh \eta$, as in equation (6.38).

Note that the function $\mathcal{L}_p^{(n)}$ is very similar to the scalar product in the form (6.38). We will soon see that this similarity is not accidental.

**Proposition 7.2.** *Let $\#\bar{x} = p$ and $\#\bar{u} = n$ so that $p \le n$. Then the completely indirect action of the product $\mathcal{T}_\kappa(\bar{x})$ on $B(\bar{u})|0\rangle$ reads*

$$\mathcal{T}_\kappa(\bar{x})B(\bar{u})|0\rangle \xrightarrow{\text{c.}-\text{indir}} \sum_{\substack{\bar{u}\mapsto\{\bar{u}_{\rm I},\bar{u}_{\rm II}\}\\ \#\bar{u}_{\rm I}=p}} \mathcal{L}_p^{(n)}(\bar{x},\bar{u}_{\rm I}|\bar{u}_{\rm II})B(\bar{x})B(\bar{u}_{\rm II})|0\rangle, \tag{7.27}$$

*that is,*

$$\mathcal{I}_p^{(n)}(\bar{u}_{\rm I}|\{\bar{x},\bar{u}_{\rm II}\}) = \mathcal{L}_p^{(n)}(\bar{x},\bar{u}_{\rm I}|\bar{u}_{\rm II}). \tag{7.28}$$

To prove this proposition, we should study some properties of the function $\mathcal{L}_p^{(n)}(\bar{x},\bar{y}|\bar{z})$ for generic complex $\bar{x}$, $\bar{y}$, and $\bar{z}$. For this, it is convenient to write down the matrix elements of the matrix in (7.26) more explicitly. Using (7.1) and (6.62), we find by straightforward calculation

$$\frac{\eta}{g(y_k,\bar{x})}\frac{\partial \tau_\kappa(y_k|\{\bar{x},\bar{z}\})}{\partial x_j} = g(y_k,x_j)g(y_k,\bar{z})\mathcal{Y}_\kappa(y_k|\{y_k,\bar{x}_j,\bar{z}\}). \tag{7.29}$$

Thus,

$$\mathcal{L}_p^{(n)}(\bar{x},\bar{y}|\bar{z}) = \Delta'(\bar{y})\Delta(\bar{x})g(\bar{y},\bar{z})\det_p\Big(g(y_k,x_j)\mathcal{Y}_\kappa(y_k|\{y_k,\bar{x}_j,\bar{z}\})\Big). \tag{7.30}$$

Immediate consequence of representation (7.30) is

$$\lim_{x_s\to\infty}\mathcal{L}_p^{(n)}(\bar{x},\bar{y}|\bar{z}) = 0, \qquad s = 1,\dots,p. \tag{7.31}$$

Indeed, let $x_p \to \infty$ (due to the symmetry of $\mathcal{L}_p^{(n)}(\bar{x},\bar{y}|\bar{z})$ over $\bar{x}$, it is enough to consider only this case). We have in the prefactor a product

$$g(x_p,\bar{x}_p) \to \left(\frac{\eta}{x_p}\right)^{p-1}, \qquad x_p \to \infty. \tag{7.32}$$

We absorb this coefficient by the determinant by multiplying the first $p-1$ rows of the matrix by $\eta/x_p$. Then we obtain

$$\mathcal{L}_p^{(n)}(\bar{x},\bar{y}|\bar{z}) \to \Delta'(\bar{y})\Delta(\bar{x}_p)g(\bar{y},\bar{z})\det_p J_{jk}, \tag{7.33}$$

where

$$\begin{aligned} J_{jk} &= \frac{\eta}{x_p}g(y_k,x_j)\mathcal{Y}_\kappa(y_k|\{y_k,\bar{x}_j,\bar{z}\}), \qquad j = 1,\dots,p-1,\\ J_{pk} &= g(y_k,x_p)\mathcal{Y}_\kappa(y_k|\{y_k,\bar{x}_p,\bar{z}\}). \end{aligned} \tag{7.34}$$

The functions $\mathcal{Y}_\kappa(y_k|\{y_k,\bar{x}_j,\bar{z}\})$ with $j < p$ are the first order polynomials in $x_p$. Hence, the first $p-1$ rows of the determinant have a finite limit at

$x_p \to \infty$. At the same time, the function $\mathcal{Y}_\kappa(y_k|\{y_k, \bar{x}_p, \bar{z}\})$ does not depend on $x_p$. Hence, the last row vanishes at $x_p \to \infty$ due to the factor $g(y_k, x_p)$.

Obviously, $\mathcal{L}_p^{(n)}(\bar{x}, \bar{y}|\bar{z})$ is a rational function of $x_p$ having simple poles at the points $x_p = y_l$, $l = 1, \dots, p$. As we just proved, this function decreases at $x_p \to \infty$. Therefore, the partial fraction decomposition of this function reads

$$\mathcal{L}_p^{(n)}(\bar{x}, \bar{y}|\bar{z}) = \frac{1}{\eta} \sum_{l=1}^{p} g(x_p, y_l) \operatorname{Res} \mathcal{L}_p^{(n)}(\bar{x}, \bar{y}|\bar{z}) \Big|_{x_p=y_l}. \tag{7.35}$$

Let us find the residues in (7.35). Due to the symmetry, we can consider the residue at $x_p = y_p$ only. Then the pole occurs in the right-lower element of the matrix. The determinant reduces to the product of this element and the corresponding minor

$$\begin{aligned} \mathcal{L}_p^{(n)}(\bar{x}, \bar{y}|\bar{z}) \Big|_{x_p=y_p} &= g(y_p, x_p)\mathcal{Y}_\kappa(y_p|\{y_p, \bar{x}_p, \bar{z}\})\Delta'(\bar{y})\Delta(\bar{x})g(\bar{y}, \bar{z}) \\ &\quad \times \det_{p-1}\Big(g(y_k, x_j)\mathcal{Y}_\kappa(y_k|\{y_k, \bar{x}_j, \bar{z}\})\Big) + \text{Reg}, \end{aligned} \tag{7.36}$$

where, as usual, Reg means regular part. Then we use obvious relations

$$\begin{aligned} \Delta'(\bar{y}) &= g(\bar{y}_p, y_p)\Delta'(\bar{y}_p), \\ \Delta(\bar{x}) &= g(x_p, \bar{x}_p)\Delta(\bar{x}_p), \\ g(\bar{y}, \bar{z}) &= g(y_p, \bar{z})g(\bar{y}_p, \bar{z}). \end{aligned} \tag{7.37}$$

Thus,

$$\Delta'(\bar{y})\Delta(\bar{x})g(\bar{y}, \bar{z}) = g(x_p, \bar{x}_p)g(y_p, \bar{z})\Delta'(\bar{y}_p)\Delta(\bar{x}_p)g(\bar{y}_p, \bar{z}'), \tag{7.38}$$

where $\bar{z}' = \{\bar{z}, y_p\}$. We also have for $j \neq p$,

$$\mathcal{Y}_\kappa(y_k|\{y_k, \bar{x}_j, \bar{z}\}) \Big|_{x_p=y_p} = \mathcal{Y}_\kappa(y_k|\{y_k, \bar{x}_{jp}, \bar{z}'\}), \tag{7.39}$$

where $\bar{x}_{jp} = \bar{x} \setminus \{x_j, x_p\}$. Thus, equation (7.36) takes the form

$$\begin{aligned} \mathcal{L}_p^{(n)}(\bar{x}, \bar{y}|\bar{z}) \Big|_{x_p=y_p} &= g(y_p, \bar{x})g(y_p, \bar{z})\mathcal{Y}_\kappa(y_p|\{y_p, \bar{x}_p, \bar{z}\})\Delta'(\bar{y}_p)\Delta(\bar{x}_p)g(\bar{y}_p, \bar{z}') \\ &\quad \times \det_{p-1}\Big(g(y_k, x_j)\mathcal{Y}_\kappa(y_k|\{y_k, \bar{x}_{jp}, \bar{z}'\})\Big) + \text{Reg}. \end{aligned} \tag{7.40}$$

The combination in the second line of (7.40) gives us $\mathcal{L}_{p-1}^{(n)}(\bar{x}_p, \bar{y}_p|\bar{z}')$, and we arrive at recursion

$$\mathcal{L}_p^{(n)}(\bar{x}, \bar{y}|\bar{z}) \Big|_{x_p=y_p} = g(y_p, \bar{x})g(y_p, \bar{z})\mathcal{Y}_\kappa(y_p|\{y_p, \bar{x}_p, \bar{z}\})\mathcal{L}_{p-1}^{(n)}(\bar{x}_p, \bar{y}_p|\bar{z}') + \text{Reg}. \tag{7.41}$$

Other residues at $x_p = y_l$ can be restored by the symmetry. Thus, equation (7.35) takes the form

$$\mathcal{L}_p^{(n)}(\bar{x}, \bar{y}|\bar{z}) = \sum_{l=1}^{p} g(y_l, \bar{x}) g(y_l, \bar{z}) \mathcal{Y}_\kappa(y_l|\{y_l, \bar{x}_p, \bar{z}\}) \mathcal{L}_{p-1}^{(n)}(\bar{x}_p, \bar{y}_l|\{\bar{z}, y_l\}). \tag{7.42}$$

The recursion obtained makes it easy to prove Proposition 7.2.

**Proof of Proposition 7.2.** We use induction over $p$. We have for $p = 1$,

$$\mathcal{L}_1^{(n)}(x, y|\bar{z}) = g(y, \bar{z}) g(y, x) \mathcal{Y}_\kappa(y|\{y, \bar{z}\}). \tag{7.43}$$

Substituting here $y = u_k$, and $\bar{z} = \bar{u}_k$ we reproduce the completely indirect action in equation (7.2). Thus, the statement of the proposition holds for $p = 1$.

Assume that it is valid for some $p-1$ with $p \geq 2$. Let $\bar{x} = \{x_1, \dots, x_p\}$. Consider the action of $\mathcal{T}_\kappa(\bar{x})$ on the Bethe vector. First, we act with $\mathcal{T}_\kappa(\bar{x}_p)$ using the induction assumption

$$\mathcal{T}_\kappa(\bar{x}_p) B(\bar{u})|0\rangle \xrightarrow{\text{c.-indir}} \sum_{\substack{\bar{u} \mapsto \{\bar{u}_{\rm I}, \bar{u}_{\rm II}\} \\ \#\bar{u}_{\rm I} = p-1}} \mathcal{L}_{p-1}^{(n)}(\bar{x}_p, \bar{u}_{\rm I}|\bar{u}_{\rm II}) B(\bar{x}_p) B(\bar{u}_{\rm II})|0\rangle. \tag{7.44}$$

Now we should apply $\mathcal{T}_\kappa(x_p)$ to this equation. It is important to emphasize that when we act with $\mathcal{T}_\kappa(x_p)$ on the vector $B(\bar{x}_p) B(\bar{u}_{\rm II})|0\rangle$ via (7.2), we are only interested in the completely indirect action of all the product $\mathcal{T}_\kappa(\bar{x})$. Therefore, we select only those terms in which $x_p$ replaces one of the parameters in the subset $\bar{u}_{\rm II}$:

$$\begin{aligned} &\mathcal{T}_\kappa(x_p) B(\bar{x}_p) B(\bar{u}_{\rm II})|0\rangle \\ &\quad = \sum_{\substack{\bar{u}_{\rm II} \mapsto \{\bar{u}_{\rm i}, \bar{u}_{\rm ii}\} \\ \#\bar{u}_{\rm i} = 1}} g(\bar{u}_{\rm i}, \bar{x}) g(\bar{u}_{\rm i}, \bar{u}_{\rm ii}) \mathcal{Y}_\kappa(\bar{u}_{\rm i}|\{\bar{x}_p, \bar{u}_{\rm II}\}) B(\bar{x}) B(\bar{u}_{\rm ii})|0\rangle + \mathfrak{Z}, \end{aligned} \tag{7.45}$$

where $\mathfrak{Z}$ does not contribute in the completely indirect action of $\mathcal{T}_\kappa(\bar{x})$. The sum is taken over partitions of the set $\bar{u}_{\rm II}$ into two subsets $\{\bar{u}_{\rm i}, \bar{u}_{\rm ii}\}$ such that $\#\bar{u}_{\rm i} = 1$. Combining equations (7.44) and (7.45) we obtain

$$\begin{aligned} \mathcal{T}_\kappa(\bar{x}) B(\bar{u})|0\rangle \xrightarrow{\text{c.-indir}} &\sum_{\substack{\bar{u} \mapsto \{\bar{u}_{\rm I}, \bar{u}_{\rm II}\} \\ \#\bar{u}_{\rm I} = p-1}} \mathcal{L}_{p-1}^{(n)}(\bar{x}_p, \bar{u}_{\rm I}|\bar{u}_{\rm II}) \sum_{\substack{\bar{u}_{\rm II} \mapsto \{\bar{u}_{\rm i}, \bar{u}_{\rm ii}\} \\ \#\bar{u}_{\rm i} = 1}} \\ &\times g(\bar{u}_{\rm i}, \bar{x}) g(\bar{u}_{\rm i}, \bar{u}_{\rm ii}) \mathcal{Y}_\kappa(\bar{u}_{\rm i}|\{\bar{x}_p, \bar{u}_{\rm II}\}) B(\bar{x}) B(\bar{u}_{\rm ii})|0\rangle. \end{aligned} \tag{7.46}$$

Here the sum actually is taken over partitions of $\bar{u}$ into three subsets

$$\begin{aligned} \mathcal{T}_\kappa(\bar{x}) B(\bar{u})|0\rangle \xrightarrow{\text{c.-indir}} &\sum_{\substack{\bar{u} \mapsto \{\bar{u}_{\rm I}, \bar{u}_{\rm i}, \bar{u}_{\rm ii}\} \\ \#\bar{u}_{\rm I} = p-1,\ \#\bar{u}_{\rm i} = 1}} g(\bar{u}_{\rm i}, \bar{x}) g(\bar{u}_{\rm i}, \bar{u}_{\rm ii}) \mathcal{Y}_\kappa(\bar{u}_{\rm i}|\{\bar{x}_p, \bar{u}_{\rm i}, \bar{u}_{\rm ii}\}) \\ &\times \mathcal{L}_{p-1}^{(n)}(\bar{x}_p, \bar{u}_{\rm I}|\{\bar{u}_{\rm i}, \bar{u}_{\rm ii}\}) B(\bar{x}) B(\bar{u}_{\rm ii})|0\rangle. \end{aligned} \tag{7.47}$$

We define now $\bar{u}_0 = \{\bar{u}_{\rm i}, \bar{u}_{\rm I}\}$. Then we recast (7.47) as follows:

$$\mathcal{T}_\kappa(\bar{x})B(\bar{u})|0\rangle \xrightarrow{\text{c.}-\text{indir}} \sum_{\substack{\bar{u}\mapsto\{\bar{u}_0,\bar{u}_{\rm ii}\}\\ \#\bar{u}_0=p}} B(\bar{x})B(\bar{u}_{\rm ii})|0\rangle$$
$$\times \sum_{\substack{\bar{u}_0\mapsto\{\bar{u}_{\rm I},\bar{u}_{\rm i}\}\\ \#\bar{u}_{\rm i}=1}} g(\bar{u}_{\rm i}, \bar{x})g(\bar{u}_{\rm i}, \bar{u}_{\rm ii})\mathcal{Y}_\kappa(\bar{u}_{\rm i}|\{\bar{x}_p, \bar{u}_{\rm i}, \bar{u}_{\rm ii}\})\mathcal{L}^{(n)}_{p-1}(\bar{x}_p, \bar{u}_{\rm I}|\{\bar{u}_{\rm i}, \bar{u}_{\rm ii}\}). \tag{7.48}$$

It remains to compare the expression in the second line of (7.48) with recursion (7.42) for $\mathcal{L}^{(n)}_p(\bar{x}, \bar{y}|\bar{z})$. We immediately arrive at

$$\sum_{\substack{\bar{u}_0\mapsto\{\bar{u}_{\rm I},\bar{u}_{\rm i}\}\\ \#\bar{u}_{\rm i}=1}} g(\bar{u}_{\rm i}, \bar{x})g(\bar{u}_{\rm i}, \bar{u}_{\rm ii})\mathcal{Y}_\kappa(\bar{u}_{\rm i}|\{\bar{x}_p, \bar{u}_{\rm i}, \bar{u}_{\rm ii}\})\mathcal{L}^{(n)}_{p-1}(\bar{x}_p, \bar{u}_{\rm I}|\{\bar{u}_{\rm i}, \bar{u}_{\rm ii}\})$$
$$= \mathcal{L}^{(n)}_p(\bar{x}, \bar{u}_0|\bar{u}_{\rm ii}). \tag{7.49}$$

Thus, equation (7.27) holds for the action of $\mathcal{T}_\kappa(\bar{x})$, and we complete the induction step. □

In the particular case $p = n$ we obtain

$$\mathcal{I}^{(n)}_n(\bar{u}|\bar{x}) = \Delta'(\bar{u})\Delta(\bar{x}) \det_n\left(\frac{\eta}{g(u_k, \bar{x})}\frac{\partial\tau_\kappa(u_k|\bar{x})}{\partial x_j}\right). \tag{7.50}$$

Replacing here $\bar{x}$ with $\bar{v}$ and using (7.9) we find

$$\langle 0|C(\bar{v})B(\bar{u})|0\rangle = d(\bar{v})\Delta'(\bar{u})\Delta(\bar{v}) \det_n\left(\frac{\eta}{g(u_k, \bar{v})}\frac{\partial\tau_\kappa(u_k|\bar{v})}{\partial v_j}\right). \tag{7.51}$$

Recall that here the dual vector $\langle 0|C(\bar{v})$ is the twisted on-shell Bethe vector, while $B(\bar{u})|0\rangle$ is the off-shell Bethe vector. Dividing equation (7.51) by the product $d(\bar{v})d(\bar{u})$, we reproduce determinant representation[4] (6.38).

### 7.1.3 *Multiple action of transfer matrices*

We now calculate the total multiple action of the twisted transfer matrices on the off-shell Bethe vector. To do this, we rewrite the equation (7.4) setting in it $\bar{w}_{\rm I} = \{\bar{u}_{\rm I}, \bar{x}_{\rm II}\}$ and $\bar{w}_{\rm II} = \{\bar{u}_{\rm II}, \bar{x}_{\rm I}\}$:

$$\mathcal{T}_\kappa(\bar{x})B(\bar{u})|0\rangle = \sum_{\substack{\bar{u}\mapsto\{\bar{u}_{\rm I},\bar{u}_{\rm II}\}\\ \bar{x}\mapsto\{\bar{x}_{\rm I},\bar{x}_{\rm II}\}\\ \#\bar{x}_{\rm II}+\#\bar{u}_{\rm I}=p}} \mathcal{A}^{(n)}_p(\{\bar{u}_{\rm I}, \bar{x}_{\rm II}\}|\{\bar{u}_{\rm II}, \bar{x}_{\rm I}\})B(\bar{x}_{\rm I})B(\bar{u}_{\rm II})|0\rangle. \tag{7.52}$$

[4] Recall that $\Delta'(\bar{u})\Delta(\bar{v}) = \Delta(\bar{u})\Delta'(\bar{v})$.

In each term of the sum, the initial parameters of the Bethe vector $\bar{u}_{\rm I}$ are replaced by the arguments of the transfer matrices $\bar{x}_{\rm I}$. Therefore, if $\#\bar{x}_{\rm I} = m$ with $m = 0, \dots, p$, then $\#\bar{u}_{\rm I} = m$. Respectively, $\#\bar{x}_{\rm II} = p - m$ and $\#\bar{u}_{\rm II} = n - m$.

Let us fix a partition in (7.52) and find the coefficient of the vector $B(\bar{x}_{\rm I})B(\bar{u}_{\rm II})|0\rangle$. For this, we reorder the product of the transfer matrices as follows: $\mathcal{T}_\kappa(\bar{x}) = \mathcal{T}_\kappa(\bar{x}_{\rm II})\mathcal{T}_\kappa(\bar{x}_{\rm I})$. Then the action of $\mathcal{T}_\kappa(\bar{x}_{\rm I})$ should be completely indirect because all $\bar{x}_{\rm I}$ must become the arguments of the resulting Bethe vector. Thus,

$$\mathcal{T}_\kappa(\bar{x}_{\rm I})B(\bar{u})|0\rangle = \sum_{\substack{\bar{u}\mapsto\{\bar{u}_{\rm i},\bar{u}_{\rm ii}\}\\ \#\bar{u}_{\rm i}=m}} \mathcal{I}_m^{(n)}(\bar{u}_{\rm i}|\{\bar{u}_{\rm ii}, \bar{x}_{\rm I}\})B(\bar{x}_{\rm I})B(\bar{u}_{\rm ii})|0\rangle + \mathfrak{Z}, \tag{7.53}$$

where $\mathfrak{Z}$ does not contribute to the desired coefficient. The sum is taken over partitions $\bar{u} \mapsto \{\bar{u}_{\rm i}, \bar{u}_{\rm ii}\}$ such that $\#\bar{u}_{\rm i} = m = \bar{x}_{\rm I}$. The action of the $\mathcal{T}_\kappa(\bar{x}_{\rm II})$ operators on $B(\bar{x}_{\rm I})B(\bar{u}_{\rm ii})|0\rangle$ either leaves the subset $\bar{u}_{\rm ii}$ unchanged or leads to the replacement of some variables from this subset by the arguments of the twisted transfer matrices. Therefore, the only contribution to the expected result is given by the partition $\bar{u}_{\rm i} = \bar{u}_{\rm I}$ and $\bar{u}_{\rm ii} = \bar{u}_{\rm II}$

$$\mathcal{T}_\kappa(\bar{x}_{\rm I})B(\bar{u})|0\rangle = \mathcal{I}_m^{(n)}(\bar{u}_{\rm I}|\{\bar{u}_{\rm II}, \bar{x}_{\rm I}\})B(\bar{x}_{\rm I})B(\bar{u}_{\rm II})|0\rangle + \mathfrak{Z}. \tag{7.54}$$

After this, when acting with the $\mathcal{T}_\kappa(\bar{x}_{\rm II})$ on $B(\bar{x}_{\rm I})B(\bar{u}_{\rm II})|0\rangle$, we should leave this vector unchanged. Hence, this action must be direct

$$\begin{aligned}\mathcal{T}_\kappa(\bar{x}_{\rm II})B(\bar{x}_{\rm I})B(\bar{u}_{\rm II})|0\rangle = \mathcal{D}_{p-m}^{(n)}(\bar{x}_{\rm II}|\{\bar{u}_{\rm II}, \bar{x}_{\rm I}\})\mathcal{I}_m^{(n)}(\bar{u}_{\rm I}|\{\bar{u}_{\rm II}, \bar{x}_{\rm I}\})\\ \times B(\bar{x}_{\rm I})B(\bar{u}_{\rm II})|0\rangle + \mathfrak{Z}.\end{aligned} \tag{7.55}$$

Thus, we find

$$\mathcal{A}_p^{(n)}(\{\bar{u}_{\rm I}, \bar{x}_{\rm II}\}|\{\bar{u}_{\rm II}, \bar{x}_{\rm I}\}) = \mathcal{D}_{p-m}^{(n)}(\bar{x}_{\rm II}|\{\bar{u}_{\rm II}, \bar{x}_{\rm I}\})\mathcal{I}_m^{(n)}(\bar{u}_{\rm I}|\{\bar{u}_{\rm II}, \bar{x}_{\rm I}\}), \tag{7.56}$$

where $m = \#\bar{u}_{\rm I} = \#\bar{x}_{\rm I}$. Taking into account (7.6) and (7.28), we have

$$\mathcal{A}_p^{(n)}(\{\bar{u}_{\rm I}, \bar{x}_{\rm II}\}|\{\bar{u}_{\rm II}, \bar{x}_{\rm I}\}) = \left(\prod_{x_j\in\bar{x}_{\rm II}} \tau_\kappa(x_j|\{\bar{u}_{\rm II}, \bar{x}_{\rm I}\})\right) \mathcal{L}_m^{(n)}(\bar{x}_{\rm I}, \bar{u}_{\rm I}|\bar{u}_{\rm II}), \tag{7.57}$$

where $\mathcal{L}_m^{(n)}$ is defined by (7.26).

**Remark 7.2.** So far, we have considered the case $p \le n$, that is, $\#\bar{x} \le \#\bar{u}$. However, it is easy to see that nothing changes in the above consideration if $\#\bar{x} > \#\bar{u}$ and we add a restriction $m \le \min(p, n)$ to the cardinalities of the subsets $\bar{x}_{\rm I}$ and $\bar{u}_{\rm I}$. Therefore, the formula (7.57) for the coefficient $\mathcal{A}_p^{(n)}(\{\bar{u}_{\rm I}, \bar{x}_{\rm II}\}|\{\bar{u}_{\rm II}, \bar{x}_{\rm I}\})$ is also valid in the case $p > n$.

Let us summarize. We showed that the scalar product of the twisted on-shell Bethe vector and the off-shell Bethe vector coincides with the completely indirect action of the twisted transfer matrix up to a normalization factor. This fact gives us a new way to calculate such scalar products. We also found the multiple action of the twisted transfer matrix on Bethe vectors. We will use this result when calculating correlation functions.

Finally, we note one more result that we got. For $\kappa = 0$, the twisted transfer matrix turns into the operator $A$. Thus, setting $\kappa = 0$ in (7.57), we obtain a formula for the action of the product of the operators $A(x_j)$ on the Bethe vector $B(\bar u)|0\rangle$. It is easy to see that

$$\begin{aligned}\mathcal{A}_p^{(n)}(\{\bar u_{\rm I}, \bar x_{\rm II}\}|\{\bar u_{\rm II}, \bar x_{\rm I}\})\Big|_{\kappa=0} &= a(\bar u_{\rm I})a(\bar x_{\rm II})f(\bar u_{\rm II}, \bar u_{\rm I})\\ &\quad\times f(\bar u_{\rm II}, \bar x_{\rm II})f(\bar x_{\rm I}, \bar x_{\rm II})\mathbb{K}_m(\bar x_{\rm I}|\bar u_{\rm I}).\end{aligned} \tag{7.58}$$

This formula can be written more compact if we go back to the sets $\bar w_{\rm I} = \{\bar u_{\rm I}, \bar x_{\rm II}\}$ and $\bar w_{\rm II} = \{\bar u_{\rm II}, \bar x_{\rm I}\}$ used in (7.4). Then

$$\mathcal{A}_p^{(n)}(\bar w_{\rm I}|\bar w_{\rm II})\Big|_{\kappa=0} = (-1)^n a(\bar w_{\rm I})f(\bar w_{\rm II}, \bar w_{\rm I})\mathbb{K}_n(\bar u|\bar w_{\rm II} + \eta). \tag{7.59}$$

Here $\bar w = \{\bar x, \bar u\}$, and we used properties (5.48), (5.49) of the highest coefficient. Then the action of the product $A(\bar x)$ reads

$$A(\bar x)B(\bar u)|0\rangle = (-1)^n \sum_{\substack{\bar w \mapsto \{\bar w_{\rm I}, \bar w_{\rm II}\}\\ \#\bar w_{\rm II} = n}} a(\bar w_{\rm I})\mathbb{K}_n(\bar u|\bar w_{\rm II} + \eta)f(\bar w_{\rm II}, \bar w_{\rm I})B(\bar w_{\rm II})|0\rangle. \tag{7.60}$$

The sum is taken over partitions $\bar w \mapsto \{\bar w_{\rm I}, \bar w_{\rm II}\}$ such that $\#\bar w_{\rm II} = n$. We invite the reader to check that for $\#\bar x = 1$, this formula reproduces the well-known action of the operator $A(x)$ on the Bethe vector obtained in Section 2.1.2.

Similarly, in the limit $\kappa \to \infty$, we obtain the multiple action of the operators $D(x_j)$:

$$D(\bar x)B(\bar u)|0\rangle = (-1)^n \sum_{\substack{\bar w \mapsto \{\bar w_{\rm I}, \bar w_{\rm II}\}\\ \#\bar w_{\rm II} = n}} d(\bar w_{\rm I})\mathbb{K}_n(\bar w_{\rm II}|\bar u + \eta)f(\bar w_{\rm I}, \bar w_{\rm II})B(\bar w_{\rm II})|0\rangle. \tag{7.61}$$

Actually, equations (7.60) and (7.61) follow from the multiple commutation relations

$$A(\bar x)B(\bar u) = (-1)^n \sum_{\substack{\bar w \mapsto \{\bar w_{\rm I}, \bar w_{\rm II}\}\\ \#\bar w_{\rm II} = n}} \mathbb{K}_n(\bar u|\bar w_{\rm II} + \eta)f(\bar w_{\rm II}, \bar w_{\rm I})B(\bar w_{\rm II})A(\bar w_{\rm I}), \tag{7.62}$$

and

$$D(\bar{x})B(\bar{u}) = (-1)^n \sum_{\substack{\bar{w}\mapsto\{\bar{w}_{\rm I},\bar{w}_{\rm II}\} \\ \#\bar{w}_{\rm II}=n}} \mathbb{K}_n(\bar{w}_{\rm II}|\bar{u}+\eta)f(\bar{w}_{\rm I},\bar{w}_{\rm II})B(\bar{w}_{\rm II})D(\bar{w}_{\rm I}). \quad (7.63)$$

Here $\bar{w} = \{\bar{x},\bar{u}\}$, $\#\bar{x} = p$, $\#\bar{u} = n$, and the sum is taken over partitions $\bar{w} \mapsto \{\bar{w}_{\rm I},\bar{w}_{\rm II}\}$ such that $\#\bar{w}_{\rm II} = n$. These multiple action formulas can be proved using double induction in $n$ and $p$ and identity (5.51) (see [N. A. Slavnov (2017); S. Belliard et al (2013a)]). Note also that formula (7.63) is obtained from (7.62) by the formal replacements $A \to D$ and $\eta \to -\eta$. This follows from the fact that the commutation relations of $D$ and $B$ (1.130) can be obtained from those of $A$ and $B$ (1.123) by the same replacements.

## 7.2 System of linear equations for scalar products

This section describes one more way to calculate scalar products of (twisted) on-shell and off-shell Bethe vectors. The basis of this method is the formula for the action of the twisted transfer matrix on the Bethe vector (7.2).

Let $\bar{u} = \{u_1,\dots,u_{n+1}\}$ be a set of arbitrary complex numbers. We draw the reader's attention that if in the previous sections the set $\bar{u}$ consisted of $n$ elements, then now it has $n+1$ elements. Let us fix some $u_j \in \bar{u}$ and write down the action of $\mathcal{T}_\kappa(u_j)$ on the vector $B(\bar{u}_j)|0\rangle$ (recall that $\bar{u}_j = \bar{u}\setminus u_j$). Then (7.2) takes the form

$$\mathcal{T}_\kappa(u_j)B(\bar{u}_j)|0\rangle = \sum_{k=1}^{n+1} g(u_k,\bar{u}_k)\mathcal{Y}_\kappa(u_k|\bar{u}_j)B(\bar{u}_k)|0\rangle. \quad (7.64)$$

Indeed, we have for $k = j$,

$$g(u_k,\bar{u}_k)\mathcal{Y}_\kappa(u_k|\bar{u}_j)B(\bar{u}_k)|0\rangle\Big|_{k=j} = \tau_\kappa(u_j|\bar{u}_j)B(\bar{u}_j)|0\rangle, \quad (7.65)$$

and we reproduce the first term in (7.2). At the same time, for $k \ne j$,

$$\begin{aligned} &g(u_k,\bar{u}_k)\mathcal{Y}_\kappa(u_k|\bar{u}_j)B(\bar{u}_k)|0\rangle\Big|_{k\ne j} \\ &\quad = g(u_k,u_j)g(u_k,\bar{u}_{kj})\mathcal{Y}_\kappa(u_k|\bar{u}_j)B(u_j)B(\bar{u}_{kj})|0\rangle, \end{aligned} \quad (7.66)$$

where $\bar{u}_{kj} = \bar{u}\setminus\{u_k,u_j\}$. We thus obtain the indirect action in (7.2).

Consider scalar products of off-shell Bethe vectors $B(\bar{u}_k)|0\rangle$ with a dual twisted on-shell Bethe vector $\langle 0|C(\bar{v})$, where $\bar{v} = \{v_1,\dots,v_n\}$. Let

$$\langle 0|C(\bar{v})B(\bar{u}_k)|0\rangle = x_k, \qquad k = 1,\dots,n+1. \quad (7.67)$$

Multiplying (7.64) from the left by $\langle 0|C(\bar{v})$, we obtain

$$\langle 0|C(\bar{v})\mathcal{T}_\kappa(u_j)B(\bar{u}_j)|0\rangle = \sum_{k=1}^{n+1} g(u_k,\bar{u}_k)\mathcal{Y}_\kappa(u_k|\bar{u}_j)x_k. \tag{7.68}$$

On the other hand, the action of the twisted transfer matrix on the twisted on-shell vector $\langle 0|C(\bar{v})$ gives eigenvalue $\tau_\kappa(u_j|\bar{v})$ multiplied by the same vector. We arrive at

$$\tau_\kappa(u_j|\bar{v})x_j = \sum_{k=1}^{n+1} g(u_k,\bar{u}_k)\mathcal{Y}_\kappa(u_k|\bar{u}_j)x_k. \tag{7.69}$$

Thus, we obtain a system of homogeneous linear equations for the scalar products $x_k$

$$\sum_{k=1}^{n+1} V_{jk}x_k = 0, \qquad j = 1,\dots,n+1, \tag{7.70}$$

where

$$V_{jk} = g(u_k,\bar{u}_k)\mathcal{Y}_\kappa(u_k|\bar{u}_j) - \delta_{jk}\tau_\kappa(u_j|\bar{v}), \qquad j,k = 1,\dots,n+1. \tag{7.71}$$

So, we have reduced the scalar product calculation to the solution of a system of linear equations, that is, to the well-studied problem of linear algebra. However, the system is homogeneous, which means that its solutions contain arbitrariness. The latter has yet to be fixed in some way. Besides, we do not yet know whether this system has solutions other than trivial ones. Nevertheless, despite all the unsolved problems listed above, we have already greatly simplified our task.

Note that we did not use the explicit form of the numerical coefficients in formula (7.2). Therefore, as soon as the action of the twisted transfer matrix on the Bethe vector is given by an expression of the form

$$\mathcal{T}_\kappa(z)B(\bar{u})|0\rangle = \alpha_0(z|\bar{u})B(\bar{u})|0\rangle + \sum_{k=1}^{n} \alpha_k(z|\bar{u})B(z)B(\bar{u}_k)|0\rangle, \tag{7.72}$$

we immediately obtain system (7.70). In this derivation, the explicit form of the functions $\alpha_s(z|\bar{u})$, $s = 0,\dots,n$, does not play any role. In turn, the action formula (7.72) is valid not only for the models discussed in this book but also, for example, in quantum integrable models with open boundaries. In these models, one can obtain a system of linear equations for scalar products in the same way. Therefore, the range of applicability of this method is vast. In this regard, it is noteworthy that such a simple method was found relatively recently [S. Belliard and N.A. Slavnov (2019b)].

It is also worth mentioning that one can derive the system (7.70) when the number of variables in the set $\bar{v}$ is not equal to $n$. Moreover, the matrix of the system still has the form (7.71). However, in this case, the system has a full rank. Therefore, it has only trivial solutions, as it should be. For example, in the case $n = 0$, the system turns into one equation

$$\big(a(u_1) + \kappa d(u_1) - \tau_\kappa(u_1|\bar{v})\big)x_1 = 0. \tag{7.73}$$

Since the expression in the parenthesis is non-vanishing (except the case $\bar{v} = \emptyset$), equation (7.73) yields $x_1 = 0$.

To study the system (7.70), we need a special matrix called the Cauchy matrix. In Chapter 12, the determinant of this matrix will play an important role in calculating the thermodynamic limit of the form factors. Therefore, before proceeding to the analysis of the system (7.70), we will stop and describe some properties of the Cauchy matrix.

### 7.2.1 *Cauchy matrix*

A matrix $\mathfrak{C}$ with the elements

$$\mathfrak{C}_{jk} = \frac{1}{x_j - y_k}, \tag{7.74}$$

is called the *Cauchy matrix*, and its determinant is called the *Cauchy determinant*. The latter is explicitly calculated.

**Proposition 7.3.** *For arbitrary complex $x_1, \dots, x_n$ and $y_1, \dots, y_n$,*

$$\det_n\left(\frac{1}{x_j - y_k}\right) = \frac{\prod_{j<k}^n (x_j - x_k)(y_k - y_j)}{\prod_{j,k=1}^n (x_j - y_k)}. \tag{7.75}$$

**Proof.** There are several relatively simple ways to prove this formula. One can, for example, use induction on $n$. One can also prove (7.75) almost without calculations. The lhs is a rational function of the variables $x_1, \dots, x_n$ and $y_1, \dots, y_n$, which has poles at the points $x_j = y_k$, $j, k = 1, \dots, n$. If $x_j = x_k$ (or $y_j = y_k$), then there are two coinciding rows in the determinant (or two columns). So the determinant has zeros at these points. If some $x_\ell$ tends to infinity while the remaining variables are fixed, then the determinant goes to zero because the $\ell$th row in the matrix tends to zero. A similar property holds if some $y_\ell \to \infty$. The listed properties are sufficient to write down the determinant in the form

$$\det_n\left(\frac{1}{x_j - y_k}\right) = C\frac{\prod_{j<k}^n (x_j - x_k)(y_k - y_j)}{\prod_{j,k=1}^n (x_j - y_k)}, \tag{7.76}$$

where a constant $C$ does not depend on either $x_1, \dots, x_n$, or $y_1, \dots, y_n$. Indeed, the poles of the determinant uniquely fix the denominator in the rhs of (7.76). The zeros at $x_j = x_k$ and $y_j = y_k$ give the numerator. It is easy to show that the determinant cannot have other zeros. Indeed, suppose that the determinant has an additional zero at $x_\ell = x_0$, where $x_\ell$ is an arbitrary element of the set $x_1, \dots, x_n$, while $x_0$ does not belong to this set. Then there would be an additional factor $(x_\ell - x_0)$ in the rhs of (7.76). In this case, the rhs does not tend to zero as $x_\ell \to \infty$. Similar considerations lead us to conclude that the determinant has no additional zeros for the parameters $y_1, \dots, y_n$. Hence, the rational functions in both sides of the formula (7.76) can differ from each other only by a constant that does not depend on either $x_1, \dots, x_n$, or $y_1, \dots, y_n$. In order to find this constant, it is sufficient to consider a special case $x_j = y_j + \varepsilon$, $j = 1, \dots, n$, and calculate the coefficients of the maximal degree $1/\varepsilon^n$. In the lhs, this coefficient is 1 (it occurs when taking the product of diagonal elements). It is easy to check that the coefficient of $1/\varepsilon^n$ in the rhs is equal to $C$. Hence, $C = 1$. □

**Corollary 7.1.** *For arbitrary $x_1, \dots, x_n$ and $y_1, \dots, y_n$,*

$$\det_n \left( \frac{1}{\sinh(x_j - y_k)} \right) = \frac{\prod_{j<k}^n \sinh(x_j - x_k) \sinh(y_k - y_j)}{\prod_{j,k=1}^n \sinh(x_j - y_k)}. \tag{7.77}$$

**Proof.** It suffices to make a change of variables $u_j = e^{2x_j}$ and $v_j = e^{2y_j}$. Then in the new variables, the determinant in (7.77) reduces to the Cauchy determinant. It remains to use (7.75) and return to the original variables. □

Both equations (7.75) and (7.77) can be written in terms of the $g$-function using our shorthand notation

$$\det_n \big(g(x_j, y_k)\big) = \frac{g(\bar{x}, \bar{y})}{\Delta'(\bar{x})\Delta(\bar{y})}. \tag{7.78}$$

It is remarkable that any minor of the matrix $g(x_j, y_k)$ also reduces to the Cauchy determinant:

$$\det_{j\neq\ell;\ k\neq m} \big(g(x_j, y_k)\big) = \frac{g(\bar{x}_\ell, \bar{y}_m)}{\Delta'(\bar{x}_\ell)\Delta(\bar{y}_m)}. \tag{7.79}$$

Therefore, we can easily calculate the inverse matrix.

**Proposition 7.4.** *Let*

$$\mathfrak{C}_{jk} = g(x_j, y_k). \tag{7.80}$$

*Then*

$$(\mathfrak{C}^{-1})_{jk} = g(x_k, y_j)\frac{g(x_k, \bar{x}_k)g(\bar{y}_j, y_j)}{g(x_k, \bar{y})g(\bar{x}, y_j)}. \tag{7.81}$$

**Proof.** As we mentioned, we can calculate the elements $(\mathfrak{C}^{-1})_{jk}$ of the inverse matrix by knowing the minors (7.79). However, we can directly prove that

$$\sum_{\ell=1}^{n}(\mathfrak{C}^{-1})_{j\ell}\, g(x_\ell, y_k) = \delta_{jk}. \tag{7.82}$$

Consider for definiteness the case of the rational $R$-matrix. Then

$$\sum_{\ell=1}^{n}(\mathfrak{C}^{-1})_{j\ell}\, g(x_\ell, y_k) = G_{jk}\frac{g(\bar{y}_j, y_j)}{g(\bar{x}, y_j)}, \tag{7.83}$$

where

$$\begin{aligned} G_{jk} &= \sum_{\ell=1}^{n} g(x_\ell, y_j)g(x_\ell, y_k)\frac{g(x_\ell, \bar{x}_\ell)}{g(x_\ell, \bar{y})} \\ &= \sum_{\ell=1}^{n}\frac{\eta}{(x_\ell - y_j)(x_\ell - y_k)}\frac{\prod_{m=1}^{n}(x_\ell - y_m)}{\prod_{m=1;\, m\neq\ell}^{n}(x_\ell - x_m)}. \end{aligned} \tag{7.84}$$

To compute $G_{jk}$, we consider an auxiliary contour integral

$$I = \frac{1}{2\pi i}\oint_{|z|=R\to\infty}\frac{\eta\, dz}{(z - y_j)(z - y_k)}\prod_{m=1}^{n}\frac{z - y_m}{z - x_m}. \tag{7.85}$$

The integral is taken over a circle of a radius $R \to \infty$ in the anticlockwise direction. The integrand behaves as $z^{-2}$ at $z \to \infty$. Hence, $I = 0$. On the other hand, the integral (7.85) equals the sum of residues in the poles within the integration contour. First, these are poles at $z = x_\ell$, $\ell = 1, \dots, n$. It is easy to see that the sum of the residues in these poles gives exactly $G_{jk}$. There are also two additional poles in the points $z = y_j$ and $z = y_k$. However, for $j \neq k$, these poles are compensated by the product of $(z - y_m)$. Only for $j = k$, we get the second order pole, while the product of $(z - y_m)$ has the first order zero only. So we get

$$I = 0 = G_{jk} + \eta\delta_{jk}\frac{\prod_{m=1;\, m\neq j}^{n}(y_j - y_m)}{\prod_{m=1}^{n}(y_j - x_m)}, \tag{7.86}$$

which implies

$$G_{jk} = \eta\delta_{jk}\frac{\prod_{m=1;\, m\neq j}^{n}(y_m - y_j)}{\prod_{m=1}^{n}(x_m - y_j)} = \delta_{jk}\frac{g(\bar{x}, y_j)}{g(\bar{y}_j, y_j)}. \tag{7.87}$$

Substituting this $G_{jk}$ in (7.83), we prove (7.82). □

**Remark 7.3.** In the case of the trigonometric $R$-matrix, the proof is also based on some auxiliary contour integral. We invite the reader to consider this case independently.

### 7.2.2 *Transformation of the system of linear equations*

We now proceed to the analysis of the system (7.70). We expect that in the case $\#\bar{v} = n$, this system has non-trivial solutions. Therefore, the condition $\operatorname{rank}(V) < n+1$ must be satisfied. To check this condition, it suffices to find a linear combination of rows of the matrix $V$, which vanishes. It is easy to do, but we will do it differently. We will try to transform the system to limit the ambiguity in its solution as much as possible. To do this, we multiply the matrix $V$ of the system (7.70) from the left by a matrix $W$ with elements

$$W_{jk} = g(u_k, w_j)\frac{g(u_k, \bar{u}_k)}{g(u_k, \bar{w})}. \tag{7.88}$$

Here $\bar{w} = \{w_1, \dots, w_{n+1}\}$ is a set of generic pairwise distinct complex numbers. This matrix is proportional to the Cauchy matrix, and hence,

$$\det_{n+1} W = \frac{\Delta(\bar{u})\Delta'(\bar{u})}{g(\bar{u}, \bar{w})} \det g(u_k, w_j) = \frac{\Delta(\bar{u})}{\Delta(\bar{w})}. \tag{7.89}$$

Thus, the matrix $W$ is nondegenerated, and the system

$$\sum_{k=1}^{n+1} \widetilde{V}_{jk} x_k = 0, \qquad \text{with} \qquad \widetilde{V} = WV, \tag{7.90}$$

is equivalent to the original system (7.70). The entries of the new matrix $\widetilde{V}$ are

$$\begin{aligned}\widetilde{V}_{jk} = g(u_k, \bar{u}_k) &\sum_{\ell=1}^{n+1} g(u_\ell, \bar{u}_\ell)\mathcal{Y}_\kappa(u_k|\bar{u}_\ell)\frac{g(u_\ell, w_j)}{g(u_\ell, \bar{w})} \\ &- g(u_k, w_j)\frac{g(u_k, \bar{u}_k)}{g(u_k, \bar{w})}\tau_\kappa(u_k|\bar{v}).\end{aligned} \tag{7.91}$$

The sum over $\ell$ in (7.91) can be computed explicitly via the standard contour integral trick. Indeed, we have

$$\begin{aligned}&\sum_{\ell=1}^{n+1} g(u_\ell, \bar{u}_\ell)\mathcal{Y}_\kappa(u_k|\bar{u}_\ell)\frac{g(u_\ell, w_j)}{g(u_\ell, \bar{w})} \\ &\quad = \sum_{\ell=1}^{n+1} g(u_\ell, \bar{u}_\ell)\frac{g(u_\ell, w_j)}{g(u_\ell, \bar{w})}\left[(-1)^n a(u_k)\frac{h(\bar{u}, u_k)}{h(u_\ell, u_k)} + \kappa d(u_k)\frac{h(u_k, \bar{u})}{h(u_k, u_\ell)}\right].\end{aligned} \tag{7.92}$$

Let us compute, for instance, the coefficient of the function $\kappa d(u_k)$. Consider a contour integral

$$I = \oint_{|z|=R\to\infty} \frac{\mathrm{d}z}{2\pi i\eta} \frac{g(z,w_j)}{h(u_k,z)} \frac{g(z,\bar{u})}{g(z,\bar{w})}. \tag{7.93}$$

This integral vanishes since the integrand behaves as $z^{-2}$ at $z\to\infty$. On the other hand, the integrand has poles in the points $z=u_\ell$, $\ell=1,\dots,n+1$, and $z=u_k+\eta$. Hence, the sum of the residues in these poles must vanish. We obtain

$$\sum_{\ell=1}^{n+1} \frac{g(u_\ell,w_j)}{h(u_k,u_\ell)} \frac{g(u_\ell,\bar{u}_\ell)}{g(u_\ell,\bar{w})} - \frac{1}{h(u_k,w_j)} \frac{h(u_k,\bar{w})}{h(u_k,\bar{u})} = 0. \tag{7.94}$$

Similarly,

$$\sum_{\ell=1}^{n+1} \frac{g(u_\ell,w_j)}{h(u_\ell,u_k)} \frac{g(u_\ell,\bar{u}_\ell)}{g(u_\ell,\bar{w})} - \frac{1}{h(w_j,u_k)} \frac{h(\bar{w},u_k)}{h(\bar{u},u_k)} = 0. \tag{7.95}$$

Using these results, we obtain after simple algebra

$$\widetilde{V}_{jk} = g(u_k,\bar{u}_k)\left\{\mathcal{Y}_\kappa(u_k|\bar{w}_j) - \frac{g(u_k,w_j)}{g(u_k,\bar{w})}\tau_\kappa(u_k|\bar{v})\right\}. \tag{7.96}$$

So far, the set $\bar{w}$ consisted of arbitrary pairwise distinct complex numbers. We now set $w_j=v_j$ for $j=1,\dots,n$. The last parameter $w_{n+1}$ remains free. Then the last row of $\widetilde{V}$ vanishes:

$$\widetilde{V}_{n+1,k} = g(u_k,\bar{u}_k)\left\{\mathcal{Y}_\kappa(u_k|\bar{v}) - \frac{1}{g(u_k,\bar{v})}\tau_\kappa(u_k|\bar{v})\right\} = 0. \tag{7.97}$$

Thus, $\det\widetilde{V}=0$, and hence, $\det V=0$. We thus proved that system (7.70) has non-trivial solutions.

In all other rows of $\widetilde{V}$ with $j\le n$, we obtain after simple algebra

$$\widetilde{V}_{jk} = \frac{g(u_k,v_j)}{g(w_{n+1},v_j)} g(u_k,\bar{u}_k)\mathcal{Y}_\kappa(u_k|\{u_k,\bar{v}_j\}), \qquad j\le n. \tag{7.98}$$

We see that the matrix elements $\widetilde{V}_{jk}$ are proportional to the Jacobian of the transfer matrix eigenvalue (see (7.29)–(7.30)):

$$\widetilde{V}_{jk} = \frac{g(u_k,\bar{u}_k)}{g(w_{n+1},v_j)} \frac{\eta}{g(u_k,\bar{v})} \frac{\partial\tau_\kappa(u_k|\bar{v})}{\partial v_j}. \tag{7.99}$$

Let us introduce an $n\times(n+1)$ matrix $S_{jk}$ with the elements

$$S_{jk} = \frac{\eta}{g(u_k,\bar{v})} \frac{\partial\tau_\kappa(u_k|\bar{v})}{\partial v_j}, \qquad j=1,\dots,n, \quad k=1,\dots,n+1. \tag{7.100}$$

Setting $x_k = y_k\Delta(\bar{u}_k)$, where $y_k$ are new unknown variables, and using obvious relation

$$g(u_k,\bar{u}_k)\Delta(\bar{u}_k) = (-1)^{n+1-k}\Delta(\bar{u}), \tag{7.101}$$

we recast system (7.90) as follows:

$$\frac{\Delta(\bar{u})}{g(w_{n+1},v_j)}\sum_{k=1}^{n+1}(-1)^{n+1-k}S_{jk}y_k = 0, \qquad j=1,\dots,n. \tag{7.102}$$

Assume that $\operatorname{rank}(S) = n$. Then we can write out a formal solution to the system by expressing $y_k$ with $k = 1,\dots,n$ in terms of $y_{n+1}$. We obtain

$$\frac{y_\ell}{\hat{S}_\ell} = \frac{y_{n+1}}{\hat{S}_{n+1}}, \qquad \ell = 1,\dots,n, \tag{7.103}$$

where $\hat{S}_\ell$ is the minor of the matrix $S$ obtained by deleting the $\ell$th column

$$\hat{S}_\ell = \det_n S_{jk}\Big|_{k\ne\ell}. \tag{7.104}$$

We have come to a key place in our analysis. Note that the structure of the matrix elements $S_{jk}$ is such that the parameter $u_\ell$ is included only in the $\ell$th column. Therefore, the minor $\hat{S}_\ell$ is independent of $u_\ell$. In addition, $y_\ell$ is independent of $u_\ell$ since $x_\ell$ and $\Delta(\bar{u}_\ell)$ are independent of this parameter by definition. Therefore, the rhs of (7.104) is independent of $u_{n+1}$. However, the lhs is independent of $u_\ell$, with $\ell$ running over $1,\dots,n$. We conclude that both sides of the relation (7.104) are independent of the parameters from the set $\bar{u}$. Hence,

$$x_\ell = \phi(\bar{v})\Delta(\bar{u}_\ell)\hat{S}_\ell, \qquad \ell = 1,\dots,n+1, \tag{7.105}$$

where $\phi(\bar{v})$ is a function of the parameters $\bar{v}$ only.

We stress that from the very beginning, we could write a formula for the solutions of system (7.70), similar to (7.103)

$$\frac{(-1)^\ell x_\ell}{\hat{V}_\ell} = \frac{(-1)^{n+1}x_{n+1}}{\hat{V}_{n+1}}, \qquad \ell = 1,\dots,n, \tag{7.106}$$

where $\hat{V}_\ell$ are the minors of the matrix $V$ obtained by deleting the $\ell$th column and, say, the $(n+1)$th row. However, the minors $\hat{V}_\ell$ still depend on $u_\ell$. Therefore, relation (7.106) does not give us any information. We simply express $x_\ell$ in terms of the unknown value $x_{n+1}$. In contrast, relation (7.103) completely fixes the dependence of the scalar product on the parameters $\bar{u}$. It is for this that we transformed the original system of equations. Now we only need to find the function $\phi(\bar{v})$.

To find this function, it suffices to consider some particular case of the scalar product in which it can be calculated most simply. For example, we can consider the case when $d(u_j) = 0$ for all $u_j \in \bar{u}$. Then formula (7.105) gives

$$\langle 0|C(\bar{v})B(\bar{u}_\ell)|0\rangle\Big|_{d(u_j)=0} = \frac{\phi(\bar{v})}{\Delta'(\bar{v})}a(\bar{u}_\ell)\mathbb{K}_n(\bar{v}|\bar{u}_\ell), \qquad \ell = 1,\dots,n+1, \tag{7.107}$$

where $\mathbb{K}_n(\bar{v}|\bar{u}_\ell)$ is the highest coefficient. On the other hand, we know that the scalar product reduces to the highest coefficient in this case (see (5.35), (5.22)):

$$\langle 0|C(\bar{v})B(\bar{u}_\ell)|0\rangle\Big|_{d(u_j)=0} = d(\bar{v})a(\bar{u}_\ell)\mathbb{K}_n(\bar{v}|\bar{u}_\ell), \qquad \ell = 1,\dots,n+1. \tag{7.108}$$

Hence,

$$\phi(\bar{v}) = d(\bar{v})\Delta'(\bar{v}). \tag{7.109}$$

So, we ultimately solved the problem of calculating the scalar product of twisted on-shell and off-shell Bethe vectors in the form of a determinant. Note that in this method framework, we mainly dealt with solving a system of linear equations. This is a typical linear algebra task, and, generally speaking, it has nothing to do with the algebraic Bethe ansatz. Only at the last stage we slightly used the $RTT$-relation to find the function $\phi(\bar{v})$. In any case, these calculations are much simpler than those that we performed in Section 5.1 or Section 7.1. Since this method applies to a broader class of models than those considered in this book, we conclude that it has a significant advantage over traditional approaches.

Another advantage of the method described above is that it sheds light on the origin of determinant representations for scalar products. Previously, the existence of such representations seemed to be an accident or even a miracle. Now we see that the determinant arises simply because the scalar products satisfy a system of linear equations. We must also require that the original system be transformed to a special form in which each parameter of the vector $B(\bar{u}_\ell)|0\rangle$ is included in only one column of the system matrix. Then the entire dependence on the parameters $\bar{u}$ is automatically expressed in terms of some determinant.

At the end of this section, we will touch on another issue. We assumed that $\mathrm{rank}(S) = n$ when deriving relation (7.103). One can argue that this is true, at least in models having physical content, in which the transfer matrix is a self-adjoint operator. Indeed, since the matrix elements $S_{jk}$

are analytic functions of the parameters $u_i$, it is sufficient to check that $\operatorname{rank}(S) = n$ at some particular point $\bar{u} = \bar{u}^0$. We can choose for such a specific point, for example, $\bar{u}_{n+1} = \bar{v}$. Then, for $\ell = n+1$, equation (7.105) gives the square of the Hamilton eigenvector norm, which cannot vanish.[5] Thus, $\hat{S}_{n+1} \neq 0$ at least for $\bar{u}_{n+1} = \bar{v}$. Since $\hat{S}_{n+1}$ is an analytic function of any $u_i$, it should be non-vanishing in some neighborhood of $\bar{u}_{n+1} = \bar{v}$. Hence, we can solve our system of equations in this neighborhood and then analytically continue the obtained solution to the whole complex plane of each variable $u_i$.

[5]It can be rigorously proved for some models that the corresponding matrix $S$ is positive definite (see remark 10.1 in Section 10.1).

# Chapter 8

# Form factors of the monodromy matrix elements

This chapter considers the form factors of the monodromy matrix entries $T_{jk}(z)$. Starting from this point, we introduce a special notation for on-shell Bethe vectors. Let

$$\boldsymbol{B}(u) = \frac{B(u)}{d(u)}, \qquad \boldsymbol{C}(u) = \frac{C(u)}{d(u)}. \tag{8.1}$$

Then we define

$$|\boldsymbol{\Phi}_n(\bar{u})\rangle = \boldsymbol{B}(\bar{u})|0\rangle, \qquad \langle\boldsymbol{\Phi}_n(\bar{u})| = \langle 0|\boldsymbol{C}(\bar{u}), \tag{8.2}$$

provided $\#\bar{u} = n$ and the parameters of the set $\bar{u}$ satisfy Bethe equations. We will also permanently deal with twisted on-shell vectors. In this case, we will write an additional superscript $(\kappa)$:

$$|\boldsymbol{\Phi}_n^{(\kappa)}(\bar{u})\rangle = \boldsymbol{B}(\bar{u})|0\rangle, \qquad \langle\boldsymbol{\Phi}_n^{(\kappa)}(\bar{u})| = \langle 0|\boldsymbol{C}(\bar{u}), \tag{8.3}$$

provided the parameters of the set $\bar{u}$ satisfy twisted Bethe equations with the twist parameter $\kappa$.

We call a *form factor* of $T_{jk}(z)$ a matrix element

$$\mathcal{F}_{n',n}^{(jk)}(\bar{v}, \bar{u}, z) = \langle\boldsymbol{\Phi}_{n'}(\bar{v})|T_{jk}(z)|\boldsymbol{\Phi}_n(\bar{u})\rangle. \tag{8.4}$$

Here $z$ is an arbitrary complex number, and both vectors are on-shell Bethe vectors. In some cases, we will also call the matrix element (8.4) a form factor if one or both vectors are twisted on-shell Bethe vectors. The subscripts $n'$ and $n$ of the function $\mathcal{F}_{n',n}^{(jk)}$ correspond to the cardinalities of the sets $\bar{v}$ and $\bar{u}$: $n' = \#\bar{v}$, $n = \#\bar{u}$. Then $n' = n - j + k$. Otherwise, the form factor (8.4) vanishes.

To calculate such form factors, we can use the formulas of the action of the operators $T_{jk}(z)$ on the (dual) Bethe vector. After that, the problem reduces to calculating scalar products in which one of the vectors is an on-shell vector. Our task is to obtain such formulas for form factors that would be convenient to use in further analysis.

In Section 8.1, we calculate the form factors of the off-diagonal elements of the monodromy matrix $B(z)$ and $C(z)$. These form factors immediately reduce to the scalar products. In Section 8.2, we study the form factors of diagonal elements. We show how they can be calculated using the twisted transfer matrix. Section 8.3 introduces the concept of a universal form factor for models with the rational $R$-matrix. Finally, in Section 8.4, we establish simple relationships between various universal form factors.

## 8.1 Form factors of the off-diagonal elements

We begin our considerations with the form factors of the operators $B(z)$ and $C(z)$. They immediately reduce to scalar products in which one of the vectors is on-shell. For example,

$$\mathcal{F}^{(12)}_{n+1,n}(\bar{v},\bar{u},z) = \langle \mathbf{\Phi}_{n+1}(\bar{v})|B(z)|\mathbf{\Phi}_n(\bar{u})\rangle. \tag{8.5}$$

Here the vector

$$B(z)|\mathbf{\Phi}_n(\bar{u})\rangle = B(z)\boldsymbol{B}(\bar{u})|0\rangle, \tag{8.6}$$

is off-shell because the union $\{\bar{u},z\}$ does not satisfy Bethe equations. However, the dual vector $\langle \mathbf{\Phi}_{n+1}(\bar{v})|$ is on-shell. Therefore, we can use determinant representation (6.34), replacing there the set $\bar{u}$ with $\bar{w}=\{\bar{u},z\}$ and $n$ with $n+1$. Taking into account the normalization (8.1), we obtain

$$\mathcal{F}^{(12)}_{n+1,n}(\bar{v},\bar{u},z) = d(z)\Delta'(\bar{v})\Delta(\bar{w})\,h(\bar{w},\bar{v})\,\det_{n+1}\mathcal{M}^{(12)}, \tag{8.7}$$

where

$$\mathcal{M}^{(12)}_{jk} = t(w_k,v_j) + (-1)^n r(w_k)t(v_j,w_k)\frac{h(\bar{v},w_k)}{h(w_k,\bar{v})}. \tag{8.8}$$

Here $w_k = u_k$ for $k=1,\dots,n$, and $w_{n+1}=z$. Therefore, using Bethe equations for the set $\bar{u}$, we transform the matrix elements $\mathcal{M}^{(12)}_{jk}$ with $k<n+1$ as follows:

$$\mathcal{M}^{(12)}_{jk} = t(u_k,v_j) - t(v_j,u_k)\frac{h(u_k,\bar{u})h(\bar{v},u_k)}{h(\bar{u},u_k)h(u_k,\bar{v})}, \qquad k<n+1. \tag{8.9}$$

In the last column, we have

$$\mathcal{M}^{(12)}_{j,n+1} = t(z,v_j) + (-1)^n r(z)t(v_j,z)\frac{h(\bar{v},z)}{h(z,\bar{v})}. \tag{8.10}$$

Recall once more that here $\#\bar{u}=n$ and $\#\bar{v}=n+1$.

The form factor of the operator $C(z)$

$$\mathcal{F}^{(21)}_{n-1,n}(\bar{v},\bar{u},z) = \langle \mathbf{\Phi}_{n-1}(\bar{v})|C(z)|\mathbf{\Phi}_n(\bar{u})\rangle, \tag{8.11}$$

is calculated in the same way. Due to the symmetry of the scalar product (5.71), it is clear that it reduces to the form factor of the operator $B(z)$. We have

$$\mathcal{F}^{(21)}_{n-1,n}(\bar{v},\bar{u},z) = d(z)\Delta(\bar{w})\Delta'(\bar{u})\; h(\bar{w},\bar{u}) \det_n \mathcal{M}^{(21)}, \tag{8.12}$$

where now $\bar{w} = \{\bar{v}, z\}$, and

$$\begin{aligned} \mathcal{M}^{(21)}_{jk} &= t(v_k,u_j) - t(u_j,v_k)\frac{h(v_k,\bar{v})h(\bar{u},v_k)}{h(\bar{v},v_k)h(v_k,\bar{u})}, \qquad k<n,\\ \mathcal{M}^{(21)}_{jn} &= t(z,u_j) + (-1)^{n-1}r(z)t(u_j,z)\frac{h(\bar{u},z)}{h(z,\bar{u})}. \end{aligned} \tag{8.13}$$

Now $\#\bar{u} = n$ and $\#\bar{v} = n-1$.

## 8.2 Form factors of the diagonal elements

Calculating the form factors of the operators $A(z)$ and $D(z)$ is a bit more complicated. First of all, we note that

$$\langle \mathbf{\Phi}_n(\bar{v})|\big(A(z)+D(z)\big)|\mathbf{\Phi}_n(\bar{u})\rangle = \delta_{\bar{v},\bar{u}}\; \tau(z|\bar{u})\langle \mathbf{\Phi}_n(\bar{u})|\mathbf{\Phi}_n(\bar{u})\rangle, \tag{8.14}$$

where $\delta_{\bar{v},\bar{u}} = 1$ if $\bar{v} = \bar{u}$, and $\delta_{\bar{v},\bar{u}} = 0$ otherwise. Hence,

$$\mathcal{F}^{(11)}_{n,n}(\bar{v},\bar{u},z) + \mathcal{F}^{(22)}_{n,n}(\bar{v},\bar{u},z) = \delta_{\bar{v},\bar{u}}\; \tau(z|\bar{u})\langle \mathbf{\Phi}_n(\bar{u})|\mathbf{\Phi}_n(\bar{u})\rangle. \tag{8.15}$$

Therefore, it is sufficient to calculate the form factors of only one of the diagonal operators, for example, the $D(z)$ operator.

The action of $D(z)$ on the vector $|\mathbf{\Phi}_n(\bar{u})\rangle$ is known. Therefore, it is not difficult to reduce this form factor to a sum of scalar products in which the dual vector remains on-shell. However, there is another way to calculate the operator $D(z)$ form factor, which we now demonstrate. This method may seem like a simple technical trick, but this trick has important consequences, which we will get acquainted with later.

Consider a twisted transfer matrix $\mathcal{T}_\kappa(z) = A(z) + \kappa D(z)$, where $\kappa$ is some complex parameter. Let us introduce

$$Q_\kappa = \langle \mathbf{\Phi}^{(\kappa)}_n(\bar{v})|\big(\mathcal{T}_\kappa(z) - \mathcal{T}(z)\big)|\mathbf{\Phi}_n(\bar{u})\rangle. \tag{8.16}$$

Here $\mathcal{T}(z)$ is the ordinary transfer matrix: $\mathcal{T}(z) = A(z) + D(z)$. The dual state $\langle \mathbf{\Phi}^{(\kappa)}_n(\bar{v})|$ is a dual twisted on-shell vector that turns into the ordinary on-shell vector at $\kappa = 1$.

On the one hand, $\mathcal{T}_\kappa(z) - \mathcal{T}(z) = (\kappa-1)D(z)$. Thus, we have

$$Q_\kappa = (\kappa-1)\langle \mathbf{\Phi}_n^{(\kappa)}(\bar v)|D(z)|\mathbf{\Phi}_n(\bar u)\rangle. \tag{8.17}$$

Differentiating over $\kappa$ at $\kappa=1$, we obtain

$$\frac{\mathrm{d}Q_\kappa}{\mathrm{d}\kappa}\Big|_{\kappa=1} = \langle \mathbf{\Phi}_n(\bar v)|D(z)|\mathbf{\Phi}_n(\bar u)\rangle. \tag{8.18}$$

In the rhs of this formula, we have set $\kappa=1$; hence, $\langle \mathbf{\Phi}_n(\bar v)|$ is an ordinary on-shell vector. Therefore, we obtain the form factor of the operator $D(z)$.

On the other hand,

$$Q_\kappa = \big(\tau_\kappa(z|\bar v) - \tau(z|\bar u)\big)\langle \mathbf{\Phi}_n^{(\kappa)}(\bar v)|\mathbf{\Phi}_n(\bar u)\rangle, \tag{8.19}$$

where $\tau_\kappa(z|\bar v)$ and $\tau(z|\bar u)$ respectively are eigenvalues of the twisted and the ordinary transfer matrices (see (6.37)). Comparing (8.18) and (8.19), we obtain

$$\langle \mathbf{\Phi}_n(\bar v)|D(z)|\mathbf{\Phi}_n(\bar u)\rangle = \frac{\mathrm{d}}{\mathrm{d}\kappa}\big(\tau_\kappa(z|\bar v) - \tau(z|\bar u)\big)\langle \mathbf{\Phi}_n^{(\kappa)}(\bar v)|\mathbf{\Phi}_n(\bar u)\rangle\Big|_{\kappa=1}. \tag{8.20}$$

Thus, to calculate the operator $D(z)$ form factor, it suffices to calculate the scalar product of the ordinary and twisted on-shell vectors and then differentiate the result over $\kappa$ at $\kappa=1$.

One can go a bit further. Consider the case when $\bar v \neq \bar u$ at $\kappa=1$. Then it is the scalar product $\langle \mathbf{\Phi}_n(\bar v)|\mathbf{\Phi}_n(\bar u)\rangle$ that vanishes at $\kappa=1$ because it is the scalar product of two different on-shell vectors. Hence, in this case, the $\kappa$-derivative in (8.20) must act only on the scalar product but not on the difference of the transfer matrix eigenvalues. In the latter, we can simply set $\kappa=1$:

$$\langle \mathbf{\Phi}_n(\bar v)|D(z)|\mathbf{\Phi}_n(\bar u)\rangle = \big(\tau(z|\bar v) - \tau(z|\bar u)\big)\frac{\mathrm{d}}{\mathrm{d}\kappa}\langle \mathbf{\Phi}_n^{(\kappa)}(\bar v)|\mathbf{\Phi}_n(\bar u)\rangle\Big|_{\kappa=1}, \tag{8.21}$$

for $\bar v \neq \bar u$.

Let now $\bar v = \bar u$ at $\kappa=1$. In another way, we can say that the parameters $\bar v$ are $\kappa$-deformations of the parameters $\bar u$. Then the difference $\tau_\kappa(z|\bar v) - \tau(z|\bar u)$ vanishes for $\kappa=1$. Therefore, the derivative over $\kappa$ in (8.20) must act only on this difference but not on the scalar product:

$$\langle \mathbf{\Phi}_n(\bar u)|D(z)|\mathbf{\Phi}_n(\bar u)\rangle = \frac{\mathrm{d}}{\mathrm{d}\kappa}\tau_\kappa(z|\bar v)\Big|_{\kappa=1}\langle \mathbf{\Phi}_n(\bar u)|\mathbf{\Phi}_n(\bar u)\rangle. \tag{8.22}$$

Note that we are dealing with the full derivative with respect to $\kappa$ in (8.21), (8.22). In other words, we must take into account the fact that in the original formula (8.16), the parameters $v_j$ depended on $\kappa$: $v_j = v_j(\kappa)$. However, in the case $\bar v \neq \bar u$, we can explicitly extract the factor $\kappa-1$ in the scalar product of the twisted on-shell and ordinary on-shell Bethe vectors in (8.21). In fact, we did this already when proving the orthogonality of (twisted) on-shell Bethe vectors in Section 6.3. Let us recall this calculation.

For instance, let us take representation (6.92), (6.93) for the scalar product of the twisted on-shell and ordinary on-shell Bethe vectors:

$$\langle \mathbf{\Phi}_n^{(\kappa)}(\bar v)|\mathbf{\Phi}_n(\bar u)\rangle = \Delta'(\bar v)\Delta(\bar u)\; h(\bar v,\bar u)\;\det_n \mathcal{M}^{(\kappa)}, \tag{8.23}$$

where

$$\mathcal{M}^{(\kappa)}_{jk} = t(v_j,u_k) + \kappa t(u_k,v_j)\frac{h(\bar u,u_k)h(u_k,\bar v)}{h(u_k,\bar u)h(\bar v,u_k)}. \tag{8.24}$$

Let

$$\nu_j = \frac{g(v_j,\bar v_j)}{g(v_j,\bar u)}, \tag{8.25}$$

we modify the last row of the matrix $\mathcal{M}^{(\kappa)}$ by adding all the other rows multiplied by the coefficients $\nu_j/\nu_n$ to it:

$$\mathcal{M}^{(\kappa;\mathrm{mod})}_{nk} = \mathcal{M}^{(\kappa)}_{nk} + \sum_{j=1}^{n-1}\frac{\nu_j}{\nu_n}\mathcal{M}^{(\kappa)}_{jk} = \frac{1}{\nu_n}\sum_{j=1}^{n}\nu_j\mathcal{M}^{(\kappa)}_{jk}. \tag{8.26}$$

The method of calculating the sum over $j$ in (8.26) is described in Section 6.3. We obtain

$$\mathcal{M}^{(\kappa;\mathrm{mod})}_{nk} = \frac{1-\kappa}{\nu_n}\frac{h(\bar u,u_k)}{h(\bar v,u_k)}. \tag{8.27}$$

Thus, the modified last row turns out to be proportional to $\kappa-1$. Hence, the determinant of the matrix $\mathcal{M}^{(\kappa)}$ is proportional to $\kappa-1$. Therefore, the $\kappa$-derivative acts only on this factor. In the rest of the expression, we simply set $\kappa=1$. Then we arrive at

$$\frac{\mathrm{d}}{\mathrm{d}\kappa}\langle \mathbf{\Phi}_n^{(\kappa)}(\bar v)|\mathbf{\Phi}_n(\bar u)\rangle\Big|_{\kappa=1} = \Delta'(\bar v)\Delta(\bar u)\; h(\bar v,\bar u)\;\det_n \mathcal{M}^{(22)}, \tag{8.28}$$

where

$$\begin{aligned} \mathcal{M}^{(22)}_{jk} &= \mathcal{M}^{(\kappa)}_{jk}\Big|_{\kappa=1}, \qquad\qquad j<n,\\ \mathcal{M}^{(22)}_{nk} &= \frac{g(\bar u,v_n)}{g(\bar v_n,v_n)}\frac{h(\bar u,u_k)}{h(\bar v,u_k)}. \end{aligned} \tag{8.29}$$

Similarly, starting with representation (6.94), we can obtain another determinant representation for the $\kappa$-derivative of the scalar product

$$\frac{\mathrm{d}}{\mathrm{d}\kappa}\langle \mathbf{\Phi}_n^{(\kappa)}(\bar v)|\mathbf{\Phi}_n(\bar u)\rangle\Big|_{\kappa=1} = \Delta'(\bar v)\Delta(\bar u)\; h(\bar v,\bar u)\;\det_n \widetilde{\mathcal{M}}^{(22)}, \tag{8.30}$$

where

$$\widetilde{\mathcal{M}}^{(22)}_{jk} = t(v_k, u_j) + t(u_j, v_k)\frac{h(v_k, \bar{v})h(\bar{u}, v_k)}{h(\bar{v}, v_k)h(v_k, \bar{u})}, \qquad j < n,$$
$$\widetilde{\mathcal{M}}^{(22)}_{nk} = \frac{g(u_n, \bar{v})}{g(u_n, \bar{u}_n)}\frac{h(v_k, \bar{v})}{h(v_k, \bar{u})}. \tag{8.31}$$

There are other, more sophisticated ways to explicitly extract the factor $\kappa - 1$ in the representations for the scalar product. We will get to know them in Section 9.4. We will also consider in Section 9.1.1 the form factor of the $D$ operator in the case $\bar{u} = \bar{v}$.

## 8.3 Universal form factor

We have already seen that form factors of the diagonal monodromy matrix entries can be presented in the form

$$\mathcal{F}^{(jj)}_{n,n}(\bar{v}, \bar{u}, z) = \big(\tau(z|\bar{v}) - \tau(z|\bar{u})\big)\mathfrak{F}^{(jj)}_{n,n}(\bar{v}, \bar{u}), \tag{8.32}$$

where a function $\mathfrak{F}^{(jj)}_{n,n}(\bar{v}, \bar{u})$ does not depend on $z$. It is easy to show that in the models with the rational $R$-matrix, such a representation is always true if $\bar{v} \neq \bar{u}$. For this, we recall the $RTT$-relation (1.116) for arbitrary matrix elements $T_{ij}(z)$ and $T_{kl}(w)$

$$[T_{ij}(z), T_{kl}(w)] = g(z, w)\big(T_{kj}(w)T_{il}(z) - T_{kj}(z)T_{il}(w)\big). \tag{8.33}$$

Observe that the rhs of (8.33) does not change under replacement $z \leftrightarrow w$. This implies that the commutator of any matrix element $T_{ij}(z)$ with the transfer matrix $\mathcal{T}(w)$ is a symmetric function of $z$ and $w$:

$$[T_{ij}(z), \mathcal{T}(w)] = [T_{ij}(w), \mathcal{T}(z)]. \tag{8.34}$$

Let us multiply (8.34) from the right by an on-shell vector $|\mathbf{\Phi}_n(\bar{u})\rangle$ and from the left by a dual on-shell vector $\langle\mathbf{\Phi}_{n'}(\bar{v})|$. We have

$$\langle\mathbf{\Phi}_{n'}(\bar{v})|[T_{ij}(z), \mathcal{T}(w)]|\mathbf{\Phi}_n(\bar{u})\rangle = \langle\mathbf{\Phi}_{n'}(\bar{v})|[T_{ij}(w), \mathcal{T}(z)]|\mathbf{\Phi}_n(\bar{u})\rangle. \tag{8.35}$$

By calculating the action of the transfer matrices on the on-shell vectors, we find

$$\begin{aligned}&\big(\tau(w|\bar{u}) - \tau(w|\bar{v})\big)\langle\mathbf{\Phi}_{n'}(\bar{v})|T_{ij}(z)|\mathbf{\Phi}_n(\bar{u})\rangle\\ &\quad= \big(\tau(z|\bar{u}) - \tau(z|\bar{v})\big)\langle\mathbf{\Phi}_{n'}(\bar{v})|T_{ij}(w)|\mathbf{\Phi}_n(\bar{u})\rangle.\end{aligned} \tag{8.36}$$

It remains to divide both sides of this equation by the differences of the transfer matrix eigenvalues. This gives us

$$\frac{\mathcal{F}^{(ij)}_{n',n}(\bar{v}, \bar{u}, z)}{\tau(z|\bar{v}) - \tau(z|\bar{u})} = \frac{\mathcal{F}^{(ij)}_{n',n}(\bar{v}, \bar{u}, w)}{\tau(w|\bar{v}) - \tau(w|\bar{u})}. \tag{8.37}$$

The lhs of (8.37) depends only on $z$, while the rhs depends only on $w$. This is possible if and only if

$$\frac{\mathcal{F}^{(ij)}_{n',n}(\bar{v},\bar{u},z)}{\tau(z|\bar{v})-\tau(z|\bar{u})}=\mathfrak{F}^{(ij)}_{n',n}(\bar{v},\bar{u}), \tag{8.38}$$

where $\mathfrak{F}^{(ij)}_{n',n}(\bar{v},\bar{u})$ does not depend on $z$.

The function $\mathfrak{F}^{(ij)}_{n',n}(\bar{v},\bar{u})$ is called a *universal form factor* of the operator $T_{ij}(z)$ [S. Pakuliak et al (2015)]. The meaning of this term is that $\mathfrak{F}^{(ij)}_{n',n}(\bar{v},\bar{u})$ does not depend on a particular model. Indeed, the form factors in different models with the rational $R$-matrix differ only by the function $r(z) = a(z)/d(z)$. It is the only function that 'remembers' the monodromy matrix of the model. However, all dependence on the parameter $z$ in (8.38) has already been extracted as the difference of the transfer matrix eigenvalues. Therefore, the function $\mathfrak{F}^{(ij)}_{n',n}(\bar{v},\bar{u})$ is model-independent.

We emphasize that only form factors in models with the rational $R$-matrix have this property. In models with the trigonometric $R$-matrix, representation (8.38) is valid only for diagonal elements. It is already incorrect for off-diagonal elements. The reason is that the commutation relations (8.33) in the case of the trigonometric $R$-matrix do not hold for all matrix elements.

### 8.3.1 *Determinant representations*

We know already that

$$\mathfrak{F}^{(22)}_{n,n}(\bar{v},\bar{u})=\frac{\mathrm{d}}{\mathrm{d}\kappa}\langle\mathbf{\Phi}^{(\kappa)}_{n}(\bar{v})|\mathbf{\Phi}_{n}(\bar{u})\rangle\Big|_{\kappa=1}, \qquad \bar{v}\neq\bar{u}, \tag{8.39}$$

$$\mathfrak{F}^{(11)}_{n,n}(\bar{v},\bar{u})=-\mathfrak{F}^{(22)}_{n,n}(\bar{v},\bar{u}), \tag{8.40}$$

and the determinant representation for $\mathfrak{F}^{(22)}_{n,n}(\bar{v},\bar{u})$ is given by (8.28) or (8.30). Let us obtain a determinant representation for $\mathfrak{F}^{(21)}_{n-1,n}(\bar{v},\bar{u})$. We apply the same method that we used to extract the factor $\kappa-1$ from the scalar product of the twisted and ordinary on-shell vectors. This time we start with the determinant of the matrix $\mathcal{M}^{(21)}$ (8.13). We write down its matrix elements in the form

$$\mathcal{M}^{(21)}_{jk}=t(w_k,u_j)-(-1)^n r(w_k)t(u_j,w_k)\frac{h(\bar{u},w_k)}{h(w_k,\bar{u})}, \tag{8.41}$$

where $w_k = v_k$ for $k = 1,\ldots,n-1$, and $w_n = z$. Our task is to explicitly extract the difference of the transfer matrix eigenvalues $\tau(z|\bar{v})-\tau(z|\bar{u})$ from

$\det \mathcal{M}^{(21)}$. For this, we add a linear combination of all the rows with $j < n$ to the last row of $\mathcal{M}^{(21)}$.

Let

$$\gamma_j = \frac{g(u_j, \bar{u}_j)}{g(u_j, \bar{v})} = \frac{\prod_{l=1}^{n-1}(u_j - v_l)}{\prod_{l=1;\, l\neq j}^{n}(u_j - u_l)}. \tag{8.42}$$

Consider the following linear combination:

$$\sum_{j=1}^{n} \mathcal{M}_{jk}^{(21)} \gamma_j = G^+(w_k) - (-1)^n r(w_k) \frac{h(\bar{u}, w_k)}{h(w_k, \bar{u})} G^-(w_k), \tag{8.43}$$

where

$$G^+(w_k) = \sum_{j=1}^{n} t(w_k, u_j)\gamma_j, \qquad G^-(w_k) = \sum_{j=1}^{n} t(u_j, w_k)\gamma_j. \tag{8.44}$$

It is easy to see that

$$G^-(w_k) = G^+(w_k)\Big|_{\eta\to-\eta}.$$

Hence, it is enough to compute $G^+(w_k)$.

Consider an auxiliary contour integral

$$I = \frac{1}{2\pi i} \oint_{|z|=R\to\infty} \frac{\eta^2\, \mathrm{d}z}{(w_k - z)(w_k - z + \eta)} \frac{\prod_{l=1}^{n-1}(z - v_l)}{\prod_{l=1}^{n}(z - u_l)}. \tag{8.45}$$

Since the integrand behaves as $z^{-3}$ at $z \to \infty$, we conclude that $I = 0$. On the other hand, the integral (8.45) equals the sum of residues in the poles within the integration contour. The sum of the residues in the poles at $z = u_j$, $j = 1, \dots, n$ gives $G^+(w_k)$. There are also two poles in $z = w_k$ and $z = w_k + \eta$. Thus, we obtain

$$I = 0 = G^+(w_k) - \eta \frac{\prod_{l=1}^{n-1}(w_k - v_l)}{\prod_{l=1}^{n}(w_k - u_l)} + \eta \frac{\prod_{l=1}^{n-1}(w_k - v_l + \eta)}{\prod_{l=1}^{n}(w_k - u_l + \eta)}. \tag{8.46}$$

This implies

$$\begin{aligned} G^+(w_k) &= \frac{g(w_k, \bar{u})}{g(w_k, \bar{v})} - \frac{h(w_k, \bar{v})}{h(w_k, \bar{u})}, \\ G^-(w_k) &= \frac{g(\bar{u}, w_k)}{g(\bar{v}, w_k)} - \frac{h(\bar{v}, w_k)}{h(\bar{u}, w_k)}. \end{aligned} \tag{8.47}$$

Using $1/g(v_k, \bar{v}) = 0$, $k = 1, \dots, n-1$ we find

$$G^+(v_k) = -\frac{h(v_k, \bar{v})}{h(v_k, \bar{u})}, \qquad G^-(v_k) = -\frac{h(\bar{v}, v_k)}{h(\bar{u}, v_k)}. \tag{8.48}$$

In the last column, we set $w_n = z$ and, after simple algebra, arrive at

$$G^+(z) = \frac{f(z,\bar u) - f(z,\bar v)}{g(z,\bar v)h(z,\bar u)}, \qquad G^-(z) = \frac{f(\bar u,z) - f(\bar v,z)}{g(\bar v,z)h(\bar u,z)}. \tag{8.49}$$

Substituting (8.48) in (8.43) for $k = 1,\dots,n-1$ we obtain

$$\sum_{j=1}^{n} \mathcal{M}^{(21)}_{jk}\gamma_j = \frac{h(v_k,\bar v)}{h(v_k,\bar u)}\Big((-1)^n r(v_k)\frac{h(\bar v,v_k)}{h(v_k,\bar v)} - 1\Big) = 0, \qquad k<n, \tag{8.50}$$

due to Bethe equations for the set $\bar v$. On the other hand, substituting (8.49) in (8.43) for $k=n$, we have

$$\begin{aligned}\sum_{j=1}^{n} \mathcal{M}^{(21)}_{jn}\gamma_j &= \frac{f(z,\bar u) + r(z)f(\bar u,z) - f(z,\bar v) - r(z)f(\bar v,z)}{g(z,\bar v)h(z,\bar u)}\\ &= \frac{\tau(z|\bar u) - \tau(z|\bar v)}{d(z)g(z,\bar v)h(z,\bar u)}.\end{aligned} \tag{8.51}$$

Now we modify the last row of the matrix $\mathcal{M}^{(21)}$ as follows:

$$\mathcal{M}^{(21;\mathrm{mod})}_{nk} = \mathcal{M}^{(21)}_{nk} + \frac{1}{\gamma_n}\sum_{j=1}^{n-1}\mathcal{M}^{(21)}_{jk}. \tag{8.52}$$

Then, using (8.50), (8.51) we find

$$\mathcal{M}^{(21;\mathrm{mod})}_{nk} = \delta_{kn}\frac{\tau(z|\bar u) - \tau(z|\bar v)}{d(z)g(z,\bar v)h(z,\bar u)}\frac{g(u_n,\bar v)}{g(u_n,\bar u_n)}. \tag{8.53}$$

Thus, in the modified last row, only the element $\mathcal{M}^{(21;\mathrm{mod})}_{nn}$ is non-zero. The determinant of the $n\times n$ matrix reduces to the product of $\mathcal{M}^{(21;\mathrm{mod})}_{nn}$ and the corresponding minor of the matrix $\mathcal{M}^{(21)}$:

$$\det_n \mathcal{M}^{(21)} = \frac{\tau(z|\bar u) - \tau(z|\bar v)}{d(z)g(z,\bar v)h(z,\bar u)}\frac{g(u_n,\bar v)}{g(u_n,\bar u_n)}\det_{n-1}\mathcal{M}^{(21)}. \tag{8.54}$$

Substituting this in (8.12), we finally arrive at

$$\mathcal{F}^{(21)}_{n-1,n}(\bar v,\bar u,z) = \Delta(\bar v)\Delta'(\bar u)\, h(\bar v,\bar u)\big(\tau(z|\bar u) - \tau(z|\bar v)\big)\frac{g(u_n,\bar v)}{g(u_n,\bar u_n)}\det_{n-1}\mathcal{M}^{(21)}, \tag{8.55}$$

where

$$\mathcal{M}^{(21)}_{jk} = t(v_k,u_j) - t(u_j,v_k)\frac{h(v_k,\bar v)h(\bar u,v_k)}{h(\bar v,v_k)h(v_k,\bar u)}, \qquad j,k = 1,\dots,n-1. \tag{8.56}$$

Thus,

$$\mathfrak{F}^{(21)}_{n-1,n}(\bar v,\bar u) = -\Delta(\bar v)\Delta'(\bar u)\, h(\bar v,\bar u)\frac{g(u_n,\bar v)}{g(u_n,\bar u_n)}\det_{n-1}\mathcal{M}^{(21)}. \tag{8.57}$$

**Remark 8.1.** At first glance, it seems that precisely the same calculations can be done in the case of the trigonometric $R$-matrix. However, this is not the case. Of course, we can introduce a trigonometric analog of the coefficients $\gamma_j$ and consider a linear combination of the form (8.43). However, we can no longer calculate them in closed form. In particular, the analog of the contour integral (8.45) is not equal to zero in the trigonometric case. We invite the reader to study this issue.

A determinant representation for the universal form factor $\mathfrak{F}^{(12)}_{n+1,n}(\bar v,\bar u)$ is derived in the same way, starting with representation (8.7). Here is the final result

$$\mathfrak{F}^{(12)}_{n+1,n}(\bar v,\bar u) = \Delta'(\bar v)\Delta(\bar u)\; h(\bar u,\bar v)\frac{g(v_{n+1},\bar u)}{g(v_{n+1},\bar v_{n+1})}\;\det_n \mathcal{M}^{(12)}, \tag{8.58}$$

where

$$\mathcal{M}^{(12)}_{jk} = t(u_k,v_j) - t(v_j,u_k)\frac{h(u_k,\bar u)h(\bar v,u_k)}{h(\bar u,u_k)h(u_k,\bar v)}, \qquad j,k=1,\dots,n. \tag{8.59}$$

Comparing representations (8.57) and (8.58), we see that

$$\mathfrak{F}^{(21)}_{n-1,n}(\bar v,\bar u) = -\mathfrak{F}^{(12)}_{n,n-1}(\bar u,\bar v). \tag{8.60}$$

This is not surprising since it immediately follows from the symmetric properties of the scalar product (5.71) that

$$\mathcal{F}^{(21)}_{n-1,n}(\bar v,\bar u,z) = \mathcal{F}^{(12)}_{n,n-1}(\bar u,\bar v,z). \tag{8.61}$$

## 8.4 Relationships between universal form factors

There exist simple relations between the universal form factors of different elements of the monodromy matrix.

**Proposition 8.1.** *Let $\bar u=\{u_1,\dots,u_n\}$ and $\bar v=\{v_1,\dots,v_n\}$. Then*

$$\mathfrak{F}^{(21)}_{n-1,n}(\bar v_n,\bar u) = \lim_{v_n\to\infty}\frac{v_n}{\eta}\mathfrak{F}^{(22)}_{n,n}(\bar v,\bar u), \tag{8.62}$$

*and*

$$\mathfrak{F}^{(12)}_{n,n-1}(\bar v,\bar u_n) = \lim_{u_n\to\infty}\frac{u_n}{\eta}\mathfrak{F}^{(22)}_{n,n}(\bar v,\bar u). \tag{8.63}$$

Relations between different form factors in models with the rational $R$-matrix were established in work [S. Pakuliak et al (2015)] using a zero modes method. We do not describe this method in this book. Therefore, we prove formulas (8.62) and (8.63) using the explicit determinant representations obtained above. We restrict ourselves to the detailed proof of (8.62). Formula (8.63) is proved similarly.

**Proof.** First of all, let us rewrite representation (8.57), replacing the set $\bar{v}$ with the set $\bar{v}_n$

$$\mathfrak{F}^{(21)}_{n-1,n}(\bar{v}_n,\bar{u}) = -\Delta(\bar{v}_n)\Delta'(\bar{u})\, h(\bar{v}_n,\bar{u})\frac{g(u_n,\bar{v}_n)}{g(u_n,\bar{u}_n)} \det_{n-1} \mathcal{M}^{(21)}. \tag{8.64}$$

Here

$$\mathcal{M}^{(21)}_{jk} = t(v_k,u_j) - t(u_j,v_k)\frac{h(v_k,\bar{v}_n)h(\bar{u},v_k)}{h(\bar{v}_n,v_k)h(v_k,\bar{u})}, \qquad j,k = 1,\dots,n-1. \tag{8.65}$$

Now we write down explicitly representations (8.39), (8.30) for the universal form factor $\mathfrak{F}^{(22)}_{n,n}(\bar{v},\bar{u})$. The prefactor reads

$$\Delta(\bar{v})\Delta'(\bar{u})\, h(\bar{v},\bar{u}) = \frac{\prod_{a,b=1}^{n}(v_a - u_b + \eta)}{\eta^n \prod_{a>b}^{n}(u_a - u_b)(v_b - v_a)}. \tag{8.66}$$

The entries of the matrix $\widetilde{\mathcal{M}}^{(22)}$ are

$$\begin{aligned}
\widetilde{\mathcal{M}}^{(22)}_{jk} &= \frac{\eta^2}{(v_k - u_j)(v_k - u_j + \eta)} + \frac{\eta^2 Y_k}{(v_k - u_j)(v_k - u_j - \eta)}, \qquad j<n,\\
\widetilde{\mathcal{M}}^{(22)}_{nk} &= \frac{\prod_{l=1}^{n-1}(u_n - u_l)}{\prod_{l=1}^{n}(u_n - v_l)} \prod_{l=1}^{n} \frac{(v_k - v_l + \eta)}{(v_k - u_l + \eta)},
\end{aligned} \tag{8.67}$$

where

$$Y_k = \prod_{l=1}^{n} \frac{(v_k - v_l + \eta)(v_k - u_l - \eta)}{(v_k - v_l - \eta)(v_k - u_l + \eta)}. \tag{8.68}$$

Let us now consider how the prefactor and the matrix elements behave in the limit $v_n \to \infty$. It is easy to see that

$$\Delta(\bar{v})\Delta'(\bar{u})\, h(\bar{v},\bar{u}) = O(v_n), \qquad v_n \to \infty, \tag{8.69}$$

and

$$\left.\begin{aligned}
\widetilde{\mathcal{M}}^{(22)}_{jk} &= O(1), \qquad j,k<n,\\
\widetilde{\mathcal{M}}^{(22)}_{jn} &= O(v_n^{-3}), \qquad j<n,\\
\widetilde{\mathcal{M}}^{(22)}_{nk} &= O(1), \qquad k<n,\\
\widetilde{\mathcal{M}}^{(22)}_{nn} &= O(v_n^{-2}),
\end{aligned}\right\} \quad v_n \to \infty. \tag{8.70}$$

Therefore, when calculating the limit of $\frac{v_n}{\eta}\mathfrak{F}^{(22)}_{n,n}$, it is convenient to multiply the last column of the matrix $\widetilde{\mathcal{M}}^{(22)}$ by $v_n^2/\eta^2$ and the prefactor by $\eta/v_n$. An elementary calculation shows that

$$\lim_{v_n\to\infty} \frac{\eta}{v_n}\Delta(\bar{v})\Delta'(\bar{u})\, h(\bar{v},\bar{u}) = \Delta'(\bar{u})\Delta(\bar{v}_n)\, h(\bar{v}_n,\bar{u}). \tag{8.71}$$

For the elements $\widetilde{\mathcal{M}}^{(22)}_{jk}$ with $j,k<n$, we obtain

$$\lim_{v_n\to\infty} \widetilde{\mathcal{M}}^{(22)}_{jk} = t(v_k,u_j) - t(u_j,v_k)\frac{h(v_k,\bar{v}_n)h(\bar{u},v_k)}{h(\bar{v}_n,v_k)h(v_k,\bar{u})}, \qquad j,k<n, \tag{8.72}$$

and we see that they coincide with the elements (8.65):

$$\lim_{v_n\to\infty} \widetilde{\mathcal{M}}^{(22)}_{jk} = \mathcal{M}^{(21)}_{jk}, \qquad j,k<n. \tag{8.73}$$

If $j<n$ and $k=n$, then

$$\lim_{v_n\to\infty} \frac{v_n^2}{\eta^2}\widetilde{\mathcal{M}}^{(22)}_{jn} = 0, \qquad j<n. \tag{8.74}$$

Finally,

$$\lim_{v_n\to\infty} \frac{v_n^2}{\eta^2}\widetilde{\mathcal{M}}^{(22)}_{nn} = -\frac{g(u_n,\bar v_n)}{g(u_n,\bar u_n)}, \tag{8.75}$$

while the elements $\widetilde{\mathcal{M}}^{(22)}_{nk}$ with $k<n$ remain finite. The determinant reduces to the limit of the element $\frac{v_n^2}{\eta^2}\widetilde{\mathcal{M}}^{(22)}_{nn}$ (8.75) multiplied by the corresponding minor. We immediately arrive at (8.64). □

To prove equation (8.63), one should use representation (8.28) for the universal form factor $\mathfrak{F}^{(22)}_{n,n}(\bar v,\bar u)$.

Relations (8.62), (8.63) allow us to obtain new determinant representations for the universal form factors. Indeed, we obtained the form factor $\mathfrak{F}^{(21)}_{n-1,n}$ by calculating the limit $v_n\to\infty$ in representation (8.30). However, we can also calculate this limit in representation (8.28). Then we obtain

$$\mathfrak{F}^{(21)}_{n-1,n}(\bar v_n,\bar u) = -\Delta'(\bar v_n)\Delta(\bar u)h(\bar u,\bar v_n)\det_n \widetilde{\mathcal{M}}^{(21)}, \tag{8.76}$$

where

$$\begin{aligned}\widetilde{\mathcal{M}}^{(21)}_{jk} &= t(u_k,v_j) - t(v_j,u_k)\frac{h(u_k,\bar u)h(\bar v_n,u_k)}{h(\bar u,u_k)h(u_k,\bar v_n)}, \qquad j<n,\\ \widetilde{\mathcal{M}}^{(21)}_{nk} &= \frac{h(u_k,\bar u)}{h(u_k,\bar v_n)}.\end{aligned} \tag{8.77}$$

Similarly, taking the limit $u_n\to\infty$ in representation (8.30), we find

$$\mathfrak{F}^{(12)}_{n,n-1}(\bar v,\bar u_n) = \Delta(\bar v)\Delta'(\bar u_n)\, h(\bar v,\bar u_n)\, \det_n \widetilde{\mathcal{M}}^{(12)}, \tag{8.78}$$

where

$$\begin{aligned}\widetilde{\mathcal{M}}^{(12)}_{jk} &= t(v_k,u_j) - t(u_j,v_k)\frac{h(v_k,\bar v)h(\bar u_n,v_k)}{h(\bar v,v_k)h(v_k,\bar u_n)}, \qquad j<n,\\ \widetilde{\mathcal{M}}^{(12)}_{nk} &= \frac{h(v_k,\bar v)}{h(v_k,\bar u_n)}.\end{aligned} \tag{8.79}$$

Several representations for the (universal) form factors are useful since it is convenient to use different formulas in different situations. Moreover, in Section 11.5, we will consider the case when different representations for the same form factor are used simultaneously in one formula.

# Chapter 9

# Form factors of local operators

This chapter is devoted to the calculation of the form factors of local operators. Let $O(x)$ be some operator that depends on the spatial variable $x$ (possibly discrete) and acts in the Hilbert space of the quantum model under consideration. We will call the form factor of this operator the matrix element of $O(x)$ in the basis of the transfer matrix eigenstates:

$$\mathcal{F}_{n',n}(\bar{v},\bar{u}|O(x)) = \langle \mathbf{\Phi}_{n'}(\bar{v})|O(x)|\mathbf{\Phi}_{n}(\bar{u})\rangle. \tag{9.1}$$

As with the form factors of the monodromy matrix elements, we will sometimes consider twisted on-shell vectors instead of the usual ones.

In spin chains, local operators are spin operators $\sigma_m^\alpha$, $\alpha = \pm, z$, acting in the local quantum space associated with the $m$th site of the chain. In the one-dimensional Bose gas model, local operators are, for example, the fields $\Psi(x)$ and $\Psi^\dagger(x)$, the density operator $\Psi^\dagger(x)\Psi(x)$, and so on.

Generally speaking, local operators can depend not only on the spatial coordinate but also on time $t$:

$$O(x,t) = e^{iHt}O(x)e^{-iHt}. \tag{9.2}$$

However, the dependence on time easily extracts into the phase factor

$$\mathcal{F}_{n',n}(\bar{v},\bar{u}|O(x,t)) = e^{it(\mathcal{E}_v-\mathcal{E}_u)}\mathcal{F}_{n',n}(\bar{v},\bar{u}|O(x)), \tag{9.3}$$

where $\mathcal{E}_v$ and $\mathcal{E}_u$ respectively are the energies of the states $\langle \mathbf{\Phi}_{n'}(\bar{v})|$ and $|\mathbf{\Phi}_{n}(\bar{u})\rangle$. They depend on the concrete models (see (2.25)–(2.27)). The dependence of form factors on the spatial coordinate is also quite simple, but this is not always obvious in advance.

In Section 9.1, we consider the form factors of local spin operators in the $XXZ$ and $XXX$ Heisenberg chains. Using the explicit solution to the quantum inverse problem, we immediately reduce them to the form factors of the monodromy matrix elements. Here we also calculate the magnetization of the spin chain. In Sections 9.2 and 9.3, we study the form factors in the one-dimensional Bose gas. We show that the form factors of local operators reduce to the universal form factors in this case. In Section 9.4, we obtain a new determinant representation for form factors. We will use this representation for calculating the thermodynamic limit.

## 9.1 Form factors of local operators in the spin chains

The quantum inverse problem formula (3.23) makes it possible to easily obtain determinant representations for the form factors of the local spin operators $\sigma_m^\alpha$, $\alpha = \pm, z$, in the $XXZ$ and $XXX$ spin chains [N. Kitanine et al (1999)]. Equivalently, we can consider form factors of the local standard basis matrices $E_m^{ij}$. It is worth mentioning that we can deal with both homogeneous and inhomogeneous chains. In any case, the form factors of local operators reduce to the form factors of the monodromy matrix elements.

Consider an inhomogeneous $XXZ$ (or $XXX$) chain. It follows from (3.23) that the local standard basis matrices $E_m^{ij}$ are given by

$$E_m^{ij} = \left(\prod_{l=1}^{m-1} \mathcal{T}(\xi_l)\right) T_{ji}(\xi_m) \left(\prod_{l=1}^{m} \mathcal{T}^{-1}(\xi_l)\right). \tag{9.4}$$

Let

$$\mathcal{F}_{n',n}(\bar{v},\bar{u}|E_m^{ij}) = \langle \mathbf{\Phi}_{n'}(\bar{v})|E_m^{ij}|\mathbf{\Phi}_n(\bar{u})\rangle. \tag{9.5}$$

Then substituting here representation (9.4), we immediately arrive at

$$\mathcal{F}_{n',n}(\bar{v},\bar{u}|E_m^{ij}) = \mathcal{F}_{n',n}^{ji}(\bar{v},\bar{u},\xi_m)\frac{\prod_{l=1}^{m-1}\tau(\xi_l|\bar{v})}{\prod_{l=1}^{m}\tau(\xi_l|\bar{u})}, \tag{9.6}$$

where $\mathcal{F}_{n',n}^{ji}(\bar{v},\bar{u},\xi_m)$ is the form factor of the operator $T_{ji}(\xi_m)$. We have already obtained determinant representations for all these form factors in the previous chapter.

Strictly speaking, in the case of form factors of the operators $E_m^{12}$ and $E_m^{21}$, we should understand equation (9.6) in the sense of the limit of $\mathcal{F}_{n',n}^{ji}(\bar{v},\bar{u},z)$ at $z \to \xi_m$. The fact is that these form factors are proportional to the function $d(z)$, and $d(\xi_m) = 0$. However, the limit is simply calculated. Omitting elementary calculations, we derive from (8.7)

$$\mathcal{F}_{n+1,n}^{(12)}(\bar{v},\bar{u},\xi_m) = a(\xi_m)\Delta(\bar{v})\Delta'(\bar{u})\, h(\bar{u},\bar{v})g(\xi_m,\bar{u})h(\bar{v},\xi_m)\, \det_{n+1}\mathcal{M}^{(12)}(\xi_m), \tag{9.7}$$

where

$$\begin{aligned}\mathcal{M}^{(12)}_{jk}(\xi_m) &= t(u_k,v_j) - t(v_j,u_k)\frac{h(u_k,\bar u)h(\bar v,u_k)}{h(\bar u,u_k)h(u_k,\bar v)}, \qquad k<n+1,\\ \mathcal{M}^{(12)}_{j,n+1}(\xi_m) &= t(v_j,\xi_m).\end{aligned} \tag{9.8}$$

In this formula, $\#\bar u = n$ and $\#\bar v = n+1$. Similarly, starting with equation (8.12), we derive

$$\mathcal{F}^{(21)}_{n-1,n}(\bar v,\bar u,\xi_m) = a(\xi_m)\Delta'(\bar v)\Delta(\bar u)\; h(\bar v,\bar u)g(\xi_m,\bar v)h(\bar u,\xi_m)\;\det_n \mathcal{M}^{(21)}(\xi_m), \tag{9.9}$$

where

$$\begin{aligned}\mathcal{M}^{(21)}_{jk}(\xi_m) &= t(v_k,u_j) - t(u_j,v_k)\frac{h(v_k,\bar v)h(\bar u,v_k)}{h(\bar v,v_k)h(v_k,\bar u)}, \qquad k<n,\\ \mathcal{M}^{(21)}_{jn}(\xi_m) &= t(u_j,\xi_m).\end{aligned} \tag{9.10}$$

Now $\#\bar u = n$ and $\#\bar v = n-1$.

Taking into account that $d(\xi_m)=0$, $m=1,\dots,N$, we have

$$\tau(\xi_m|\bar w) = a(\xi_m)f(\bar w,\xi_m), \tag{9.11}$$

for any set $\bar w$. Then

$$\begin{aligned}\mathcal{F}_{n+1,n}(\bar v,\bar u|E^{21}_m) &= \Delta(\bar v)\Delta'(\bar u)\frac{\prod_{l=1}^{m-1} f(\bar v,\xi_l)}{\prod_{l=1}^{m} f(\bar u,\xi_l)}\\ &\quad\times h(\bar u,\bar v)g(\xi_m,\bar u)h(\bar v,\xi_m)\;\det_{n+1}\mathcal{M}^{(12)}(\xi_m),\end{aligned} \tag{9.12}$$

and

$$\begin{aligned}\mathcal{F}_{n-1,n}(\bar v,\bar u|E^{12}_m) &= \Delta'(\bar v)\Delta(\bar u)\frac{\prod_{l=1}^{m-1} f(\bar v,\xi_l)}{\prod_{l=1}^{m} f(\bar u,\xi_l)}\\ &\quad\times h(\bar v,\bar u)g(\xi_m,\bar v)h(\bar u,\xi_m)\;\det_{n}\mathcal{M}^{(21)}(\xi_m).\end{aligned} \tag{9.13}$$

Form factors of diagonal matrices $E^{jj}_m$ reduce to the form factors of the monodromy matrix diagonal elements. If $\bar v \ne \bar u$, then

$$\mathcal{F}_{n,n}(\bar v,\bar u|E^{11}_m) = -\mathcal{F}_{n,n}(\bar v,\bar u|E^{22}_m), \tag{9.14}$$

and

$$\begin{aligned}\mathcal{F}_{n,n}(\bar v,\bar u|E^{22}_m) &= \Delta'(\bar v)\Delta(\bar u)\frac{\prod_{l=1}^{m-1} f(\bar v,\xi_l)}{\prod_{l=1}^{m} f(\bar u,\xi_l)}\\ &\quad\times h(\bar v,\bar u)\big(f(\bar v,\xi_m)-f(\bar u,\xi_m)\big)\;\det_n\mathcal{M}^{(22)},\end{aligned} \tag{9.15}$$

where $\mathcal{M}^{(22)}$ is given by (8.29). Instead of this matrix, we can also take the matrix $\widetilde{\mathcal{M}}^{(22)}$ (8.31).

In the case of the homogeneous $XXX$ chain, we can reduce the form factors of local operators to the universal form factors. Setting all $\xi_l = \eta/2$ in (9.6) we obtain

$$\mathcal{F}_{n',n}(\bar{v}, \bar{u}|E_m^{ij}) = \mathcal{F}^{ji}_{n',n}(\bar{v}, \bar{u}, \tfrac{\eta}{2})\frac{\tau^{m-1}(\frac{\eta}{2}|\bar{v})}{\tau^m(\frac{\eta}{2}|\bar{u})}. \tag{9.16}$$

Let us introduce operators

$$\boldsymbol{E}_s^{ij} = \sum_{m=1}^{s} E_m^{ij}. \tag{9.17}$$

Then

$$\mathcal{F}_{n',n}(\bar{v}, \bar{u}|\boldsymbol{E}_s^{ij}) = \mathcal{F}^{ji}_{n',n}(\bar{v}, \bar{u}, \tfrac{\eta}{2})\sum_{m=1}^{s}\frac{\tau^{m-1}(\frac{\eta}{2}|\bar{v})}{\tau^m(\frac{\eta}{2}|\bar{u})}. \tag{9.18}$$

Summing up the geometric progression, we get

$$\mathcal{F}_{n',n}(\bar{v}, \bar{u}|\boldsymbol{E}_s^{ij}) = \frac{\mathcal{F}^{ji}_{n',n}(\bar{v}, \bar{u}, \frac{\eta}{2})}{\tau(\frac{\eta}{2}|\bar{v}) - \tau(\frac{\eta}{2}|\bar{u})}\left[\left(\frac{\tau(\frac{\eta}{2}|\bar{v})}{\tau(\frac{\eta}{2}|\bar{u})}\right)^s - 1\right]. \tag{9.19}$$

Thus, we have arrived at the universal form factors

$$\mathcal{F}_{n',n}(\bar{v}, \bar{u}|\boldsymbol{E}_s^{ij}) = \mathfrak{F}^{ji}_{n',n}(\bar{v}, \bar{u})\left[\left(\frac{f(\bar{v}, \frac{\eta}{2})}{f(\bar{u}, \frac{\eta}{2})}\right)^s - 1\right], \tag{9.20}$$

where we used (9.11). Form factors of the local operators are now expressed as a difference

$$\mathcal{F}_{n',n}(\bar{v}, \bar{u}|E_s^{ij}) = \mathcal{F}_{n',n}(\bar{v}, \bar{u}|\boldsymbol{E}_s^{ij}) - \mathcal{F}_{n',n}(\bar{v}, \bar{u}|\boldsymbol{E}_{s-1}^{ij}). \tag{9.21}$$

### 9.1.1 *Magnetization*

Recall that when studying the form factors of the monodromy matrix diagonal elements, we can deal with two different cases: $\bar{v} \neq \bar{u}$ and $\bar{v} = \bar{u}$. So far, we have considered the first of them. Let us now turn to case $\bar{v} = \bar{u}$. Then the form factor is given by formula (8.22), in which we need to calculate the full $\kappa$-derivative of the twisted transfer matrix eigenvalue. In particular, it is necessary to differentiate solutions of the twisted Bethe equations $v_j = v_j(\kappa)$. Let us describe how to do this.

Consider the magnetization $\langle\sigma^z\rangle$ of a state $|\boldsymbol{\Phi}_n(\bar{u})\rangle$ of the homogeneous $XXZ$ chain. It is defined as follows:

$$\langle\sigma^z\rangle = \frac{\langle\boldsymbol{\Phi}_n(\bar{u})|\sigma^z_m|\boldsymbol{\Phi}_n(\bar{u})\rangle}{\langle\boldsymbol{\Phi}_n(\bar{u})|\boldsymbol{\Phi}_n(\bar{u})\rangle}. \tag{9.22}$$

The site number $m$ is arbitrary. The magnetization does not depend on $m$ due to periodic boundary conditions.

**Proposition 9.1.** *Let $|\mathbf{\Phi}_n(\bar{u})\rangle$ be an on-shell Bethe vector. Then*

$$\langle \sigma^z \rangle = 1 - \frac{2n}{N}, \tag{9.23}$$

*where $N$ is the length of the chain.*

**Proof.** Using $\sigma^z_m = 1 - 2E^{22}_m$ and (9.6) we have

$$\langle \sigma^z \rangle = 1 - 2\frac{\mathcal{F}^{22}_{n,n}(\bar{u}, \bar{u}, \frac{\eta}{2})}{\tau(\frac{\eta}{2}|\bar{u})\langle \mathbf{\Phi}_n(\bar{u})|\mathbf{\Phi}_n(\bar{u})\rangle}. \tag{9.24}$$

As we expected, magnetization does not depend on the site number $m$.

Let $\bar{v}(\kappa) = \{v_1(\kappa)\dots, v_n(\kappa)\}$ be a solution of twisted Bethe equations such that $v_j(1) = u_j$, $j = 1, \dots, n$. Then due to (8.22), we obtain

$$\langle \sigma^z \rangle = 1 - 2\frac{\mathrm{d}}{\mathrm{d}\kappa} \log \tau_\kappa(\tfrac{\eta}{2}|\bar{v})\Big|_{\kappa=1}. \tag{9.25}$$

Let us compute this logarithmic derivative of the twisted transfer matrix eigenvalue. Since $d(\frac{\eta}{2}) = 0$, we have

$$\tau_\kappa(\tfrac{\eta}{2}|\bar{v}) = a(\tfrac{\eta}{2}) f(\bar{v}, \tfrac{\eta}{2}). \tag{9.26}$$

Then

$$\frac{\mathrm{d}}{\mathrm{d}\kappa} \log \tau_\kappa(\tfrac{\eta}{2}|\bar{v})\Big|_{\kappa=1} = \sum_{j=1}^{n} \frac{\partial \log \tau_\kappa(\frac{\eta}{2}|\bar{u})}{\partial u_j} \frac{\mathrm{d}v_j}{\mathrm{d}\kappa}\Big|_{\kappa=1} = \sum_{j=1}^{n} U_j \frac{\mathrm{d}v_j}{\mathrm{d}\kappa}\Big|_{\kappa=1}, \tag{9.27}$$

where

$$U_j = \frac{\partial \log f(u_j, \frac{\eta}{2})}{\partial u_j}. \tag{9.28}$$

To find the derivatives $\mathrm{d}v_j/\mathrm{d}\kappa$ at $\kappa = 1$, we consider twisted Bethe equations

$$\frac{a(v_j)}{d(v_j)} = \kappa(-1)^{n-1} \prod_{k=1}^{n} \frac{\sinh(\eta + v_j - v_k)}{\sinh(\eta - v_j + v_k)}, \quad \text{where} \quad \frac{a(v_j)}{d(v_j)} = f^N(v_j, \tfrac{\eta}{2}). \tag{9.29}$$

Taking the logarithm of (9.29), we obtain

$$N \log f(v_j, \tfrac{\eta}{2}) = \sum_{k=1}^{n} \log \frac{\sinh(\eta + v_j - v_k)}{\sinh(\eta - v_j + v_k)} + \log \kappa + 2\pi i I_j, \tag{9.30}$$

where $I_j$ are (half)integers. Now we take the $\kappa$-derivative of (9.30) at $\kappa = 1$:

$$NU_j \frac{\mathrm{d}v_j}{\mathrm{d}\kappa}\Big|_{\kappa=1} = \sum_{k=1}^{n} \mathcal{K}(u_j - u_k) \left(\frac{\mathrm{d}v_k}{\mathrm{d}\kappa} - \frac{\mathrm{d}v_j}{\mathrm{d}\kappa}\right)\Big|_{\kappa=1} + 1, \tag{9.31}$$

where

$$\mathcal{K}(u) = -\frac{\mathrm{d}}{\mathrm{d}u}\log\frac{\sinh(\eta+u)}{\sinh(\eta-u)} = \frac{\sinh 2\eta}{\sinh(u-\eta)\sinh(u+\eta)}. \tag{9.32}$$

The system of equations (9.31) can be written as

$$\sum_{k=1}^{n}\mathcal{N}_{jk}\frac{\mathrm{d}v_k}{\mathrm{d}\kappa}\Big|_{\kappa=1} = 1, \tag{9.33}$$

where the matrix $\mathcal{N}_{jk}$ is

$$\mathcal{N}_{jk} = \delta_{jk}\left(NU_j + \sum_{l=1}^{n}\mathcal{K}(u_j-u_l)\right) - \mathcal{K}(u_j-u_k). \tag{9.34}$$

We have already encountered this matrix when calculating the norm of (twisted) on-shell Bethe vectors (see (6.77)). Note that this matrix is symmetric: $\mathcal{N}_{jk} = \mathcal{N}_{kj}$. The system (9.33) has a formal solution

$$\frac{\mathrm{d}v_j}{\mathrm{d}\kappa}\Big|_{\kappa=1} = \sum_{l=1}^{n}\mathcal{N}^{-1}_{jl} = \sum_{l=1}^{n}\mathcal{N}^{-1}_{lj}. \tag{9.35}$$

It also follows from (9.34) that

$$\sum_{k=1}^{n}\mathcal{N}_{jk} = NU_j. \tag{9.36}$$

Substituting (9.36) and (9.35) into (9.27) we find

$$\frac{\mathrm{d}}{\mathrm{d}\kappa}\log\tau_\kappa(\tfrac{\eta}{2}|\bar{v})\Big|_{\kappa=1} = \frac{1}{N}\sum_{j=1}^{n}\sum_{k=1}^{n}\sum_{l=1}^{n}\mathcal{N}^{-1}_{lj}\mathcal{N}_{jk} = \frac{n}{N}. \tag{9.37}$$

Substituting this in (9.25), we complete the proof. □

## 9.2 Form factors of local operators in the one-dimensional Bose gas

We now proceed to the calculation of the form factors of local operators in the one-dimensional Bose gas. We could use the formulas of the quantum inverse problem for the lattice QNLSE model [T. Oota (2004)] and then take the continuous limit. In this case, the calculations mostly are repeated those already done. However, we will consider another method, which is based on the composite model (see Section 4.2). In some cases, this approach is more cumbersome. Ultimately, however, we will see that the form factors of local operators in the one-dimensional Bose gas are nevertheless reduced to the universal form factors of the monodromy matrix elements.

Let us start with a small reminder of the contents of Section 1.6.2. For now, we limit ourselves to the basic formulas that we need to calculate the form factors. We will return to the one-dimensional Bose gas again in Chapter 10. We also refer the reader to works [V. E. Korepin et al (1993); M. Gaudin (1983); E. H. Lieb and W. Liniger (1963); E. H. Lieb (1963); F. A. Berezin et al (1964); C. N. Yang and C. P. Yang (1969)] for a more detailed description of the model.

The Hamiltonian of the quantum nonlinear Schrödinger equation has the form (1.76)

$$H = \int_0^L \big(\partial_x\Psi^\dagger(x)\partial_x\Psi(x) + c\Psi^\dagger(x)\Psi^\dagger(x)\Psi(x)\Psi(x)\big)\, \mathrm{d}x, \tag{9.38}$$

where $\Psi(x)$ and $\Psi^\dagger(x)$ are canonical Bose fields, and $c$ is a coupling constant. It is related to the $R$-matrix constant $\eta$ by $c = -i\eta$. We focus on the repulsive case $c > 0$. We also recall the momentum operator

$$P = -\frac{i}{2}\int_0^L \Big(\Psi^\dagger(x)\partial_x\Psi(x) - \big(\partial_x\Psi^\dagger(x)\big)\Psi(x)\Big)\, \mathrm{d}x, \tag{9.39}$$

that commutes with the Hamiltonian. On-shell Bethe vectors are common eigenvectors of $H$ and $P$ with

$$\begin{aligned} P|\mathbf{\Phi}_n(\bar u)\rangle &= \left(\sum_{k=1}^n u_k\right)|\mathbf{\Phi}_n(\bar u)\rangle,\\ H|\mathbf{\Phi}_n(\bar u)\rangle &= \left(\sum_{k=1}^n u_k^2\right)|\mathbf{\Phi}_n(\bar u)\rangle. \end{aligned} \tag{9.40}$$

These formulas can be obtained by applying decomposition (1.79) to the transfer matrix eigenvalue.

It follows from (9.39) that

$$\partial_x\Psi(x) = -i[P,\Psi(x)], \qquad \partial_x\Psi^\dagger(x) = -i[P,\Psi^\dagger(x)], \tag{9.41}$$

and thus,

$$\Psi(x) = e^{-ixP}\Psi(0)e^{ixP}, \qquad \Psi^\dagger(x) = e^{-ixP}\Psi^\dagger(0)e^{ixP}. \tag{9.42}$$

To formulate the one-dimensional Bose gas model in the language of the QISM, we introduce a lattice approximation. Namely, we chose $N$ points $x_1,\dots,x_N$ in the interval $[0,L]$, so that $x_n = \Delta n$, where $\Delta = L/N$ is the lattice step. With each lattice site, we associate the $L$-operator (1.74), in which we set

$$\rho_n^+ = \sqrt{c}, \qquad \rho_n^- = \sqrt{c}\left(1 + \frac{c\Delta^2}{4}\psi_n^\dagger\psi_n\right). \tag{9.43}$$

The operators $\psi_n$ and $\psi_n^\dagger$ are lattice approximations of the original Bose fields and satisfy the relations

$$[\psi_n, \psi_m^\dagger] = \frac{\delta_{nm}}{\Delta}. \tag{9.44}$$

The monodromy matrix appears in the continuous limit $\Delta \to 0$ in the standard expression $T(u) = L_N(u) \dots L_1(u)$.

### 9.2.1 *Field form factors*

To calculate form factors of the fields $\Psi(x)$ and $\Psi^\dagger(x)$, we use the method proposed in [A. G. Izergin et al (1987); T. Kojima et al (1997a)]. Consider a form factor of the field $\Psi(x)$:

$$\mathcal{F}_{n-1,n}(\bar{v}, \bar{u}|\Psi(x)) = \langle \mathbf{\Phi}_{n-1}(\bar{v})|\Psi(x)|\mathbf{\Phi}_n(\bar{u})\rangle. \tag{9.45}$$

Here $\bar{v} = \{v_1, \dots, v_{n-1}\}$ and $\bar{u} = \{u_1, \dots, u_n\}$, otherwise the form factor vanishes. It follows from (9.42) and (9.40) that

$$\mathcal{F}_{n-1,n}(\bar{v}, \bar{u}|\Psi(x)) = e^{-ix\mathcal{P}} \mathcal{F}_n(\bar{v}, \bar{u}|\Psi(0)), \tag{9.46}$$

where

$$\mathcal{P} = \sum_{j=1}^{n-1} v_j - \sum_{j=1}^{n} u_j. \tag{9.47}$$

For calculating the form factor of the field $\Psi(0)$, we consider the lattice QNLSE model as a composite model with

$$T^{(2)}(u) = L_N(u) \dots L_2(u), \qquad T^{(1)}(u) = L_1(u). \tag{9.48}$$

Using (1.74), we obtain

$$\begin{aligned} B(u) &= A^{(2)}(u)B^{(1)}(u) + B^{(2)}(u)D^{(1)}(u) \\ &= -i\sqrt{c}\Delta A^{(2)}(u)\psi_1^\dagger(u) + B^{(2)}(u)\left(1 + \frac{iu\Delta}{2} + \frac{c\Delta^2}{2}\psi_1^\dagger\psi_1\right). \end{aligned} \tag{9.49}$$

Then

$$[\psi_1, B(u)] = -i\sqrt{c}A^{(2)}(u) + \frac{c\Delta}{2}B^{(2)}(u)\psi_1. \tag{9.50}$$

In the continuous limit $\Delta \to 0$, we have $\psi_1 \to \Psi(0)$ and $A^{(2)}(u) \to A(u)$. Thus, we arrive at[1]

$$[\Psi(0), B(u)] = -i\sqrt{c}A(u). \tag{9.51}$$

Similarly, one can calculate

$$[C(u), \Psi^\dagger(0)] = i\sqrt{c}D(u). \tag{9.52}$$

Now we can compute $\Psi(0)$ action on a Bethe vector $B(\bar{u})|0\rangle$. We use the standard method of the algebraic Bethe ansatz. First, we should find a general structure of the result. Due to $\Psi(x)|0\rangle = 0$, we have

$$\begin{aligned}\Psi(0)B(\bar{u})|0\rangle &= \sum_{k=1}^{n} B(u_1)\dots B(u_{k-1})[\Psi(0), B(u_k)]B(u_{k+1})\dots B(u_n)|0\rangle \\ &= -i\sqrt{c}\sum_{k=1}^{n} B(u_1)\dots B(u_{k-1})A(u_k)B(u_{k+1})\dots B(u_n)|0\rangle.\end{aligned} \tag{9.53}$$

Due to commutation relation of the $A$ and $B$ operators (1.123), we conclude that the final result has the following structure:

$$\Psi(0)B(\bar{u})|0\rangle = -i\sqrt{c}\sum_{j=1}^{n} \Lambda_j(\bar{u})a(u_j)B(\bar{u}_j)|0\rangle. \tag{9.54}$$

where $\Lambda_j(\bar{u})$ are numerical coefficients to be determined.

Due to the symmetry of the Bethe vector over $\bar{u}$, it is enough to find $\Lambda_1(\bar{u})$. We present the result in the form

$$\Psi(0)B(\bar{u})|0\rangle = -i\sqrt{c}\Lambda_1(\bar{u})a(u_1)B(\bar{u}_1)|0\rangle + \mathfrak{Z}, \tag{9.55}$$

where $\mathfrak{Z}$ contains all other contributions. Then we must use the commutation relation of $\Psi(0)$ and $B(u_1)$ in the very beginning:

$$\Psi(0)B(\bar{u})|0\rangle = [\Psi(0), B(u_1)]B(\bar{u}_1)|0\rangle + \mathfrak{Z}. \tag{9.56}$$

Otherwise, we obtain a term

$$B(u_1)\Psi(0)B(\bar{u}_1)|0\rangle,$$

and the final result will contain $B(u_1)$. Thus, we arrive at

$$\Psi(0)B(\bar{u})|0\rangle = -i\sqrt{c}A(u_1)B(\bar{u}_1)|0\rangle + \mathfrak{Z}. \tag{9.57}$$

[1] Calculating the continuous limit was studied in more detail in [K. K. Kozlowski (2011)].

After this, we should move $A(u_1)$ to the right, keeping its argument. Otherwise, we do not obtain $a(u_1)$. This means that we should use only the first scheme in the commutation relation (1.123). This gives us

$$\Psi(0)B(\bar{u})|0\rangle = -i\sqrt{c}a(u_1)f(\bar{u}_1, u_1)B(\bar{u}_1)|0\rangle + \mathfrak{Z}. \tag{9.58}$$

Hence,

$$\Lambda_1(\bar{u}) = f(\bar{u}_1, u_1), \tag{9.59}$$

and due to the symmetry over $\bar{u}$, we find

$$\Lambda_\ell(\bar{u}) = f(\bar{u}_\ell, u_\ell), \qquad \ell = 1, \dots, n. \tag{9.60}$$

Thus, the action of $\Psi(0)$ on an on-shell Bethe vector $|\mathbf{\Phi}_n(\bar{u})|0\rangle$ is

$$\Psi(0)|\mathbf{\Phi}_n(\bar{u})|0\rangle = -i\sqrt{c}\sum_{\ell=1}^{n} r(u_\ell)f(\bar{u}_\ell, u_\ell)\boldsymbol{B}(\bar{u}_\ell)|0\rangle, \tag{9.61}$$

and we immediately reduce the form factor (9.46) to a sum of scalar products

$$\mathcal{F}_{n-1,n}(\bar{v}, \bar{u}|\Psi(0)) = -i\sqrt{c}\sum_{\ell=1}^{n} r(u_\ell)f(\bar{u}_\ell, u_\ell)\langle\mathbf{\Phi}_{n-1}(\bar{v})|\boldsymbol{B}(\bar{u}_\ell)|0\rangle. \tag{9.62}$$

In the obtained scalar products, the dual vector is still on-shell. Therefore, we can use determinant representations (6.35) for these scalar products

$$\mathcal{F}_{n-1,n}(\bar{v}, \bar{u}|\Psi(0)) = -i\sqrt{c}\Delta'(\bar{v})\sum_{\ell=1}^{n} r(u_\ell)f(\bar{u}_\ell, u_\ell)h(\bar{u}_\ell, \bar{v})\Delta(\bar{u}_\ell)\det_{k\neq\ell}\mathcal{M}_j(u_k), \tag{9.63}$$

where

$$\mathcal{M}_j(u_k) = t(u_k, v_j)\left(1 - r(u_k)\frac{f(\bar{v}_j, u_k)}{f(u_k, \bar{v}_j)}\right). \tag{9.64}$$

We deliberately wrote the matrix elements as $\mathcal{M}_j(u_k)$ to emphasize that the $k$th column of the matrix is associated with the parameter $u_k$. In formula (9.63), we have determinants of matrices of size $(n-1) \times (n-1)$ in which the $\ell$th column is omitted.

It is easy to see that

$$f(\bar{u}_\ell, u_\ell)h(\bar{u}_\ell, \bar{v})\Delta(\bar{u}_\ell) = (-1)^{\ell-1}\frac{h(\bar{u}, u_\ell)}{h(u_\ell, \bar{v})}h(\bar{u}, \bar{v})\Delta(\bar{u}). \tag{9.65}$$

Then (9.63) takes the form

$$\begin{aligned}\mathcal{F}_{n-1,n}(\bar{v}, \bar{u}|\Psi(0)) &= -i\sqrt{c}\Delta'(\bar{v})\Delta(\bar{u})h(\bar{u}, \bar{v}) \\ &\quad\times\sum_{\ell=1}^{n}(-1)^{\ell-1}r(u_\ell)\frac{h(\bar{u}, u_\ell)}{h(u_\ell, \bar{v})}\det_{k\neq\ell}\mathcal{M}_j(u_k).\end{aligned} \tag{9.66}$$

Let us introduce an $n \times n$ matrix

$$\begin{aligned}\mathcal{M}^{(\Psi)}_{jk} &= \mathcal{M}_j(u_k), \qquad k < n, \\ \mathcal{M}^{(\Psi)}_{nk} &= (-1)^{n-1}r(u_k)\frac{h(\bar{u}, u_k)}{h(u_k, \bar{v})}.\end{aligned} \tag{9.67}$$

It becomes obvious that the sum in the second line of (9.66) is nothing but the expansion of the determinant of the matrix $\mathcal{M}^{(\Psi)}_{jk}$ along the last row. Hence,

$$\mathcal{F}_{n-1,n}(\bar{v},\bar{u}|\Psi(0)) = -i\sqrt{c}\Delta'(\bar{v})\Delta(\bar{u})h(\bar{u},\bar{v})\det_n \mathcal{M}^{(\Psi)}. \tag{9.68}$$

It remains to substitute Bethe equations

$$r(u_k) = (-1)^{n-1}\frac{h(u_k,\bar{u})}{h(\bar{u},u_k)}$$

in the matrix elements $\mathcal{M}^{(\Psi)}_{jk}$, and we arrive at

$$\begin{aligned} \mathcal{M}^{(\Psi)}_{jk} &= t(u_k,v_j) - t(v_j,u_k)\frac{h(u_k,\bar{u})h(\bar{v},u_k)}{h(\bar{u},u_k)h(u_k,\bar{v})}, \qquad j<n,\\ \mathcal{M}^{(\Psi)}_{nk} &= \frac{h(u_k,\bar{u})}{h(u_k,\bar{v})}. \end{aligned} \tag{9.69}$$

Comparing the obtained determinant representation with representation (8.76), (8.77), we conclude that

$$\mathcal{F}_{n-1,n}(\bar{v},\bar{u}|\Psi(0)) = i\sqrt{c}\,\mathfrak{F}^{(21)}_{n-1,n}(\bar{v},\bar{u}), \tag{9.70}$$

where $\mathfrak{F}_{n-1,n}(\bar{v},\bar{u})$ is the universal form factor of the operator $C(z)$.

The form factor of the field $\Psi^\dagger(0)$ can be found precisely in the same manner. The result is quite predictable

$$\mathcal{F}_{n+1,n}(\bar{v},\bar{u}|\Psi^\dagger(0)) = i\sqrt{c}\,\mathfrak{F}^{(12)}_{n+1,n}(\bar{v},\bar{u}). \tag{9.71}$$

It is also easy to check that for real[2] $\bar{u}$ and $\bar{v}$,

$$\mathcal{F}^*_{n-1,n}(\bar{v},\bar{u}|\Psi(0)) = \mathcal{F}_{n,n-1}(\bar{u},\bar{v}|\Psi^\dagger(0)). \tag{9.72}$$

This property follows from $C(u) = B^\dagger(u)$ for $u\in\mathbb{R}$. However, relationship (9.72) can also be checked directly using explicit formulas for the universal form factors. To do this, we set $\eta = -ic$, $c>0$. Then for real $u$ and $v$,

$$g^*(u,v) = g(v,u), \qquad h^*(u,v) = h(v,u), \qquad t^*(u,v) = t(v,u).$$

It is also easy to check that

$$\Big(\mathfrak{F}^{(21)}_{n-1,n}(\bar{v},\bar{u})\Big)^* = \mathfrak{F}^{(21)}_{n-1,n}(\bar{v},\bar{u}), \qquad \Big(\mathfrak{F}^{(12)}_{n,n-1}(\bar{v},\bar{u})\Big)^* = \mathfrak{F}^{(12)}_{n,n-1}(\bar{v},\bar{u}). \tag{9.73}$$

Then using (9.70), (9.71), and (8.60), we arrive at (9.72).

[2]In Section 10.1, we prove that Bethe equations have only real solutions in the one-dimensional Bose gas with $c>0$.

## 9.3 Calculation of generating function

In this section, we calculate a form factor of the density operator $\Psi^\dagger(x)\Psi(x)$. This operator is an analog of the spin operator $\sigma_k^z$ in the $XXZ$ chain. Therefore, it is again convenient for us to use the twisted on-shell vectors.

Let

$$\hat{q}(x) = \int_0^x \Psi^\dagger(y)\Psi(y)\,\mathrm{d}y. \tag{9.74}$$

Consider a generating function

$$\mathsf{Q}_n^{(\alpha)}(\bar{v}|\bar{u}) = \langle \mathbf{\Phi}_n^{(\kappa_2)}(\bar{v})|e^{\alpha\hat{q}(x)}|\mathbf{\Phi}_n^{(\kappa_1)}(\bar{u})\rangle, \qquad \kappa_\ell = e^{\beta_\ell}, \quad \ell = 1,2. \tag{9.75}$$

Here $\alpha$ is a complex number. The states $\langle \mathbf{\Phi}_n^{(\kappa_2)}(\bar{v})|$ and $|\mathbf{\Phi}_n^{(\kappa_1)}(\bar{u})\rangle$ are twisted on-shell Bethe vectors with the twist parameters $\kappa_2$ and $\kappa_1$. The density form factor is obtained by differentiating the generating function at special values of $\alpha$ and $\beta_\ell$ (see below).

To calculate the generating function (9.75), we use a composite model. Namely, we associate partial monodromy matrices $T^{(1)}(u)$ and $T^{(2)}(u)$ respectively with the intervals $[0,x]$ and $]x,L]$. The total monodromy matrix is then related to the interval $[0,L]$.

We use formulas (4.12) and (4.27), which express total Bethe vectors in terms of partial ones. However, first, we need to rewrite these formulas using normalized partial operators

$$\boldsymbol{B}^{(j)}(u) = \frac{B^{(j)}(u)}{d^{(j)}(u)}, \qquad \boldsymbol{C}^{(j)}(u) = \frac{C^{(j)}(u)}{d^{(j)}(u)}, \qquad j = 1,2. \tag{9.76}$$

Then equations (4.12) and (4.27) take the form

$$\boldsymbol{B}(\bar{u})|0\rangle = \sum_{\bar{u}\mapsto\{\bar{u}_{\rm I},\bar{u}_{\rm II}\}} r^{(2)}(\bar{u}_{\rm I})\boldsymbol{B}^{(1)}(\bar{u}_{\rm I})|0\rangle^{(1)}\boldsymbol{B}^{(2)}(\bar{u}_{\rm II})|0\rangle^{(2)} f(\bar{u}_{\rm II},\bar{u}_{\rm I}), \tag{9.77}$$

and

$$\langle 0|\boldsymbol{C}(\bar{v}) = \sum_{\bar{v}\mapsto\{\bar{v}_{\rm I},\bar{v}_{\rm II}\}} r^{(1)}(\bar{v}_{\rm II})\langle 0|^{(1)}\boldsymbol{C}^{(1)}(\bar{v}_{\rm I})\langle 0|^{(2)}\boldsymbol{C}^{(2)}(\bar{v}_{\rm II})\, f(\bar{v}_{\rm I},\bar{v}_{\rm II}), \tag{9.78}$$

where

$$r^{(j)}(u) = \frac{a^{(j)}(u)}{d^{(j)}(u)}, \qquad j = 1,2. \tag{9.79}$$

Substituting these expressions in the generating function, we obtain

$$\begin{aligned}\mathsf{Q}_n^{(\alpha)}(\bar{v}|\bar{u}) = &\sum_{\substack{\bar{u}\mapsto\{\bar{u}_{\rm I},\bar{u}_{\rm II}\}\\ \bar{v}\mapsto\{\bar{v}_{\rm I},\bar{v}_{\rm II}\}}} r^{(1)}(\bar{v}_{\rm II})r^{(2)}(\bar{u}_{\rm I})f(\bar{v}_{\rm I},\bar{v}_{\rm II})f(\bar{u}_{\rm II},\bar{u}_{\rm I})\langle 0|^{(1)}\boldsymbol{C}^{(1)}(\bar{v}_{\rm I})\, e^{\alpha\hat{q}(x)}\\ &\times \boldsymbol{B}^{(1)}(\bar{u}_{\rm I})|0\rangle^{(1)}\langle 0|^{(2)}\boldsymbol{C}^{(2)}(\bar{v}_{\rm II})\boldsymbol{B}^{(2)}(\bar{u}_{\rm II})|0\rangle^{(2)}.\end{aligned} \tag{9.80}$$

The sum is taken over partitions $\bar{u} \mapsto \{\bar{u}_{\rm I}, \bar{u}_{\rm II}\}$ and $\bar{v} \mapsto \{\bar{v}_{\rm I}, \bar{v}_{\rm II}\}$ such that $\#\bar{u}_{\rm I} = \#\bar{v}_{\rm I}$. We also took into account that the operator $\hat{q}(x)$ does not act on vectors $\boldsymbol{B}^{(2)}(\bar{u})|0\rangle^{(2)}$.

The operator $\hat{q}(x)$ coincides with the number of particles operator on the interval $[0, x]$ (see (1.80)). Therefore,

$$\hat{q}(x)\boldsymbol{B}^{(1)}(\bar{u})|0\rangle^{(1)} = n\boldsymbol{B}^{(1)}(\bar{u})|0\rangle^{(1)}, \tag{9.81}$$

for arbitrary set $\bar{u}$ of the cardinality $n$. Then

$$\begin{aligned} \mathsf{Q}_n^{(\alpha)}(\bar{v}|\bar{u}) = \sum_{\substack{\bar{u}\mapsto\{\bar{u}_{\rm I},\bar{u}_{\rm II}\}\\ \bar{v}\mapsto\{\bar{v}_{\rm I},\bar{v}_{\rm II}\}}} & r^{(1)}(\bar{v}_{\rm II})r^{(2)}(\bar{u}_{\rm I})f(\bar{v}_{\rm I},\bar{v}_{\rm II})f(\bar{u}_{\rm II},\bar{u}_{\rm I})e^{\alpha n_{\rm I}}\langle 0|^{(1)}\boldsymbol{C}^{(1)}(\bar{v}_{\rm I}) \\ & \times \boldsymbol{B}^{(1)}(\bar{u}_{\rm I})|0\rangle^{(1)}\langle 0|^{(2)}\boldsymbol{C}^{(2)}(\bar{v}_{\rm II})\boldsymbol{B}^{(2)}(\bar{u}_{\rm II})|0\rangle^{(2)}, \end{aligned} \tag{9.82}$$

where $n_{\rm I} = \#\bar{u}_{\rm I}$.

Let

$$\ell(u) = r^{(1)}(u) = e^{-ixu}. \tag{9.83}$$

Then due to the twisted Bethe equations for the set $\bar{u}$, we have

$$r^{(2)}(\bar{u}_{\rm I})f(\bar{u}_{\rm II},\bar{u}_{\rm I}) = \frac{r(\bar{u}_{\rm I})f(\bar{u}_{\rm II},\bar{u}_{\rm I})}{r^{(1)}(\bar{u}_{\rm I})} = e^{\beta_1 n_{\rm I}}\frac{f(\bar{u}_{\rm I},\bar{u}_{\rm II})}{\ell(\bar{u}_{\rm I})}. \tag{9.84}$$

Equation (9.82) then takes the form

$$\begin{aligned} \mathsf{Q}_n^{(\alpha)}(\bar{v}|\bar{u}) = \sum_{\substack{\bar{u}\mapsto\{\bar{u}_{\rm I},\bar{u}_{\rm II}\}\\ \bar{v}\mapsto\{\bar{v}_{\rm I},\bar{v}_{\rm II}\}}} & \frac{\ell(\bar{v}_{\rm II})}{\ell(\bar{u}_{\rm I})}f(\bar{v}_{\rm I},\bar{v}_{\rm II})f(\bar{u}_{\rm I},\bar{u}_{\rm II})e^{(\alpha+\beta_1)n_{\rm I}} \\ & \times \boldsymbol{S}^{(1)}_{n_{\rm I}}(\bar{v}_{\rm I}|\bar{u}_{\rm I})\boldsymbol{S}^{(2)}_{n_{\rm II}}(\bar{v}_{\rm II}|\bar{u}_{\rm II}). \end{aligned} \tag{9.85}$$

Here $\boldsymbol{S}^{(j)}_m(\bar{v}|\bar{u})$ are scalar products of the partial Bethe vectors:

$$\boldsymbol{S}^{(j)}_m(\bar{v}|\bar{u}) = \langle 0|^{(j)}\boldsymbol{C}^{(j)}(\bar{v})\boldsymbol{B}^{(j)}(\bar{u})|0\rangle^{(j)}, \qquad j=1,2, \qquad m = n_{\rm I}, n_{\rm II}. \tag{9.86}$$

For their calculation, we can use formula (5.70), replacing the functions $r(z)$ with the functions $r^{(1)}(z) = \ell(z)$ in the case of the scalar product $\boldsymbol{S}^{(1)}_{n_{\rm I}}$, and $r(z)$ with $r^{(2)}(z)$ for the scalar product $\boldsymbol{S}^{(2)}_{n_{\rm II}}$. We will have new partitions into new subsets. Because of a large number of such subsets, we will reject the use of Roman numerals and mark the subsets with bold superscripts. Then we obtain for the scalar product $\boldsymbol{S}^{(1)}_{n_{\rm I}}$

$$\begin{aligned} \boldsymbol{S}^{(1)}_{n_{\rm I}}(\bar{v}_{\rm I}|\bar{u}_{\rm I}) = \sum_{\substack{\bar{u}_{\rm I}\mapsto\{\bar{u}^{\mathbf{1}},\bar{u}^{\mathbf{3}}\}\\ \bar{v}_{\rm I}\mapsto\{\bar{v}^{\mathbf{1}},\bar{v}^{\mathbf{3}}\}}} & \ell(\bar{u}^{\mathbf{3}})\ell(\bar{v}^{\mathbf{1}}) \\ & \times \mathbb{K}_{n_1}(\bar{u}^{\mathbf{1}}|\bar{v}^{\mathbf{1}})\mathbb{K}_{n_3}(\bar{v}^{\mathbf{3}}|\bar{u}^{\mathbf{3}})\, f(\bar{v}^{\mathbf{3}},\bar{v}^{\mathbf{1}})f(\bar{u}^{\mathbf{1}},\bar{u}^{\mathbf{3}}). \end{aligned} \tag{9.87}$$

Here the sum is taken over all possible partitions $\bar{u}_{\rm I} \mapsto \{\bar{u}^{\bf 1}, \bar{u}^{\bf 3}\}$ and $\bar{v}_{\rm I} \mapsto \{\bar{v}^{\bf 1}, \bar{v}^{\bf 3}\}$, such that $\#\bar{u}^{\bf 1} = \#\bar{v}^{\bf 1} = n_1$, $\#\bar{u}^{\bf 3} = \#\bar{v}^{\bf 3} = n_3$ and $n_1 + n_3 = n_{\rm I}$.

Similarly,

$$\boldsymbol{S}^{(2)}_{n_{\rm II}}(\bar{v}_{\rm II}|\bar{u}_{\rm II}) = \sum_{\substack{\bar{u}_{\rm II}\mapsto\{\bar{u}^{\bf 2},\bar{u}^{\bf 4}\} \\ \bar{v}_{\rm II}\mapsto\{\bar{v}^{\bf 2},\bar{v}^{\bf 4}\}}} r^{(2)}(\bar{u}^{\bf 4}) r^{(2)}(\bar{v}^{\bf 2}) \times \mathbb{K}_{n_2}(\bar{u}^{\bf 2}|\bar{v}^{\bf 2})\mathbb{K}_{n_4}(\bar{v}^{\bf 4}|\bar{u}^{\bf 4}) f(\bar{v}^{\bf 4}, \bar{v}^{\bf 2}) f(\bar{u}^{\bf 2}, \bar{u}^{\bf 4}). \tag{9.88}$$

Here the notation is similar to the one in (9.87).

Now we use the fact that the original vectors $\langle \boldsymbol{\Phi}^{(\kappa_2)}_n(\bar{v})|$ and $|\boldsymbol{\Phi}^{(\kappa_1)}_n(\bar{u})\rangle$ were the twisted on-shell vectors. Therefore, we can express the function $r^{(2)}(z)$ in terms of the function $\ell(z)$ in (9.88). Due to the twisted Bethe equations, we have

$$r^{(2)}(\bar{u}^{\bf 4}) = \frac{r(\bar{u}^{\bf 4})}{\ell(\bar{u}^{\bf 4})} = \frac{e^{n_4\beta_1}}{\ell(\bar{u}^{\bf 4})} \frac{f(\bar{u}^{\bf 4}, \bar{u}^{\bf 1}) f(\bar{u}^{\bf 4}, \bar{u}^{\bf 2}) f(\bar{u}^{\bf 4}, \bar{u}^{\bf 3})}{f(\bar{u}^{\bf 1}, \bar{u}^{\bf 4}) f(\bar{u}^{\bf 2}, \bar{u}^{\bf 4}) f(\bar{u}^{\bf 3}, \bar{u}^{\bf 4})}, \tag{9.89}$$

and

$$r^{(2)}(\bar{v}^{\bf 2}) = \frac{r(\bar{v}^{\bf 2})}{\ell(\bar{v}^{\bf 2})} = \frac{e^{n_2\beta_2}}{\ell(\bar{v}^{\bf 2})} \frac{f(\bar{v}^{\bf 2}, \bar{v}^{\bf 1}) f(\bar{v}^{\bf 2}, \bar{v}^{\bf 3}) f(\bar{v}^{\bf 2}, \bar{v}^{\bf 4})}{f(\bar{v}^{\bf 1}, \bar{v}^{\bf 2}) f(\bar{v}^{\bf 3}, \bar{v}^{\bf 2}) f(\bar{v}^{\bf 4}, \bar{v}^{\bf 2})}. \tag{9.90}$$

Then the scalar product $\boldsymbol{S}^{(2)}_{n_{\rm II}}(\bar{v}_{\rm II}|\bar{u}_{\rm II})$ takes the form

$$\boldsymbol{S}^{(2)}_{n_{\rm II}}(\bar{v}_{\rm II}|\bar{u}_{\rm II}) = \sum_{\substack{\bar{u}_{\rm II}\mapsto\{\bar{u}^{\bf 2},\bar{u}^{\bf 4}\} \\ \bar{v}_{\rm II}\mapsto\{\bar{v}^{\bf 2},\bar{v}^{\bf 4}\}}} \frac{e^{n_2\beta_2+n_4\beta_1}}{\ell(\bar{u}^{\bf 4})\ell(\bar{v}^{\bf 2})} \mathbb{K}_{n_2}(\bar{u}^{\bf 2}|\bar{v}^{\bf 2})\mathbb{K}_{n_4}(\bar{v}^{\bf 4}|\bar{u}^{\bf 4}) \times \frac{f(\bar{u}^{\bf 4}, \bar{u}^{\bf 1}) f(\bar{u}^{\bf 4}, \bar{u}^{\bf 2}) f(\bar{u}^{\bf 4}, \bar{u}^{\bf 3}) f(\bar{v}^{\bf 2}, \bar{v}^{\bf 1}) f(\bar{v}^{\bf 2}, \bar{v}^{\bf 3}) f(\bar{v}^{\bf 2}, \bar{v}^{\bf 4})}{f(\bar{u}^{\bf 1}, \bar{u}^{\bf 4}) f(\bar{u}^{\bf 3}, \bar{u}^{\bf 4}) f(\bar{v}^{\bf 1}, \bar{v}^{\bf 2}) f(\bar{v}^{\bf 3}, \bar{v}^{\bf 2})}. \tag{9.91}$$

The last thing that remains for us is to express the products $\ell(\bar{v}_{\rm II})$, $\ell(\bar{u}_{\rm I})$, $f(\bar{v}_{\rm I}, \bar{v}_{\rm II})$, and $f(\bar{u}_{\rm I}, \bar{u}_{\rm II})$ in (9.85) in terms of products over new subsets:

$$\begin{aligned} &e^{n_{\rm I}(\alpha+\beta_1)} \frac{\ell(\bar{v}_{\rm II})}{\ell(\bar{u}_{\rm I})} f(\bar{v}_{\rm I}, \bar{v}_{\rm II}) f(\bar{u}_{\rm I}, \bar{u}_{\rm II}) \\ &\quad = e^{(n_1+n_3)(\alpha+\beta_1)} \frac{\ell(\bar{v}^{\bf 2})\ell(\bar{v}^{\bf 4})}{\ell(\bar{u}^{\bf 1})\ell(\bar{u}^{\bf 3})} f(\bar{v}^{\bf 1}, \bar{v}^{\bf 2}) f(\bar{v}^{\bf 1}, \bar{v}^{\bf 4}) f(\bar{v}^{\bf 3}, \bar{v}^{\bf 2}) f(\bar{v}^{\bf 3}, \bar{v}^{\bf 4}) \\ &\qquad \times f(\bar{u}^{\bf 1}, \bar{u}^{\bf 2}) f(\bar{u}^{\bf 1}, \bar{u}^{\bf 4}) f(\bar{u}^{\bf 3}, \bar{u}^{\bf 2}) f(\bar{u}^{\bf 3}, \bar{u}^{\bf 4}). \end{aligned} \tag{9.92}$$

Now we need to substitute formulas (9.87), (9.91), and (9.92) into representation (9.85). All of these formulas are long enough. Therefore, one should be careful. However, nothing but accuracy here is required, and as a result, we obtain

$$\mathsf{Q}_n^{(\alpha)}(\bar v|\bar u) = \sum e^{(n_1+n_3)\alpha+(n_1+n_3+n_4)\beta_1+n_2\beta_2}\frac{\ell(\bar v^{\mathbf{1}})\ell(\bar v^{\mathbf{4}})}{\ell(\bar u^{\mathbf{1}})\ell(\bar u^{\mathbf{4}})}\,F_u\,F_v$$
$$\times\,\mathbb{K}_{n_1}(\bar u^{\mathbf{1}}|\bar v^{\mathbf{1}})\mathbb{K}_{n_3}(\bar v^{\mathbf{3}}|\bar u^{\mathbf{3}})\mathbb{K}_{n_2}(\bar u^{\mathbf{2}}|\bar v^{\mathbf{2}})\mathbb{K}_{n_4}(\bar v^{\mathbf{4}}|\bar u^{\mathbf{4}}), \qquad (9.93)$$

where

$$\begin{aligned}F_u &= f(\bar u^{\mathbf{1}},\bar u^{\mathbf{2}})f(\bar u^{\mathbf{1}},\bar u^{\mathbf{3}})f(\bar u^{\mathbf{3}},\bar u^{\mathbf{2}})f(\bar u^{\mathbf{4}},\bar u^{\mathbf{1}})f(\bar u^{\mathbf{4}},\bar u^{\mathbf{2}})f(\bar u^{\mathbf{4}},\bar u^{\mathbf{3}}),\\ F_v &= f(\bar v^{\mathbf{2}},\bar v^{\mathbf{1}})f(\bar v^{\mathbf{2}},\bar v^{\mathbf{4}})f(\bar v^{\mathbf{1}},\bar v^{\mathbf{4}})f(\bar v^{\mathbf{2}},\bar v^{\mathbf{3}})f(\bar v^{\mathbf{3}},\bar v^{\mathbf{4}})f(\bar v^{\mathbf{3}},\bar v^{\mathbf{1}}).\end{aligned} \qquad (9.94)$$

The sum in (9.93) is taken over partitions of the original sets $\bar v$ and $\bar u$ into four subsets each:

$$\bar u \mapsto \{\bar u^{\mathbf{1}},\bar u^{\mathbf{2}},\bar u^{\mathbf{3}},\bar u^{\mathbf{4}}\}, \qquad \bar v \mapsto \{\bar v^{\mathbf{1}},\bar v^{\mathbf{2}},\bar v^{\mathbf{3}},\bar v^{\mathbf{4}}\}. \qquad (9.95)$$

The restrictions are imposed only on the cardinalities of the subsets, namely, $\#\bar v^{\boldsymbol{j}} = \#\bar u^{\boldsymbol{j}} = n_j$, $j = 1,2,3,4$. By the way, let us recall once again that these restrictions are implicitly indicated in (9.93) in the form of subscripts of the highest coefficients. The latter are defined only under the condition that the number of elements to the left of the vertical line is equal to the number of elements to the vertical line's right.

Actually, all the calculations are already done, and we only need to see the result. For this, we introduce new subsets:

$$\begin{aligned}\bar u_{\rm i} &= \{\bar u^{\mathbf{1}},\bar u^{\mathbf{4}}\}, &\qquad \bar u_{\rm ii} &= \{\bar u^{\mathbf{2}},\bar u^{\mathbf{3}}\},\\ \bar v_{\rm i} &= \{\bar v^{\mathbf{1}},\bar v^{\mathbf{4}}\}, &\qquad \bar v_{\rm ii} &= \{\bar v^{\mathbf{2}},\bar v^{\mathbf{3}}\}.\end{aligned} \qquad (9.96)$$

Then, it is easy to see that

$$\begin{aligned}F_u &= f(\bar u_{\rm i},\bar u_{\rm ii})f(\bar u^{\mathbf{3}},\bar u^{\mathbf{2}})f(\bar u^{\mathbf{4}},\bar u^{\mathbf{1}}),\\ F_v &= f(\bar v_{\rm ii},\bar v_{\rm i})f(\bar v^{\mathbf{1}},\bar v^{\mathbf{4}})f(\bar v^{\mathbf{2}},\bar v^{\mathbf{3}}).\end{aligned} \qquad (9.97)$$

As a result, we obtain

$$\mathsf{Q}_n^{(\alpha)}(\bar v|\bar u) = e^{\beta_1 n}\sum_{\substack{\bar u\mapsto\{\bar u_{\rm i},\bar u_{\rm ii}\}\\ \bar v\mapsto\{\bar v_{\rm i},\bar v_{\rm ii}\}}}\frac{\ell(\bar v_{\rm i})}{\ell(\bar u_{\rm i})}\,f(\bar u_{\rm i},\bar u_{\rm ii})f(\bar v_{\rm ii},\bar v_{\rm i})e^{n_{\rm ii}\alpha}$$
$$\times\, G_1(\bar v_{\rm i}|\bar u_{\rm i})\,G_2(\bar v_{\rm ii}|\bar u_{\rm ii}), \qquad (9.98)$$

and functions $G_1(\bar v_{\rm i}|\bar u_{\rm i})$ and $G_2(\bar v_{\rm ii}|\bar u_{\rm ii})$ are given as sums over partitions into additional subsets:

$$G_1(\bar v_{\rm i}|\bar u_{\rm i}) = \sum_{\substack{\bar u_{\rm i}\mapsto\{\bar u^{\mathbf{1}},\bar u^{\mathbf{4}}\}\\ \bar v_{\rm i}\mapsto\{\bar v^{\mathbf{1}},\bar v^{\mathbf{4}}\}}} e^{n_1\alpha}\mathbb{K}_{n_1}(\bar u^{\mathbf{1}}|\bar v^{\mathbf{1}})\mathbb{K}_{n_4}(\bar v^{\mathbf{4}}|\bar u^{\mathbf{4}})f(\bar v^{\mathbf{1}},\bar v^{\mathbf{4}})f(\bar u^{\mathbf{4}},\bar u^{\mathbf{1}}),$$
$$(9.99)$$

and

$$G_2(\bar v_{\rm ii}|\bar u_{\rm ii}) = \sum_{\substack{\bar u_{\rm ii}\mapsto\{\bar u^{\bf 2},\bar u^{\bf 3}\}\\ \bar v_{\rm ii}\mapsto\{\bar v^{\bf 2},\bar v^{\bf 3}\}}} e^{n_2(\beta_2-\beta_1-\alpha)}\mathbb{K}_{n_3}(\bar v^{\bf 3}|\bar u^{\bf 3})\mathbb{K}_{n_2}(\bar u^{\bf 2}|\bar v^{\bf 2}) \times f(\bar v^{\bf 2},\bar v^{\bf 3})f(\bar u^{\bf 3},\bar u^{\bf 2}). \tag{9.100}$$

In other words, the initial summation over partitions (9.95) is now organized into several stages: first, we have partitions $\bar u\mapsto\{\bar u_{\rm i},\bar u_{\rm ii}\}$ and $\bar v\mapsto\{\bar v_{\rm i},\bar v_{\rm ii}\}$, and then each of the resulting subsets is divided once again into two subsets.

The sums (9.99) and (9.100) were calculated in proposition 6.7. Indeed, they are different from the function $\boldsymbol{S}^{(\kappa)}_n(\bar v|\bar u)$ (6.91) only by the labels of the subsets. Hence,

$$\begin{aligned} G_1(\bar v_{\rm i}|\bar u_{\rm i}) &= \boldsymbol{S}^{(\kappa)}_{n_{\rm i}}(\bar v_{\rm i}|\bar u_{\rm i}), \qquad \kappa=e^{\alpha},\\ G_2(\bar v_{\rm ii}|\bar u_{\rm ii}) &= \boldsymbol{S}^{(\kappa)}_{n_{\rm ii}}(\bar v_{\rm ii}|\bar u_{\rm ii}), \qquad \kappa=e^{\beta_2-\beta_1-\alpha}. \end{aligned} \tag{9.101}$$

Equation (9.98) takes the form

$$\mathsf{Q}^{(\alpha)}_n(\bar v|\bar u) = e^{\beta_1 n}\sum_{\substack{\bar u\mapsto\{\bar u_{\rm i},\bar u_{\rm ii}\}\\ \bar v\mapsto\{\bar v_{\rm i},\bar v_{\rm ii}\}}} \frac{\ell(\bar v_{\rm i})}{\ell(\bar u_{\rm i})} f(\bar u_{\rm i},\bar u_{\rm ii})f(\bar v_{\rm ii},\bar v_{\rm i})e^{n_{\rm ii}\alpha} \times \boldsymbol{S}^{(\kappa)}_{n_{\rm i}}(\bar v_{\rm i}|\bar u_{\rm i})\Big|_{\kappa=e^{\alpha}}\boldsymbol{S}^{(\kappa)}_{n_{\rm ii}}(\bar v_{\rm ii}|\bar u_{\rm ii})\Big|_{\kappa=e^{\beta_2-\beta_1-\alpha}}. \tag{9.102}$$

Let now $\alpha=\beta_2-\beta_1$. Then $G_2(\bar v_{\rm ii}|\bar u_{\rm ii})=\delta_{n_{\rm ii},0}$ due to (6.100). This implies $\bar v_{\rm ii}=\bar u_{\rm ii}=\emptyset$. Thus, $\bar v_{\rm i}=\bar v$, $\bar u_{\rm i}=\bar u$ and $n_{\rm i}=n$. We arrive at

$$G_1(\bar v|\bar u) = \boldsymbol{S}^{(\kappa)}_n(\bar v|\bar u), \qquad \kappa=e^{\beta_2-\beta_1}. \tag{9.103}$$

Hence, we obtain

$$\mathsf{Q}^{(\alpha)}_n(\bar v|\bar u) = e^{n\beta_1}\frac{\ell(\bar v)}{\ell(\bar u)}\boldsymbol{S}^{(\kappa)}_n(\bar v|\bar u), \qquad \kappa=e^{\beta_2-\beta_1}, \tag{9.104}$$

for $\alpha=\beta_2-\beta_1$.

Thus, all the numerous and very cumbersome sums over the partitions are eventually reduced to the scalar product of the twisted and the ordinary on-shell vectors, for which we have a compact representation in the form of a determinant.

Setting $\beta_1=0$ and $\beta_2=\beta$ we find

$$\langle\boldsymbol{\Phi}^{(\kappa)}_n(\bar v)|e^{\beta\hat q(x)}|\boldsymbol{\Phi}_n(\bar u)\rangle = \frac{\ell(\bar v)}{\ell(\bar u)}\boldsymbol{S}^{(\kappa)}_n(\bar v|\bar u) = \frac{\ell(\bar v)}{\ell(\bar u)}\langle\boldsymbol{\Phi}^{(\kappa)}_n(\bar v)|\boldsymbol{\Phi}_n(\bar u)\rangle, \tag{9.105}$$

for $\kappa = e^{\beta}$. Taking the $\beta$-derivative at $\beta = 0$ and differentiating over $x$, we finally arrive at

$$\langle \mathbf{\Phi}_n(\bar{v})|\Psi^{\dagger}(x)\Psi(x)|\mathbf{\Phi}_n(\bar{u})\rangle = \frac{\partial}{\partial x}\frac{\partial}{\partial \beta}\frac{\ell(\bar{v})}{\ell(\bar{u})}\langle \mathbf{\Phi}_n^{(\kappa)}(\bar{v})|\mathbf{\Phi}_n(\bar{u})\rangle\Big|_{\beta=0}. \tag{9.106}$$

Here in the lhs, both states are ordinary on-shell vectors. Thus, this is the form factor of the density operator $\Psi^{\dagger}(x)\Psi(x)$:

$$\mathcal{F}_{n,n}(\bar{v},\bar{u}|\Psi^{\dagger}(x)\Psi(x)) = \frac{\partial}{\partial x}\frac{\partial}{\partial \beta}\frac{\ell(\bar{v})}{\ell(\bar{u})}\langle \mathbf{\Phi}_n^{(\kappa)}(\bar{v})|\mathbf{\Phi}_n(\bar{u})\rangle\Big|_{\beta=0}. \tag{9.107}$$

As usual, one should distinguish two cases: (i) $\bar{v} \neq \bar{u}$ at $\beta = 0$; (ii) $\bar{v} = \bar{u}$ at $\beta = 0$. In the first case, the $\beta$-derivative in (9.106) acts only on the scalar product. Then due to (8.39) and (9.83), we obtain

$$\mathcal{F}_{n,n}(\bar{v},\bar{u}|\Psi^{\dagger}(x)\Psi(x)) = \left(\frac{\partial}{\partial x}\prod_{k=1}^{n} e^{ix(u_k-v_k)}\right)\mathfrak{F}_{n,n}^{(22)}(\bar{v},\bar{u}), \tag{9.108}$$

where $\mathfrak{F}_{n,n}^{(22)}(\bar{v},\bar{u})$ is the universal form factor of the operator $D(z)$.

In case (ii), the $\beta$-derivative acts only on the ratio $\ell(\bar{v})/\ell(\bar{u})$. We can easily calculate this derivative since the twisted Bethe equations in the one-dimensional Bose gas have the form

$$e^{-iLv_j} = -e^{\beta}\prod_{k=1}^{n}\frac{v_j - v_k - ic}{v_j - v_k + ic}. \tag{9.109}$$

It is clear that if $v_k = u_k$ at $\beta = 0$ for $k = 1,\dots,n$, then $v_k = u_k + i\beta/L$. Then we immediately arrive at

$$\langle \mathbf{\Phi}_n(\bar{u})|\Psi^{\dagger}(x)\Psi(x)|\mathbf{\Phi}_n(\bar{u})\rangle = \frac{n}{L}\langle \mathbf{\Phi}_n(\bar{u})|\mathbf{\Phi}_n(\bar{u})\rangle, \tag{9.110}$$

which is entirely analogous to the result (9.37) that we obtained when calculating the magnetization.

Setting $\beta_2 = 0$, $\beta_1 = \beta$ in (9.104) we obtain one more representation

$$\langle \mathbf{\Phi}_n(\bar{v})|e^{-\beta\hat{q}(x)}|\mathbf{\Phi}_n^{(\kappa)}(\bar{u})\rangle = e^{n\beta}\frac{\ell(\bar{v})}{\ell(\bar{u})}\boldsymbol{S}_n^{(\kappa^{-1})}(\bar{v}|\bar{u}), \tag{9.111}$$

where $\kappa = e^{\beta}$. Using proposition 6.7, one can easily check that

$$\kappa^n \boldsymbol{S}_n^{(\kappa^{-1})}(\bar{v}|\bar{u}) = \boldsymbol{S}_n^{(\kappa)}(\bar{u}|\bar{v}). \tag{9.112}$$

Substituting this in (9.111) and replacing $\bar{v} \leftrightarrow \bar{u}$, we obtain

$$\langle \mathbf{\Phi}_n(\bar{u})|e^{-\beta\hat{q}(x)}|\mathbf{\Phi}_n^{(\kappa)}(\bar{v})\rangle = \frac{\ell(\bar{u})}{\ell(\bar{v})}\boldsymbol{S}_n^{(\kappa)}(\bar{v}|\bar{u}) = \frac{\ell(\bar{u})}{\ell(\bar{v})}\langle \mathbf{\Phi}_n^{(\kappa)}(\bar{v})|\mathbf{\Phi}_n(\bar{u})\rangle, \tag{9.113}$$

where $\kappa = e^{\beta}$.

Let us summarize. The calculation of the generating function within the framework of the composite model is much more cumbersome than the calculation based on the quantum inverse problem. However, in this calculation, we do not use any additional information about the representation of the $RTT$ algebra. All required for this calculation is the operator $\hat{q}(x)$ action on the partial Bethe vector (see (9.81)). All other calculations are universal. Note also that there is a complete analogy between the final results for the form factors of local operators in a one-dimensional Bose gas and the $XXX$ chain. In both cases, the form factors of local operators are expressed in terms of the universal form factors.

## 9.4 One more determinant representation

The determinant formulas for form factors obtained above are not very convenient for calculating the thermodynamic limit, in which the number of Bethe parameters tends to infinity. In this section, we will transform them into a new form, which we will use later in calculating the thermodynamic limit of form factors. The idea of the transformation is to explicitly extract the Cauchy determinant (7.78) in representations for form factors. We will consider in detail a concrete example of such a transformation in the case of the scalar product of on-shell and twisted on-shell Bethe vectors for the models with the rational $R$-matrix. Determinant representations for other form factors transform similarly.

Consider a scalar product of twisted on-shell and ordinary on-shell Bethe vectors

$$\boldsymbol{S}_n^{(\kappa)}(\bar{v}|\bar{u}) = \langle \boldsymbol{\Phi}_n^{(\kappa)}(\bar{v})|\boldsymbol{\Phi}_n(\bar{u})\rangle. \tag{9.114}$$

We begin with representation (6.93)

$$\boldsymbol{S}_n^{(\kappa)}(\bar{v}|\bar{u}) = \Delta'(\bar{v})\Delta(\bar{u})h(\bar{v},\bar{u}) \det_n \mathcal{M}^{(\kappa)}. \tag{9.115}$$

Here

$$\mathcal{M}_{jk}^{(\kappa)} = t(v_j,u_k) + \kappa t(u_k,v_j)\frac{V^-(u_k)}{V^+(u_k)}, \tag{9.116}$$

where

$$V^+(z) = \frac{h(\bar{v},z)}{h(\bar{u},z)}, \qquad V^-(z) = \frac{h(z,\bar{v})}{h(z,\bar{u})}. \tag{9.117}$$

**Proposition 9.2.** *The scalar product $\boldsymbol{S}_n^{(\kappa)}(\bar{v}|\bar{u})$ has the following determinant representation:*

$$\boldsymbol{S}_n^{(\kappa)}(\bar{v}|\bar{u}) = \frac{(1-\kappa)f(\bar{v},\bar{u})}{V^+(\omega)-\kappa V^-(\omega)} \prod_{a=1}^{n}\left(1-\kappa\frac{V^-(u_a)}{V^+(u_a)}\right) \det_n \big(\delta_{jk} - U_{jk}(\omega)\big). \tag{9.118}$$

*Here $\omega$ is an arbitrary complex number. The entries $U_{jk}$ are*

$$U_{jk}(\omega) = \frac{g(u_k, \bar{u}_k)}{g(u_k, \bar{v})} \left( \frac{K_\kappa(u_j, u_k) - K_\kappa(u_j, \omega)}{\kappa V^-(u_k) - V^+(u_k)} \right), \tag{9.119}$$

*where*

$$K_\kappa(u, v) = \frac{1}{h(u, v)} + \frac{\kappa}{h(v, u)}. \tag{9.120}$$

**Proof.** We first prove equation (9.118) for $\omega = u_n$. For this, we transform $\det_n \mathcal{M}^{(\kappa)}$ as follows:

$$\det_n \mathcal{M}^{(\kappa)} = \frac{\det_n(\mathcal{A}\mathcal{M}^{(\kappa)})}{\det_n \mathcal{A}}, \tag{9.121}$$

where

$$\mathcal{A}_{jk} = \frac{g(v_k, \bar{v}_k)}{g(v_k, \bar{u})} \widetilde{\mathcal{A}}_{jk}, \qquad \text{and} \qquad \widetilde{\mathcal{A}}_{jk} = \begin{cases} g(v_k, u_j), & j \neq n, \\ 1, & j = n. \end{cases} \tag{9.122}$$

**Remark 9.1.** Here we are dealing with the rare case when the formula for $\widetilde{\mathcal{A}}_{jk}$ does not generalize literally to the models with the trigonometric $R$-matrix. In the latter case, we should put

$$\widetilde{\mathcal{A}}_{jk} = \begin{cases} \coth(v_k - u_j), & j \neq n, \\ 1, & j = n. \end{cases} \tag{9.123}$$

All further considerations nevertheless do not change.

**Remark 9.2.** We could put $\widetilde{\mathcal{A}}_{jk} = g(v_k, u_j)$ in all rows of the matrix $\widetilde{\mathcal{A}}$, including $j = n$. Then we get another representation for the scalar product, which is slightly different from (9.118). The choice of the matrix $\widetilde{\mathcal{A}}$ in the form (9.122) is explained by the fact that we want to extract the factor $\kappa - 1$ explicitly. In the future, this will make it easier for us to differentiate over $\kappa$ at the point $\kappa = 1$.

Let us find $\det_n \mathcal{A}$. Obviously,

$$\det_n \mathcal{A} = \frac{\Delta'(\bar{v})\Delta(\bar{v})}{g(\bar{v}, \bar{u})} \det_n \widetilde{\mathcal{A}}. \tag{9.124}$$

To compute $\det_n \widetilde{\mathcal{A}}$, we subtract the last column of $\widetilde{\mathcal{A}}_{jk}$ from all previous columns. Then we obtain zeros in the last row of this matrix everywhere except the element $\widetilde{\mathcal{A}}_{nn} = 1$. Thus,

$$\det_n \widetilde{\mathcal{A}} = \det_{n-1} \Big(g(v_k, u_j) - g(v_n, u_j)\Big). \tag{9.125}$$

Using

$$g(v_k, u_j) - g(v_n, u_j) = \frac{g(v_k, u_j)g(v_n, u_j)}{g(v_n, v_k)},$$

we reduce (9.125) to the Cauchy determinant (7.78)

$$\det_n \widetilde{\mathcal{A}} = \frac{g(v_n, \bar{u}_n)}{g(v_n, \bar{v}_n)} \det_{n-1} g(v_k, u_j) = \frac{g(v_n, \bar{u}_n)}{g(v_n, \bar{v}_n)} \frac{g(\bar{v}_n, \bar{u}_n)}{\Delta(\bar{v}_n)\Delta'(\bar{u}_n)}. \tag{9.126}$$

Combining (9.126) and (9.124), we find

$$\det_n \mathcal{A} = \frac{\Delta'(\bar{v})}{\Delta'(\bar{u})} \frac{g(\bar{u}_n, u_n)}{g(\bar{v}, u_n)}. \tag{9.127}$$

Now we should calculate the product $\mathcal{A}\mathcal{M}^{(\kappa)}$. For this, we use the standard contour integral method. In particular, we find for $j < n$:

$$\begin{aligned} \sum_{\ell=1}^{n} \mathcal{A}_{j\ell} t(u_k, v_\ell) &= \delta_{jk} \frac{g(u_k, \bar{v})}{g(u_k, \bar{u}_k)} - \frac{1}{h(u_k, u_j)V^-(u_k)}, \\ \sum_{\ell=1}^{n} \mathcal{A}_{j\ell} t(v_\ell, u_k) &= -\delta_{jk} \frac{g(u_k, \bar{v})}{g(u_k, \bar{u}_k)} - \frac{1}{h(u_j, u_k)V^+(u_k)}, \end{aligned} \tag{9.128}$$

and for $j = n$:

$$\begin{aligned} \sum_{\ell=1}^{n} \mathcal{A}_{j\ell} t(u_k, v_\ell) &= -\frac{1}{V^-(u_k)}, \\ \sum_{\ell=1}^{n} \mathcal{A}_{j\ell} t(v_\ell, u_k) &= \frac{1}{V^+(u_k)}. \end{aligned} \tag{9.129}$$

Let us give a brief comment on how these formulas were obtained. Let us derive, for instance, the first equation (9.128). For this, we consider a contour integral

$$I = \oint_{|z|=R\to\infty} \frac{\eta^2 \mathrm{d}z}{2\pi i(u_k - z + \eta)(u_k - z)(z - u_j)} \prod_{p=1}^{n} \frac{z - u_p}{z - v_p}. \tag{9.130}$$

This integral vanishes because the integrand behaves as $z^{-3}$ at $z \to \infty$. On the other hand, it is equal to the sum of the residues within the integration contour. The sum of the residues at $z = v_\ell$, $\ell = 1, \dots, n$ gives the lhs of the first equation (9.128). One more pole occurs at $z = u_k + \eta$. Finally, there is a simple pole at $z = u_k$ for $k = j$. The sum of all these residues vanishes, which implies the first equation (9.128).

All other sums in (9.128), (9.129) can be calculated similarly. Using these formulas, we find

$$\begin{aligned}(\mathcal{AM}^{(\kappa)})_{jk} &= \delta_{jk}\frac{g(u_k,\bar v)}{g(u_k,\bar u_k)}\left(\kappa\frac{V^-(u_k)}{V^+(u_k)}-1\right) - \frac{K_\kappa(u_j,u_k)}{V^+(u_k)}, \qquad j<n,\\ (\mathcal{AM}^{(\kappa)})_{nk} &= \frac{1-\kappa}{V^+(u_k)}.\end{aligned} \tag{9.131}$$

Now we can subtract the last column $(\mathcal{AM}^{(\kappa)})_{jn}$ multiplied by $V^+(u_n)/V^+(u_k)$ from all previous columns. Then the elements of the last row vanish except the element $(\mathcal{AM}^{(\kappa)})_{nn} = (1-\kappa)/V^+(u_n)$. We obtain

$$\begin{aligned}\det_n(\mathcal{AM}^{(\kappa)}) = \frac{1-\kappa}{V^+(u_n)}\det_{n-1}\Bigg[&\delta_{jk}\frac{g(u_k,\bar v)}{g(u_k,\bar u_k)}\left(\kappa\frac{V^-(u_k)}{V^+(u_k)}-1\right)\\ &- \frac{K_\kappa(u_j,u_k)-K_\kappa(u_j,u_n)}{V^+(u_k)}\Bigg].\end{aligned} \tag{9.132}$$

Finally, extracting the factor of the $\delta_{jk}$, we arrive at

$$\begin{aligned}\det_n(\mathcal{AM}^{(\kappa)}) = &\frac{1-\kappa}{V^+(u_n)}\frac{g(\bar u_n,\bar v)}{\Delta'(\bar u)\Delta(\bar u_n)}\\ &\times\prod_{a=1}^{n-1}\left(\kappa\frac{V^-(u_a)}{V^+(u_a)}-1\right)\det_{n-1}\big(\delta_{jk}-U_{jk}(u_n)\big).\end{aligned} \tag{9.133}$$

Using (9.115) and (9.127), we obtain

$$\boldsymbol{S}_n^{(\kappa)}(\bar v|\bar u) = \frac{1-\kappa}{V^+(u_n)}f(\bar v,\bar u)\prod_{a=1}^{n-1}\left(1-\kappa\frac{V^-(u_a)}{V^+(u_a)}\right)\det_n\big(\delta_{jk}-U_{jk}(u_n)\big). \tag{9.134}$$

This formula coincides with (9.118) for $\omega = u_n$. Note that we have replaced the determinant of the $(n-1)\times(n-1)$ matrix in (9.133) with the determinant of the $n\times n$ matrix in (9.134). This can be done because $U_{nn}(u_n) = 0$.

Let us prove now that a combination

$$\frac{\det_n\big(\delta_{jk}-U_{jk}(\omega)\big)}{\kappa V^-(\omega)-V^+(\omega)},$$

does not depend on $\omega$. For this, we transform the determinant of the matrix $\delta_{jk}-U_{jk}(\omega)$ as follows.

Let

$$\gamma_j = \frac{g(u_j, \bar{u}_j)}{g(u_j, \bar{v})}. \tag{9.135}$$

Using the contour integral method, we easily find

$$\begin{aligned} \sum_{j=1}^{n} \frac{\gamma_j}{h(w, u_j)} &= V^-(w) - 1, \\ \sum_{j=1}^{n} \frac{\gamma_j}{h(u_j, w)} &= 1 - V^+(w), \end{aligned} \tag{9.136}$$

for arbitrary complex $w$. Using these formulas, we obtain

$$\sum_{j=1}^{n} \gamma_j U_{jk}(\omega) = \gamma_k \left(1 - \frac{\kappa V^-(\omega) - V^+(\omega)}{\kappa V^-(u_k) - V^+(u_k)}\right). \tag{9.137}$$

Hence, adding to the last row $\delta_{nk} - U_{nk}$ all other rows multiplied by $\gamma_j/\gamma_n$, we arrive at

$$\delta_{nk} - U_{nk}(\omega) + \sum_{j=1}^{n-1} \frac{\gamma_j}{\gamma_n}(\delta_{jk} - U_{jk}(\omega)) = \frac{c_k}{\gamma_n}\big(\kappa V^-(\omega) - V^+(\omega)\big), \tag{9.138}$$

where

$$c_k = \frac{\gamma_k}{\kappa V^-(u_k) - V^+(u_k)}. \tag{9.139}$$

Hence,

$$\det_n \big(\delta_{jk} - U_{jk}(\omega)\big) = \frac{\kappa V^-(\omega) - V^+(\omega)}{\gamma_n} \det_n \mathcal{B}_{jk}, \tag{9.140}$$

where

$$\begin{aligned} \mathcal{B}_{jk} &= \delta_{jk} - c_k\big(K_\kappa(u_j, u_k) - K_\kappa(u_j, \omega)\big), \qquad j < n, \\ \mathcal{B}_{nk} &= c_k. \end{aligned} \tag{9.141}$$

Now we can subtract the last column multiplied by $c_k/c_n$ from all previous columns. This gives

$$\det_n \mathcal{B}_{jk} = c_n \det_n \widetilde{\mathcal{B}}_{jk}, \tag{9.142}$$

where

$$\widetilde{\mathcal{B}}_{jk} = \delta_{jk} - c_k\big(K_\kappa(u_j, u_k) - K_\kappa(u_j, u_n)\big). \tag{9.143}$$

Combining all together, we arrive at

$$\begin{aligned} &\frac{\det_n \Big(\delta_{jk} - c_k\big(K_\kappa(u_j, u_k) - K_\kappa(u_j, \omega)\big)\Big)}{\kappa V^-(\omega) - V^+(\omega)} \\ &\qquad = \frac{\det_n \Big(\delta_{jk} - c_k\big(K_\kappa(u_j, u_k) - K_\kappa(u_j, u_n)\big)\Big)}{\kappa V^-(u_n) - V^+(u_n)}. \end{aligned} \tag{9.144}$$

We see that the combination in the rhs of (9.144) does not depend on $\omega$. Thus, equation (9.118) is proved. □

Similarly, starting with representation (6.95), one can prove

$$\boldsymbol{S}_n^{(\kappa)}(\bar v|\bar u) = \frac{(1-\kappa)f(\bar v,\bar u)}{\big(V^-(\omega)\big)^{-1} - \kappa\big(V^+(\omega)\big)^{-1}} \prod_{a=1}^{n}\left(1-\kappa\frac{V^-(v_a)}{V^+(v_a)}\right) \times \det_n\big(\delta_{jk} + \tilde U_{jk}(\omega)\big), \tag{9.145}$$

where the entries $\tilde U_{jk}$ are

$$\tilde U_{jk}(\omega) = \frac{g(v_k,\bar v_k)}{g(v_k,\bar u)} \frac{K_\kappa(v_k,v_j) - K_\kappa(\omega,v_j)}{\kappa\big(V^+(v_k)\big)^{-1} - \big(V^-(v_k)\big)^{-1}}. \tag{9.146}$$

Due to relationships (9.70), (8.39), and (8.62), we easily obtain a new determinant representation for the field form factor

$$\mathcal{F}_{n-1,n}(\bar v,\bar u|\Psi(0)) = \frac{-i\sqrt{c}f(\bar v,\bar u)}{V^+(\omega)+V^-(\omega)} \prod_{a=1}^{n}\left(1+\frac{V^-(u_a)}{V^+(u_a)}\right) \times \det_n\big(\delta_{jk} - U_{jk}^{(\Psi)}(\omega)\big). \tag{9.147}$$

Here the notation is basically the same as in (9.118). In particular, $\omega$ is an arbitrary complex number, and the entries $U_{jk}^{(\Psi)}$ are

$$U_{jk}^{(\Psi)}(\omega) = \frac{g(u_k,\bar u_k)}{g(u_k,\bar v)} \frac{K_1(u_j,u_k) - K_1(u_j,\omega)}{V^-(u_k)+V^+(u_k)}, \tag{9.148}$$

where

$$K_1(u,v) = K_\kappa(u,v)\Big|_{\kappa=1}. \tag{9.149}$$

The functions $V^\pm(z)$ are still given by (9.117). However, it should be remembered that $\#\bar u = \#\bar v = n$ in the case of the scalar product, while $\#\bar u = n$ and $\#\bar v = n-1$ in the case of the field form factor. The difference between the two $V^\pm(z)$ becomes obvious if we abandon the shorthand notation for the products

$$\begin{aligned} V^\pm(z) &= \prod_{j=1}^{n}\frac{z-v_j\mp\eta}{z-u_j\mp\eta}, && \text{for the scalar product,}\\ V^\pm(z) &= \frac{\prod_{j=1}^{n-1}(z-v_j\mp\eta)}{\prod_{j=1}^{n}(z-u_j\mp\eta)}, && \text{for the field form factor.}\end{aligned} \tag{9.150}$$

Relationships between form factors of type (8.62), (8.63) exist only for models with the rational $R$-matrix. In particular, they are valid for the one-dimensional Bose gas and the $XXX$ chain. However, in the $XXZ$ chain, we cannot obtain the form factors of the $\sigma^\pm$ operators by taking the limit in the form factor of the $\sigma^z$ operator. Therefore, to extract the Cauchy determinant from representations (8.7), (8.12), we have to repeat the procedure described above. We hope that it will not be difficult for the reader to do this on their own. Representations of type (9.147) for the $\sigma^{\pm}$ form factors in the $XXZ$ chain can be found in [N. Kitanine et al (2011a)].

# Chapter 10

# Thermodynamic limit

Before proceeding directly to calculating correlation functions, we describe the thermodynamic limit of the models considered above. In the thermodynamic limit, the system size goes to infinity. At the same time, the magnetization of the spin chain (or the average density of the one-dimensional Bose gas) remains finite.

We consider the thermodynamic limit for two reasons. First, large systems are of the most significant interest from the point of view of physics. Secondly, if we are dealing, for example, with a spin chain of finite length, then, in fact, we already have everything necessary to calculate the correlation functions. In particular, we have explicit determinant representations for the form factors of local spin operators. We should only solve the Bethe equations numerically and substitute the roots found in explicit formulas. After that, we can calculate the correlation functions, for example, using the form factor expansion, which we will consider in Section 11.5 and Chapter 12. Such numerical calculations were carried out in [D. Biegel et al (2002, 2003); M. Karbach et al (2002); J. Sato et al (2204); P. Calabrese and J.-S. Caux (2007); J.-S. Caux and P. Calabrese (2006); J.-S. Caux et al (2007, 2003, 2005); J.-S. Caux and J.-M. Maillet (2005); R. Hagemans et al (2005); R. G. Pereira et al (2006, 2008); J. Sirker et al (2008)].

The thermodynamic limit allows us to go a little further without resorting to the help of a computer. Then, instead of Bethe equations, we get an integral equation for the distribution density of Bethe roots. Respectively, the sums over the Bethe roots turn into integrals, which in some cases can be explicitly calculated. In this sense, the thermodynamic limit is an additional condition that simplifies the problem.

In this chapter, we mainly focus on the thermodynamic limit of the ground state of the one-dimensional Bose gas. In this case, we can rigorously prove most statements. However, it should be noted that rigorous proofs often require quite complicated mathematical tools in the study of the thermodynamic limit. Therefore, sometimes we will use qualitative arguments at the level of physical intuition in our reasoning.

In Section 10.1, we study the properties of the Bethe equations solutions in a one-dimensional Bose gas with repulsive interaction. In Section 10.2, we construct the ground state of this model in the thermodynamic limit. Here we derive an integral equation for the distribution density. Section 10.3 is devoted to excited states. Here we deviate from the traditional interpretation and consider the twisted on-shell Bethe vectors as excited states. From the point of view of calculating correlation functions, this approach is often more convenient. At the same time, we can always go back to the traditional approach by setting the twist parameter equal to 1. In Section 10.4, we introduce an additional term proportional to the particles number operator into the Hamiltonian of the quantum nonlinear Schrödinger equation. This allows us to consider excitations of a more general type. Section 10.5 describes a nonlinear identity between the dressed charge and the shift function at the Fermi boundaries. In Section 10.6, we consider the thermodynamic limit of the $XXZ$ chain in the massless regime. We mainly focus on some differences between this model and the one-dimensional bosons. In particular, we consider the case of a zero external magnetic field when the integral equation for the distribution density is explicitly solved in elementary functions. Finally, in Section 10.7, we give examples of calculating the thermodynamic limits of sums, products, and determinants.

The reader can learn more about the thermodynamic limit of the ground state of the one-dimensional bosons in the original works [E. H. Lieb and W. Liniger (1963); E. H. Lieb (1963); E. Lieb and D. Mattis (1961)]. Thermodynamics at a finite temperature was studied in [C. N. Yang and C. P. Yang (1969)]. The thermodynamic limit of the $XXZ$ chain was studied in works [L. Hultén (1939); C. N. Yang and C. P. Yang (1966a,b); R. J. Baxter (1982); L. D. Faddeev and L. A. Takhtajan (1984); E. V. Gusev (1980, 1985, 1986); R. K. Bullough et al (1990); M. Takahashi (1971, 1973); M. Takahashi and M. Suzuki (1972); T. C. Dorlas and M. Samsonov (2009); K. K. Kozlowski (2018b); H. Duminil-Copin et al (2020)]. Both models are described in sufficient detail in monographs [M. Gaudin (1983); V. E. Korepin et al (1993); H.-P. Eckle (2019)].

## 10.1 Bethe equations in the one-dimensional Bose gas

This section studies some properties of the solutions to the Bethe equations in the model of the one-dimensional Bose gas. We will need them to study the thermodynamic limit. From now on, we consider only the positive coupling constant $c > 0$, that is, the repulsive case.

Bethe equations in the one-dimensional Bose gas have the form

$$e^{iLu_j} = -\prod_{k=1}^{N} \frac{u_j - u_k + ic}{u_j - u_k - ic}, \qquad j = 1, \dots, N. \tag{10.1}$$

Recall that $L$ is the length of the interval on which the Hamiltonian of the model (1.76) is initially defined. The roots of the Bethe equations correspond to the on-shell Bethe vectors $|\mathbf{\Phi}_N(\bar{u})\rangle$, $\bar{u} = \{u_1, \dots, u_N\}$. Note that starting from this section, we denote the number of Bethe parameters by the capital $N$. A small $n$ will usually denote the number of excitations. Our goal is to study the limit in which $N, L \to \infty$, while their ratio $N/L = D$ remains a fixed finite value.

We have already noted in Section 4.3.2 that in the framework of the coordinate Bethe ansatz, equations (10.1) correspond to periodic boundary conditions. We can also consider the one-dimensional Bose gas with quasi-periodic boundary conditions $\Psi(x + L) = e^{i\alpha}\Psi(x)$, $\Im\alpha = 0$. In this case, we are dealing with twisted Bethe equations of the form

$$e^{iLu_j} = -e^{i\alpha} \prod_{k=1}^{N} \frac{u_j - u_k + ic}{u_j - u_k - ic}, \qquad j = 1, \dots, N. \tag{10.2}$$

In both cases, the Hamiltonian eigenvalues $E$ are given by

$$E = \sum_{j=1}^{N} u_j^2. \tag{10.3}$$

It is easy to see that every solution of system (10.2) can be obtained by shifting by $\alpha/L$ the solutions of system (10.1).

**Proposition 10.1.** *If* $c > 0$, *then all the roots of the system* (10.1) *are real.*

**Proof.** Let $u_{\max}$ be the root of the system (10.1) with the maximum imaginary part. Consider the equation corresponding to this root and take the absolute value of both sides

$$\left|e^{iLu_{\max}}\right| = \prod_{k=1}^{N} \left|\frac{u_{\max} - u_k + ic}{u_{\max} - u_k - ic}\right|. \tag{10.4}$$

Obviously, the rhs of this equation is greater than or equal to 1. Looking at the lhs of (10.4), we conclude that $\Im u_{\max} \leq 0$. Similarly, we prove that $\Im u_{\min} \geq 0$, where $u_{\min}$ is the root of the system (10.1) with the minimum imaginary part. Hence, we arrive at the statement of the proposition. □

Let us write down Bethe equations in a logarithmic form, as we did in (6.78)

$$Lu_j + \sum_{k=1}^{N} \theta(u_j - u_k) = 2\pi I_j. \tag{10.5}$$

Here

$$\theta(u) = i \log\left(\frac{ic+u}{ic-u}\right) = 2\arctan\frac{u}{c}, \qquad \theta(0) = 0, \tag{10.6}$$

and $I_j$ are integer or half-integer numbers (depending on whether $N$ is odd or even). Note that $\theta(u)$ is a monotonically increasing function: $\theta(u_2) > \theta(u_1)$ for $u_2 > u_1$.

**Proposition 10.2.** *[C. N. Yang and C. P. Yang (1969)] System* (10.5) *has a unique solution at* $I_j$ *fixed.*

**Proof.** Consider a functional

$$Y = \frac{L}{2}\sum_{j=1}^{N} u_j^2 - 2\pi \sum_{j=1}^{N} I_j u_j + \frac{1}{2}\sum_{j,k=1}^{N} \theta_1(u_j - u_k), \tag{10.7}$$

where

$$\theta_1(u) = \int_0^u \theta(v)\, dv. \tag{10.8}$$

We prove that the functional has a unique minimum. The extremum condition $\partial Y/\partial u_j = 0$ coincides with the system (10.5). Consider a matrix of second derivatives

$$\frac{\partial^2 Y}{\partial u_j \partial u_k} = \delta_{jk}\left(L + \sum_{a=1}^{N} \mathcal{K}(u_j - u_a)\right) - \mathcal{K}(u_j - u_k), \tag{10.9}$$

where

$$\mathcal{K}(u) = \theta'(u) = \frac{2c}{u^2 + c^2}. \tag{10.10}$$

Let $\bar{x} = \{x_1, \dots, x_N\}$ be a set of real numbers. Then we obtain for quadratic form

$$\sum_{j,k=1}^{N} \frac{\partial^2 Y}{\partial u_j \partial u_k} x_j x_k = L\sum_{j=1}^{N} x_j^2 + \sum_{j>k}^{N} \mathcal{K}(u_j - u_k)(x_j - x_k)^2 \geq 0. \tag{10.11}$$

Thus, the matrix (10.9) is positive definite. Hence, the functional $Y$ is convex. It behaves as $Lu_j^2/2$ at $u_j \to \infty$, $j = 1, \dots, N$. Thus, $Y$ has a unique minimum. □

**Remark 10.1.** In Section 7.2.2 (see the footnote on page 151), we noted that in some models, it is possible to rigorously prove that the principal minor of the matrix $S$ (7.100) is positive definite for $\bar{u} = \bar{v}$. It is easy to check that this minor coincides with the matrix (10.9) in the model of one-dimensional bosons.

**Remark 10.2.** The proof of proposition 10.2 implies the positivity of the norm of on-shell Bethe vectors in the one-dimensional Bose gas model (see the general formula (6.69) and the more detailed expression (10.181)).

Let us now make some estimates for the roots of the system (10.5).

**Proposition 10.3.** *If $I_j > I_k$, then $u_j > u_k$. If $I_j = I_k$, then $u_j = u_k$.*

**Proof.** Subtracting the equations corresponding to $I_j$ and $I_k$, we obtain

$$L(u_j - u_k) + \sum_{l=1}^{N} \big(\theta(u_j - u_l) - \theta(u_k - u_l)\big) = 2\pi(I_j - I_k). \tag{10.12}$$

Since $\theta(u)$ monotonically increases, the lhs has the same sign as the first term $L(u_j - u_k)$. □

Due to proposition 10.3, the integers $I_j$ in equations (10.5) should be pairwise distinct. Otherwise, the Bethe equations have coinciding roots. Then the corresponding vector Bethe is not an on-shell vector.

**Proposition 10.4.** *At $I_j > I_k$, the following estimates are valid:*

$$\frac{2\pi(I_j - I_k)}{L} \geq u_j - u_k \geq \frac{2\pi(I_j - I_k)}{L(1 + 2D/c)}, \tag{10.13}$$

*where $D = N/L$ is an average density.*

**Proof.** It follows from (10.12) that

$$u_j - u_k \leq \frac{2\pi(I_j - I_k)}{L}. \tag{10.14}$$

To prove the second inequality, we note that for real $u$, the function $\mathcal{K}(u)$ defined in (10.10) satisfies a condition

$$0 < \mathcal{K}(u) \leq \frac{2}{c}. \tag{10.15}$$

Then for $u_j > u_k$,

$$\theta(u_j - u_l) - \theta(u_k - u_l) = \int_{u_k}^{u_j} \mathcal{K}(v - u_l)\,\mathrm{d}v \leq \frac{2}{c}(u_j - u_k). \tag{10.16}$$

Substituting this estimate into (10.12), we find

$$\frac{2\pi(I_j - I_k)}{L} = u_j - u_k + \frac{1}{L}\sum_{l=1}^{N}\left(\theta(u_j - u_l) - \theta(u_k - u_l)\right)$$
$$\leq (u_j - u_k)\left(1 + 2\frac{D}{c}\right). \tag{10.17}$$

□

**Proposition 10.5.** *Let integers $I_j$ in (10.5) be such that $I_j = -I_{N+1-j}$. Then $u_j = -u_{N+1-j}$.*

**Proof.** It follows from (10.5) that

$$-Lu_j + \sum_{k=1}^{N}\theta(-u_j + u_k) = -2\pi I_j, \tag{10.18}$$

where we used $\theta(-x) = -\theta(x)$. Setting $u_l = -v_{N+1-l}$, $l = 1, \dots, N$, and using $I_j = -I_{N+1-j}$ we obtain

$$Lv_{N+1-j} + \sum_{k=1}^{N}\theta(v_{N+1-j} - v_k) = 2\pi I_{N+1-j}. \tag{10.19}$$

Due to the uniqueness of the solution of system (10.5), we conclude that $v_{N+1-j} = u_{N+1-j}$. □

To conclude this section, we consider the limit $c \to \infty$. Such a system is called impenetrable bosons (Tonks–Girardeau gas) [L. Tonks (1936); M. Girardeau (1960); M. Girardeau et al (2001)]. It is equivalent to free fermions[1] [D. B. Creamer et al (1980)]. In this limit, $\theta(u) = 0$, and the system (10.5) has an explicit solution $u_j = 2\pi I_j/L$. We can also apply perturbation theory to study the properties of solutions of the Bethe equations for sufficiently large $c$. For example, the first correction to the solution for $u_j = 2\pi I_j/L$ has the form

$$u_j = \frac{2\pi I_j}{L}\left(1 - \frac{2D}{c}\right) + \frac{4\pi}{cL^2}\sum_{k=1}^{N} I_k + O(c^{-2}), \qquad D = \frac{N}{L}. \tag{10.20}$$

The limit $c \to \infty$ allows us to obtain good qualitative ideas of the description of the spectrum in the general case. Therefore, we will often use it.

[1]Quantum particles in $(1 + 1)$-dimensional models can have both boson and fermion properties because the spin-statistics theorem does not work for such models.

## 10.2 Ground state in the thermodynamic limit

### 10.2.1 *Ground state density*

Consider a state parameterized by $N$ Bethe roots. The energy of this state is given by (10.3). The ground state of the system corresponds to the minimum value of $E$. Thus, we need to find (half)integers $I_j$ such that the roots of the system (10.5) minimize the sum (10.3). We can easily solve this problem for impenetrable bosons ($c \to \infty$). We obtain

$$E = \frac{4\pi^2}{L^2} \sum_{j=1}^{N} I_j^2. \tag{10.21}$$

Recall that here all (half)integers $I_j$ are pairwise distinct. It is evident that under such conditions, the minimum energy occurs if $I_j$ are arranged symmetrically around zero and $I_{j+1} - I_j = 1$:

$$I_j = j - \frac{N+1}{2}, \qquad j = 1, \dots, N. \tag{10.22}$$

It is easy to check that in the first order of the perturbation theory in $1/c$ (see (10.20)), the minimum energy occurs with the same choice of $I_j$. It seems quite reasonable to assume that the same is true in the general case. We cannot prove this rigorously at $N$ finite. However, we will prove that this is indeed the case in the thermodynamic limit.

The thermodynamic limit corresponds to the limit $N, L \to \infty$ at a fixed average density $N/L = D = const$. We denote this limit by the symbol Lim. Consider the solution of system (10.5) for $I_j$ given by (10.22). To distinguish this state from others, we will denote, in this case, $I_j = n_j$. It follows from proposition 10.5 that $u_{N+1-j} = -u_j$. Then it follows from proposition 10.4 that

$$\begin{aligned} &\frac{2\pi}{L} \geq u_{j+1} - u_j \geq \frac{2\pi}{L(1+2D/c)}, \\ &\frac{\pi(N-1)}{L} \geq u_N \geq \frac{\pi(N-1)}{L(1+2D/c)}. \end{aligned} \tag{10.23}$$

Thus, in the thermodynamic limit, the parameters $u_j$ densely fill an interval $[-q, q]$, where $q = \operatorname{Lim} u_N$. We call this interval the *Fermi zone*. Since $u_{j+1} - u_j \sim 1/L$ in the thermodynamic limit, it is natural to describe the Bethe equations solution by a distribution density.

Let

$$\rho(u_j) = \operatorname{Lim} \frac{1}{L(u_{j+1} - u_j)}. \tag{10.24}$$

Then for arbitrary integrable on $[-q,q]$ function $f(u)$, we have

$$\begin{aligned}\mathrm{Lim}\,\frac{1}{L}\sum_{j=1}^{N} f(u_j) &= \mathrm{Lim}\,\frac{1}{L}\sum_{j=1}^{N} f(u_j)\frac{u_{j+1}-u_j}{u_{j+1}-u_j}\\ &= \mathrm{Lim}\sum_{j=1}^{N} f(u_j)\rho(u_j)(u_{j+1}-u_j).\end{aligned} \tag{10.25}$$

The sum in the rhs is nothing but an approximation of the Riemann integral. We obtain

$$\mathrm{Lim}\,\frac{1}{L}\sum_{j=1}^{N} f(u_j) = \int_{-q}^{q} f(u)\rho(u)\,du. \tag{10.26}$$

Thus, the function $\rho(u)$ has the meaning of the distribution density of Bethe roots. It is normalized to the average density $D = N/L$. Indeed, setting $f(u) = 1$ in (10.26), we obtain

$$D = \int_{-q}^{q} \rho(u)\,du. \tag{10.27}$$

Consider now the difference of two Bethe equations (10.5) with $n_{j+1}$ and $n_j$

$$L(u_{j+1}-u_j) + \sum_{l=1}^{N}\big(\theta(u_{j+1}-u_l) - \theta(u_j-u_l)\big) = 2\pi. \tag{10.28}$$

For $L$ large enough, we can replace the finite difference $\theta(u_{j+1}-u_l)-\theta(u_j-u_l)$ with the derivative

$$L(u_{j+1}-u_j) + \sum_{l=1}^{N}\mathcal{K}(u_j-u_l)\Big((u_{j+1}-u_j) + O(L^{-2})\Big) = 2\pi, \tag{10.29}$$

where $\mathcal{K}(u) = \theta'(u)$ is given by (10.10). Finally, dividing (10.29) by $L(u_{j+1}-u_j)$ and replacing the sum with an integral via (10.26), we arrive at an integral equation for the distribution density

$$2\pi\rho(u) = 1 + \int_{-q}^{q}\mathcal{K}(u-v)\rho(v)\,dv. \tag{10.30}$$

This equation was obtained in [E. H. Lieb and W. Liniger (1963)].

Consider an integral operator $\mathbf{I} - \frac{1}{2\pi}\mathbf{K}$ acting in $L^2([-q,q])$ as follows:

$$\begin{aligned}{[(\mathbf{I} - \tfrac{1}{2\pi}\mathbf{K})f](u)} &= f(u) - \frac{1}{2\pi}\int_{-q}^{q}\mathcal{K}(u-v)f(v)\,dv,\\ {[f(\mathbf{I} - \tfrac{1}{2\pi}\mathbf{K})](v)} &= f(v) - \frac{1}{2\pi}\int_{-q}^{q}\mathcal{K}(u-v)f(u)\,du.\end{aligned} \tag{10.31}$$

Then we can write down equation (10.30) in the form

$$[(\mathbf{I} - \tfrac{1}{2\pi}\mathbf{K})\rho](u) = \frac{1}{2\pi}. \tag{10.32}$$

**Proposition 10.6.** *Let $\phi(u)$ be an eigenfunction of the operator $\frac{1}{2\pi}\mathbf{K}$ with an eigenvalue $\lambda$:*

$$\lambda\phi(u) = \frac{1}{2\pi}\int_{-q}^{q} \mathcal{K}(u-v)\phi(v)\,dv. \tag{10.33}$$

*The eigenvalues of the operator $\frac{1}{2\pi}\mathbf{K}$ are bounded: $|\lambda| < 1$.*

**Proof.** Let $\|\phi\| = \sup_{u\in[-q,q]} |\phi(u)|$. Then

$$|\lambda|\,|\phi(u)| \le \frac{\|\phi\|}{2\pi}\int_{-q}^{q} \mathcal{K}(u-v)\,dv. \tag{10.34}$$

In its turn

$$\frac{1}{2\pi}\int_{-q}^{q} \mathcal{K}(u-v)\,dv < \frac{1}{2\pi}\int_{-\infty}^{\infty} \mathcal{K}(u-v)\,dv = 1. \tag{10.35}$$

Hence,

$$|\lambda|\,|\phi(u)| < \|\phi\|. \tag{10.36}$$

The kernel of $\mathcal{K}(u-v)$ is continuous and bounded for $u, v \in [-q, q]$. Therefore, the eigenfunctions of operator $\frac{1}{2\pi}\mathbf{K}$ are continuous [V. I. Smirnov (1964)]. Hence, there exists $u_0 \in [-q, q]$ such that $\|\phi\| = |\phi(u_0)|$. Setting $u = u_0$ in (10.36), we get $|\lambda| < 1$. Due to the kernel $\mathcal{K}(u-v)$ continuity, its eigenvalues have a single limit point 0. Therefore, $|\lambda|$ cannot tend to 1 from below. □

It follows from proposition 10.6 that $\|\frac{1}{2\pi}\mathbf{K}\| < 1$. Hence, the series

$$\mathbf{R} = \sum_{n=1}^{\infty} \left(\tfrac{1}{2\pi}\mathbf{K}\right)^n \tag{10.37}$$

is absolutely convergent, and operator $\mathbf{I} - \frac{1}{2\pi}\mathbf{K}$ is invertible:

$$(\mathbf{I} + \mathbf{R})(\mathbf{I} - \tfrac{1}{2\pi}\mathbf{K}) = (\mathbf{I} - \tfrac{1}{2\pi}\mathbf{K})(\mathbf{I} + \mathbf{R}) = \mathbf{I}. \tag{10.38}$$

Thus, integral equation (10.32) has a unique solution.

**Proposition 10.7.** *Let $f(u)$ solve equation*

$$[(\mathbf{I} - \tfrac{1}{2\pi}\mathbf{K})f](u) = g(u). \tag{10.39}$$

*If function $g(u)$ is even (odd), then $f(u)$ also is even (odd).*

**Proof.** To prove this statement, it is enough to notice that $\mathcal{K}(u) = \mathcal{K}(-u)$. □

Finally, using (10.23), we also can obtain the following estimates:

$$\frac{1}{2\pi}\left(1 + 2\frac{D}{c}\right) \ge \rho(u) \ge \frac{1}{2\pi}, \qquad u \in [-q, q]. \tag{10.40}$$

### 10.2.2 *Ground state counting function*

Integral equation (10.30) determines the density of the ground state in the thermodynamic limit $N, L \to \infty$. If $N$ and $L$ are large but finite, then the density function $\rho(u)$ has corrections. Similar corrections occur to the energy and momentum of excited states. In some cases, when calculating correlation functions, these corrections play a significant role. Therefore, we give another derivation of the equation for the ground state density, which allows us to consider the finite-size corrections.

Let $\bar{u} = \{u_1, \dots, u_N\}$ be the roots of the Bethe equations corresponding to the (half)integers $n_j$ (10.22). We introduce a *counting function* $\hat{\xi}_u(z)$ of the ground state by

$$\hat{\xi}_u(z) = \frac{z}{2\pi} + \frac{1}{2\pi L}\sum_{k=1}^{N} \theta(z - u_k) + \frac{N+1}{2L}. \tag{10.41}$$

Due to Bethe equations,

$$\hat{\xi}_u(u_j) = \frac{j}{L}, \qquad j = 1, \dots, N. \tag{10.42}$$

Since $\theta(z)$ is a monotonically increasing bounded function, the counting function $\hat{\xi}_u(z)$ is also a monotonically increasing function such that $\hat{\xi}_u(z) \to \pm\infty$ as $z \to \pm\infty$. Therefore, there exists an inverse function $\hat{\xi}_u^{-1}(z)$.

The derivative of the function $\hat{\xi}_u(z)$ is naturally called the ground state distribution density

$$\hat{\rho}_u(z) = \hat{\xi}_u'(z) = \frac{1}{2\pi} + \frac{1}{2\pi L}\sum_{k=1}^{N} \mathcal{K}(z - u_k). \tag{10.43}$$

**Remark 10.3.** The definitions of the counting function and density are given for finite $N$ and $L$. In this case, we use the hat symbol. The subscript shows that the functions $\hat{\xi}_u(z)$ and $\hat{\rho}_u(z)$ are defined by the set of roots $\bar{u}$. We will use the notation $\xi(z)$ and $\rho(z)$ (without the hat and the subscript) for the thermodynamic limits of these functions. We will follow a similar rule for the counting function of excited states, their density, and the shift function (see below).

Obviously,

$$\operatorname{Lim} \hat{\rho}_u(u_j) = \operatorname{Lim} \frac{\hat{\xi}_u(u_{j+1}) - \hat{\xi}_u(u_j)}{u_{j+1} - u_j} = \operatorname{Lim} \frac{1}{L(u_{j+1} - u_j)}. \tag{10.44}$$

We see that the limit of $\hat{\rho}_u(u_j)$ defined by (10.43) coincides with the limit of the ground state distribution density introduced by (10.24). On the other hand, we can compute the sum in (10.43) via the Euler–Maclaurin formula

$$\frac{1}{L}\sum_{a\leq j<b} f\left(\tfrac{j}{L}\right) = \int_{a/L}^{b/L} f(x)\,\mathrm{d}x + \sum_{k-1}^{m} \frac{B_k}{L^k k!} f^{(k-1)}(x)\Big|_{a/L}^{b/L} + O(L^{-m-1}), \tag{10.45}$$

where $f(x)$ is a function that is $m$ times continuously differentiable on an interval $[a/L, b/L]$, and $B_k$ are Bernoulli numbers. In fact, we will need only the integral term and the first correction:

$$\frac{1}{L}\sum_{j=1}^{N} f\left(\tfrac{j}{L}\right) = \int_{\frac{1}{L}}^{\frac{N+1}{L}} f(x)\,\mathrm{d}x - \frac{1}{2L}\Big(f\left(\tfrac{N+1}{L}\right) - f\left(\tfrac{1}{L}\right)\Big) + O(L^{-2}). \tag{10.46}$$

To eliminate the term outside the integral, we can rewrite this formula in the form

$$\frac{1}{L}\sum_{j=1}^{N} f\left(\tfrac{j}{L}\right) = \int_{\frac{1}{2L}}^{\frac{N+1/2}{L}} f(x)\,\mathrm{d}x + O(L^{-2}). \tag{10.47}$$

Now we take advantage of the fact that the Fermi boundary $q$ is defined only in the limit $q = \operatorname{Lim} u_N$. Therefore, we have some freedom in choosing the exact (prelimit) value of $q$. Using this freedom, we define $q$ as follows:

$$\hat{\xi}_u(q) = \frac{N}{L} + \frac{1}{2L}, \qquad \hat{\xi}_u(-q) = \frac{1}{2L}. \tag{10.48}$$

Then we have

$$\frac{1}{L}\sum_{j=1}^{N} f(u_j) = \frac{1}{L}\sum_{j=1}^{N} f\left(\hat{\xi}_u^{-1}\left(\tfrac{j}{L}\right)\right) = \int_{\frac{1}{2L}}^{\frac{N+1/2}{L}} f\left(\hat{\xi}_u^{-1}(x)\right)\mathrm{d}x + O(L^{-2}). \tag{10.49}$$

Setting $x = \hat{\xi}_u(u)$ we arrive at

$$\frac{1}{L}\sum_{j=1}^{N} f(u_j) = \int_{-q}^{q} f(u)\hat{\rho}_u(u)\,\mathrm{d}u + O(L^{-2}). \tag{10.50}$$

Applying this formula to (10.43), we obtain

$$\hat{\rho}_u(z) = \frac{1}{2\pi} + \frac{1}{2\pi}\int_{-q}^{q} \mathcal{K}(z-u)\hat{\rho}_u(u)\,\mathrm{d}u + O(L^{-2}). \tag{10.51}$$

In the thermodynamic limit, we reproduce integral equation (10.30).

Using the counting function turns out to be extremely useful when calculating sums (products) containing singularities. In this case, a direct application of the Euler–Maclaurin formula is impossible. First, we get the divergent integral. Secondly, all the correction terms actually turn out to be of the same order. Methods based on the counting function allows us to overcome these difficulties. We will take a closer look at this situation in Chapter 12.

## 10.3 Excited states

### 10.3.1 *Counting function of excited states*

We assumed that a state characterized by the (half)integers $n_j = j - (N+1)/2$ is the ground state. In other words, this is a physical vacuum. Any state characterized by a different set of (half)integers $I_j$ is an excited state. Let us fix $n$ integer numbers $h_1, \dots, h_n$ such that $h_k \in \{1, \dots, N\}$ and $n$ integer numbers $p_1, \dots, p_n$ such that $p_k \notin \{1, \dots, N\}$. Let

$$\begin{aligned} I_j &= j - \frac{N+1}{2}, \qquad && j \ne h_1, \dots, h_n, \\ I_{h_k} &= p_k - \frac{N+1}{2}, \qquad && k = 1, \dots, n. \end{aligned} \tag{10.52}$$

We say that this state has $n$ particles and $n$ holes. Any excited state with $N$ Bethe parameters can be created in this way by choosing different $n$ and different sets of $h_k$ and $p_k$. The integers $p_1, \dots, p_n$ (resp. $h_1, \dots, h_n$) are called *quantum numbers* of particles (resp. holes). We call excitations the whole set of particles and holes.

This is the standard approach to the description of excited states. However, when calculating correlation functions, it is convenient to deal with solutions of twisted Bethe equations. As we have already seen, the roots of the twisted Bethe equations have an additional shift relative to the roots of the ordinary Bethe equations. This shift plays the role of a kind of regularization. For example, when calculating the form factors of the monodromy matrix diagonal elements, we should especially consider the case when both states coincide. If one of the states is a twisted on-shell vector, then such a match never occurs. Therefore, if we consider twisted on-shell vectors as excited states, we can obtain universal formulas for such form factors. A similar effect occurs when calculating correlation functions. Therefore, from the very beginning, we will consider the twisted on-shell Bethe vectors as excited states.

Consider a solution of twisted Bethe equations $\bar{v} = \{v_1, \dots, v_N\}$ corresponding to the (half)integers (10.52)

$$Lv_j + \sum_{k=1}^{N} \theta(v_j - v_k) = 2\pi I_j + \alpha, \qquad j = 1, \dots, N, \tag{10.53}$$

where $\alpha$ is a real number. Then the twist parameter is $\kappa = e^{-i\alpha}$. Define a counting function $\hat{\xi}_v(z)$:

$$\hat{\xi}_v(z,\alpha) = \frac{z}{2\pi} + \frac{1}{2\pi L}\sum_{k=1}^{N}\theta(z-v_k) + \frac{N+1}{2L} - \frac{\alpha}{2\pi L}. \tag{10.54}$$

Here we explicitly indicated that the counting function depends on $\alpha$. However, in most cases, we will omit this dependency and write simply $\hat{\xi}_v(z)$.

It is easy to see that

$$\begin{aligned} \hat{\xi}_v(v_j) &= \frac{j}{L}, \qquad j \ne h_1,\dots,h_n, \\ \hat{\xi}_v(v_{h_k}) &= \frac{p_k}{L}, \qquad k = 1,\dots,n. \end{aligned} \tag{10.55}$$

**Remark 10.4.** In a particular case, we can consider a state without particles and holes, i.e., $I_j = n_j$, $j = 1,\dots,N$. We call this state an *$\alpha$-twisted ground state*. In our interpretation, it also becomes one of the excited states.

Similarly to the counting function of the ground state, the function $\hat{\xi}_v(z)$ is a monotonically increasing function such that $\hat{\xi}_v(z) \to \pm\infty$ as $z \to \pm\infty$. Therefore, there exists an inverse function $\hat{\xi}_v^{-1}(z)$. We call a *momentum of a particle* the value $\lambda_{p_k} = \hat{\xi}_v^{-1}(p_k/L)$. By virtue of (10.55), $\lambda_{p_k} = v_{h_k}$. We call a *momentum of a hole* the value $\lambda_{h_k} = \hat{\xi}_v^{-1}(h_k/L)$. The values $\lambda_{h_k}$ do not belong to the set $\bar{v}$.

Let us define a set $\bar{v}' = \{v'_1,\dots,v'_N\}$ such that

$$\hat{\xi}_v(v'_j) = \frac{j}{L}, \qquad j = 1,\dots,N. \tag{10.56}$$

Then $v'_j = v_j$ for $j \notin \{h_1,\dots,h_n\}$ and $v'_j = \lambda_{h_k}$ for $j = h_k$.

We can now match any integer $m$ with a 'vacancy' $\hat{\xi}_v^{-1}(m/L)$. We say that a vacancy is occupied if $\hat{\xi}_v^{-1}(m/L) \in \bar{v}$, and the vacancy is free otherwise. Any free vacancy such that $m \in \{1,\dots,N\}$ corresponds to a hole. An occupied vacancy corresponds to a particle if $m \notin \{1,\dots,N\}$. The function $\hat{\xi}_v(z)$ defines the total distribution of vacancies.

The excited state distribution density $\hat{\rho}_v(z)$ is defined similarly to the ground state density

$$\hat{\rho}_v(z) = \hat{\xi}'_v(z) = \frac{1}{2\pi} + \frac{1}{2\pi L}\sum_{k=1}^{N}\mathcal{K}(z-v_k). \tag{10.57}$$

### 10.3.2 *Shift function*

In the limit $c \to \infty$ (impenetrable bosons), the function $\theta(z)$ vanishes. Then twisted Bethe equations (10.53) take the form

$$Lv_j = 2\pi I_j + \alpha, \qquad j = 1,\dots,N, \qquad c \to \infty. \tag{10.58}$$

If $j \notin \{h_1, \ldots, h_n\}$, then $I_j = n_j$. Hence, the corresponding root $v_j$ coincides with the root of the $\alpha$-twisted ground state. In particular, $v_j = u_j$ for $j \neq \{h_1, \ldots, h_n\}$ and $\alpha = 0$, where $u_j$ are the ground state roots. However, this property is a distinctive feature of the impenetrable bosons. It occurs because the system of twisted equations decouples in the limit $c \to \infty$. With a finite coupling constant, this property disappears. For $N$ and $L$ large enough, the Bethe roots $v_j$ of the excited state (which do not correspond to the momenta of particles) are slightly shifted with respect to the ground state roots $u_j$: $v_j - u_j \sim 1/L$, $j \neq \{h_1, \ldots, h_n\}$. This shift remains non-vanishing even for $\alpha = 0$. The rigorous proof of this statement is quite spacious, so we do not give it here. However, the reader can easily see this by using the explicit solution (10.20) that considers the first correction in the $1/c$ expansion. The value of the shift $v_j - u_j$ is determined by a *shift function*, which plays an important role in calculating the thermodynamic limit of form factors and correlation functions.

Consider an excited state characterized by the quantum numbers $\{p_1, \ldots, p_n\}$ and $\{h_1, \ldots, h_n\}$. Define a shift function $\hat{F}(z)$ as

$$\hat{F}(z) = L\big(\hat{\xi}_u(z) - \hat{\xi}_v(z)\big). \tag{10.59}$$

Here $\hat{\xi}_u(z)$ and $\hat{\xi}_v(z)$ respectively are the counting functions of the ground state and the excited state under consideration.

For $j \notin \{h_1, \ldots, h_n\}$, we have $\hat{\xi}_v(v_j) = j/L$ due to the definition of $\hat{\xi}_v(z)$. On the other hand,

$$\hat{\xi}_u(v_j) = \hat{\xi}_u(u_j) + (v_j - u_j)\hat{\xi}_u'(u_j) + O(L^{-2}) = \frac{j}{L} + (v_j - u_j)\hat{\rho}_u(u_j) + O(L^{-2}), \tag{10.60}$$

and we arrive at

$$v_j - u_j = \frac{\hat{F}(v_j)}{L\hat{\rho}_u(u_j)} + O(L^{-2}) = \frac{\hat{F}(u_j)}{L\hat{\rho}_u(u_j)} + O(L^{-2}). \tag{10.61}$$

Thus, this function characterizes the shift of the roots $v_j$ with respect to the roots $u_j$ for $j \neq h_k$.

In the thermodynamic limit, the shift function satisfies an integral equation similar to equation (10.32). To derive this integral equation, we substitute explicit expressions for the counting functions in (10.59):

$$2\pi\hat{F}(z) = \alpha + \sum_{k=1}^{N} \big(\theta(z - u_k) - \theta(z - v_k)\big). \tag{10.62}$$

Using auxiliary parameters $\bar{v}'$ (10.56), we recast this equation as follows:

$$2\pi\hat{F}(z) = \alpha + \sum_{k=1}^{N}\big(\theta(z-u_k)-\theta(z-v_k')\big) + \sum_{j=1}^{n}\big(\theta(z-\lambda_{h_j})-\theta(u-\lambda_{p_j})\big). \tag{10.63}$$

Developing the difference

$$\theta(z-u_k)-\theta(z-v_k') = (v_k'-u_k)\mathcal{K}(z-u_k)+O(L^{-2}), \tag{10.64}$$

and using (10.61), we obtain

$$F(z) - \frac{1}{2\pi}\int_{-q}^{q}\mathcal{K}(z-v)F(v)\,\mathrm{d}v = \frac{\alpha}{2\pi} + \frac{1}{2\pi}\sum_{j=1}^{n}\big(\theta(z-\lambda_{h_j})-\theta(u-\lambda_{p_j})\big). \tag{10.65}$$

Here $F(z)$ means the limiting value of $\hat{F}(z)$. Thus, the shift function satisfies the integral equation with the same integral operator $\mathbf{I}-\frac{1}{2\pi}\mathbf{K}$, which arises in the equation for the density. Therefore, equation (10.65) is uniquely solvable.

We see that the shift function parametrically depends on the momenta of the particles and the holes: $F(z) = F(z|\{\lambda_{p_j},\lambda_{h_j}\})$. As a rule, we will omit this dependence to avoid too heavy notation.

It is convenient to introduce a function $\phi(u,v)$ as the solution to the integral equation

$$\phi(u,v) - \frac{1}{2\pi}\int_{-q}^{q}\mathcal{K}(u-w)\phi(w,v)\,\mathrm{d}w = \frac{1}{2\pi}\theta(u-v). \tag{10.66}$$

Then

$$F(u) = \alpha\rho(u) + \sum_{j=1}^{n}\big(\phi(u,\lambda_{h_j})-\phi(u,\lambda_{p_j})\big). \tag{10.67}$$

It is easy to express the density $\rho(u)$ in terms of a special shift function. To do this, we introduce a function $\chi(u) = \rho(u)-1/2\pi$. Then equation (10.30) for the density implies

$$\chi(u) - \frac{1}{2\pi}\int_{-q}^{q}\mathcal{K}(u-v)\left(\chi(v)+\tfrac{1}{2\pi}\right)\mathrm{d}v = 0. \tag{10.68}$$

Using $\mathcal{K}(u)=\theta'(u)$ we obtain

$$\chi(u) - \frac{1}{2\pi}\int_{-q}^{q}\mathcal{K}(u-v)\chi(v)\,\mathrm{d}v = \frac{\theta(u+q)-\theta(u-q)}{4\pi^2}. \tag{10.69}$$

Hence,

$$\chi(u) = \frac{1}{2\pi}\big(\phi(u,-q)-\phi(u,q)\big) = \frac{F(u|q,-q)}{2\pi}\Big|_{\alpha=0}, \tag{10.70}$$

and we finally obtain

$$2\pi\rho(u) = 1 + \phi(u,-q) - \phi(u,q) = 1 + F(u|q,-q)\Big|_{\alpha=0}. \tag{10.71}$$

This relationship has a simple physical interpretation. Let $\bar{v}$ be a solution of the twisted Bethe equations corresponding to the $\alpha$-twisted ground state. Then the shift function is

$$F(u) = \alpha\rho(u). \tag{10.72}$$

On the other hand, the twisted Bethe equations turn into the ordinary Bethe equations at $\alpha = 2\pi$. Then the solution $\bar{v}$ corresponds to (half)integers $m_j = n_j + 1$ due to (10.53). Hence, we can consider this state as an excited state with one particle and one hole at $\alpha = 0$. The corresponding quantum numbers are $p = N + 1$ and $h = 1$. This means that the momenta of the particle and the hole respectively go to the right and left Fermi boundaries: $\lambda_p = q$, $\lambda_h = -q$. Hence, we obtain due to (10.67),

$$F(u|q,-q)\Big|_{\alpha=0} = \phi(u,-q) - \phi(u,q). \tag{10.73}$$

It remains to notice that

$$\hat{\xi}_v(u,0) - \hat{\xi}_v(u,\alpha) = \frac{\alpha}{2\pi L}, \tag{10.74}$$

which implies (10.71).

This example shows that we can treat some excited states at $\alpha = 0$ as special cases of twisted on-shell Bethe vectors. In Section 12.6, we will see that these states determine the main contributions to the asymptotic behavior of correlation functions.

Concluding this section, we note that it follows from the finiteness of the shift function that the thermodynamic limits of counting functions $\hat{\xi}_u(z)$ and $\hat{\xi}_v(z)$ coincide: $\operatorname{Lim}\hat{\xi}_u(z) = \operatorname{Lim}\hat{\xi}_v(z) = \xi(z)$. Hence, the limits of the distribution densities also coincide: $\operatorname{Lim}\hat{\rho}_u(z) = \operatorname{Lim}\hat{\rho}_v(z) = \rho(z)$. For this reason, we omit the subscript at the thermodynamic limit of the counting function $\xi(z)$ and the density $\rho(z)$.

### 10.3.3 *Energy and momentum of the excited states*

We proceed now to the calculation of the energy of the excited states. To do this, we introduce new functions and find their properties.

Consider an auxiliary function $\tilde{\varepsilon}(u)$ which solves an equation

$$[(\mathbf{I} - \tfrac{1}{2\pi}\mathbf{K})\tilde{\varepsilon}](u) = u^2. \tag{10.75}$$

Obviously, $\tilde{\varepsilon}(u) = \tilde{\varepsilon}(-u)$ (see proposition 10.7). Besides, by virtue of the absolute convergence of the series (10.37) and the condition $\mathcal{K}(u-v) > 0$ for $-q \le u, v \le q$, we conclude that $\tilde{\varepsilon}(u) \ge 0$. Define a new function

$$[(\mathbf{I} - \tfrac{1}{2\pi}\mathbf{K})\varepsilon](u) = u^2 - \mu, \tag{10.76}$$

where $\mu$ is a positive constant. Then $\varepsilon(u) = \tilde{\varepsilon}(u) - 2\pi\mu\rho(u)$. Since $\rho(u)$ is even function and $\rho(u) > 0$, there exists $\mu$ such that $\varepsilon(\pm q) = 0$.

**Proposition 10.8.** *Let $\varepsilon(u)$ satisfy equation* (10.76), *where $\mu$ is such that $\varepsilon(\pm q) = 0$. Then $\varepsilon(u) < 0$ for $|u| < q$, and $\varepsilon(u) > 0$ for $|u| > q$.*

**Proof.** Let us differentiate equation (10.76) with respect to $u$. Using the fact that the kernel $K$ depends on the difference, we can integrate by parts and take into account $\varepsilon(\pm q) = 0$. We obtain

$$[(\mathbf{I} - \tfrac{1}{2\pi}\mathbf{K})\varepsilon'](u) = 2u. \tag{10.77}$$

The function $\varepsilon'(u)$ is odd. We prove that it is positive for $u > 0$. To do this, we define a sequence of functions $\varepsilon'_n(u)$ by

$$\varepsilon'_{n+1}(u) = 2u + \frac{1}{2\pi}\int_{-q}^{q} \mathcal{K}(u-v)\varepsilon'_n(v)\,\mathrm{d}v, \qquad \varepsilon'_1(u) = 2u. \tag{10.78}$$

Due to the absolute convergence of the series (10.37), this sequence of functions converges to the solution of equation (10.77). On the other hand, let some $\varepsilon'_n(u) > 0$ for $u > 0$. Then

$$\varepsilon'_{n+1}(u) = 2u + \frac{1}{2\pi}\int_{0}^{q} \big(\mathcal{K}(u-v) - \mathcal{K}(u+v)\big)\varepsilon'_n(v)\,\mathrm{d}v. \tag{10.79}$$

Since $\mathcal{K}(u-v) - \mathcal{K}(u+v) > 0$ for $u, v > 0$, both terms in the rhs of (10.79) are positive. Hence, $\varepsilon'_{n+1}(u) > 0$.

Thus, the function $\varepsilon'(u)$ has a single zero at $u = 0$. Hence, $\varepsilon(u)$ has a single extremum at $u = 0$. This extremum is a minimum because $\varepsilon(u) \to u^2 - \mu$ at $u \to \pm\infty$. Finally, given that $\varepsilon(\pm q) = 0$, we arrive at the proof of the proposition. □

Now we are ready to calculate the energy of an excited state. Let this state have one hole and one particle for simplicity. Then

$$\mathcal{E}_{\rm ex} = \sum_{j=1}^{N}(v_j^2 - u_j^2) = \lambda_p^2 - \lambda_h^2 + \sum_{j=1}^{N}(v'^2_j - u_j^2), \tag{10.80}$$

where $v'_j$ are given by (10.56). Using the fact that

$$v'^2_j - u_j^2 = 2u_j(v'_j - u_j) + O(L^{-2}),$$

we obtain

$$\mathcal{E}_{\text{ex}} = \lambda_p^2 - \lambda_h^2 + \sum_{j=1}^{N} 2u_j \left( \frac{\hat{F}(u_j)}{L\hat{\rho}_u(u_j)} + O(L^{-2}) \right). \tag{10.81}$$

where we used (10.61). Replacing the sum by the integral, we arrive at

$$\mathcal{E}_{\text{ex}} = \lambda_p^2 - \lambda_h^2 + \int_{-q}^{q} 2uF(u)\, du. \tag{10.82}$$

We now use equation (10.77) and present $2u$ as the result of the action of the integral operator $\mathbf{I} - \frac{1}{2\pi}\mathbf{K}$ on the function $\varepsilon'(v)$:

$$\int_{-q}^{q} 2uF(u)\, du = \int_{-q}^{q}\int_{-q}^{q} \left( \delta(u-v) - \frac{1}{2\pi}\mathcal{K}(u-v) \right) \varepsilon'(v)F(u)\, du\, dv. \tag{10.83}$$

Using the symmetry of $\mathbf{I} - \frac{1}{2\pi}\mathbf{K}$ and applying it to the shift function, we obtain

$$\int_{-q}^{q} 2uF(u)\, du = \frac{1}{2\pi}\int_{-q}^{q} (\alpha + \theta(v-\lambda_h) - \theta(v-\lambda_p))\varepsilon'(v)\, dv. \tag{10.84}$$

Integrating this equation by parts and using $\varepsilon(\pm q) = 0$ we find

$$\int_{-q}^{q} 2uF(u)\, du = \frac{1}{2\pi}\int_{-q}^{q} (\mathcal{K}(v-\lambda_p) - \mathcal{K}(v-\lambda_h))\varepsilon(v)\, dv. \tag{10.85}$$

Then (10.76) yields

$$\int_{-q}^{q} 2uF(u)\, du = \varepsilon(\lambda_p) - \varepsilon(\lambda_h) - \lambda_p^2 + \lambda_h^2. \tag{10.86}$$

Substituting this into (10.82), we immediately find

$$\mathcal{E}_{\text{ex}} = \varepsilon(\lambda_p) - \varepsilon(\lambda_h). \tag{10.87}$$

Due to proposition 10.8, $\varepsilon(\lambda_p) \geq 0$ because $|\lambda_p| \geq q$ and $\varepsilon(\lambda_h) \leq 0$ because $|\lambda_h| \leq q$. We see that the energy of the excited state is non-negative.

For $L$ and $N$ large but finite, the energy is strictly positive since $|\lambda_p| > q$ and $|\lambda_h| < q$. Therefore, the state constructed in Section 10.2 is indeed the ground state. However, the momenta of particles and holes can coincide with the Fermi boundaries in the thermodynamic limit. For example, it is easy to show that if $p = N + m$, where $m > 0$ and remains finite in the thermodynamic limit, then $\lambda_p - q = O(L^{-1})$. Thus, the spectrum of the Hamiltonian turns out to be gapless.

It is easy to show that if we take a state with several holes and particles, then due to (10.67), the energy of such excitation is equal to

$$\mathcal{E}_{\text{ex}} = \sum_{k=1}^{n} \left( \varepsilon(\lambda_{p_k}) - \varepsilon(\lambda_{h_k}) \right). \tag{10.88}$$

We see that the energy is positive if at least one particle or hole is separated from the Fermi boundaries. If all the particles and holes fall on the Fermi boundaries, then the energy of this excitation vanishes.

Due to (9.40), the momentum of an excited state is

$$\mathcal{P}_{\text{ex}} = \sum_{j=1}^{N}(v_j - u_j). \tag{10.89}$$

It can be calculated similarly. However, we can do this more simply using that $\theta(u)$ is an odd function. This implies

$$\sum_{j=1}^{m}\sum_{k=1}^{m}\theta(w_j - w_k) = 0, \tag{10.90}$$

for any set $w_1, \dots, w_m$. Then due to (10.41), we have

$$\sum_{j=1}^{N} u_j = 2\pi\sum_{j=1}^{N}\left(\hat{\xi}_u(u_j) - \frac{N+1}{2L}\right) = \frac{2\pi}{L}\sum_{j=1}^{N}\left(j - \frac{N+1}{2}\right) = 0. \tag{10.91}$$

Similarly, if an excited state is characterized by the quantum numbers $p_1, \dots, p_n$ and $h_1, \dots, h_n$, equation (10.54) yields

$$\sum_{j=1}^{N} v_j = \frac{\alpha N}{L} + 2\pi\sum_{j=1}^{N}\left(\hat{\xi}_v(v_j) - \frac{N+1}{2L}\right) = \alpha D + \frac{2\pi}{L}\sum_{k=1}^{n}(p_k - h_k), \tag{10.92}$$

leading to

$$\mathcal{P}_{\text{ex}} = \alpha D + 2\pi\sum_{k=1}^{n}\left(\hat{\xi}_v(\lambda_{p_k}) - \hat{\xi}_v(\lambda_{h_k})\right) = \alpha D + 2\pi\sum_{k=1}^{n}\int_{\lambda_{h_k}}^{\lambda_{p_k}}\hat{\rho}_v(v)\,dv. \tag{10.93}$$

The term $\alpha D$ corresponds to the general shift of the excited state due to the twist. The second term in (10.93) corresponds to a change in momentum due to the presence of particles and holes. If we define a dressed momentum $k(u)$ as

$$k(u) = 2\pi\int_0^u \rho(v)\,dv, \tag{10.94}$$

then the excitation momentum looks similar to the excitation energy (10.88):

$$\mathcal{P}_{\text{ex}} = \alpha D + \sum_{j=1}^{n}\left(k(\lambda_{p_j}) - k(\lambda_{h_j})\right). \tag{10.95}$$

It is easy to see that the dressed momentum is equal to the counting function up to the factor of $2\pi$ and an additive constant:

$$k(u) = 2\pi\xi(u) - \pi D. \tag{10.96}$$

The additive constant can be found from the relations

$$k(q) = 2\pi \int_0^q \rho(v)\,dv = \pi \int_{-q}^q \rho(v)\,dv = \pi D, \tag{10.97}$$

and

$$\xi(q) = \mathrm{Lim}\,\hat{\xi}_u(u_N) = \mathrm{Lim}\,\frac{N}{L} = D. \tag{10.98}$$

The value $k_F = k(q)$ is called the *Fermi momentum*. It is related to the average density by $k_F = \pi D$.

## 10.4 Chemical potential

So far, we have only considered excited states in which the total number of Bethe parameters $\#\bar{v} = N$ was the same as in the ground state. However, we can also consider excitations in which $\#\bar{v} = N'$ and $N' \neq N$. We deal with such states, for example, when calculating the field form factors. It is convenient not to fix the number of the Bethe parameters in the ground state but slightly modify the original model. For this, we add to the original Hamiltonian of the quantum nonlinear Schrödinger equation a term proportional to the operator of the number of particles

$$H \to H - \mu Q, \tag{10.99}$$

where $Q$ is given by (1.80). It is natural to call $\mu$ a *chemical potential*. Thus, we can say that we are considering a grand canonical ensemble. Since $[H, Q] = 0$, the new Hamiltonian eigenfunctions do not change and are still parameterized by solutions of the Bethe equations. However, the eigenvalues now have the form

$$E = \sum_{j=1}^{N} (u_j^2 - \mu). \tag{10.100}$$

Now, for small values of $u$, there are negative contributions to the energy. The minimum energy achieves when all vacancies with $u^2 - \mu < 0$ are filled. In this case, the number of the Bethe parameters in the ground state $N$ (and hence the density $D$) becomes a function of the chemical potential $\mu$. As before, in the thermodynamic limit, the roots of the Bethe equations $u_j$ fill a finite interval $[-q, q]$. However, the value of $q$ is now determined by the equation for the energy (10.76) with the additional condition $\varepsilon(\pm q) = 0$. Thus, $q$ is implicitly but unambiguously fixed as a function of $c$ and $\mu$. The distribution density $\rho(u)$ and the average density $D$ also become functions of $c$ and $\mu$.

It becomes quite natural to consider such excitations that occur due to changes in the number of the Bethe parameters in the ground state. If $\#\bar{v} = N'$, then the counting function $\hat{\xi}_v(z)$ of the excited state has the form

$$\hat{\xi}_v(z) = \frac{z}{2\pi} + \frac{1}{2\pi L}\sum_{k=1}^{N'}\theta(z - v_k) + \frac{N'+1}{2L} - \frac{\alpha}{2\pi L}. \tag{10.101}$$

We will consider only those states in which $\Delta N = N' - N = \pm 1$.

Excited states are still characterized by quantum numbers of holes $h_1, \dots, h_n$ and particles $p_1, \dots, p_n$, but now $h_k \in \{1, \dots, N'\}$ and $p_k \notin \{1, \dots, N'\}$.

**Remark 10.5.** The question of which quantum numbers correspond to particles and holes is a matter of convention. A more traditional approach is to treat the quantum numbers of holes and particles so that they satisfy conditions $h_k \in \{1, \dots, N\}$ and $p_k \notin \{1, \dots, N\}$. Then the excited state always has one extra particle for $\Delta N = 1$ or one extra hole for $\Delta N = -1$. From a physical point of view, this approach looks more natural. In turn, our interpretation of quantum numbers is more convenient for calculating correlation functions. In particular, in all excited states, the number of particles is equal to the number of holes.

The definition of the shift function and the derivation of the integral equation for it do not change. As a result, we come to

$$\begin{aligned}[(\mathbf{I} - \tfrac{1}{2\pi}\mathbf{K})F](u) &= \frac{\alpha}{2\pi} - \frac{\Delta N}{2}\Big(1 + \tfrac{1}{\pi}\theta(z-q)\Big)\\ &\quad + \frac{1}{2\pi}\sum_{k=1}^{n}\big(\theta(u - \lambda_{h_k}) - \theta(u - \lambda_{p_k})\big),\end{aligned} \tag{10.102}$$

for a state with $n$ particles and $n$ holes. The solution of this equation is given in terms of the function $\phi(u, v)$ (10.66) and the distribution density $\rho(u)$ (10.30):

$$F(u) = (\alpha - \Delta N\pi)\rho(u) - \Delta N\phi(u, q) + \sum_{k=1}^{n}\big(\phi(u, \lambda_{h_k}) - \phi(u, \lambda_{p_k})\big). \tag{10.103}$$

The energy of an excited state is now defined as

$$\mathcal{E}_{\text{ex}} = \sum_{j=1}^{N'}(v_j^2 - \mu) - \sum_{j=1}^{N}(u_j^2 - \mu). \tag{10.104}$$

Further calculation of the thermodynamic limit of this expression completely repeats the derivation of Section 10.3.3. We allow the reader to check that the final result is still given by (10.88).

## 10.5 Dressed charge and shift function

In Section 10.3, we established an identity between the distribution density and the special shift function corresponding to the particle and hole at the Fermi boundaries (see (10.71)). In this Section, we obtain another identity that is already nonlinear.

We introduce a function $Z(u)$, which is the unique solution to the integral equation

$$Z(u) - \frac{1}{2\pi}\int_{-q}^{q} \mathcal{K}(u-v)Z(v)\,\mathrm{d}v = 1. \tag{10.105}$$

The function $Z(u)$ is called a *dressed charge*. This terminology came from the massive Thirring model, in which the value $Z(u)$ gives the observed charge of an elementary excitation. Comparing equations (10.105) and (10.30), we see that $Z(u) = 2\pi\rho(u)$. However, this property is a feature of the one-dimensional boson model. In Section 10.6, we will see that in the $XXZ$ Heisenberg chain, the distribution density and the dressed charge are no longer related in any way.

Consider the integral equation (10.66) for the function $\phi(u,v)$. An important feature of this equation is that the kernel of the integral operator $\mathcal{K}(u)$ equals the derivative of the inhomogeneous term: $\mathcal{K}(u) = \theta'(u)$. This relation leads to a rather unexpected identity for the solution of this equation.

**Proposition 10.9.** *Let $\phi(u,v)$ be defined by equation* (10.66)

$$\phi(u,v) - \frac{1}{2\pi}\int_{-q}^{q} \mathcal{K}(u-w)\phi(w,v)\,\mathrm{d}w = \frac{1}{2\pi}\theta(u-v). \tag{10.106}$$

*Then it satisfies the following identity:*

$$\Big(1-\phi(q,q)\Big)\Big(1+\phi(-q,-q)\Big) + \phi(q,-q)\phi(-q,q) = 1. \tag{10.107}$$

**Proof.** We use the fact that the integral operator $\mathbf{I} - \frac{1}{2\pi}\mathbf{K}$ is invertible. Then the resolvent $\mathcal{R}(u,v)$ satisfies the following equation:

$$\mathcal{R}(u,v) - \frac{1}{2\pi}\int_{-q}^{q} \mathcal{K}(u-w)\mathcal{R}(w,v)\,\mathrm{d}w = \frac{1}{2\pi}\mathcal{K}(u-v). \tag{10.108}$$

We prove that the lhs of (10.107) does not depend on $q$. For this, we take a full $q$-derivative of (10.107). In its turn, the full derivative can be expressed in terms of the partial ones:

$$\frac{\mathrm{d}\phi(q,\pm q)}{\mathrm{d}q} = \left(\frac{\partial}{\partial u} \pm \frac{\partial}{\partial v} + \frac{\partial}{\partial q}\right)\phi(u,v)\Big|_{\substack{u=q\\ v=\pm q}},$$
$$\frac{\mathrm{d}\phi(-q,\pm q)}{\mathrm{d}q} = \left(-\frac{\partial}{\partial u} \pm \frac{\partial}{\partial v} + \frac{\partial}{\partial q}\right)\phi(u,v)\Big|_{\substack{u=-q\\ v=\pm q}}. \tag{10.109}$$

Differentiating (10.106) over $v$, we find

$$\frac{\partial\phi(u,v)}{\partial v} - \frac{1}{2\pi}\int_{-q}^{q}\mathcal{K}(u-w)\frac{\partial\phi(w,v)}{\partial v}\,\mathrm{d}w = -\frac{1}{2\pi}\mathcal{K}(u-v). \tag{10.110}$$

Comparing this equation with (10.108), we conclude that

$$\frac{\partial\phi(u,v)}{\partial v} = -\mathcal{R}(u,v). \tag{10.111}$$

Differentiating (10.106) over $u$, we get

$$\frac{\partial\phi(u,v)}{\partial u} - \frac{1}{2\pi}\int_{-q}^{q}\frac{\partial\mathcal{K}(u-w)}{\partial u}\phi(w,v)\,\mathrm{d}w = \frac{1}{2\pi}\mathcal{K}(u-v). \tag{10.112}$$

Using $\partial\mathcal{K}(u-w)/\partial u = -\partial\mathcal{K}(u-w)/\partial w$ and integrating by parts, we obtain

$$\frac{\partial\phi(u,v)}{\partial u} - \frac{1}{2\pi}\int_{-q}^{q}\mathcal{K}(u-w)\frac{\partial\phi(w,v)}{\partial w}\,\mathrm{d}w$$
$$= \frac{1}{2\pi}\Big(\mathcal{K}(u-v) - \mathcal{K}(u-q)\phi(q,v) + \mathcal{K}(u+q)\phi(-q,v)\Big). \tag{10.113}$$

Thus,

$$\frac{\partial\phi(u,v)}{\partial u} = \mathcal{R}(u,v) - \mathcal{R}(u,q)\phi(q,v) + \mathcal{R}(u,-q)\phi(-q,v). \tag{10.114}$$

Finally, we should compute the $q$-derivative of $\phi(u,v)$. Differentiating (10.106) over $q$, we find

$$\frac{\partial\phi(u,v)}{\partial q} - \frac{1}{2\pi}\int_{-q}^{q}\mathcal{K}(u-w)\frac{\partial\phi(w,v)}{\partial q}\,\mathrm{d}w$$
$$= \frac{1}{2\pi}\Big(\mathcal{K}(u-q)\phi(q,v) + \mathcal{K}(u+q)\phi(-q,v)\Big). \tag{10.115}$$

This yields

$$\frac{\partial\phi(u,v)}{\partial q} = \mathcal{R}(u,q)\phi(q,v) + \mathcal{R}(u,-q)\phi(-q,v). \tag{10.116}$$

Using these results, we obtain

$$\begin{aligned}
\frac{\mathrm{d}\phi(q,q)}{\mathrm{d}q} &= 2\mathcal{R}(q,-q)\phi(-q,q),\\
\frac{\mathrm{d}\phi(-q,-q)}{\mathrm{d}q} &= 2\mathcal{R}(-q,q)\phi(q,-q),\\
\frac{\mathrm{d}\phi(q,-q)}{\mathrm{d}q} &= 2\mathcal{R}(q,-q)\big(1+\phi(-q,-q)\big),\\
\frac{\mathrm{d}\phi(-q,q)}{\mathrm{d}q} &= -2\mathcal{R}(-q,q)\big(1-\phi(q,q)\big).
\end{aligned} \tag{10.117}$$

It remains to calculate the $q$-derivative of (10.107). We provide the reader with the opportunity to do this simple exercise and make sure that the derivative vanishes. Finally, we note that the equality (10.107) becomes obvious for $q=0$, since $\phi(u,v)=\theta(u-v)$. □

We draw attention to the fact that we did not use the function $\theta(u)$ explicit form in our proof. It is only important that the kernel $\mathcal{K}(u-v)$ depends on the difference of the arguments, and $\mathcal{K}(u)=\theta'(u)$. In our particular case, we have an additional property $\theta(-u)=-\theta(u)$. This implies $\phi(-u,-v)=-\phi(u,v)$ due to proposition 10.7. Then identity (10.107) takes the form

$$\Big(1-\phi(q,q)-\phi(q,-q)\Big)\Big(1-\phi(q,q)+\phi(q,-q)\Big)=1. \tag{10.118}$$

Equality (10.118) allows us to obtain a new expression for the dressed charge at the Fermi boundaries in terms of the function $\phi$. Let us write down (10.71) in the form

$$Z(u)=1+\phi(u,-q)-\phi(u,q). \tag{10.119}$$

Setting here $u=q$ we obtain

$$Z(q)=1+\phi(q,-q)-\phi(q,q). \tag{10.120}$$

Then, due to (10.118), we have

$$\frac{1}{Z(q)}=1-\phi(q,-q)-\phi(q,q). \tag{10.121}$$

This nonlinear relationship between solutions of linear integral equations may seem surprising. One should remember, however, that equality (10.121) is not a functional identity. It is valid only at points when the function $\phi(u,v)$ arguments take the values $\pm q$.

## 10.6 Thermodynamic limit of the $XXZ$ chain

We now describe the thermodynamic limit in the $XXZ$ Heisenberg chain. In many respects, the analysis is similar to the analysis considered above for the model of one-dimensional bosons. However, there are some differences. We mainly focus on the latter. In particular, we do not present derivations of integral equations for the density of the ground state, the shift function, and so on. The reader can find a more detailed exposition in books [M. Gaudin (1983); V. E. Korepin et al (1993); H.-P. Eckle (2019)].

First of all, we slightly modify the Hamiltonian of the $XXZ$ chain compared to the formula (1.91). Let

$$H = H_0 - \mu S_z, \qquad [H_0, S_z] = 0. \tag{10.122}$$

Here

$$H_0 = \sum_{k=1}^{L} \Big( \sigma_k^x \sigma_{k+1}^x + \sigma_k^y \sigma_{k+1}^y + \Delta(\sigma_k^z \sigma_{k+1}^z - \mathbf{1}) \Big), \tag{10.123}$$

and

$$S_z = \frac{1}{2} \sum_{k=1}^{L} \sigma_k^z. \tag{10.124}$$

Now the length of the chain is denoted by $L$ (similarly to the length of the interval in the one-dimensional Bose gas). With a sufficiently large length of the chain, the physical properties should not depend on the parity of $L$. So for simplicity, we choose $L$ to be even. We also subtracted the identity operator from the product $\sigma_k^z \sigma_{k+1}^z$. This is done so that the Hamiltonian $H_0$ would have zero eigenvalue on the vacuum $|0\rangle$. Finally, we added the operator of the third component of the total spin $S_z$ to the Hamiltonian. The role of this operator is completely analogous to the role of the operator $Q$ in (10.99). Since $[H_0, S_z] = 0$, modification (10.122) does not change the eigenvectors of $H_0$. However, it shifts the eigenvalues. The factor $\mu$, in this case, has the meaning of an external magnetic field directed along the $z$-axis.

There exist the following similarity transformations:

$$\begin{aligned} \left( \prod_{m=1}^{L} \sigma_m^x \right) H(\mu) \left( \prod_{m=1}^{L} \sigma_m^x \right) &= H(-\mu), \\ \left( \prod_{m=1}^{L/2} \sigma_{2m}^z \right) H_0(\Delta) \left( \prod_{m=1}^{L/2} \sigma_{2m}^z \right) &= -H_0(-\Delta). \end{aligned} \tag{10.125}$$

Thanks to the first transformation, it is sufficient to consider only the case $\mu \geq 0$. The second transformation will be discussed a bit later.

We now turn to the description of the thermodynamic limit. In the $XXZ$ model, the vacuum vector $|0\rangle$ is a state with all spins up. Spin-down particles can be interpreted as excitations above the vacuum. The eigenstate, which depends on $N$ roots of the Bethe equations $\bar{u} = \{u_1, \ldots, u_N\}$, is a state containing $N$ spin-down particles. The chain length $L$ and $N$ go to infinity at a fixed ratio $N/L$ in the thermodynamic limit. This ratio is related to the magnetization of the chain $\langle \sigma^z \rangle = 1 - 2N/L$ (see (9.23)).

**Remark 10.6.** The thermodynamic limit should be distinguished from the scaling limit in which the model becomes continuous. In the thermodynamic limit of the $XXZ$ chain, the system remains discrete, and the distance between the sites of the chain does not vanish. Generally speaking, the thermodynamic limit can be considered both in discrete models (spin chains) and in continuous models (one-dimensional Bose gas).

The ground state of the Hamiltonian of the $XXZ$ chain at zero external magnetic field strongly depends on the value of the anisotropy parameter $\Delta$. At $\Delta > 1$, the ground state is antiferromagnetic. The lowest excited state is separated from the ground state by a finite gap. Therefore, this phase is called massive. For $-1 < \Delta \leq 1$, the ground state is also antiferromagnetic, but there is no gap in the spectrum. Therefore, this phase is called massless. Finally, for $\Delta \leq -1$, the ground state is ferromagnetic. We stress that this classification is valid for the Hamiltonian $H_0$. Adding the term $-\mu S_z$ makes the situation more involved. For example, with a sufficiently large $\mu$, the spectral gap can disappear, and the massive phase transforms into the massless one. Similarly, the massless phase can go into the ferromagnetic phase. We restrict ourselves to studying the thermodynamic limit in the massless phase for magnetic field values less than critical (that is, when $0 < N/L < 1$). Moreover, we do not consider the value $\Delta = 1$, though many formulas remain valid in this case.

**Remark 10.7.** When classifying different phases of the $XXZ$ chain, the second transformation (10.125) should be taken into account. Some authors prefer to have a common minus sign in the Hamiltonian. This leads to a change in the sign of the anisotropy parameter $\Delta$. Then, the value $\Delta < -1$ corresponds to the massive phase, and $\Delta \geq 1$ corresponds to the ferromagnetic phase. The possibility of replacing $\Delta$ with $-\Delta$ also leads to the fact that the equations describing the thermodynamic limit may differ slightly from one author to another.

When describing the $XXZ$ chain, we usually used a parameter $\eta$, so as $\cosh\eta = \Delta$. Since we consider the case $|\Delta| < 1$, the parameter $\eta$ turns out to be purely imaginary. Therefore, it is convenient to change variables $i\eta = \zeta$, taking $\zeta$ as a real parameter belonging to the interval $0 < \zeta < \pi$.

Then the system of Bethe equations reads

$$\left(\frac{\sinh(u_j + \frac{i\zeta}{2})}{\sinh(u_j - \frac{i\zeta}{2})}\right)^L = -\prod_{k=1}^{N} \frac{\sinh(u_j - u_k + i\zeta)}{\sinh(u_j - u_k - i\zeta)}. \tag{10.126}$$

Unlike the one-dimensional boson model, this system has both real and complex conjugate roots [A. A. Vladimirov (1986)]. Solutions that correspond to complex conjugate roots are usually called strings. The presence of such strings significantly complicates the analysis of excited states. A detailed description of string excitations is beyond the scope of this book, so we refer the reader to works [M. Fowler and X. Zotos (1981); M. Takahashi (1971, 1973); M. Takahashi and M. Suzuki (1972)]. In the ground state, all the roots of the system (10.126) are real.

To proceed to the logarithmic form of Bethe equations, we introduce

$$\begin{aligned} p_0(u) &= i\log\left(\frac{\sinh(\frac{i\zeta}{2} + u)}{\sinh(\frac{i\zeta}{2} - u)}\right), \qquad && p_0(0) = 0, \\ \theta(u) &= i\log\left(\frac{\sinh(i\zeta + u)}{\sinh(i\zeta - u)}\right), \qquad && \theta(0) = 0. \end{aligned} \tag{10.127}$$

The function $p_0(u)$ plays the role of the bare momentum of the elementary excitation. In the model of one-dimensional bosons, the roots of the Bethe equations $u_j$ themselves were the bare momenta.

An analog of the logarithmic Bethe equations (10.5) is

$$Lp_0(u_j) - \sum_{k=1}^{N} \theta(u_j - u_k) = 2\pi I_j, \tag{10.128}$$

where $I_j$ are (half)integers. As before, the ground state corresponds to a set $I_j \equiv n_j = j - \frac{N+1}{2}$. One can show that in this case, and for $\mu > 0$, the roots of the system fill a finite interval of the real axis $[-q, q]$. In the thermodynamic limit, the roots condense and can be described by a distribution density $\rho(u)$. Similarly to (10.41), we introduce a counting function

$$\hat{\xi}_u(u) = \frac{p_0(u)}{2\pi} - \frac{1}{2\pi L}\sum_{k=1}^{N} \theta(u - u_k) + \frac{N+1}{2L}, \tag{10.129}$$

and its derivative $\hat{\rho}_u(u) = \hat{\xi}'(u)$

$$\hat{\rho}_u(u) = \frac{p_0'(u)}{2\pi} - \frac{1}{2\pi L}\sum_{k=1}^{N} \mathcal{K}(u - u_k). \tag{10.130}$$

Here

$$\begin{aligned} p_0'(u) &= \frac{\mathrm{d}p_0(u)}{\mathrm{d}u} = \frac{\sin\zeta}{\sinh(u - \frac{i\zeta}{2})\sinh(u + \frac{i\zeta}{2})}, \\ \mathcal{K}(u) &= \frac{\sin 2\zeta}{\sinh(u - i\zeta)\sinh(u + i\zeta)}. \end{aligned} \tag{10.131}$$

In the thermodynamic limit, the function $\hat{\rho}_u(u)$ turns into a distribution density of the ground state $\rho(u)$. Formula (10.130) turns into an integral equation for the distribution density

$$\rho(u) + \frac{1}{2\pi}\int_{-q}^{q} \mathcal{K}(u - v)\rho(v)\,\mathrm{d}v = \frac{1}{2\pi}p_0'(u). \tag{10.132}$$

Note that, unlike equation (10.30), the integral operator has a different sign. This difference is not fundamental and is explained by choosing the initial form of the Hamiltonian $H_0$. As we already explained above, the Hamiltonian (10.123) admits a similarity transformation, leading to the change in the sign of $\Delta$. This corresponds to a change $\zeta \to \pi - \zeta$, which leads to a change in the sign of the kernel $\mathcal{K}(u)$.

Another distinguishing feature of the ground state in the $XXZ$ chain is that the parameters $u_j$ fill the real axis at zero magnetic field. More precisely, in the limit $N, L \to \infty$, the roots $u_j$ belong to the interval $[-\frac{\zeta}{\pi}\log L, \frac{\zeta}{\pi}\log L]$. Near the interval ends, the distance between adjacent roots is not small but of order $\log L$. However, in the bulk of the interval, the difference between neighboring roots is of order $1/L$. Thus, their distribution can be described by some density. The case of zero magnetic field is also described by equation (10.132), in which one should put $q \to \infty$. We will consider this case in more detail in the next section.

In Section 6.2, we introduced a function $t(u, v)$ (see (6.22)). In the case of the trigonometric $R$-matrix, it has the following form:

$$t(u, v) = \frac{-\sin^2\zeta}{\sinh(u - v)\sinh(u - v - i\zeta)}. \tag{10.133}$$

Obviously,

$$p_0'(u) = -\frac{t(u, -\frac{i\zeta}{2})}{\sin\zeta}. \tag{10.134}$$

Then we can introduce a so-called *inhomogeneous density* by

$$\rho(u,z) + \frac{1}{2\pi}\int_{-q}^{q} \mathcal{K}(u-v)\rho(v,z)\,\mathrm{d}v = -\frac{t(u,z)}{2\pi\sin\zeta}. \tag{10.135}$$

We will meet this function in Chapter 11 when computing correlation functions in the $XXZ$ chain. For $z = -\frac{i\zeta}{2}$, inhomogeneous density turns into the distribution density $\rho(u)$.

Particle-hole excitations are described entirely analogous to the model of the one-dimensional Bose gas. A counting function of a twisted excited state is defined as

$$\hat{\xi}_v(z) = \frac{p_0(z)}{2\pi} - \frac{1}{2\pi L}\sum_{k=1}^{N'}\theta(z-v_k) + \frac{N'+1}{2L} - \frac{\alpha}{2\pi L}, \tag{10.136}$$

where $N' = \#\bar{v}$. A shift function is still defined by (10.59). In the thermodynamic limit, the shift function of a state with $n$ particles and $n$ holes satisfies an integral equation

$$\begin{aligned} F(u) + \frac{1}{2\pi}\int_{-q}^{q} \mathcal{K}(u-v)F(v)\,\mathrm{d}v = \frac{\alpha}{2\pi} + \frac{\Delta N}{2}\Big(\tfrac{1}{\pi}\theta(z-q) - 1\Big) \\ + \frac{1}{2\pi}\sum_{k=1}^{n}\big(\theta(u-\lambda_{p_k}) - \theta(u-\lambda_{h_k})\big). \end{aligned} \tag{10.137}$$

The parameters $\lambda_{p_k}$ and $\lambda_{h_k}$ in the $XXZ$ model are usually called *quasi-momenta* or *rapidities* of particles and holes.

The solution of equation (10.137) is

$$F(u) = (\alpha - \Delta N\pi)\frac{Z(u)}{2\pi} + \Delta N\phi(u,q) + \sum_{k=1}^{n}\big(\phi(u,\lambda_{p_k}) - \phi(u,\lambda_{h_k})\big), \tag{10.138}$$

where $\phi(u,v)$ solves an equation similar to (10.66)

$$\phi(u,v) + \frac{1}{2\pi}\int_{-q}^{q} \mathcal{K}(u-w)\phi(w,v)\,\mathrm{d}w = \frac{1}{2\pi}\theta(u-v). \tag{10.139}$$

The function $Z(u)$ is the dressed charge. It satisfies an integral equation similar to (10.105)

$$Z(u) + \frac{1}{2\pi}\int_{-q}^{q} \mathcal{K}(u-v)Z(v)\,\mathrm{d}v = 1. \tag{10.140}$$

Unlike the one-dimensional Bose gas model, we see that now the solution to this equation has nothing to do with the distribution density (10.132).

At the same time, the solutions of equations (10.139) and (10.140) are still connected by identities similar to (10.120) and (10.121)

$$\begin{aligned} Z(q) &= 1 - \phi(q,-q) + \phi(q,q), \\ \frac{1}{Z(q)} &= 1 + \phi(q,-q) + \phi(q,q). \end{aligned} \tag{10.141}$$

The difference in the function $\phi(u,v)$ sign is explained by the opposite sign of the integral kernel $\mathcal{K}(u-v)$.

### 10.6.1 *Zero magnetic field*

For $q \to \infty$, we can explicitly solve equation (10.132) using the Fourier transform. Let

$$\begin{aligned} \mathring{\rho}(\lambda) &= \int_{-\infty}^{\infty} e^{iu\lambda} \rho(u)\, du, \\ \mathring{\mathcal{K}}(\lambda) &= \frac{1}{2\pi} \int_{-\infty}^{\infty} e^{iu\lambda} \mathcal{K}(u)\, du, \\ \mathring{p}_0'(\lambda) &= \frac{1}{2\pi} \int_{-\infty}^{\infty} e^{iu\lambda} p_0'(u)\, du. \end{aligned} \tag{10.142}$$

Multiplying equation (10.132) at $q \to \infty$ by $e^{iu\lambda}$ and integrating over $u$, we obtain

$$\mathring{\rho}(\lambda) + \frac{1}{2\pi} \int_{-\infty}^{\infty} e^{i(u-v)\lambda} e^{iv\lambda} \mathcal{K}(u-v) \rho(v)\, du\, dv = \mathring{p}_0'(\lambda). \tag{10.143}$$

Setting $u - v = u'$ in the double integral, we immediately find

$$\mathring{\rho}(\lambda)\big(1 + \mathring{\mathcal{K}}(\lambda)\big) = \mathring{p}_0'(\lambda), \tag{10.144}$$

which implies

$$\mathring{\rho}(\lambda) = \frac{\mathring{p}_0'(\lambda)}{1 + \mathring{\mathcal{K}}(\lambda)}. \tag{10.145}$$

Thus, to solve equation (10.132), one should find Fourier transforms $\mathring{\mathcal{K}}(\lambda)$ and $\mathring{p}_0'(\lambda)$. They can be easily computed:

$$\begin{aligned} \mathring{\mathcal{K}}(\lambda) &= \frac{\sinh \lambda(\frac{\pi}{2} - \zeta)}{\sinh \frac{\pi\lambda}{2}}, \\ \mathring{p}_0'(\lambda) &= \frac{\sinh \frac{\lambda}{2}(\pi - \zeta)}{\sinh \frac{\pi\lambda}{2}}. \end{aligned} \tag{10.146}$$

Then, due to (10.145),

$$\overset{\circ}{\rho}(\lambda) = \frac{1}{2\cosh\frac{\zeta\lambda}{2}}. \tag{10.147}$$

Making the inverse Fourier transform, we finally obtain

$$\rho(u) = \frac{1}{2\zeta\cosh\frac{\pi u}{\zeta}}. \tag{10.148}$$

We see that the distribution density decreases exponentially as $u \to \pm\infty$. This behavior is consistent with the fact that the distance between the neighboring roots of the Bethe equations is of the order of $\log L$ on the wings of their distribution.

For the zero external magnetic field, we also obtain

$$D \equiv \frac{N}{L} = \int_{-\infty}^{\infty} \rho(u)\,\mathrm{d}u = \overset{\circ}{\rho}(0) = \frac{1}{2}. \tag{10.149}$$

Thus, the magnetization $1 - 2N/L$ of the ground state vanishes.

Finally, equation (10.135) for the inhomogeneous density also becomes explicitly solvable for the zero magnetic field. We have

$$\rho(u, z) = \frac{i}{2\zeta\sinh\frac{\pi}{\zeta}(u-z)}, \tag{10.150}$$

provided $-\zeta < \Im z < 0$.

### 10.6.2 *Dressed charge at the Fermi boundary*

In Section 12.6, we will see that the asymptotics of correlation functions in the Bethe ansatz solvable models depend on the dressed charge values at the Fermi boundary. In the $XXZ$ model at zero magnetic field, the dressed charge function has a curious feature, which we will now discuss.

Let

$$\overset{\circ}{Z}(\lambda) = \int_{-\infty}^{\infty} e^{iu\lambda} Z(u)\,\mathrm{d}u. \tag{10.151}$$

Then using the method described above, we obtain

$$\overset{\circ}{Z}(\lambda) = \frac{2\pi\delta(\lambda)}{1 + \overset{\circ}{\mathcal{K}}(\lambda)}. \tag{10.152}$$

Here we took advantage of the fact that the Fourier transform of 1 gives $2\pi\delta(\lambda)$. Then

$$Z(u) = \frac{1}{1 + \overset{\circ}{\mathcal{K}}(0)} = \frac{\pi}{2(\pi - \zeta)}. \tag{10.153}$$

Thus, the dressed charge at zero magnetic field is equal to the constant value.

It seems that then the function $Z(u)$ value at the Fermi boundaries should be equal to the same constant (10.153). However, this is not the case. As follows from equation (10.140), the function $Z(u)$ depends parametrically on $q$: $Z(u) = Z(u,q)$. To find the value of the dressed charge at the Fermi boundary at zero magnetic field, we must first put $u = q$ and only then send $q$ to infinity. Instead, solving equation (10.140), we considered the limit $q \to \infty$ at a finite $u$.

A careful analysis of equation (10.140) shows that [N. M. Bogolyubov et al (1987); N. Kitanine et al (2009a); N. A. Slavnov (2010)]

$$\lim_{q\to\infty} Z(q,q) = \sqrt{\lim_{u\to\infty}\lim_{q\to\infty} Z(u,q)} = \sqrt{\frac{\pi}{2(\pi-\zeta)}}. \tag{10.154}$$

However, this analysis requires using the Riemann–Hilbert problem methods (see, e.g., [N. Kitanine et al (2009a)]) and is beyond the scope of this book. Therefore, we will limit ourselves to illustrating equation (10.154) in the first order of perturbation theory near the point[2] $\zeta = \pi/2$.

Let $\zeta = \pi/2 + \epsilon$. Then the integral kernel $\mathcal{K}(u-v)$ is

$$\mathcal{K}(u-v) = \frac{-2\epsilon}{\cosh^2(u-v)} + O(\epsilon^2). \tag{10.155}$$

We are looking for the solution to the integral equation (10.140) in the form $Z(u,q) = 1 + \epsilon Z_1(u,q) + O(\epsilon^2)$. Substituting this expansion into (10.140) and using (10.155), we immediately arrive at

$$Z_1(u,q) = \frac{1}{\pi}\int_{-q}^{q} \frac{dv}{\cosh^2(u-v)}. \tag{10.156}$$

Thus, we find

$$Z(u,q) = 1 + \frac{\epsilon}{\pi}\big(\tanh(q-u) + \tanh(q+u)\big) + O(\epsilon^2). \tag{10.157}$$

Taking the limit $q \to \infty$ at $u$ being arbitrary finite, we obtain

$$\lim_{q\to\infty} Z(u,q) = 1 + \frac{2\epsilon}{\pi} + O(\epsilon^2), \tag{10.158}$$

which is consistent with (10.153). On the other hand, setting $u = q$ in (10.157) and sending $q \to \infty$, we have

$$\lim_{q\to\infty} Z(q,q) = 1 + \frac{\epsilon}{\pi} + O(\epsilon^2). \tag{10.159}$$

This result agrees with (10.154).

[2]The $XXZ$ becomes equivalent to the free fermions at $\Delta = 0$ (see Section 11.6).

## 10.7 Examples of calculating the thermodynamic limit

This section considers examples of calculating the thermodynamic limit of sums, products, and determinants. As a rule, we use the Euler–Maclaurin formula (10.45) for this. Most often, it is sufficient to limit ourselves to the integral term of this formula.

### 10.7.1 *Thermodynamic limit of sums and products*

Let $f(u)$ be some smooth function on the interval $[-q, q]$. Consider a sum

$$\mathcal{J}_1 = \sum_{j=1}^{N} \big(f(v_j) - f(u_j)\big). \tag{10.160}$$

Here $\bar{u} = \{u_1, \dots, u_N\}$ are the roots of Bethe equations corresponding to the ground state, and $\bar{v} = \{v_1, \dots, v_N\}$ are the roots of twisted Bethe equations corresponding to the $\alpha$-twisted ground state. Setting $v_j = u_j + \epsilon_j$, where $\epsilon_j$ is of order $1/L$, we have

$$\mathcal{J}_1 = \sum_{j=1}^{N} \big(\epsilon_j f'(u_j) + O(L^{-2})\big). \tag{10.161}$$

Using (10.61), we find

$$\operatorname{Lim} \mathcal{J}_1 = \int_{-q}^{q} f'(u) F(u)\, du. \tag{10.162}$$

Here the shift function corresponds to the $\alpha$-twisted ground state.

Similarly, we can consider a double sum. Let

$$\mathcal{J}_2 = \sum_{j,k=1}^{N} \big(f(u_j, v_k) + f(v_j, u_k) - f(v_j, v_k) - f(u_j, u_k)\big), \tag{10.163}$$

where $f(u, v)$ is differentiable over $u$ and over $v$. Setting again $v_j = u_j + \epsilon_j$, we obtain

$$\mathcal{J}_2 = \sum_{j,k=1}^{N} \left(-\epsilon_j \epsilon_k \frac{\partial^2 f(u_j, u_k)}{\partial u_j \partial u_k} + O(L^{-3})\right), \tag{10.164}$$

and thus,

$$\operatorname{Lim} \mathcal{J}_2 = -\int_{-q}^{q} \frac{\partial^2 f(u, v)}{\partial u \partial v} F(u) F(v)\, du\, dv. \tag{10.165}$$

Let now the set $\bar{v}$ correspond to the state with $n$ particles with momenta $\{\lambda_{p_1},\dots,\lambda_{p_n}\}$ and $n$ holes with momenta $\{\lambda_{h_1},\dots,\lambda_{h_n}\}$. Then we have for the sum $\mathcal{J}_1$

$$\mathcal{J}_1=\sum_{j=1}^{N}\big(f(v'_j)-f(u_j)\big)+\sum_{k=1}^{n}\big(f(\lambda_{p_k})-f(\lambda_{h_k})\big), \tag{10.166}$$

where $v'_j$ are defined in (10.56). Using (10.162) for the first sum in (10.166), we obtain

$$\operatorname{Lim}\mathcal{J}_1=\int_{-q}^{q}f'(u)F(u)\,du+\sum_{k=1}^{n}\big(f(\lambda_{p_k})-f(\lambda_{h_k})\big). \tag{10.167}$$

Although the integral terms in formulas (10.162) and (10.167) look the same, it should be remembered that the shift functions in them are different. In (10.167), the shift function corresponds to the excited state with the given values of the particle and hole momenta.

Note that formula (10.167) can be written more compactly if we introduce a modified shift function by

$$F_{\rm mod}(u)=F(u)\chi_{[-q,q]}(u)+\sum_{j=1}^{n}\Big(\chi_{]-\infty,\lambda_{p_j}]}(u)-\chi_{]-\infty,\lambda_{h_j}]}(u)\Big). \tag{10.168}$$

Here $\chi_{[a,b]}$ is the characteristic function of the interval $[a,b]$. Then the limit of the sum $\mathcal{J}_1$ takes the form (10.162) provided the integration interval $[-q,q]$ is replaced with the real axis:

$$\operatorname{Lim}\mathcal{J}_1=\int_{-\infty}^{\infty}f'(u)F_{\rm mod}(u)\,du. \tag{10.169}$$

Using this trick, we can easily calculate the limit of the sum $\mathcal{J}_2$ (10.163)

$$\operatorname{Lim}\mathcal{J}_2=-\int_{-\infty}^{\infty}\frac{\partial^2 f(u,v)}{\partial u\partial v}F_{\rm mod}(u)F_{\rm mod}(v)\,du\,dv. \tag{10.170}$$

Substituting here (10.168) we find

$$\begin{aligned}\operatorname{Lim}\mathcal{J}_2&=\sum_{j,k=1}^{n}\big(f(\lambda_{h_j},\lambda_{p_k})+f(\lambda_{p_j},\lambda_{h_k})-f(\lambda_{p_j},\lambda_{p_k})-f(\lambda_{h_j},\lambda_{h_k})\big)\\&+\sum_{j=1}^{n}\int_{-q}^{q}F(u)\frac{\partial}{\partial u}\big(f(\lambda_{h_j},u)+f(u,\lambda_{h_j})-f(\lambda_{p_j},u)-f(u,\lambda_{p_j})\big)\,du\\&-\int_{-q}^{q}\frac{\partial^2 f(u,v)}{\partial u\partial v}F(u)F(v)du\,dv.\end{aligned} \tag{10.171}$$

The thermodynamic limits of products easily reduce to the limits of sums. Let, for example,

$$\mathcal{P}_1 = \prod_{j=1}^{N} \frac{f(v_j)}{f(u_j)}. \tag{10.172}$$

Taking the logarithm, we immediately reduce this product to the sum (10.166)

$$\operatorname{Lim} \log \mathcal{P}_1 = \int_{-\infty}^{\infty} \frac{\mathrm{d} \log f(u)}{\mathrm{d}u} F_{\mathrm{mod}}(u)\, \mathrm{d}u. \tag{10.173}$$

Thus, we obtain

$$\operatorname{Lim} \mathcal{P}_1 = \prod_{j=1}^{n} \frac{f(\lambda_{p_j})}{f(\lambda_{h_j})} \exp\left\{ \int_{-q}^{q} \frac{\mathrm{d} \log f(u)}{\mathrm{d}u} F(u)\, \mathrm{d}u \right\}. \tag{10.174}$$

Similarly, the limits of double products are calculated.

**Remark 10.8.** When deriving formula (10.174), we assumed that $f(u) \neq 0$ for $u \in [-q, q]$. In this case, the integral in formula (10.174) is well defined. If this condition is not fulfilled, then the thermodynamic limit of such a product may not exist or be calculated in a more complex way. We will consider examples of such products in Section 12.3.2.

### 10.7.2 *Definition of Fredholm determinant*

Let us now consider the thermodynamic limit of determinants. We need to do this since the form factors of local operators are expressed in terms of determinants of matrices of finite size. However, in the thermodynamic limit, the size of these matrices goes to infinity. Therefore, the determinants of matrices turn into the Fredholm determinants of integral operators. First of all, we recall the definition of the Fredholm determinant.

Assume that a two-variable function $V(u, v)$ is bounded in a rectangle $a \leq u, v \leq b$. Consider an integral operator $\mathbf{I} + z\mathbf{V}$ acting in $L^2([a, b])$ as follows:

$$\begin{aligned} [(\mathbf{I} + z\mathbf{V})f](u) &= f(u) + z \int_a^b V(u, v) f(v)\, \mathrm{d}v, \\ [f(\mathbf{I} + z\mathbf{V})](v) &= f(v) + z \int_a^b V(u, v) f(u)\, \mathrm{d}u. \end{aligned} \tag{10.175}$$

Here $z$ is a complex number.

A Fredholm determinant of the integral operator $\mathbf{I}+z\mathbf{V}$ is a series of the form

$$\det(\mathbf{I}+z\mathbf{V})=\sum_{n=0}^{\infty}\frac{z^n}{n!}\int_a^b \det_n\begin{pmatrix}V(u_1,u_1)&\cdots&V(u_1,u_n)\\ \dots&\ddots&\vdots\\ V(u_n,u_1)&\cdots&V(u_n,u_n)\end{pmatrix}du_1\dots du_n. \tag{10.176}$$

By definition, the term corresponding to $n=0$ is equal to 1.

The Fredholm determinant is a generalization of the concept of the matrix determinant. It arises when the matrix size goes to infinity.

One can prove (see, e.g., [V. I. Smirnov (1964)]) that under the conditions stated above, the series (10.176) absolutely converges for any finite $z$. Thus, the Fredholm determinant is an entire function of $z$.

Sometimes it is more convenient to use another formula for the Fredholm determinant

$$\det(\mathbf{I}+z\mathbf{V})=\exp\Big\{\operatorname{tr}\log(\mathbf{I}+z\mathbf{V})\Big\}. \tag{10.177}$$

Expanding the logarithm into the power series of $z$, we obtain

$$\det(\mathbf{I}+z\mathbf{V})=\exp\Big\{\sum_{n=1}^{\infty}(-1)^{n-1}\frac{z^n}{n}\operatorname{tr}\mathbf{V}^n\Big\}. \tag{10.178}$$

Note that when decomposing the logarithm into the series, we assume that $|z|$ is small enough so that $\|z\mathbf{V}\|<1$. Then the series in (10.178) absolutely converges. After that, using the Fredholm determinant's analyticity, we can arrange an analytic continuation to any point in the complex plane of $z$.

If we expand the exponent in the rhs of (10.178) into a power series in $z$, then we reproduce series (10.176). We suggest that the reader independently verify this by calculating the first few terms in the decomposition.

**Remark 10.9.** Formulas similar to (10.176)–(10.178) also hold for the usual matrix determinant. In particular,

$$\det_N(\delta_{jk}+zV_{jk})=\exp\Big\{\sum_{n=1}^{\infty}(-1)^{n-1}\frac{z^n}{n}\sum_{j_1,\dots,j_n=1}^{N}V_{j_1,j_2}V_{j_2,j_3}\dots V_{j_n,j_1}\Big\}. \tag{10.179}$$

**Remark 10.10.** The Fredholm determinant concept naturally generalizes to integral operators that act on an infinite interval (for example, on the real axis) or on some contour in the complex plane. In this case, it is necessary to impose additional requirements on the kernel $V(u,v)$ that ensure the convergence of the integrals in $\operatorname{tr}\mathbf{V}^n$.

### 10.7.3 *Norm of on-shell Bethe vectors in the thermodynamic limit*

We have calculated the norm of on-shell Bethe vectors in Section 6.4 (see (6.76), (6.77)). In the one-dimensional Bose gas, these formulas take the form

$$\langle \mathbf{\Phi}_N(\bar{u})|\mathbf{\Phi}_N(\bar{u})\rangle = c^N \left( \prod_{\substack{j,k=1\\ j\neq k}}^{N} \frac{u_j - u_k - ic}{u_j - u_k} \right) \det_N \mathcal{N}_{jk}, \tag{10.180}$$

where

$$\mathcal{N}_{jk} = \delta_{jk}\Big(L + \sum_{m=1}^{N} \mathcal{K}(u_j - u_m)\Big) - \mathcal{K}(u_j - u_k), \tag{10.181}$$

and $\mathcal{K}(u)$ is given by (10.10). Let the parameters $\bar{u}$ correspond to the ground state. Consider the behavior of $\det \mathcal{N}$ in the thermodynamic limit.

Due to (10.43), we have

$$L + \sum_{m=1}^{N} \mathcal{K}(u_j - u_m) = 2\pi L \hat{\rho}_u(u_j). \tag{10.182}$$

Then

$$\det_N \mathcal{N}_{jk} = \det_N (\delta_{jk} - V_{jk}) \prod_{a=1}^{N} \big(2\pi L \hat{\rho}_u(u_a)\big), \tag{10.183}$$

where

$$V_{jk} = \frac{\mathcal{K}(u_j - u_k)}{2\pi L \hat{\rho}_u(u_k)}. \tag{10.184}$$

Let us calculate the traces of the powers of the matrix $V$ in the thermodynamic limit. We have

$$\operatorname{tr} V = \sum_{j=1}^{N} V_{jj} = \sum_{j=1}^{N} \frac{\mathcal{K}(0)}{2\pi L \hat{\rho}_u(u_j)} \to \frac{1}{2\pi} \int_{-q}^{q} \mathcal{K}(0)\, \mathrm{d}u = \operatorname{tr}\left(\tfrac{1}{2\pi}\mathbf{K}\right), \tag{10.185}$$

$$\begin{aligned} \operatorname{tr} V^2 &= \sum_{j,k=1}^{N} V_{jk} V_{kj} = \sum_{j,k=1}^{N} \frac{\mathcal{K}(u_j - u_k)\mathcal{K}(u_k - u_j)}{4\pi^2 L^2 \hat{\rho}_u(u_j)\hat{\rho}_u(u_k)} \\ &\to \frac{1}{4\pi^2} \int_{-q}^{q} \mathcal{K}(u - v)\mathcal{K}(v - u)\, \mathrm{d}u\, \mathrm{d}v = \operatorname{tr}\left(\tfrac{1}{2\pi}\mathbf{K}\right)^2, \end{aligned} \tag{10.186}$$

and so on. Thus,

$$\operatorname{tr} V^n \to \operatorname{tr}\left(\tfrac{1}{2\pi}\mathbf{K}\right)^n, \qquad n = 1, 2, \ldots, \tag{10.187}$$

in the thermodynamic limit. Due to equations (10.178) and (10.179), we conclude that the matrix determinant in (10.183) turns to the Fredholm determinant

$$\operatorname{Lim} \det_N \left( \delta_{jk} - \frac{\mathcal{K}(u_j - u_k)}{2\pi L \hat{\rho}_u(u_k)} \right) = \det\left(\mathbf{I} - \tfrac{1}{2\pi}\mathbf{K}\right). \tag{10.188}$$

The product $\prod_{a=1}^{N} \left(2\pi L \hat{\rho}_u(u_a)\right)$ has no thermodynamic limit. However, it disappears in the normalized expressions for form factors and correlation functions.

Consider now the norm of an excited state with a finite number of particles and holes. Let this state be parameterized by the set of Bethe parameters $\bar{v} = \{v_1, \ldots, v_N\}$. Then

$$\langle \mathbf{\Phi}_N^{(\kappa)}(\bar{v}) | \mathbf{\Phi}_N^{(\kappa)}(\bar{v}) \rangle = e^{-i\alpha N} c^N \left( \prod_{\substack{j,k=1 \\ j \neq k}}^{N} \frac{v_j - v_k - ic}{v_j - v_k} \right) \det_N \mathcal{N}_{jk}^{(\mathrm{ex})}. \tag{10.189}$$

Here

$$\begin{aligned} \det_N \mathcal{N}_{jk}^{(\mathrm{ex})} &= \det_N \left( \delta_{jk} \Big( L + \sum_{m=1}^{N} \mathcal{K}(v_j - v_m) \Big) - \mathcal{K}(v_j - v_k) \right) \\ &= \det_N \left( \delta_{jk} - V_{jk}^{\mathrm{ex}} \right) \prod_{a=1}^{N} \left( 2\pi L \hat{\rho}_v(v_a) \right), \end{aligned} \tag{10.190}$$

where now

$$V_{jk}^{\mathrm{ex}} = \frac{\mathcal{K}(v_j - v_k)}{2\pi L \hat{\rho}_v(v_k)}. \tag{10.191}$$

The entries of the matrix $V^{\mathrm{ex}}$, which depend on the momenta of holes and particles, may differ significantly from the corresponding matrix elements of the matrix $V$ (10.184). However, with a finite number of holes and particles, the contribution of these elements to the traces of the powers of the matrix $V^{\mathrm{ex}}$ is of order $O(L^{-1})$. To illustrate this statement, consider, for instance, the difference $\operatorname{tr}(V^{\mathrm{ex}} - V)$:

$$\begin{aligned} \operatorname{tr}(V^{\mathrm{ex}} - V) &= \frac{\mathcal{K}(0)}{2\pi L} \sum_{k=1}^{N} \left( \frac{1}{\hat{\rho}_v(v_k)} - \frac{1}{\hat{\rho}_u(u_k)} \right) \\ &= \frac{\mathcal{K}(0)}{2\pi L} \sum_{k=1}^{N} \left( \frac{1}{\hat{\rho}_v(v_k')} - \frac{1}{\hat{\rho}_u(u_k)} \right) \\ &\quad + \frac{\mathcal{K}(0)}{2\pi L} \sum_{j=1}^{n} \left( \frac{1}{\hat{\rho}_v(\lambda_{p_j})} - \frac{1}{\hat{\rho}_v(\lambda_{h_j})} \right), \end{aligned} \tag{10.192}$$

where $v'_k$ are given by (10.56). The contribution from the finite sum over the momenta of particles and holes vanishes as it has the prefactor $1/L$. At the same time, the difference $1/\hat{\rho}_v(v'_k) - 1/\hat{\rho}_u(u_k)$ is a value of order $1/L$. Therefore, the terms in the first sum in the rhs of (10.192) are of order $1/L^2$. Hence, this contribution also vanishes, and we arrive at

$$\operatorname{Lim} \operatorname{tr} V^{\text{ex}} = \operatorname{Lim} \operatorname{tr} V = \operatorname{tr}\left(\frac{1}{2\pi}\mathbf{K}\right). \tag{10.193}$$

Similarly, we can prove that $\operatorname{tr}(V^{\text{ex}})^k = \operatorname{tr}(V)^k$ in the thermodynamic limit for any $k > 1$. Thus, we eventually arrive at

$$\operatorname{Lim} \det_N \left(\delta_{jk} - \frac{\mathcal{K}(v_j - v_k)}{2\pi L \hat{\rho}_v(v_k)}\right) = \det\left(\mathbf{I} - \tfrac{1}{2\pi}\mathbf{K}\right). \tag{10.194}$$

### 10.7.4 *Form factors in the thermodynamic limit*

Let us give another example of the Fredholm determinant. This is a typical determinant that arises when studying the thermodynamic limit of form factors.

Let $G(z, z')$ be an analytical function if $z$ and $z'$ belong to some neighborhood of the Fermi zone $[-q, q]$. Consider an $N \times N$ matrix $G$ of the form

$$G_{jk} = G(u_j, u_k) \frac{\prod_{l=1}^{N}(u_k - v_l)}{\prod_{l=1;\, l\neq k}^{N}(u_k - u_l)}. \tag{10.195}$$

Here parameters $\bar{u}$ correspond to the ground state, and parameters $\bar{v}$ correspond to an excited state.

**Proposition 10.10.** *Under the condition formulated above,*

$$\det_N(\delta_{jk} + G_{jk}) = \det\left(\mathbf{I} + \tfrac{1}{2\pi i}\mathbf{G}\right). \tag{10.196}$$

*Here the integral operator acts on a contour* $\Gamma([-q, q])$ *as follows:*

$$[(\mathbf{I} + \tfrac{1}{2\pi i}\mathbf{G})f](z) = f(z) + \int\limits_{\Gamma([-q,q])} G(z, z') \left(\prod_{l=1}^{N} \frac{z' - v_l}{z' - u_l}\right) f(z') \,\frac{\mathrm{d}z'}{2\pi i}. \tag{10.197}$$

*The integration contour* $\Gamma([-q, q])$ *surrounds the interval* $[-q, q]$ *and belongs to the domain of analyticity of the function* $G(z, z')$ *(see Fig. 10.1).*

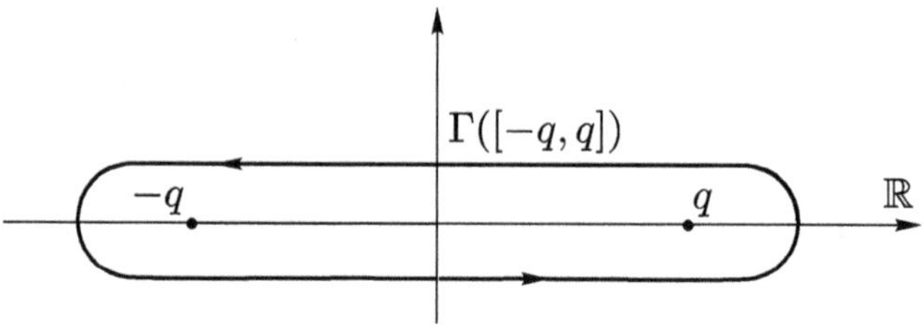


Fig. 10.1 Contour $\Gamma([-q,q])$.

**Proof.** It is sufficient to calculate the traces of $\mathbf{G}^k$, $k=1,2,\ldots$. We have

$$\operatorname{tr}\tfrac{1}{2\pi i}\mathbf{G}=\int\limits_{\Gamma([-q,q])}G(z,z)\left(\prod_{l=1}^{N}\frac{z-v_l}{z-u_l}\right)\frac{\mathrm{d}z}{2\pi i}. \tag{10.198}$$

The integral is equal to the sum of the residues in the points $z=u_k$, $k=1,\ldots,N$. Thus,

$$\operatorname{tr}\tfrac{1}{2\pi i}\mathbf{G}=\sum_{k=1}^{N}G(u_k,u_k)\frac{\prod_{l=1}^{N}(u_k-v_l)}{\prod_{l=1;\,l\neq k}^{N}(u_k-u_l)}=\operatorname{tr}G. \tag{10.199}$$

Similarly,

$$\operatorname{tr}\left(\tfrac{1}{2\pi i}\mathbf{G}\right)^2=\int\limits_{\Gamma([-q,q])}G(z_1,z_2)G(z_2,z_1)\left(\prod_{a=1}^{2}\prod_{l=1}^{N}\frac{z_a-v_l}{z_a-u_l}\right)\frac{\mathrm{d}z_1\,\mathrm{d}z_2}{(2\pi i)^2}. \tag{10.200}$$

Computing again the integral by the residues, we obtain

$$\operatorname{tr}\left(\tfrac{1}{2\pi i}\mathbf{G}\right)^2 =\sum_{k_1,k_2=1}^{N}G(u_{k_1},u_{k_2})G(u_{k_2},u_{k_1})\prod_{a=1}^{2}\frac{\prod_{l=1}^{N}(u_{k_a}-v_l)}{\prod_{l=1;\,l\neq k_a}^{N}(u_{k_a}-u_l)}=\operatorname{tr}G^2, \tag{10.201}$$

and so on. It becomes clear that

$$\operatorname{tr}\left(\tfrac{1}{2\pi i}\mathbf{G}\right)^k=\operatorname{tr}G^k,\qquad k=1,2,\ldots, \tag{10.202}$$

and proposition 10.10 is proved. □

It is important to emphasize that at this stage, we have not yet considered the thermodynamic limit. Equality (10.196) is exact for any finite $N$. To proceed to the thermodynamic limit, we need to calculate the limit of the product over $l$ in (10.197). Let the set $\bar{v}$ contain $n$ particles with momenta $\lambda_{p_k}$ and $n$ holes with momenta $\lambda_{h_k}$, $k = 1, \dots, n$. Then using (10.174), we obtain

$$\prod_{l=1}^{N} \frac{z' - v_l}{z' - u_l} = \left( \prod_{k=1}^{n} \frac{z' - \lambda_{p_k}}{z' - \lambda_{h_k}} \right) e^{-\tilde{F}(z')}, \tag{10.203}$$

where $\tilde{F}(z')$ is the Cauchy transform of the thermodynamic limit of the shift function:

$$\tilde{F}(z') = \int_{-q}^{q} \frac{F(u)\,\mathrm{d}u}{z' - u}. \tag{10.204}$$

Pay attention that $z' \in \Gamma([-q, q])$, while $u \in [-q, q]$. Therefore, $|z' - u|$ is bounded from below by some positive value. Hence, the Cauchy transform $\tilde{F}(z')$ is well defined.

Thus, we finally arrive at

$$\operatorname*{Lim}_{N} \det(\delta_{jk} + G_{jk}) = \det\left(\mathbf{I} + \tfrac{1}{2\pi i}\mathbf{G}\right), \tag{10.205}$$

where the integral operator acts on the contour $\Gamma([-q, q])$ as follows:

$$\begin{aligned} &[(\mathbf{I} + \tfrac{1}{2\pi i}\mathbf{G})f](z) \\ &\quad = f(z) + \int\limits_{\Gamma([-q,q])} G(z, z') e^{-\tilde{F}(z')} \left( \prod_{k=1}^{n} \frac{z' - \lambda_{p_k}}{z' - \lambda_{h_k}} \right) f(z') \frac{\mathrm{d}z'}{2\pi i}. \end{aligned} \tag{10.206}$$

# Chapter 11

# Multiple integral representations for correlation functions

The technique developed in the previous chapters allows us to proceed to the calculating correlation functions of quantum integrable systems. Even within the framework of the algebraic Bethe Ansatz, there are many approaches to this problem. We confine ourselves to two specific examples. However, these examples clearly illustrate the main methods and techniques.

In this chapter, we study multiple integral representations for the correlation functions of the $XXZ$ spin chain. For the first time, they were obtained in [M. Jimbo et al (1992); M. Jimbo and T. Miwa (1995); M. Jimbo and T. Miwa (1996)] using an approach based on the $q$-deformed Knizhnik–Zamolodchikov equations arising from the representation theory of quantum affine algebras. The first multiple integral representations described the density matrix elements on the finite segment of the chain at zero temperature and zero external magnetic field. These results were then generalized in the algebraic Bethe ansatz framework to the case of a nonzero external magnetic field [N. Kitanine et al (2000)]. Subsequently, new integral representations were obtained for two-point functions [N. Kitanine et al (2002a,c, 2004, 2005a,b,c,d,e)] and correlation functions at a finite temperature [F. Göhmann et al (2004, 2005a,b,d, 2010); F. Göhmann and J. Suzuki (2010)].

The multiple integral representations proved to be very convenient for the explicit calculation of correlation functions in the case of a small lattice distance between spin operators [G. Kato et al (2003, 2004); K. Sakai et al (2003); H. E. Boos and V. E. Korepin (2001, 2002); H. E. Boos et al (2003, 2008)]. It was later shown that the multiple integrals for the elements of the inhomogeneous density matrix are factorized [H. Boos et al (2006a,b, 2007, 2009); M. Jimbo et al (2009); H. E. Boos et al (2007, 2006); J. Damerau et al (2007); H. E. Boos and F. Göhmann (2009)]. However, it is not straightforward to evaluate the explicit homogeneous limit in these expressions when the lattice distance is large. The long-distance asymptotic analysis of the multiple integral representations was performed in [N. Kitanine et al (2002d, 2009b)]. However, the method turned out to be too complicated.

Multiple integral representations were also derived for the correlation functions of the one-dimensional Bose gas [N. Kitanine et al (2007); A. Seel et al (2007)]. They were used for calculating the long-distance asymptotics of the two-point functions at zero and finite temperature in [N. Kitanine et al (2009b); K. K. Kozlowski et al (2011a,b); K. K. Kozlowski and V. Terras (2011); K. K. Kozlowski (2015)].

Let us now describe the content of this chapter. In Section 11.1, we formulate the problem of computing two-point functions of local operators in the algebraic Bethe ansatz framework. In Section 11.2, we specify this problem to the case of local spin operators of the $XXZ$ chain. Section 11.3 introduces a special generating function that allows us to calculate the correlation function of the third spin components. Here we calculate this generating function for the chain of finite length. To do this, we use the formula for the multiple action of the twisted transfer matrix obtained in Section 7.1.3. The final result is given as a sum over the partitions of the Bethe parameters and inhomogeneities. The thermodynamic limit of this formula is calculated in Section 11.4. The sum over the partitions of inhomogeneities turns into contour integrals of the Cauchy type, and the sum over the partitions of the Bethe parameters turns into integrals over the Fermi zone. As a result, we obtain a multiple integral representation for the generating function in the thermodynamic limit. In Section 11.5, we reproduce the same result, but in a completely different way. Namely, we compute the generating function using the form factor expansion. In Section 11.6, we study the resulting multiple integral representation in the particular case of the $XXZ$ chain when the anisotropy parameter $\Delta = 0$. In this case, the model is equivalent to free fermions. All multiple integrals are easily calculated, and we get an explicit answer for the correlation function of the third spin components. In Section 11.7, we consider another particular case of the generating function called the emptiness formation probability. We calculate the corresponding multiple integral and obtain an exact result with a special value of the anisotropy parameter $\Delta = 1/2$.

In conclusion, we note that many different multiple integral representations for the correlation functions of spin chains are known today. We give only one example of such a representation. The reader can find other integral formulas for correlation functions in the works cited above.

## 11.1 Two-point correlation functions via the algebraic Bethe ansatz

Let $\mathfrak{O}$ be an operator acting in the quantum space $\mathcal{H}$ and corresponding to some observable. Then the correlation function of this operator at a finite temperature $T$ is defined as follows:

$$\langle \mathfrak{O} \rangle_T = \mathbf{Z}^{-1} \sum_n \langle \omega_n | \mathfrak{O} | \omega_n \rangle e^{-E_n/T}, \tag{11.1}$$

where $\mathbf{Z}$ is a partition function

$$\mathbf{Z} = \sum_n e^{-E_n/T}. \tag{11.2}$$

The sum is taken over all possible normalized eigenstates $|\omega_n\rangle$ of the Hamiltonian, $E_n$ being the corresponding energy in these formulas. At zero temperature, only those terms that correspond to the minimum energy survive in the sum. Assuming that the ground state is not degenerate, we obtain

$$\langle \mathfrak{O} \rangle = \langle \omega_0 | \mathfrak{O} | \omega_0 \rangle. \tag{11.3}$$

Here $|\omega_0\rangle$ is the normalized ground state of the Hamiltonian. The algebraic Bethe ansatz allows one to consider correlation functions at both finite and zero temperatures. However, for simplicity, we restrict ourselves to the study of the case (11.3). Besides, we will consider the operators $\mathfrak{O}$ of the form

$$\mathfrak{O} = \mathcal{O}^\dagger(x_1, t_1)\mathcal{O}(x_2, t_2). \tag{11.4}$$

Here $\mathcal{O}(x, t)$ is some local operator (Bose field, spin operator, etc.) that depends on the spatial and temporal coordinates. The corresponding correlation functions are called two-point functions. Thus, our goal is to calculate an expectation value

$$\langle \mathcal{O}^\dagger(x_1, t_1)\mathcal{O}(x_2, t_2) \rangle = \langle \omega_0 | \mathcal{O}^\dagger(x_1, t_1)\mathcal{O}(x_2, t_2) | \omega_0 \rangle, \tag{11.5}$$

in the thermodynamic limit.

We briefly describe the main strategy for calculating correlation functions in the algebraic Bethe ansatz framework using a spin chain example. At the initial stage, we consider an average

$$\langle\!\langle \mathcal{O}^\dagger(x_1, t_1)\mathcal{O}(x_2, t_2) \rangle\!\rangle = \frac{\langle \mathbf{\Phi}_N(\bar{u}) | \mathcal{O}^\dagger(x_1, t_1)\mathcal{O}(x_2, t_2) | \mathbf{\Phi}_N(\bar{u}) \rangle}{\langle \mathbf{\Phi}_N(\bar{u}) | \mathbf{\Phi}_N(\bar{u}) \rangle}, \tag{11.6}$$

where $|\mathbf{\Phi}_N(\bar{u})\rangle$ is an on-shell Bethe vector corresponding to some solution of the Bethe equations $\bar{u} = \{u_1, \dots, u_N\}$. Here and further, double angle brackets mean that we are considering the normalized average value for a finite number of Bethe parameters $N$ and a finite number of chain sites $L$. We compute this average value using the algebraic Bethe ansatz methods. After that, we choose a Bethe equation solution corresponding to the minimum energy and calculate the limit $N, L \to \infty$ at $D = N/L = const.$ The advantage of this approach is that the average value (11.6) is a well-defined quantity. In particular, for a finite chain length, the norm of the on-shell vector $|\mathbf{\Phi}_N(\bar{u})\rangle$ is finite. Therefore, at the first stage of calculations, we can obtain an exact result for the correlation function in terms of the Bethe equations roots $\bar{u}$. We consider the limit $N, L \to \infty$ after the average (11.6) is calculated. In other words, we calculate the limit of a certain scalar function but not the limit in which the number of local quantum spaces $\mathbb{C}^2$ tends to infinity. In fairness, it should also be mentioned that the exact answers for correlation functions for finite $N$ and $L$ are often rather complex. Therefore, calculating the thermodynamic limit is a non-trivial problem.

There are at least two ways to calculate the average value (11.6) within the algebraic Bethe ansatz framework. First, we can calculate the action of operators $\mathcal{O}^\dagger\mathcal{O}$, for example, on the vector $|\mathbf{\Phi}_N(\bar{u})\rangle$. If the result of this action is a linear combination of off-shell Bethe vectors, then it remains to calculate the scalar products of these off-shell vectors with the dual on-shell Bethe vector $\langle\mathbf{\Phi}_N(\bar{u})|$. In the previous sections, we obtained explicit determinant representations for these scalar products. This method is more often used to calculate time-independent correlation functions, although it can also be generalized when the operators depend on time [N. Kitanine et al (2005d)].

Second, we can insert the identity operator between $\mathcal{O}^\dagger$ and $\mathcal{O}$ as the sum of projectors to the eigenstates of the Hamiltonian

$$\begin{aligned}&\langle\langle\mathcal{O}^\dagger(x_1,t_1)\mathcal{O}(x_2,t_2)\rangle\rangle\\ &\quad=\sum_{N'}\sum_{\bar{v}}\frac{\langle\mathbf{\Phi}_N(\bar{u})|\mathcal{O}^\dagger(x_1,t_1)|\mathbf{\Phi}_{N'}(\bar{v})\rangle\langle\mathbf{\Phi}_{N'}(\bar{v})|\mathcal{O}(x_2,t_2)|\mathbf{\Phi}_N(\bar{u})\rangle}{\langle\mathbf{\Phi}_{N'}(\bar{v})|\mathbf{\Phi}_{N'}(\bar{v})\rangle\langle\mathbf{\Phi}_N(\bar{u})|\mathbf{\Phi}_N(\bar{u})\rangle}. \qquad (11.7)\end{aligned}$$

Here the sum is taken over solutions of the Bethe equations $\bar{v} = \{v_1, \dots, v_{N'}\}$. Then we get the form factors of local operators $\mathcal{O}^\dagger$ and $\mathcal{O}$, for which we also have explicit formulas in the form of determinants. Thus, the problem reduces to summing form factors over the solutions of the Bethe equations. This way is especially appropriate for calculating dynamic (i.e., time-dependent) correlation functions since the dependence of each form factor on time is immediately separated as a simple phase factor.

Let us also say a few words about other methods for calculating correlation functions. We have mentioned already that the algebraic Bethe ansatz allows us to study correlation functions at a finite temperature. In particular, one can reduce the summation over the Hamiltonian eigenstates to averaging over some effective state

$$\langle \mathcal{O}^\dagger(x_1,t_1)\mathcal{O}(x_2,t_2)\rangle_T = \langle \omega_T|\mathcal{O}^\dagger(x_1,t_1)\mathcal{O}(x_2,t_2)|\omega_T\rangle. \tag{11.8}$$

For example, in the one-dimensional Bose gas, $|\omega_T\rangle$ is one of the eigenstates corresponding to thermodynamic equilibrium at a finite temperature [C. N. Yang and C. P. Yang (1969)]. In the thermodynamic limit, the result of averaging does not depend on the choice of a particular state [N. M. Bogolyubov and V. E. Korepin (1985)]. A similar average can be obtained in spin chains using a quantum transfer matrix method [A. Klümper (1993, 2004)]. Then $|\omega_T\rangle$ is the eigenstate of the quantum transfer matrix corresponding to the leading eigenvalue. In both cases, we always start calculating the correlation functions with the normalized average (11.6). For finite $N$ and $L$, this stage is the same as for the zero temperature case since $|\mathbf{\Phi}_N(\bar{u})\rangle$ is an arbitrary on-shell Bethe vector. The differences arise only in the thermodynamic limit since it depends on the distribution of the roots of the Bethe equations $\bar{u} = \{u_1,\dots,u_N\}$.

Another approach was proposed in [V. E. Korepin (1987, 1989)]. This method uses auxiliary quantum operators, which are called *dual fields*. Using this method, we can reduce multiple integral representations to a more compact form. Namely, the correlation functions are calculated as an average in the auxiliary Fock space of a certain determinant, which, in turn, depends on the dual fields. This approach is described in sufficient detail in the book [V. E. Korepin et al (1993)]. Several interesting results have been obtained using this method [F. H. L. Essler et al (1995a,b, 1997); V. E. Korepin et al (1994); T. Kojima et al (1997a,b); V. E. Korepin and N. A. Slavnov (1991, 1997a,b, 1998)]. The use of dual fields allows us to obtain very compact expressions for correlation functions. However, strictly speaking, these expressions are intermediate results in which the procedure for averaging dual fields still needs to be performed. The latter is a challenging task [A. R. Its and N. A. Slavnov (1999); N. A. Slavnov (1999a,b)].

Finally, numerical methods for calculating correlation functions have become widely used in recent years [D. Biegel et al (2002, 2003);

M. Karbach et al (2002); J. Sato et al (2204); P. Calabrese and J.-S. Caux (2007); J.-S. Caux and P. Calabrese (2006); J.-S. Caux et al (2007, 2003, 2005); J.-S. Caux and J.-M. Maillet (2005); R. Hagemans et al (2005); R. G. Pereira et al (2006, 2008); J. Sirker et al (2008)]. They are based on the form factor expansion (11.7). In numerical analysis, determinant representations for the form factors of local operators play a key role. The time required to compute the determinant of the $N \times N$ matrix increases polynomial-wise with the growth of $N$. Therefore, a computer can easily cope with this task even for sufficiently large $N$. The Bethe equations can also be solved numerically. Of course, it is not realistic to calculate the sum of the form factors over all the Hamiltonian eigenstates since there are too many. However, the main contribution to the correlation functions is provided by a relatively small number of states. The sum of the corresponding form factors gives an approximate result; however, it is easy to control the accuracy of this result.

## 11.2 Two-point functions in the $XXZ$ chain

Our main task is to obtain integral representations for two-point correlation functions of local spin operators in the ground state of the $XXZ$ chain. To do this, we consider the following expectation value:

$$\langle\!\langle \sigma^\alpha_{k_1}\sigma^\alpha_{k_2}\rangle\!\rangle = \frac{\langle \mathbf{\Phi}_N(\bar u)|\sigma^\alpha_{k_1}\sigma^\alpha_{k_2}|\mathbf{\Phi}_N(\bar u)\rangle}{\langle \mathbf{\Phi}_N(\bar u)|\mathbf{\Phi}_N(\bar u)\rangle}, \qquad \alpha = x, y, z. \tag{11.9}$$

Here $|\mathbf{\Phi}_N(\bar u)\rangle$ is an on-shell Bethe vector, and we assume that $k_2 > k_1$ for definiteness. We need to calculate the average (11.9) using the algebraic Bethe ansatz methods at the first stage. After that, we should require that the parameters $\bar u = \{u_1, \dots, u_N\}$ correspond to the ground state and proceed to the thermodynamic limit. A bit later, it will become clear that it is convenient to consider an inhomogeneous $XXZ$ chain at the first stage. It is also convenient to switch from the Pauli matrices to the standard basis matrices $E^{ij}_k$:

$$\langle\!\langle E^{ij}_{k_1}E^{ji}_{k_2}\rangle\!\rangle = \frac{\langle \mathbf{\Phi}_N(\bar u)|E^{ij}_{k_1}E^{ji}_{k_2}|\mathbf{\Phi}_N(\bar u)\rangle}{\langle \mathbf{\Phi}_N(\bar u)|\mathbf{\Phi}_N(\bar u)\rangle}, \qquad i,j = 1,2. \tag{11.10}$$

It is clear that the initial expectation values (11.9) are expressed in terms of simple linear combinations of averages (11.10).

Using the solution of the quantum inverse problem (9.4), we present (11.10) as

$$\langle\!\langle E^{ij}_{k_1} E^{ji}_{k_2} \rangle\!\rangle = \frac{1}{\langle \mathbf{\Phi}_N(\bar{u}) | \mathbf{\Phi}_N(\bar{u}) \rangle} \langle \mathbf{\Phi}_N(\bar{u})| \left( \prod_{l=1}^{k_1-1} \mathcal{T}(\xi_l) \right) T_{ji}(\xi_{k_1})$$

$$\times \left( \prod_{l=k_1+1}^{k_2-1} \mathcal{T}(\xi_l) \right) T_{ij}(\xi_{k_2}) \left( \prod_{l=1}^{k_2} \mathcal{T}^{-1}(\xi_l) \right) |\mathbf{\Phi}_N(\bar{u})\rangle. \quad (11.11)$$

The actions of the transfer matrices products on the on-shell vectors $\langle \mathbf{\Phi}_N(\bar{u})|$ and $|\mathbf{\Phi}_N(\bar{u})\rangle$ give the products of the corresponding eigenvalues

$$\langle\!\langle E^{ij}_{k_1} E^{ji}_{k_2} \rangle\!\rangle = \left( \langle \mathbf{\Phi}_N(\bar{u}) | \mathbf{\Phi}_N(\bar{u}) \rangle \prod_{l=k_1}^{k_2} \tau(\xi_l|\bar{u}) \right)^{-1}$$

$$\times \langle \mathbf{\Phi}_N(\bar{u}) | T_{ji}(\xi_{k_1}) \left( \prod_{l=k_1+1}^{k_2-1} \mathcal{T}(\xi_l) \right) T_{ij}(\xi_{k_2}) |\mathbf{\Phi}_N(\bar{u})\rangle. \quad (11.12)$$

Now we need to act with the remaining operators, for example, on the vector $|\mathbf{\Phi}_N(\bar{u})\rangle$. As a result, we get some linear combination of off-shell Bethe vectors. After that, we just need to calculate the scalar products of these off-shell vectors with the dual on-shell vector $\langle \mathbf{\Phi}_N(\bar{u})|$. We have explicit determinant representations for such scalar products.

This is the general scheme for calculating the expectation value (11.10). The main difficulty lies in the fact that we have to act with the product of the transfer matrices $\prod_{l=k_1+1}^{k_2-1} \mathcal{T}(\xi_l)$ on the vector $T_{ij}(\xi_{k_2})|\mathbf{\Phi}_N(\bar{u})\rangle$. This vector is no longer an on-shell vector but a linear combination of off-shell vectors. Therefore, we need to use formulas (7.52), (7.57) for the multiple action of the transfer matrices. As a result, the final formula is quite cumbersome since it contains sums over partitions of many parameters. The main task is to write down these sums in a form convenient for taking the thermodynamic limit.

Let us make a few general comments. The multiple action formula (7.52) is defined for any values of the transfer matrix arguments. In particular, it can be used when the arguments of transfer matrices coincide. However, this is not very convenient since we have to deal with indeterminate forms. It is for this reason that we consider the inhomogeneous model from the very beginning. Note, however, that formula (11.11) includes only inhomogeneities at sites $l$ that satisfy a condition $k_1 \le l \le k_2$. Therefore, we can put $\xi_l = -i\zeta/2$ for $l < k_1$ and $l > k_2$. It does not affect the calculations at all. Note also that the expectation value (11.12) depends only on the difference $k_2 - k_1$ in the homogeneous limit. This is quite natural since we imposed periodic boundary conditions. Therefore, we can set $k_1 = 1$ without loss of generality.

When transfer matrices $\mathcal{T}(\xi_l)$ act on the state $T_{ij}(\xi_{k_2})|\mathbf{\Phi}_N(\bar{u})\rangle$, we obtain off-shell Bethe vectors that depend on inhomogeneities. Then the normalization of eigenstates to the product of the functions $d$ (8.2) becomes inconvenient, since $d(\xi_l) = 0$ for any $\xi_l$. Therefore, we recast (11.10) in terms of original on-shell Bethe vectors $\langle 0|C(\bar{u})$ and $B(\bar{u})|0\rangle$:

$$\langle\langle E_{k_1}^{ij} E_{k_2}^{ji} \rangle\rangle = \frac{\langle 0|C(\bar{u})\; E_{k_1}^{ij} E_{k_2}^{ji}\; B(\bar{u})|0\rangle}{\langle 0|C(\bar{u})B(\bar{u})|0\rangle}. \tag{11.13}$$

To conclude this section, we note that this method can be generalized to calculate the time-dependent (dynamical) correlation functions

$$\langle E_{k_1}^{ij}(t_1) E_{k_2}^{ji}(t_2)\rangle, \qquad \text{where} \qquad E_k^{ab}(t) = e^{iHt} E_k^{ab} e^{-iHt}. \tag{11.14}$$

In this case, it is necessary to find the action of the evolution operator $e^{iH(t_2-t_1)}$ on the off-shell Bethe vectors. It arises in a special limit of the multiple action of transfer matrices. Details can be found in [N. Kitanine et al (2005d)].

## 11.3 Generating function

Let us now consider the method described above in more detail using the example of the two-point correlation function of the third components of spin

$$\langle\langle \sigma_1^z \sigma_{n+1}^z \rangle\rangle = \frac{\langle 0|C(\bar{u})\; \sigma_1^z \sigma_{n+1}^z\; B(\bar{u})|0\rangle}{\langle 0|C(\bar{u})B(\bar{u})|0\rangle}. \tag{11.15}$$

We consider an inhomogeneous $XXZ$ chain. Therefore, we do not yet require that the parameters $\bar{u} = \{u_1, \dots, u_N\}$ correspond to the ground state. We will impose this condition later. So far, $B(\bar{u})|0\rangle$ is an arbitrary on-shell Bethe vector.

As with the form factors of the $\sigma_k^z$ operator, it is convenient to introduce a generating function. To do this, consider a family of operators

$$\mathcal{Q}_{m,n}^{(\kappa)} = \prod_{k=m}^{n} \left(E_k^{11} + \kappa E_k^{22}\right), \tag{11.16}$$

parameterized by two integers $m$ and $n$ ($m \le n$) and a complex number $\kappa$. It is easy to see that

$$\begin{aligned} &\mathcal{Q}_{m,n+1}^{(\kappa)} - \mathcal{Q}_{m,n}^{(\kappa)} - \mathcal{Q}_{m+1,n+1}^{(\kappa)} + \mathcal{Q}_{m+1,n}^{(\kappa)} \\ &\quad = (\kappa-1)^2 E_m^{22} \left( \prod_{k=m+1}^{n} \left(E_k^{11} + \kappa E_k^{22}\right) \right) E_{n+1}^{22}, \end{aligned} \tag{11.17}$$

and hence,

$$\frac{1}{2}\frac{\mathrm{d}^2}{\mathrm{d}\kappa^2}\Big[\mathcal{Q}^{(\kappa)}_{m,n+1}-\mathcal{Q}^{(\kappa)}_{m,n}-\mathcal{Q}^{(\kappa)}_{m+1,n+1}+\mathcal{Q}^{(\kappa)}_{m+1,n}\Big]_{\kappa=1}=E^{22}_{m}E^{22}_{n+1}. \tag{11.18}$$

We now introduce a generating function

$$Q^{(\kappa)}_{n;N}=\frac{\langle 0|C(\bar u)\mathcal{Q}^{(\kappa)}_{1,n}B(\bar u)|0\rangle}{\langle 0|C(\bar u)B(\bar u)|0\rangle}, \tag{11.19}$$

where we set $m=1$ for the operator $\mathcal{Q}^{(\kappa)}_{m,n}$. This is always possible due to the translation invariance of the correlation function. Then it follows from (11.18) that

$$\langle\!\langle E^{22}_1E^{22}_{n+1}\rangle\!\rangle=\frac{1}{2}\frac{\mathrm{d}^2}{\mathrm{d}\kappa^2}\Big[Q^{(\kappa)}_{n+1;N}-2Q^{(\kappa)}_{n;N}+Q^{(\kappa)}_{n-1;N}\Big]_{\kappa=1}. \tag{11.20}$$

From this, we easily find the correlation function (11.15), since

$$\langle\!\langle \sigma^z_1\sigma^z_{n+1}\rangle\!\rangle=4\langle\!\langle E^{22}_1E^{22}_{n+1}\rangle\!\rangle+2\langle\!\langle\sigma^z\rangle\!\rangle-1. \tag{11.21}$$

Here $\langle\!\langle\sigma^z\rangle\!\rangle$ is a magnetization

$$\langle\!\langle\sigma^z\rangle\!\rangle=\frac{\langle 0|C(\bar u)\;\sigma^z_n\;B(\bar u)|0\rangle}{\langle 0|C(\bar u)B(\bar u)|0\rangle}. \tag{11.22}$$

We had computed it in Section 9.1.1 (see (9.23)): $\langle\!\langle\sigma^z\rangle\!\rangle=1-\frac{2N}{L}$. However, it can also be obtained from the generating function:

$$\langle\!\langle\sigma^z\rangle\!\rangle=1-2\frac{\mathrm{d}}{\mathrm{d}\kappa}\Big[Q^{(\kappa)}_{n+1;N}-Q^{(\kappa)}_{n;N}\Big]_{\kappa=1}. \tag{11.23}$$

The advantage of this approach is that the operator $\mathcal{Q}^{(\kappa)}_{1,n}$ has a very simple expression in terms of the monodromy matrix elements. Indeed, let $\bar\xi=\{\xi_1,\dots,\xi_n\}$. Using the quantum inverse problem formula (9.4), we easily find that

$$\mathcal{Q}^{(\kappa)}_{1,n}=\mathcal{T}_\kappa(\bar\xi)\mathcal{T}^{-1}(\bar\xi), \tag{11.24}$$

where $\mathcal{T}_\kappa(z)$ is the twisted transfer matrix with the twist parameter $\kappa$. Then we obtain

$$Q^{(\kappa)}_{n;N}=\frac{\langle 0|C(\bar u)\mathcal{T}_\kappa(\bar\xi)B(\bar u)|0\rangle}{\langle 0|C(\bar u)B(\bar u)|0\rangle\prod_{k=1}^{n}\tau(\xi_k|\bar u)}. \tag{11.25}$$

Thus, the problem reduces to calculating the action of the product of $\mathcal{T}_\kappa(\bar\xi)$ on the vector $B(\bar u)|0\rangle$. This action was calculated in Section 7.1 (see (7.52) and (7.57)):

$$\mathcal{T}_\kappa(\bar\xi)B(\bar u)|0\rangle = \sum_{m=0}^{n} \sum_{\substack{\bar u\mapsto\{\bar u_{\rm I},\bar u_{\rm II}\}\\ \bar\xi\mapsto\{\bar\xi_{\rm I},\bar\xi_{\rm II}\}\\ \#\bar\xi_{\rm I}=\#\bar u_{\rm I}=m}} \left(\prod_{\xi_j\in\bar\xi_{\rm II}} \tau_\kappa(\xi_j|\{\bar u_{\rm II},\bar\xi_{\rm I}\})\right) \times\mathcal{L}_m^{(N)}(\bar\xi_{\rm I},\bar u_{\rm I}|\bar u_{\rm II})B(\bar\xi_{\rm I})B(\bar u_{\rm II})|0\rangle, \tag{11.26}$$

where

$$\mathcal{L}_m^{(N)}(\bar x,\bar y|\bar w) = \Delta'(\bar y)\Delta(\bar x)g(\bar y,\bar w)\det_m\Big(g(y_k,x_j)\mathcal{Y}_\kappa(y_k|\{y_k,\bar x_j,\bar w\})\Big). \tag{11.27}$$

Recall that here $\#\bar x = \#\bar y = m$ and $\#\bar w = N-m$.

Since $d(\xi_j)=0$ for any $\xi_j$, we have

$$\tau_\kappa(\xi_j|\bar w) = \tau(\xi_j|\bar w) = a(\xi_j)f(\bar w,\xi_j), \tag{11.28}$$

for any set $\bar w$. Therefore,

$$\begin{aligned} &\prod_{\xi_j\in\bar\xi_{\rm II}} \tau_\kappa(\xi_j|\{\bar u_{\rm II},\bar\xi_{\rm I}\}) = a(\bar\xi_{\rm II})f(\bar\xi_{\rm I},\bar\xi_{\rm II})f(\bar u_{\rm II},\bar\xi_{\rm II}),\\ &\prod_{\xi_j\in\bar\xi} \tau(\xi_j|\bar u) = a(\bar\xi_{\rm I})f(\bar u,\bar\xi). \end{aligned} \tag{11.29}$$

Substituting (11.26) and (11.29) in (11.25) we obtain

$$Q^{(\kappa)}_{n;N} = \sum_{m=0}^{n} \sum_{\substack{\bar u\mapsto\{\bar u_{\rm I},\bar u_{\rm II}\}\\ \bar\xi\mapsto\{\bar\xi_{\rm I},\bar\xi_{\rm II}\}\\ \#\bar\xi_{\rm I}=\#\bar u_{\rm I}=m}} \frac{f(\bar\xi_{\rm I},\bar\xi_{\rm II})\mathcal{L}_m^{(N)}(\bar\xi_{\rm I},\bar u_{\rm I}|\bar u_{\rm II})}{a(\bar\xi_{\rm I})f(\bar u_{\rm I},\bar\xi)f(\bar u_{\rm II},\bar\xi_{\rm I})}\,\frac{S_N(\bar u|\{\bar\xi_{\rm I},\bar u_{\rm II}\})}{S_N(\bar u|\bar u)}, \tag{11.30}$$

where $S_N(\bar u|\bar v)$ is the scalar product (5.1)

$$S_N(\bar u|\bar v) = \langle 0|C(\bar u)B(\bar v)|0\rangle. \tag{11.31}$$

Equation (11.30) is the result of calculating the generating function for the $XXZ$ chain of finite length. Now our task is to calculate the thermodynamic limit in this expression.

## 11.4 Multiple integral representation for generating function

### 11.4.1 *Replacement of sums with integrals*

To calculate the thermodynamic limit of the generating function $Q^{(\kappa)}_{n;N}$, we should require that the roots of the Bethe equations $\bar u$ correspond to the ground state. Strictly speaking, to do this, we first need to go to the homogeneous limit. Indeed, there is no state corresponding to the minimum energy in the inhomogeneous $XXZ$ chain.

**Remark 11.1.** If necessary, we can consider an 'almost homogeneous' $XXZ$ chain. This means that we treat the parameters $\bar{\xi} = \{\xi_1, \dots, \xi_n\}$ as arbitrary complex numbers and assume the other inhomogeneities $\{\xi_{n+1}, \dots, \xi_L\}$ to be equal to $-i\zeta/2$. Since $L-n \to \infty$ in the thermodynamic limit and $n$ remains finite, such an almost homogeneous model coincides with the homogeneous one in the limit. In this case, as the ground state, we choose a solution of the Bethe equations that goes to the true ground state in the limit $\xi_1, \dots, \xi_n \to -i\zeta/2$. We can easily use this approach in further calculations, but this is not necessary. We will only assume that all inhomogeneities are in a sufficiently small neighborhood of the point $-i\zeta/2$.

Let us recall the standard functions that we permanently used:

$$g(u,v) = \frac{-i\sin\zeta}{\sinh(u-v)}, \qquad f(u,v) = \frac{\sinh(u-v-i\zeta)}{\sinh(u-v)},$$

$$h(u,v) = \frac{\sinh(u-v-i\zeta)}{-i\sin\zeta}, \qquad t(u,v) = \frac{-\sin^2\zeta}{\sinh(u-v)\sinh(u-v-i\zeta)}. \tag{11.32}$$

We will continue to use this notation to facilitate the formulas. We will also use the convention on the shorthand notation for the products if it does not lead to misunderstanding.

### 11.4.2 *Sum over partitions of inhomogeneities*

The replacement of the sum over partitions of the parameters $\bar{\xi}$ by an integral of Cauchy type carries out using the following proposition.

**Proposition 11.1.** *Let $\xi_j \in \mathcal{D}_r$, $j = 1, \dots, n$, where $\mathcal{D}_r$ is a circle centered at $-i\zeta/2$ and radius $r < \zeta/2$. Let $F(\bar{z}) = F(z_1, \dots, z_m)$ be a symmetric function of variables $\bar{z} = \{z_1, \dots, z_m\}$, where $m \le n$. Assume also that $F(\bar{z})$ is holomorphic in $\mathcal{D}_r$ for each $z_k$, $k = 1, \dots, m$. Then*

$$\sum_{\substack{\bar{\xi}\mapsto\{\bar{\xi}_{\rm I},\bar{\xi}_{\rm II}\} \\ \#\bar{\xi}_{\rm I}=m}} f(\bar{\xi}_{\rm I},\bar{\xi}_{\rm II})F(\bar{\xi}_{\rm I}) = \frac{1}{(2\pi\sin\zeta)^m m!}\oint_{\partial\mathcal{D}_r} \frac{F(\bar{z})f(\bar{z},\bar{\xi})\,\mathrm{d}\bar{z}}{h(\bar{z},\bar{z})\Delta(\bar{z})\Delta'(\bar{z})}, \tag{11.33}$$

*where integration is carried out in the positive direction along the boundary of $\mathcal{D}_r$, and $\mathrm{d}\bar{z} = \mathrm{d}z_1 \dots \mathrm{d}z_m$.*

We used the shorthand notation in the integrand (11.33). Let us also give this integrand in a detailed form:

$$\frac{F(\bar{z})f(\bar{z},\bar{\xi})\,\mathrm{d}\bar{z}}{h(\bar{z},\bar{z})\Delta(\bar{z})\Delta'(\bar{z})} = F(z_1,\dots,z_m)\left(\prod_{k=1}^{m}\prod_{j=1}^{n}\frac{\sinh(z_k-\xi_j-i\zeta)}{\sinh(z_k-\xi_j)}\right)$$
$$\times\left(\prod_{\substack{a,b=1\\ a\neq b}}^{m}\frac{\sinh(z_a-z_b)}{\sinh(z_a-z_b-i\zeta)}\right)\prod_{k=1}^{m}\mathrm{d}z_k. \tag{11.34}$$

**Proof.** First of all, we note that under the conditions of the proposition, the only singularities of the integrand in the domain $\mathcal{D}_r$ are the simple poles at the points $z_k=\xi_j$, $k=1,\dots,m$, $j=1,\dots,n$. The poles at points $z_a=z_b\pm i\zeta$ are outside the integration contours because $|z_a-z_b|<\zeta$ for $z_a,z_b\in\partial\mathcal{D}_r$. Then the integral in (11.33) is the sum of the residues at $z_k=\xi_j$.

We use induction over $m$. Using $h(z,z)=1$, we obtain for $m=1$

$$\frac{1}{2\pi\sin\zeta}\oint_{\partial\mathcal{D}_r}F(\bar{z})f(z,\bar{\xi})\,\mathrm{d}z=\sum_{j=1}^{n}F(\xi_j)f(\xi_j,\bar{\xi}_j). \tag{11.35}$$

This equation coincides with (11.33) for $\#\bar{\xi}_{\rm I}=1$.

Assume that (11.33) holds for $\#\bar{\xi}_{\rm I}=m-1$. Then we can present the $m$-fold integral as follows:

$$\oint_{\partial\mathcal{D}_r}\frac{F(\bar{z})f(\bar{z},\bar{\xi})\,\mathrm{d}\bar{z}}{h(\bar{z},\bar{z})\Delta(\bar{z})\Delta'(\bar{z})}=\oint_{\partial\mathcal{D}_r}f(z_m,\bar{\xi})\,\mathrm{d}z_m$$
$$\times\oint_{\partial\mathcal{D}_r}\frac{F(\bar{z})}{f(z_m,\bar{z}_m)f(\bar{z}_m,z_m)}\frac{f(\bar{z}_m,\bar{\xi})\,\mathrm{d}\bar{z}_m}{h(\bar{z}_m,\bar{z}_m)\Delta(\bar{z}_m)\Delta'(\bar{z}_m)}, \tag{11.36}$$

where we used $g(u,v)h(u,v)=f(u,v)$. Using induction assumption, we can calculate the $(m-1)$-fold integral in the second line of (11.36). We obtain

$$\frac{1}{(2\pi\sin\zeta)^m m!}\oint_{\partial\mathcal{D}_r}\frac{F(\bar{z})f(\bar{z},\bar{\xi})\,\mathrm{d}\bar{z}}{h(\bar{z},\bar{z})\Delta(\bar{z})\Delta'(\bar{z})}$$
$$=\frac{1}{2\pi m\sin\zeta}\sum_{\substack{\bar{\xi}\mapsto\{\bar{\xi}_{\rm I},\bar{\xi}_{\rm II}\}\\ \#\bar{\xi}_{\rm I}=m-1}}f(\bar{\xi}_{\rm I},\bar{\xi}_{\rm II})\oint_{\partial\mathcal{D}_r}F(\{\bar{\xi}_{\rm I},z_m\})\frac{f(z_m,\bar{\xi}_{\rm II})}{f(\bar{\xi}_{\rm I},z_m)}\,\mathrm{d}z_m. \tag{11.37}$$

The remaining integral is equal to the sum of the residues in the poles at the points $z_m = \xi_j$, where $\xi_j \in \bar{\xi}_{\text{II}}$. We also calculate it using equation (11.33) and arrive at

$$\begin{aligned}&\frac{1}{(2\pi\sin\zeta)^m m!}\oint_{\partial\mathcal{D}_r}\frac{F(\bar{z})f(\bar{z},\bar{\xi})\,\mathrm{d}\bar{z}}{h(\bar{z},\bar{z})\Delta(\bar{z})\Delta'(\bar{z})}\\&\quad=\frac{1}{m}\sum_{\substack{\bar{\xi}\mapsto\{\bar{\xi}_{\text{I}},\bar{\xi}_{\text{II}}\}\\ \#\bar{\xi}_{\text{I}}=m-1}} f(\bar{\xi}_{\text{I}},\bar{\xi}_{\text{II}})\sum_{\substack{\bar{\xi}_{\text{II}}\mapsto\{\bar{\xi}_{\text{i}},\bar{\xi}_{\text{ii}}\}\\ \#\bar{\xi}_{\text{i}}=1}} F(\{\bar{\xi}_{\text{I}},\bar{\xi}_{\text{i}}\})\frac{f(\bar{\xi}_{\text{i}},\bar{\xi}_{\text{ii}})}{f(\bar{\xi}_{\text{I}},\bar{\xi}_{\text{i}})}.\end{aligned}\tag{11.38}$$

Using $f(\bar{\xi}_{\text{I}},\bar{\xi}_{\text{II}}) = f(\bar{\xi}_{\text{I}},\bar{\xi}_{\text{i}})f(\bar{\xi}_{\text{I}},\bar{\xi}_{\text{ii}})$, we obtain

$$\begin{aligned}&\frac{1}{(2\pi\sin\zeta)^m m!}\oint_{\partial\mathcal{D}_r}\frac{F(\bar{z})f(\bar{z},\bar{\xi})\,\mathrm{d}\bar{z}}{h(\bar{z},\bar{z})\Delta(\bar{z})\Delta'(\bar{z})}\\&\quad=\frac{1}{m}\sum_{\substack{\bar{\xi}\mapsto\{\bar{\xi}_{\text{I}},\bar{\xi}_{\text{i}},\bar{\xi}_{\text{ii}}\}\\ \#\bar{\xi}_{\text{I}}=m-1;\#\bar{\xi}_{\text{i}}=1}} F(\{\bar{\xi}_{\text{I}},\bar{\xi}_{\text{i}}\})f(\bar{\xi}_{\text{I}},\bar{\xi}_{\text{ii}})f(\bar{\xi}_{\text{i}},\bar{\xi}_{\text{ii}}).\end{aligned}\tag{11.39}$$

Let $\{\bar{\xi}_{\text{I}},\bar{\xi}_{\text{i}}\} = \bar{\xi}_0$. Then the sum over partitions in (11.39) can be presented in the form

$$\begin{aligned}&\frac{1}{m}\sum_{\substack{\bar{\xi}\mapsto\{\bar{\xi}_{\text{I}},\bar{\xi}_{\text{i}},\bar{\xi}_{\text{ii}}\}\\ \#\bar{\xi}_{\text{I}}=m-1;\#\bar{\xi}_{\text{i}}=1}} F(\{\bar{\xi}_{\text{I}},\bar{\xi}_{\text{i}}\})f(\bar{\xi}_{\text{I}},\bar{\xi}_{\text{ii}})f(\bar{\xi}_{\text{i}},\bar{\xi}_{\text{ii}})\\&\quad=\frac{1}{m}\sum_{\substack{\bar{\xi}\mapsto\{\bar{\xi}_0,\bar{\xi}_{\text{ii}}\}\\ \#\bar{\xi}_0=m}}\sum_{\substack{\bar{\xi}_0\mapsto\{\bar{\xi}_{\text{I}},\bar{\xi}_{\text{i}}\}\\ \#\bar{\xi}_{\text{i}}=1}} F(\bar{\xi}_0)f(\bar{\xi}_0,\bar{\xi}_{\text{ii}}).\end{aligned}\tag{11.40}$$

The sum over partitions $\bar{\xi}_0 \mapsto \{\bar{\xi}_{\text{I}},\bar{\xi}_{\text{i}}\}$ gives $m$ since nothing depends on $\bar{\xi}_{\text{I}}$ and $\bar{\xi}_{\text{i}}$. Thus, we finally arrive at

$$\frac{1}{(2\pi\sin\zeta)^m m!}\oint_{\partial\mathcal{D}_r}\frac{F(\bar{z})f(\bar{z},\bar{\xi})\,\mathrm{d}\bar{z}}{h(\bar{z},\bar{z})\Delta(\bar{z})\Delta'(\bar{z})}=\sum_{\substack{\bar{\xi}\mapsto\{\bar{\xi}_0,\bar{\xi}_{\text{ii}}\}\\ \#\bar{\xi}_0=m}} F(\bar{\xi}_0)f(\bar{\xi}_0,\bar{\xi}_{\text{ii}}),\tag{11.41}$$

which coincides with (11.33) up to the labels of the subsets. □

It is easy to see that the sum over the partitions of inhomogeneities in formula (11.30) satisfies all the conditions of proposition 11.1. The entire dependence on the parameters from the set $\bar{\xi}_{\text{II}}$ is found only in the product $f(\bar{\xi}_{\text{I}},\bar{\xi}_{\text{II}})$, as in (11.33). The remaining part depends only on the parameters from the set $\bar{\xi}_{\text{I}}$, and we can denote it as $F(\bar{\xi}_{\text{I}})$. Since we intend to go to the homogeneous limit, we can assume that the inhomogeneities locate in an arbitrarily small neighborhood $\mathcal{D}_r$ of the point $-i\zeta/2$. Finally, the poles of the function $F(\bar{\xi}_{\text{I}})$ are at the points $\xi_j = u_k$, $j = 1, \dots, n$, $k = 1, \dots, N$. Since $u_k$ are real in the ground state, these points are separated from $-i\zeta/2$. Therefore, we can always choose the integration contours so that the only singularities within the contours are the points $\xi_j$.

Thus, applying proposition 11.1 to (11.30) we obtain

$$Q_{n;N}^{(\kappa)} = \sum_{m=0}^{n} \sum_{\substack{\bar{u} \mapsto \{\bar{u}_{\text{I}}, \bar{u}_{\text{II}}\} \\ \#\bar{z} = \#\bar{u}_{\text{I}} = m}} \oint_{\bar{\xi}} \frac{d\bar{z}}{(2\pi \sin\zeta)^m m!} \frac{S_N(\bar{u}|\{\bar{z}, \bar{u}_{\text{II}}\})}{S_N(\bar{u}|\bar{u})}$$
$$\times \frac{f(\bar{z}, \bar{\xi})}{f(\bar{u}_{\text{I}}, \bar{\xi})} \frac{\mathcal{L}_m^{(N)}(\bar{z}, \bar{u}_{\text{I}}|\bar{u}_{\text{II}})}{a(\bar{z}) f(\bar{u}_{\text{II}}, \bar{z}) h(\bar{z}, \bar{z}) \Delta(\bar{z}) \Delta'(\bar{z})}. \tag{11.42}$$

The integration contours surround the points $\bar{\xi}$ and do not contain any other singularities of the integrand. Note that now we can proceed to the homogeneous limit and set $\xi_k = -i\zeta/2$, $k = 1, \dots, n$. Let

$$\varphi(z) = f(z, -\tfrac{i\zeta}{2}) = \frac{\sinh(z - \frac{i\zeta}{2})}{\sinh(z + \frac{i\zeta}{2})}. \tag{11.43}$$

Then we arrive at

$$Q_{n;N}^{(\kappa)} = \sum_{m=0}^{n} \sum_{\substack{\bar{u} \mapsto \{\bar{u}_{\text{I}}, \bar{u}_{\text{II}}\} \\ \#\bar{z} = \#\bar{u}_{\text{I}} = m}} \oint_{-\frac{i\zeta}{2}} \frac{d\bar{z}}{(2\pi \sin\zeta)^m m!} \frac{S_N(\bar{u}|\{\bar{z}, \bar{u}_{\text{II}}\})}{S_N(\bar{u}|\bar{u})}$$
$$\times \left( \frac{\prod_{j=1}^{m} \varphi(z_j)}{\prod_{u_j \in \bar{u}_{\text{I}}} \varphi(u_j)} \right)^n \frac{\mathcal{L}_m^{(N)}(\bar{z}, \bar{u}_{\text{I}}|\bar{u}_{\text{II}})}{a(\bar{z}) f(\bar{u}_{\text{II}}, \bar{z}) h(\bar{z}, \bar{z}) \Delta(\bar{z}) \Delta'(\bar{z})}. \tag{11.44}$$

### 11.4.3 *Sum over partitions of Bethe parameters*

The replacement of the sum over partitions of the set $\bar{u}$ by a multiple integral occurs only in the thermodynamic limit. Let the roots of the system of Bethe equations $\bar{u}$ correspond to the ground state. Consider a sum of the form

$$\sum_{\substack{\bar{u} \mapsto \{\bar{u}_{\text{I}}, \bar{u}_{\text{II}}\} \\ \#\bar{u}_{\text{I}} = m}} F(\bar{u}_{\text{I}}) \prod_{u_j \in \bar{u}_{\text{I}}} \frac{1}{L \hat{\rho}_u(u_j)}, \tag{11.45}$$

where $F$ is a symmetric function of $m$ variables, integrable in the $m$-dimensional cube $u_j \in [-q, q]$, $j = 1, \dots, m$. It is essential that this function depends only on the subset $\bar{u}_{\text{I}}$ but not on the subset $\bar{u}_{\text{II}}$. We also assume that $F$ vanishes as soon as any two arguments coincide. A function $\hat{\rho}_u(u)$ is defined by equation (10.130), and it goes to the distribution density (10.132) in the thermodynamic limit.

It is easy to see that

$$\sum_{\substack{\bar u\mapsto\{\bar u_{\rm I},\bar u_{\rm II}\}\\ \#\bar u_{\rm I}=m}} F(\bar u_{\rm I})\prod_{u_j\in\bar u_{\rm I}}\frac{1}{L\hat\rho_u(u_j)}$$
$$=\frac{1}{m!}\sum_{u_1\in\bar u}\cdots\sum_{u_m\in\bar u}F(u_1,\dots,u_m)\prod_{j=1}^{m}\frac{1}{L\hat\rho_u(u_j)}. \qquad (11.46)$$

Here in the rhs, each variable $u_k$ independently runs through the set $\bar u$. Indeed, due to the condition imposed on the function $F$, there remain only those terms in which all $u_k$ are distinct in the rhs of the formula (11.46). Those terms that differ only by permuting the arguments are killed by the factor $1/m!$. What remains is the sum over partitions of the set $\bar u$ into subsets $\bar u_{\rm I}$ and $\bar u_{\rm II}$.

In its turn, the sum in the rhs of (11.46) becomes an $m$-fold integral in the thermodynamic limit. Therefore,

$$\sum_{\substack{\bar u\mapsto\{\bar u_{\rm I},\bar u_{\rm II}\}\\ \#\bar u_{\rm I}=m}} F(\bar u_{\rm I})\prod_{u_j\in\bar u_{\rm I}}\frac{1}{L\hat\rho_u(u_j)}\xrightarrow{N,L\to\infty}\frac{1}{m!}\int_{-q}^{q}F(\bar u)\,\mathrm{d}\bar u, \qquad (11.47)$$

where $\mathrm{d}\bar u=\mathrm{d}u_1\dots\mathrm{d}u_m$.

**Remark 11.2.** In fact, the requirement that $F(u_1,\dots,u_m)$ vanishes for coinciding arguments is not crucial. If we refuse this requirement, additional terms arise in the rhs of (11.46), which are not in the sum over the partitions. However, it is easy to check that the contributions of these terms disappear in the thermodynamic limit. For example, we recommend that the reader consider the case $m=2$ and evaluate the contribution to the sum (11.46) with $u_1=u_2$.

We cannot immediately use formula (11.47) to replace the sum of the partitions of the Bethe parameters $\bar u$ in formula (11.44) with a multiple integral. The fact is that the terms of this sum depend not only on the subset $\bar u_{\rm I}$ but also on the subset $\bar u_{\rm II}$. So, first of all, we need to get rid of the subset $\bar u_{\rm II}$.

#### 11.4.3.1 *Transformation of $\mathcal{L}^{(N)}_m$*

Let us start with the transformation of the function $\mathcal{L}^{(N)}_m$

$$\mathcal{L}^{(N)}_m(\bar z,\bar u_{\rm I}|\bar u_{\rm II})=\Delta'(\bar z)\Delta(\bar u_{\rm I})g(\bar u_{\rm I},\bar u_{\rm II})\det_m\Big(g(u^{\rm I}_k,z_j)\mathcal{Y}_\kappa(u^{\rm I}_k|\{u^{\rm I}_k,\bar z_j,\bar u_{\rm II}\})\Big). \qquad (11.48)$$

Here $u^{\text{I}}_k$ denotes the $k$th element of the subset $\bar{u}_{\text{I}}$. Substituting here

$$\mathcal{Y}_\kappa(u^{\text{I}}_k|\{u^{\text{I}}_k,\bar{z}_j,\bar{u}_{\text{II}}\}) = (-1)^N a(u^{\text{I}}_k)h(\bar{z}_j,u^{\text{I}}_k)h(\bar{u}_{\text{II}},u^{\text{I}}_k) + \kappa d(u^{\text{I}}_k)h(u^{\text{I}}_k,\bar{z}_j)h(u^{\text{I}}_k,\bar{u}_{\text{II}}), \tag{11.49}$$

and using Bethe equations for the set $\bar{u}$ we find

$$\mathcal{L}^{(N)}_m(\bar{z},\bar{u}_{\text{I}}|\bar{u}_{\text{II}}) = d(\bar{u}_{\text{I}})\Delta'(\bar{z})\Delta(\bar{u}_{\text{I}})f(\bar{u}_{\text{I}},\bar{u}_{\text{II}})h(\bar{z},\bar{u}_{\text{I}}) \det_m \mathcal{M}^{(\kappa)}(\bar{z},\bar{u}_{\text{I}}), \tag{11.50}$$

where

$$\mathcal{M}^{(\kappa)}_{jk}(\bar{z},\bar{u}_{\text{I}}) = t(z_j,u^{\text{I}}_k) + \kappa t(u^{\text{I}}_k,z_j)\frac{h(\bar{u}_{\text{I}},u^{\text{I}}_k)h(u^{\text{I}}_k,\bar{z})}{h(u^{\text{I}}_k,\bar{u}_{\text{I}})h(\bar{z},u^{\text{I}}_k)}. \tag{11.51}$$

Note that the matrix $\mathcal{M}^{(\kappa)}$ had already occurred when calculating the scalar product of the twisted on-shell and ordinary on-shell vectors (see (6.93)). One of the important properties of this matrix is that $\det \mathcal{M}^{(\kappa)} = 0$ at $\kappa = 1$.

#### 11.4.3.2 *Ratio of scalar products*

We now proceed to calculate the ratio of two scalar products. To simplify the notation, we consider a special partition of the Bethe parameters, in which $\bar{u}_{\text{I}} = \{u_1,\dots,u_m\}$ and $\bar{u}_{\text{II}} = \{u_{m+1},\dots,u_N\}$. The case of a general partition reduces to this special partition (see below).

The norm of the (twisted) on-shell Bethe vector was calculated in Section 6.4. We should only consider the new normalization of the scalar product and set $\kappa = 1$, $\eta = -i\zeta$ in (6.76) and (6.77). We also use

$$\frac{\mathrm{d}}{\mathrm{d}z}\log r(z)\Big|_{z=u_k} = \frac{iL\sin\zeta}{\sinh(u_k-\frac{i\zeta}{2})\sinh(u_k+\frac{i\zeta}{2})} = iLp_0'(u_k), \tag{11.52}$$

where $p_0'(u)$ is defined by (10.131). Then we obtain

$$S_N(\bar{u}|\bar{u}) = (-\sin\zeta)^N d^2(\bar{u})\Delta(\bar{u})\Delta'(\bar{u})h(\bar{u},\bar{u}) \det_N \Omega(\bar{u}|\bar{u}), \tag{11.53}$$

where

$$\Omega_{jk}(\bar{u}|\bar{u}) = \delta_{jk}\left(Lp_0'(u_k) - \sum_{l=1}^N \mathcal{K}(u_k-u_l)\right) + \mathcal{K}(u_k-u_j), \tag{11.54}$$

and $\mathcal{K}(u)$ is defined by (10.131).

Consider now $S_N(\bar{u}|\{\bar{z},\bar{u}_{\text{II}}\})$. The scalar product of a dual twisted on-shell vector $\langle 0|C(\bar{v})$ and an arbitrary off-shell vector $B(\bar{u})|0\rangle$ is given by equation (6.34). We should replace there $\bar{v} \leftrightarrow \bar{u}$, set $\kappa = 1$, and multiply by the product of $d(\bar{u})d(\bar{v})$. Then

$$
\begin{aligned}
S_N(\bar{u}|\bar{v}) = d(\bar{u})\Delta'(\bar{u})\Delta(\bar{v}) \det_N \big[d(v_k)t(v_k,u_j)h(v_k,\bar{u}) \\
+ (-1)^{N-1}a(v_k)t(u_j,v_k)h(\bar{u},v_k)\big].
\end{aligned} \tag{11.55}
$$

It is convenient to rewrite this equation as follows:

$$
S_N(\bar{u}|\bar{v}) = (-\sin\zeta)^N d(\bar{u})a(\bar{v})\Delta'(\bar{u})\Delta(\bar{v})h(\bar{u},\bar{v}) \det_N \widetilde{\Omega}(\bar{u}|\bar{v}), \tag{11.56}
$$

where

$$
\widetilde{\Omega}_{jk}(\bar{u}|\bar{v}) = \frac{(-1)^N t(v_k,u_j)}{r(v_k)\sin\zeta} \frac{h(v_k,\bar{u})}{h(\bar{u},v_k)} - \frac{t(u_j,v_k)}{\sin\zeta}. \tag{11.57}
$$

Then it is easy to check that

$$
\widetilde{\Omega}_{jk}(\bar{u}|\bar{v})\Big|_{v_k=u_k} = \Omega_{jk}(\bar{u}|\bar{u}). \tag{11.58}
$$

Now we should substitute $\bar{v} = \{\bar{z}, \bar{u}_{\rm II}\}$ in (11.56). We omit the details of this simple calculation and give only the final result

$$
\begin{aligned}
S_N(\bar{u}|\{\bar{z},\bar{u}_{\rm II}\}) = (-\sin\zeta)^N a(\bar{z})a(\bar{u}_{\rm II})d(\bar{u})\Delta'(\bar{u})\Delta(\bar{u}_{\rm II})\Delta(\bar{z})g(\bar{u}_{\rm II},\bar{z}) \\
\times h(\bar{u},\bar{z})h(\bar{u},\bar{u}_{\rm II}) \det_N \widetilde{\Omega}(\bar{u}|\{\bar{z},\bar{u}_{\rm II}\}),
\end{aligned} \tag{11.59}
$$

where

$$
\begin{aligned}
&\widetilde{\Omega}_{jk}(\{\bar{z},\bar{u}_{\rm II}\}|\bar{u}) = -\frac{t(u_j,z_k)}{\sin\zeta}, \qquad k = 1,\dots,m, \\
&\widetilde{\Omega}_{j,k+m}(\{\bar{z},\bar{u}_{\rm II}\}|\bar{u}) = \Omega_{jk}(\bar{u}|\bar{u}), \qquad k = 1,\dots,N-m.
\end{aligned} \tag{11.60}
$$

Due to (11.58), the last $N-m$ columns of the $\widetilde{\Omega}$ matrix coincide with the ones of the matrix $\Omega$. In the first $m$ columns, we put $d(z_k) = 0$. Indeed, the function $d(z)$ has zero of order $L$ at the point $z = -i\zeta/2$. On the other hand, the result of integration for each variable $z_k$ in (11.44) is equal to the residue of the integrand at the pole of order $n$ at $z_k = -i\zeta/2$. Since $L > n$, it is clear that all contributions proportional to $d(z_k)$ vanish. Therefore, we can put $d(z_k) = 0$ in advance. This does not affect the final result in any way.

Thus, we arrive at the following representation for the ratio of the scalar products:

$$
\frac{S_N(\bar{u}|\{\bar{z},\bar{u}_{\rm II}\})}{S_N(\bar{u}|\bar{u})} = \frac{a(\bar{z})a(\bar{u}_{\rm II})\Delta(\bar{z})h(\bar{u},\bar{z})g(\bar{u}_{\rm II},\bar{z})}{d(\bar{u})\Delta(\bar{u}_{\rm I})h(\bar{u},\bar{u}_{\rm I})g(\bar{u}_{\rm II},\bar{u}_{\rm I})} \frac{\det_N \widetilde{\Omega}(\bar{u}|\{\bar{z},\bar{u}_{\rm II}\})}{\det_N \Omega(\bar{u}|\bar{u})}. \tag{11.61}
$$

Substituting (11.61) and (11.50) in representation (11.44) we obtain

$$Q^{(\kappa)}_{n;N} = \sum_{m=0}^{n} \sum_{\substack{\bar{u}\mapsto\{\bar{u}_{\rm I},\bar{u}_{\rm II}\}\\ \#\bar{z}=\#\bar{u}_{\rm I}=m}} \oint_{-\frac{i\zeta}{2}} \frac{d\bar{z}}{(2\pi\sin\zeta)^m m!} \left(\frac{\prod_{j=1}^{m}\varphi(z_j)}{\prod_{u_j\in\bar{u}_{\rm I}}\varphi(u_j)}\right)^n$$

$$\times W_m(\bar{z},\bar{u}_{\rm I}) \det_m \mathcal{M}^{(\kappa)}(\bar{z},\bar{u}_{\rm I}) \frac{\det_N \widetilde{\Omega}(\bar{u}|\{\bar{z},\bar{u}_{\rm II}\})}{\det_N \Omega(\bar{u}|\bar{u})}, \tag{11.62}$$

where

$$W_m(\bar{z},\bar{u}_{\rm I}) = \frac{h(\bar{z},\bar{u}_{\rm I})h(\bar{u}_{\rm I},\bar{z})}{h(\bar{z},\bar{z})h(\bar{u}_{\rm I},\bar{u}_{\rm I})}. \tag{11.63}$$

To derive (11.62), we also used Bethe equations in the form

$$\frac{a(\bar{u}_{\rm II})}{d(\bar{u}_{\rm II})} = \frac{f(\bar{u}_{\rm II},\bar{u}_{\rm I})}{f(\bar{u}_{\rm I},\bar{u}_{\rm II})}. \tag{11.64}$$

Now the subset $\bar{u}_{\rm II}$ is only present in the matrix $\widetilde{\Omega}$. We calculate the ratio of the determinants of the matrices $\widetilde{\Omega}$ and $\Omega$ using

$$\frac{\det_N \widetilde{\Omega}(\bar{u}|\{\bar{z},\bar{u}_{\rm II}\})}{\det_N \Omega(\bar{u}|\bar{u})} = \det_N \Big( \big(\Omega(\bar{u}|\bar{u})\big)^{-1} \widetilde{\Omega}(\bar{u}|\{\bar{z},\bar{u}_{\rm II}\}) \Big). \tag{11.65}$$

Since the columns of these matrices with numbers $k = m+1,\dots,N$ coincide, we get $(\Omega^{-1}\widetilde{\Omega})_{jk} = \delta_{jk}$ in these columns. As a result, the determinant of the $N\times N$ matrix reduces to the determinant of an $m\times m$ matrix:

$$\frac{\det_N \widetilde{\Omega}(\bar{u}|\{\bar{z},\bar{u}_{\rm II}\})}{\det_N \Omega(\bar{u}|\bar{u})} = \det_m \mathcal{R}, \tag{11.66}$$

where the entries of the matrix $\mathcal{R}$ satisfy the following equations:

$$\sum_{l=1}^{N} \Omega_{jl}(\bar{u}|\bar{u})\mathcal{R}_{lk} = -\frac{t(u_j,z_k)}{\sin\zeta}, \qquad j,k = 1,\dots,m. \tag{11.67}$$

We can solve equations (11.67) in the thermodynamic limit in terms of inhomogeneous density (10.135). First of all, we recall the definition of the ground state density (10.130):

$$2\pi\hat{\rho}_u(u_k) = p_0'(u_k) - \frac{1}{L}\sum_{l=1}^{N} \mathcal{K}(u_k-u_l). \tag{11.68}$$

Then we can write down the entries $\Omega_{jk}$ as

$$\Omega_{jk}(\bar{u}|\bar{u}) = 2\pi L\hat{\rho}_u(u_k)\left(\delta_{jk} + \frac{\mathcal{K}(u_k-u_j)}{2\pi L\hat{\rho}_u(u_k)}\right). \tag{11.69}$$

Let us look for the solution of (11.67) in the form

$$\mathcal{R}_{jk} = \frac{\rho(u_j, z_k)}{L\hat{\rho}_u(u_j)}, \tag{11.70}$$

where $\rho(u_j, z_k)$ is a new unknown function. Then (11.67) takes the following form:

$$\sum_{l=1}^{N} \left(\delta_{jl} + \frac{\mathcal{K}(u_l - u_j)}{2\pi L\hat{\rho}_u(u_l)}\right) \rho(u_l, z_k) = -\frac{t(u_j, z_k)}{2\pi \sin\zeta}, \qquad j, k = 1, \dots, m. \tag{11.71}$$

Obviously, in the thermodynamic limit, this system of equations turns into an integral equation

$$\rho(u, z_k) + \frac{1}{2\pi}\int_{-q}^{q} \mathcal{K}(u - v)\rho(v, z_k)\,\mathrm{d}v = -\frac{t(u, z_k)}{2\pi \sin\zeta}, \tag{11.72}$$

which coincides with equation (10.135) for the inhomogeneous density. Thus,

$$\mathrm{Lim}\left(\prod_{l=1}^{m} L\hat{\rho}_u(u_l)\right) \frac{\det_N \widetilde{\Omega}(\bar{u}|\{\bar{z}, \bar{u}_{\mathrm{II}}\})}{\det_N \Omega(\bar{u}|\bar{u})} = \det_m \rho(u_j, z_k). \tag{11.73}$$

It remains to recall that we considered the special partition $\bar{u}_{\mathrm{I}} = \{u_1, \dots, u_m\}$ and $\bar{u}_{\mathrm{II}} = \{u_{m+1}, \dots, u_N\}$. If we deal with a general partition, we only need to make one change. Namely, we should reorder the rows and columns of the matrix $\Omega(\bar{u}|\bar{u})$ so that the first $m$ rows and columns correspond to the parameters of the subset $\bar{u}_{\mathrm{I}}$, and the remaining rows and columns correspond to the parameters of the subset $\bar{u}_{\mathrm{II}}$. Then

$$\mathrm{Lim}\left(\prod_{u_l \in \bar{u}_{\mathrm{I}}}^{m} L\hat{\rho}_u(u_l)\right) \frac{\det_N \widetilde{\Omega}(\bar{u}|\{\bar{z}, \bar{u}_{\mathrm{II}}\})}{\det_N \Omega(\bar{u}|\bar{u})} = \det_m \rho(u^{\mathrm{I}}_j, z_k), \tag{11.74}$$

where, as usual, $u^{\mathrm{I}}_j$ means the $j$th element from the subset $\bar{u}_{\mathrm{I}}$.

Thus, we arrive at

$$Q^{(\kappa)}_{n;N} = \sum_{m=0}^{n} \sum_{\substack{\bar{u} \mapsto \{\bar{u}_{\mathrm{I}}, \bar{u}_{\mathrm{II}}\} \\ \#\bar{z} = \#\bar{u}_{\mathrm{I}} = m}} \oint_{-\frac{i\zeta}{2}} \frac{\mathrm{d}\bar{z}}{(2\pi \sin\zeta)^m m!} \left(\frac{\prod_{j=1}^{m} \varphi(z_j)}{\prod_{u_j \in \bar{u}_{\mathrm{I}}} L\hat{\rho}_u(u_j)\varphi(u_j)}\right)^n \\ \times W_m(\bar{z}, \bar{u}_{\mathrm{I}}) \det_m \mathcal{M}^{(\kappa)}(\bar{z}, \bar{u}_{\mathrm{I}}) \det_m \rho(u^{\mathrm{I}}_j, z_k). \tag{11.75}$$

In this expression, we have neglected the corrections of order $O(L^{-2})$ that arise when replacing the sum (11.71) with the integral (11.72). Now all the functions in the expression (11.75) depend on the subset $\bar{u}_{\mathrm{I}}$ but do not depend on the subset $\bar{u}_{\mathrm{II}}$. Hence, we can use equation (11.47). We obtain

$$Q_n^{(\kappa)} = \sum_{m=0}^{n} \oint \frac{\mathrm{d}\bar{z}}{(2\pi \sin\zeta)^m (m!)^2} \int_{-q}^{q} \mathrm{d}\bar{u} \prod_{j=1}^{m} \varphi^n(z_j)\varphi^{-n}(u_j) \times W_m(\bar{z},\bar{u}) \det_m \mathcal{M}^{(\kappa)}(\bar{z},\bar{u}) \det_m \rho(u_j,z_k), \tag{11.76}$$

where $Q_n^{(\kappa)} = \operatorname{Lim} Q_{n;N}^{(\kappa)}$, and $\bar{u} = \{u_1,\dots,u_m\}$. The integration contours for the variables $z_j$ surround the point $-i\zeta/2$ and do not contain any other integrand singularities. Equation (11.76) gives us the required integral representation for the generating function $Q_n^{(\kappa)}$ in the thermodynamic limit.

For the zero external magnetic field $\mu = 0$, the inhomogeneous density $\rho(u_j, z_k)$ has an explicit representation (10.150). Besides, it is necessary to send the boundary of the Fermi zone $q$ to infinity. Then

$$Q_n^{(\kappa)} = \sum_{m=0}^{n} \oint \frac{\mathrm{d}\bar{z}}{(2\pi \sin\zeta)^m (m!)^2} \int_{-\infty}^{\infty} \mathrm{d}\bar{u} \prod_{j=1}^{m} \varphi^n(z_j)\varphi^{-n}(u_j) W_m(\bar{z},\bar{u}) \det_m \times \mathcal{M}^{(\kappa)}(\bar{z},\bar{u}) \det_m \Big[\frac{i}{2\zeta \sinh\frac{\pi}{\zeta}(u_j - z_k)}\Big], \qquad \mu = 0. \tag{11.77}$$

#### 11.4.3.3 *Two particular cases*

It is interesting to see how equation (11.76) works in simple particular cases, where the result is known in advance. In particular, operator $\mathcal{Q}_{m,n}^{(\kappa)}$ (11.16) turns into the identity operator at $\kappa = 1$. The average of the latter should be equal to 1. Formula (11.76) does give this result because

$$\det_m \mathcal{M}^{(\kappa)}(\bar{z},\bar{u})\Big|_{\kappa=1} = \delta_{m0}, \tag{11.78}$$

as we already mentioned.

Another obvious particular case is $n = 1$. Then

$$\mathcal{Q}_{1,1}^{(\kappa)} = E_1^{11} + \kappa E_1^{22} = 1 + \frac{\kappa - 1}{2}(1 - \sigma_1^z). \tag{11.79}$$

Thus, the non-trivial part of the generating function $Q_1^{(\kappa)}$ reduces to the magnetization $\langle \sigma^z \rangle$. We calculated the latter in Section 9.1.1: $\langle \sigma^z \rangle = 1 - 2D$ (see (9.23)). Here

$$D = \frac{N}{L} = \int_{-q}^{q} \rho(u)\,\mathrm{d}u, \tag{11.80}$$

and $\rho(u)$ is the ground state distribution density (10.132). Hence, we should obtain

$$Q_1^{(\kappa)} = 1 + (\kappa - 1)D. \tag{11.81}$$

We suggest that the reader check that representation (11.76) does give this result for $n = 1$.

Concluding this section, we note that we can obtain an integral representation for the generating function $Q_n^{(\kappa)}$ in the $XXX$ chain in the same way. However, this result can be extracted directly from representation (11.76) in the limit $\zeta \to 0$. To do this, we need to make the change of integration variables $u_j \to \zeta u_j$, $z_j \to \zeta z_j$, and let $\zeta$ go to zero.

## 11.5 Generating function via form factor expansion

In this section, we show that multiple integral representation (11.76) for the generating function $Q_n^{(\kappa)}$ can be obtained in a completely different way. Namely, we will use the form factor expansion. The advantage of this method is that we do not need to calculate a sufficiently complex multiple action of the twisted transfer matrix $\mathcal{T}_\kappa(\bar{\xi})$ on the Bethe vector. This method also allows us to obtain several multiple integral representations for the generating function. Some of them have not yet been investigated. They may be more suitable for calculating the asymptotics of correlation functions.

The idea of this method is elementary. We start with representations (11.24) and insert between $\mathcal{T}_\kappa(\bar{\xi})$ and $\mathcal{T}^{-1}(\bar{\xi})$ a projector on the states with $N$ spins down in the form

$$\mathbb{P}_N = \sum_{\bar{v}} \frac{|\mathbf{\Phi}_N^{(\kappa)}(\bar{v})\rangle\langle\mathbf{\Phi}_N^{(\kappa)}(\bar{v})|}{\langle\mathbf{\Phi}_N^{(\kappa)}(\bar{v})|\mathbf{\Phi}_N^{(\kappa)}(\bar{v})\rangle}. \tag{11.82}$$

Here the sum is taken over all admissible solutions of twisted Bethe equations $\bar{v} = \{v_1, \dots, v_N\}$ with $\#\bar{v} = N$. Then representation (11.25) takes the form

$$Q_{n;N}^{(\kappa)} = \sum_{\bar{v}} \frac{\langle\mathbf{\Phi}_N(\bar{u})|\mathcal{T}_\kappa(\bar{\xi})|\mathbf{\Phi}_N^{(\kappa)}(\bar{v})\rangle\langle\mathbf{\Phi}_N^{(\kappa)}(\bar{v})|\mathcal{T}^{-1}(\bar{\xi})|\mathbf{\Phi}_N(\bar{u})\rangle}{\langle\mathbf{\Phi}_N(\bar{u})|\mathbf{\Phi}_N(\bar{u})\rangle\langle\mathbf{\Phi}_N^{(\kappa)}(\bar{v})|\mathbf{\Phi}_N^{(\kappa)}(\bar{v})\rangle}. \tag{11.83}$$

Let $|\kappa|$ be small enough. We will also assume that all inhomogeneities $\{\xi_1, \dots, \xi_L\}$ are pairwise distinct and belong to a small neighborhood of the point $-i\zeta/2$. Then, according to the results of Section 2.4.2, all admissible roots of the twisted Bethe equations are in a small neighborhood of the point $i\zeta/2$ (see proposition 2.1), and the set of twisted on-shell vectors forms a basis. Hence, $\mathbb{P}_N$ is indeed the identity operator in the subspace of states with $N$ spins down.

The actions of $\mathcal{T}_\kappa(\bar{\xi})$ and $\mathcal{T}^{-1}(\bar{\xi})$ respectively on the states $|\mathbf{\Phi}_N^{(\kappa)}(\bar{v})\rangle$ and $|\mathbf{\Phi}_N(\bar{u})\rangle$ are easily calculated

$$\begin{aligned}\mathcal{T}_\kappa(\bar{\xi})|\mathbf{\Phi}_N^{(\kappa)}(\bar{v})\rangle &= \left(\prod_{k=1}^{n} \tau_\kappa(\xi_k|\bar{v})\right)|\mathbf{\Phi}_N^{(\kappa)}(\bar{v})\rangle,\\ \mathcal{T}^{-1}(\bar{\xi})|\mathbf{\Phi}_N(\bar{u})\rangle &= \left(\prod_{k=1}^{n} \tau^{-1}(\xi_k|\bar{u})\right)|\mathbf{\Phi}_N(\bar{u})\rangle.\end{aligned} \tag{11.84}$$

Recall that $\bar{\xi} = \{\xi_1, \dots, \xi_n\}$. Using $d(\xi_k) = 0$ for all $k = 1, \dots, n$, we obtain

$$Q_{n;N}^{(\kappa)} = \sum_{\bar{v}} \frac{f(\bar{v},\bar{\xi})}{f(\bar{u},\bar{\xi})} \frac{\langle \mathbf{\Phi}_N(\bar{u})|\mathbf{\Phi}_N^{(\kappa)}(\bar{v})\rangle\langle \mathbf{\Phi}_N^{(\kappa)}(\bar{v})|\mathbf{\Phi}_N(\bar{u})\rangle}{\langle \mathbf{\Phi}_N(\bar{u})|\mathbf{\Phi}_N(\bar{u})\rangle\langle \mathbf{\Phi}_N^{(\kappa)}(\bar{v})|\mathbf{\Phi}_N^{(\kappa)}(\bar{v})\rangle}. \tag{11.85}$$

### 11.5.1 *Sums over solutions of twisted Bethe equations*

So, the generating function reduces to normalized scalar products of twisted and ordinary on-shell Bethe vectors, for which we have explicit determinant representations. However, now we need to calculate the sum over the solutions of the twisted Bethe equations. We cannot find the roots $v_j$ explicitly, but it is not necessary. The sum (11.85) can be reduced to a multiple contour integral of the Cauchy type.

To illustrate the main idea, we consider a toy example of summing a rational function over the roots of an algebraic equation. Let us have a polynomial $P_n(z)$ of degree $n$, and let $x_1, \dots, x_n$ be the roots of this polynomial:

$$P_n(x_j) = 0, \qquad j = 1, \dots, n. \tag{11.86}$$

Suppose we want to calculate a sum

$$S(a) = \sum_{j=1}^{n} \frac{1}{x_j - a}, \tag{11.87}$$

where $a \neq x_j$, $j = 1, \dots, n$. It is easy to see that the sum (11.87) can be written as a contour integral

$$S(a) = \oint_{\Gamma(\bar{x})} \frac{\mathrm{d}z}{2\pi i} \frac{P_n'(z)}{(z-a)P_n(z)}. \tag{11.88}$$

Here $\Gamma(\bar{x})$ is a positively oriented closed contour that surrounds the roots of $\bar{x}$ and does not contain the point $z = a$. The integral (11.88) calculation by the residues in the poles within the contour returns us to the sum (11.87). However, we can take the integral also by residues outside the contour. The integrand behaves as $z^{-2}$ at $z \to \infty$. Thus, the residue at infinity is zero. Therefore, the integral over the original contour is equal to the integral over a closed contour around $a$ with the opposite sign:

$$S(a) = -\oint_{\Gamma(a)} \frac{dz}{2\pi i} \frac{P_n'(z)}{(z-a)P_n(z)} = -\frac{P_n'(a)}{P_n(a)}. \tag{11.89}$$

Thus, we have calculated the sum $S(a)$ without solving the equation $P_n(z) = 0$.

The method described above naturally generalizes to a sum over the roots of a system of equations. In particular, let us consider the system of twisted Bethe equations in the form (6.63)

$$\mathcal{Y}_\kappa(v_k|\bar{v}) = 0, \qquad k = 1, \dots, N, \tag{11.90}$$

where $\mathcal{Y}_\kappa(z|\bar{v})$ is given by (6.62) (see also (6.71) for a more explicit form). We can formulate a problem of summing a rational function $\phi(\bar{z})$ over all possible solutions $\bar{v}$. Then

$$\sum_{\bar{v}} \phi(\bar{v}) = \frac{1}{N!} \oint_{\Gamma(\bar{v})} \phi(\bar{z}) \det_N \left( \frac{\partial \mathcal{Y}_\kappa(z_k|\bar{z})}{\partial z_j} \right) \prod_{a=1}^{N} \frac{dz_a}{2\pi i \mathcal{Y}_\kappa(z_a|\bar{z})}. \tag{11.91}$$

Here $\Gamma(\bar{v})$ is a multidimensional closed contour (the boundary of a polydisc) that encompasses all solutions of the system (11.90). However, all other singularities of the integrand should be outside $\Gamma(\bar{v})$. The factor $1/N!$ appears because the result of integration includes the sets of roots $\bar{v}$ that differ only by a permutation of the parameters $v_k$. We should not distinguish such sets in the sum because they correspond to the same Bethe vector.

It is remarkable that the integrand now contains the Jacobian of the system (11.90) (analogous to the derivative $P_n'(z)$ in formula (11.88)). At the same time, the norm of a twisted on-shell vector is proportional to the same Jacobian (see (6.70)). Therefore, this Jacobian cancels in normalized expressions of type (11.85).

Thus, instead of the sum over all admissible solutions of the twisted Bethe equations (11.90), we can write down the following integral representation for the generating function:

$$Q_{n;N}^{(\kappa)} = \frac{1}{N!} \oint_{\Gamma(\bar{v})} \frac{f(\bar{z},\bar{\xi})}{f(\bar{u},\bar{\xi})} \prod_{a=1}^{N} \frac{dz_a}{2\pi \sin\zeta \, \mathcal{Y}_\kappa(z_a|\bar{z})} \times \frac{d(\bar{z}) (\langle \mathbf{\Phi}_N^{(\kappa)}(\bar{z}) | \mathbf{\Phi}_N(\bar{u}) \rangle)^2}{\langle \mathbf{\Phi}_N(\bar{u}) | \mathbf{\Phi}_N(\bar{u}) \rangle \Delta'(\bar{z}) \Delta(\bar{z})}. \tag{11.92}$$

Here we already used (6.70) for the norm of the twisted on-shell Bethe vector $\langle \mathbf{\Phi}_N^{(\kappa)}(\bar{v})|\mathbf{\Phi}_N^{(\kappa)}(\bar{v})\rangle$. We also used the fact that

$$\langle \mathbf{\Phi}_N(\bar{u})|\mathbf{\Phi}_N^{(\kappa)}(\bar{v})\rangle = \langle \mathbf{\Phi}_N^{(\kappa)}(\bar{v})|\mathbf{\Phi}_N(\bar{u})\rangle. \tag{11.93}$$

Since all admissible solutions lie in a small neighborhood of the point $i\zeta/2$, we can assume that the integration contour for each variable $z_j$ is a circle of small radius centered in $i\zeta/2$. At the same time, we should ensure that all other poles of the integrand are outside these contours. Then we can try to calculate the integral by the residues in these 'external' singularities.

There is another problem that we need to solve. Namely, we should get rid of the spurious solutions of the twisted Bethe equations. Indeed, if we formally calculate the integral (11.92) by the residues in the roots of the system (11.90), we obtain the sum over *all* solutions of the twisted Bethe equations. However, we know that not all of them correspond to on-shell vectors. We will return to this question later, but in the meantime, we will discuss singularities of the integrand in (11.92).

### 11.5.2 *Poles of the integrand*

The poles outside the integration contours depend on which representation we choose for the scalar product. We have several determinant formulas for them. Moreover, since we have two scalar products in formula (11.92), we can choose different representations for each of them.

We will use representations (6.92) and (6.94). Replacing there the parameters $v_k$ with $z_k$, we have

$$\langle \mathbf{\Phi}_N^{(\kappa)}(\bar{z})|\mathbf{\Phi}_N(\bar{u})\rangle = F(\bar{z},\bar{u}) \det_N \mathcal{M}^{(\kappa)}(\bar{z},\bar{u}), \tag{11.94}$$

$$\langle \mathbf{\Phi}_N^{(\kappa)}(\bar{z})|\mathbf{\Phi}_N(\bar{u})\rangle = F(\bar{z},\bar{u}) \det_N \widetilde{\mathcal{M}}^{(\kappa)}(\bar{z},\bar{u}). \tag{11.95}$$

Here

$$F(\bar{z},\bar{u}) = \Delta'(\bar{z})\Delta(\bar{u})\, h(\bar{z},\bar{u}), \tag{11.96}$$

and the entries of $\mathcal{M}^{(\kappa)}(\bar{z},\bar{u})$ and $\widetilde{\mathcal{M}}^{(\kappa)}(\bar{z},\bar{u})$ are

$$\mathcal{M}_{jk}^{(\kappa)}(\bar{z},\bar{u}) = t(z_j,u_k) + \kappa\, t(u_k,z_j)\frac{h(\bar{u},u_k)h(u_k,\bar{z})}{h(u_k,\bar{u})h(\bar{z},u_k)}, \tag{11.97}$$

$$\widetilde{\mathcal{M}}_{jk}^{(\kappa)}(\bar{z},\bar{u}) = t(z_k,u_j) - (-1)^N r(z_k)\, t(u_j,z_k)\frac{h(\bar{u},z_k)}{h(z_k,\bar{u})}. \tag{11.98}$$

We draw the reader's attention to the matrix elements (11.98). If the set $\bar{z}$ coincides with some solution of the twisted Bethe equations $\bar{v}$, then

$r(z_k) = r(v_k)$, and we can express this function in terms of the ratio of the $h$-functions. Then we reproduce the matrix elements (6.95). Thus, representation (11.95) is not, in general, a scalar product of the twisted and ordinary on-shell vectors. However, we can use this representation quite well since the set $\bar{z}$ coincides with one of the solutions $\bar{v}$ of the twisted Bethe equations as a result of integration.

For convenience, let us rewrite formulas (11.96)–(11.98) once more, but this time explicitly. Then the prefactor $F(\bar{z},\bar{u})$ reads

$$F(\bar{z},\bar{u}) = \frac{\prod_{a=1}^{N}\prod_{b=1}^{N}\sinh(z_a - u_b - i\zeta)}{(-i\sin\zeta)^N \prod_{a>b}^{N}\sinh(u_a - u_b)\sinh(z_b - z_a)}. \tag{11.99}$$

The matrix elements $\mathcal{M}^{(\kappa)}_{jk}(\bar{z},\bar{u})$ are

$$\begin{aligned}\mathcal{M}^{(\kappa)}_{jk}(\bar{z},\bar{u}) = &\frac{-(\sin\zeta)^2}{\sinh(z_j - u_k)\sinh(z_j - u_k - i\zeta)}\\ &+ \frac{-\kappa(\sin\zeta)^2 V(u_k)}{\sinh(u_k - z_j)\sinh(u_k - z_j - i\zeta)},\end{aligned} \tag{11.100}$$

where

$$V(u_k) = \prod_{a=1}^{N}\frac{\sinh(u_a - u_k - i\zeta)\sinh(u_k - z_a - i\zeta)}{\sinh(u_k - u_a - i\zeta)\sinh(z_a - u_k - i\zeta)}. \tag{11.101}$$

Finally, the matrix elements $\widetilde{\mathcal{M}}^{(\kappa)}_{jk}(\bar{z},\bar{u})$ have the following form:

$$\begin{aligned}\widetilde{\mathcal{M}}^{(\kappa)}_{jk}(\bar{z},\bar{u}) = &\frac{-(\sin\zeta)^2}{\sinh(z_k - u_j)\sinh(z_k - u_j - i\zeta)}\\ &+ \frac{(-1)^N(\sin\zeta)^2\tilde{V}(z_k)}{\sinh(u_j - z_k)\sinh(u_j - z_k - i\zeta)},\end{aligned} \tag{11.102}$$

where

$$\tilde{V}(z_k) = \left(\prod_{b=1}^{L}\frac{\sinh(z_k - \xi_b - i\zeta)}{\sinh(z_k - \xi_b)}\right)\left(\prod_{a=1}^{N}\frac{\sinh(u_a - z_k - i\zeta)}{\sinh(z_k - u_a - i\zeta)}\right). \tag{11.103}$$

Now it is easy to see that in representation (11.94), the scalar product is a rational hyperbolic function of variables $z_k$ with poles in the points $z_k = u_j$, $j,k = 1,\dots,N$. Since the set $\bar{u}$ corresponds to the ground state, all parameters $u_j$ are real. Therefore, they are separated from the point $i\zeta/2$ and, hence, are outside the integration contours. In representation (11.95), the scalar product is a rational function of variables $z_k$ with poles in the points $d(z_k) = 0$, $k = 1,\dots,N$, that is, $z_k = \xi_j$, $j = 1,\dots,L$. These poles are also outside the integration contours. At the same time, the matrix $\widetilde{\mathcal{M}}^{(\kappa)}_{jk}$ is not singular in the points $z_k = u_j$ due to the Bethe equations for the parameters $\bar{u}$.

Note also that in addition to the factor $(\langle \mathbf{\Phi}_N^{(\kappa)}(\bar{z})|\mathbf{\Phi}_N(\bar{u})\rangle)^2$, the integrand in (11.92) also contains the product $f(\bar{z},\bar{\xi})$ and the product $d(\bar{z})$. The first one has poles at the points $z_k = \xi_j$, $k = 1,\dots,N$, $j = 1,\dots,n$, and the second vanishes at $z_k = \xi_j$, $k = 1,\dots,N$, $j = 1,\dots,L$. Thus, the possible poles of the integrand that lie outside the integration contours are inhomogeneities $\bar{\xi}$ and the Bethe parameters $\bar{u}$ that correspond to the ground state.

**Remark 11.3.** Now it becomes clear why we use the form factor expansion over twisted on-shell vectors. Thanks to this, we can separate the solutions of the twisted Bethe equations from the parameters $\bar{u}$, which are one of the solutions of the ordinary Bethe equations. Therefore, there exist integration contours such that they surround only zeros of the functions $\mathcal{Y}_\kappa(z_a|\bar{z})$, and all other singularities of the integrand are outside the contours. We could not choose integration contours that satisfy these conditions when summing over the usual Bethe equations solutions.

So, we have three possibilities to write down the integrand (11.92): (i) choose representation (11.94) for both scalar products; (ii) choose representation (11.95) for both scalar products; (iii) choose representation (11.94) for one scalar product and representation (11.95) for the other. In the first case, we obtain double poles at the points $u_1,\dots,u_N$. In the second case, we obtain double poles at points $\xi_1,\dots,\xi_n$ and simple poles at points $\xi_{n+1},\dots,\xi_L$. Finally, in the third case, we get simple poles at points $\xi_1,\dots,\xi_n$ and $u_1,\dots,u_N$. We choose the third possibility for reasons of simplicity of calculating residues in the first order poles. Then we have

$$Q^{(\kappa)}_{n;N} = \frac{1}{N!}\oint_{\Gamma(\bar{v})} \frac{f(\bar{z},\bar{\xi})}{f(\bar{u},\bar{\xi})} \frac{d(\bar{z})d(\bar{u})h^2(\bar{z},\bar{u})}{\det_N\left(\frac{\partial \mathcal{Y}(u_k|\bar{u})}{\partial u_j}\right)} \\ \times \det_N \mathcal{M}^{(\kappa)}(\bar{z},\bar{u}) \det_N \widetilde{\mathcal{M}}^{(\kappa)}(\bar{z},\bar{u}) \prod_{a=1}^{N} \frac{i\, dz_a}{2\pi \sin^2\zeta\, \mathcal{Y}_\kappa(z_a|\bar{z})}. \tag{11.104}$$

Here we also used (6.70) (at $\kappa = 1$) for the square of the ground state norm $\langle \mathbf{\Phi}_N(\bar{u})|\mathbf{\Phi}_N(\bar{u})\rangle$. Representation (11.104) was first obtained in [N. Kitanine et al (2005c)].

### 11.5.3 *Calculating the integral*

We can now calculate the integral (11.104) by the residues in the points $\xi_1, \dots, \xi_n$ and $u_1, \dots, u_N$. Let $\mathcal{I}(\bar{z}, \bar{u})$ denote the integrand in (11.104). Observe that $\mathcal{I}(\bar{z}, \bar{u})$ is $i\pi$-periodic in each $z_j$ and decreases as $e^{-2|z_j|}$ at $z_j \to \pm\infty$. Therefore, the integral over each $z_j$ along the boundary of any horizontal strip of width $i\pi$ is equal to zero. Hence,

$$\begin{aligned}\oint_{\Gamma(\bar{v})} \mathcal{I}(\bar{z}, \bar{u})\,\mathrm{d}z_j &= \oint_{\text{Strip}} \mathcal{I}(\bar{z}, \bar{u})\,\mathrm{d}z_j - \oint_{\Gamma(\bar{\xi})\cup\Gamma(\bar{u})} \mathcal{I}(\bar{z}, \bar{u})\,\mathrm{d}z_j \\ &= -\oint_{\Gamma(\bar{\xi})\cup\Gamma(\bar{u})} \mathcal{I}(\bar{z}, \bar{u})\,\mathrm{d}z_j. \end{aligned} \tag{11.105}$$

Here the contour $\Gamma(\bar{\xi}) \cup \Gamma(\bar{u})$ encloses the points $\bar{\xi}$ and $\bar{u}$, and the integral with the subscript 'Strip' means the integral over the above-mentioned strip of width $i\pi$, which is equal to zero. Thus, the $N$-fold integral (11.104) turns into

$$\begin{aligned} Q^{(\kappa)}_{n;N} &= \frac{(-1)^N}{N!} \oint_{\Gamma(\bar{\xi})\cup\Gamma(\bar{u})} \frac{f(\bar{z}, \bar{\xi})}{f(\bar{u}, \bar{\xi})} \frac{d(\bar{z})d(\bar{u})h^2(\bar{z}, \bar{u})}{\det_N\left(\frac{\partial \mathcal{Y}(u_k|\bar{u})}{\partial u_j}\right)} \\ &\quad \times \det_N \mathcal{M}^{(\kappa)}(\bar{z}, \bar{u}) \det_N \widetilde{\mathcal{M}}^{(\kappa)}(\bar{z}, \bar{u}) \prod_{a=1}^{N} \frac{i\,\mathrm{d}z_a}{2\pi \sin^2\zeta\, \mathcal{Y}_\kappa(z_a|\bar{z})}. \end{aligned} \tag{11.106}$$

For each integration variable $z_j$, the integral over the contour $\Gamma(\bar{\xi})\cup\Gamma(\bar{u})$ can be written as the sum of two integrals: over the contour $\Gamma(\bar{\xi})$ (around the points $\bar{\xi}$) and the contour $\Gamma(\bar{u})$ (around the points $\bar{u}$). Since the integrand in (11.106) is symmetric with respect to the variables $\bar{z}$, the total $N$-fold integral is then written as

$$\begin{aligned} &\frac{(-1)^N}{N!} \oint_{\Gamma(\bar{\xi})\cup\Gamma(\bar{u})} \mathcal{I}(\bar{z}, \bar{u}) \prod_{j=1}^{N} \mathrm{d}z_j \\ &= \sum_{m=0}^{N} \frac{(-1)^N}{m!(N-m)!} \oint_{\Gamma(\bar{\xi})} \left(\prod_{j=1}^{m} \mathrm{d}z_j\right) \oint_{\Gamma(\bar{u})} \left(\prod_{j=m+1}^{N} \mathrm{d}z_j\right) \mathcal{I}(\bar{z}, \bar{u}). \end{aligned} \tag{11.107}$$

Here, for the integration variables $z_1, \dots, z_m$, the contours encircle the points $\bar{\xi}$, and for the remaining variables $z_{m+1}, \dots, z_N$ the integration is carried out along contours enclosing the points $\bar{u}$. Note that the number of poles within the contour $\Gamma(\bar{\xi})$ equals $n$ (these are the points $\xi_1, \dots, \xi_n$). In addition, the integrand has the second order zero if two variables $z_j$ and

$z_k$ coincide. Therefore, calculating the residues at the points $\xi_1, \dots, \xi_n$, we cannot put two different $z_j$ and $z_k$ equal to the same inhomogeneity $\xi_\ell$. It follows that in the integrals over the contours $\Gamma(\bar\xi)$, the number of integration variables $z_1, \dots, z_n$ can not exceed the number of the poles $\xi_1, \dots, \xi_n$, that is, $m \le n$. Otherwise, the $m$-fold integral over the contours $\Gamma(\bar\xi)$ vanishes. Thus, the sum over $m$ in (11.107) is actually limited to $m \le n$.

For the time being, let us leave the integrals over the contours $\Gamma(\bar\xi)$ and pass to the integrals over the contours $\Gamma(\bar u)$. In these integrals, we also cannot put two different $z_j$ and $z_k$ equal to the same value $u_\ell$. Therefore, the integral over the contours $\Gamma(\bar u)$ is equal to the sum of the residues at the points $z_{m+k} = u_{j_k}$, $k = 1, \dots, N-m$, where the sum is taken over all possible sets $u_{j_1}, \dots, u_{j_{N-m}}$. In other words, this integral generates a sum over partitions of the complete set $\bar u$ into two subsets: $\bar u^{\mathrm{I}}$ and $\bar u^{\mathrm{II}}$, where $\#\bar u^{\mathrm{I}} = m$ and $\#\bar u^{\mathrm{II}} = N-m$. The residues are taken at the points $u_j \in \bar u^{\mathrm{II}}$. Taking into account the fact that we can do all possible permutations of the elements in the subset $\bar u^{\mathrm{II}}$, the sum over the partitions $\bar u \mapsto \{\bar u^{\mathrm{I}}, \bar u^{\mathrm{II}}\}$ should be multiplied by $(N-m)!$. Thus, we obtain

$$\frac{(-1)^N}{N!} \oint_{\Gamma(\bar\xi)\cup\Gamma(\bar u)} \mathcal{I}(\bar z, \bar u) \prod_{j=1}^{N} \mathrm{d}z_j$$
$$= \sum_{m=0}^{n} \frac{(-1)^N}{m!} \oint_{\Gamma(\bar\xi)} \prod_{j=1}^{m} \mathrm{d}z_j \sum_{\bar u \mapsto \{\bar u^{\mathrm{I}}, \bar u^{\mathrm{II}}\}} (2\pi i)^{N-m} \operatorname*{Res}_{z_{m+1}=u_1^{\mathrm{II}}} \cdots \operatorname*{Res}_{z_N = u_{N-m}^{\mathrm{II}}} \mathcal{I}(\bar z, \bar u). \tag{11.108}$$

Similarly, the integrals over the contours $\Gamma(\bar\xi)$ give the sum over partitions of the parameters $\xi_1, \dots, \xi_n$. However, we will leave these integrals as they are now. We only note that in the remaining integrand, we can put $d(z_j) = 0$, $j = 1, \dots, m$. Indeed, after taking the integral, each variable $z_j$ becomes equal to one of the inhomogeneities $\xi_\ell$. But $d(\xi_\ell) = 0$, so we can put $d(z_j) = 0$ in advance. In other words, all terms proportional to $d(z_j)$ are holomorphic within the contours $\Gamma(\bar\xi)$. Therefore, they give zero contribution and can be discarded.

All that remains to be done is to compute the residues at the points $z_{m+s} = u_s^{\mathrm{II}}$, $s = 1, \dots, N-m$. However, before doing this, we slightly transform the integrand. We are going to reproduce equation (11.62). Therefore, it is convenient to rewrite the integrand in (11.106) in terms of the matrices $\Omega(\bar u|\bar u)$ (11.54) and $\widetilde\Omega(\bar u|\bar v)$ (11.57).

It is easy to see that

$$\frac{\partial \mathcal{Y}(u_k|\bar u)}{\partial u_j} = -i d(u_k) h(u_k, \bar u) \Omega_{jk}(\bar u|\bar u), \tag{11.109}$$

and

$$\widetilde{\mathcal{M}}_{jk}^{(\kappa)}(\bar{z},\bar{u}) = (-1)^N \sin\zeta\, r(z_k)\frac{h(\bar{u},z_k)}{h(z_k,\bar{u})}\widetilde{\Omega}_{jk}(\bar{u}|\bar{z}). \tag{11.110}$$

Let us also introduce a function $\widetilde{\mathcal{Y}}_\kappa(z_a|\bar{z})$ by

$$\widetilde{\mathcal{Y}}_\kappa(z_a|\bar{z}) = \frac{(-1)^N\mathcal{Y}_\kappa(z_a|\bar{z})}{a(z_k)h(\bar{z},z_k)} = 1 + (-1)^N\kappa\frac{d(z_k)h(z_k,\bar{z})}{a(z_k)h(\bar{z},z_k)}. \tag{11.111}$$

Then we recast the integral (11.106) as follows:

$$Q_{n;N}^{(\kappa)} = \frac{1}{N!}\oint_{\Gamma(\bar{\xi})\cup\Gamma(\bar{u})} \frac{f(\bar{z},\bar{\xi})}{f(\bar{u},\bar{\xi})}\frac{W_N(\bar{z},\bar{u})\det_N\mathcal{M}^{(\kappa)}(\bar{z},\bar{u})}{\prod_{a=1}^N\big(2\pi\sin\zeta\,\widetilde{\mathcal{Y}}_\kappa(z_a|\bar{z})\big)} \times\frac{\det_N\widetilde{\Omega}(\bar{u}|\bar{z})}{\det_N\Omega(\bar{u}|\bar{u})}\prod_{a=1}^N \mathrm{d}z_a, \tag{11.112}$$

where

$$W_N(\bar{z},\bar{u}) = \frac{h(\bar{z},\bar{u})h(\bar{u},\bar{z})}{h(\bar{z},\bar{z})h(\bar{u},\bar{u})}. \tag{11.113}$$

Now we are ready to calculate the integral in (11.112). The determinant of the matrix $\mathcal{M}^{(\kappa)}(\bar{z},\bar{u})$ has simple poles at $z_j = u_k$, $j,k=1,\dots,N$. Using (11.97) or (11.100), we immediately obtain

$$\operatorname*{Res}_{z_N=u_N}\frac{\det_N\mathcal{M}^{(\kappa)}(\bar{z},\bar{u})}{\widetilde{\mathcal{Y}}_\kappa(z_N|\bar{z})} = -i\sin\zeta\det_{N-1}\mathcal{M}^{(\kappa)}(\bar{z}_N,\bar{u}_N). \tag{11.114}$$

It is convenient to set $\{z_1,\dots,z_m\} = \bar{z}^{\mathrm{I}}$ and $\{z_{m+1},\dots,z_N\} = \bar{z}^{\mathrm{II}}$. Then we should consider the residues in the points $z_j^{\mathrm{II}} = u_j^{\mathrm{II}}$, $j=1,\dots,N-m$. It follows from (11.114) that

$$(2\pi i)^{N-m}\operatorname*{Res}_{z_1^{\mathrm{II}}=u_1^{\mathrm{II}}}\cdots\operatorname*{Res}_{z_{N-m}^{\mathrm{II}}=u_{N-m}^{\mathrm{II}}}\frac{\det_N\mathcal{M}^{(\kappa)}(\bar{z},\bar{u})}{\prod_{a=1}^N\big(2\pi\sin\zeta\,\widetilde{\mathcal{Y}}_\kappa(z_a|\bar{z})\big)} = \frac{\det_m\mathcal{M}^{(\kappa)}(\bar{z}^{\mathrm{I}},\bar{u}^{\mathrm{I}})}{\prod_{a=1}^m\big(2\pi\sin\zeta\,\widetilde{\mathcal{Y}}_\kappa(z_a|\{\bar{z}^{\mathrm{I}},\bar{u}^{\mathrm{II}}\})\big)}. \tag{11.115}$$

It is also easy to see that

$$W_N(\bar{z},\bar{u})\Big|_{\bar{z}^{\mathrm{II}}=\bar{u}^{\mathrm{II}}} = W_m(\bar{z}^{\mathrm{I}},\bar{u}^{\mathrm{I}}) = \frac{h(\bar{z}^{\mathrm{I}},\bar{u}^{\mathrm{I}})h(\bar{u}^{\mathrm{I}},\bar{z}^{\mathrm{I}})}{h(\bar{z}^{\mathrm{I}},\bar{z}^{\mathrm{I}})h(\bar{u}^{\mathrm{I}},\bar{u}^{\mathrm{I}})}. \tag{11.116}$$

Thus, taking into account (11.108), we arrive at

$$Q_{n;N}^{(\kappa)} = \sum_{m=0}^n\frac{1}{m!}\oint_{\Gamma(\bar{\xi})}\sum_{\bar{u}\mapsto\{\bar{u}^{\mathrm{I}},\bar{u}^{\mathrm{II}}\}}\frac{f(\bar{z},\bar{\xi})}{f(\bar{u},\bar{\xi})}\frac{W_m(\bar{z}^{\mathrm{I}},\bar{u}^{\mathrm{I}})\det_m\mathcal{M}^{(\kappa)}(\bar{z}^{\mathrm{I}},\bar{u}^{\mathrm{I}})}{\prod_{a=1}^m\big(2\pi\sin\zeta\,\widetilde{\mathcal{Y}}_\kappa(z_a|\{\bar{z}^{\mathrm{I}},\bar{u}^{\mathrm{II}}\})\big)} \times\frac{\det_N\widetilde{\Omega}(\bar{u}|\{\bar{z}^{\mathrm{I}},\bar{u}^{\mathrm{II}}\})}{\det_N\Omega(\bar{u}|\bar{u})}\prod_{a=1}^m\mathrm{d}z_a. \tag{11.117}$$

Now we can put $d(z_a) = 0$ for $a = 1, \dots, m$. Then using (11.111), we find

$$\widetilde{\mathcal{Y}}_\kappa(z_a|\{\bar{z}^{\mathrm{I}}, \bar{u}^{\mathrm{II}}\}) = 1, \qquad \text{for} \qquad d(z_a) = 0, \qquad a = 1, \dots, m. \tag{11.118}$$

Hence,

$$Q^{(\kappa)}_{n;N} = \sum_{m=0}^{n} \frac{1}{m!} \oint_{\Gamma(\bar{\xi})} \sum_{\bar{u} \mapsto \{\bar{u}^{\mathrm{I}}, \bar{u}^{\mathrm{II}}\}} \frac{f(\bar{z}, \bar{\xi})}{f(\bar{u}, \bar{\xi})} W_m(\bar{z}^{\mathrm{I}}, \bar{u}^{\mathrm{I}}) \det_m \mathcal{M}^{(\kappa)}(\bar{z}^{\mathrm{I}}, \bar{u}^{\mathrm{I}})$$
$$\times \frac{\det_N \widetilde{\Omega}(\bar{u}|\{\bar{z}^{\mathrm{I}}, \bar{u}^{\mathrm{II}}\})}{\det_N \Omega(\bar{u}|\bar{u})} \prod_{a=1}^{m} \frac{dz_a}{2\pi \sin\zeta}. \tag{11.119}$$

Proceeding to the homogeneous limit $\xi_k = -i\zeta/2$, $k = 1, \dots, n$, we reproduce (11.62).

Now we can get rid of all restrictions on the parameter $\kappa$. Recall that we assumed that $|\kappa|$ was small enough at the very beginning. This requirement ensured the completeness of the system of twisted on-shell vectors. Besides, all solutions $\bar{v}$ of the twisted Bethe equations turned out to be in a small vicinity of the point $i\zeta/2$. Therefore, they have been separated from the Bethe roots $\bar{u}$ and the inhomogeneities. Thus, equation (11.119) is obtained under the condition $|\kappa| < \kappa_0$, where $\kappa_0$ is small enough. Note, however, that in formula (11.119), the parameter $\kappa$ is included only in the determinant of the matrix $\mathcal{M}^{(\kappa)}$. Hence, the integrand is a polynomial in $\kappa$ of degree $m$ for any fixed $m$. If we know a polynomial in the neighborhood $|\kappa| < \kappa_0$, we know it in the whole complex plane. Thus, in formula (11.119), the parameter $\kappa$ can take any value, including $\kappa = 1$.

In the thermodynamic limit, formula (11.119) turns into the multiple integral representation (11.76). Thus, we have shown that the multiple integral representation for the generating function can be obtained using the form factor expansion. The same approach also can be applied to dynamical (time-dependent) correlation functions [N. Kitanine et al (2005d)].

### 11.5.4 *Spurious solutions*

We have proved that the integral representation (11.92) reduces to the formula (11.62). It means that integral (11.92) takes into account only admissible solutions of the twisted Bethe equations that correspond to the eigenvectors of the twisted transfer matrix. All spurious solutions, for some reason, give vanishing contribution.

It is easy to understand that solutions in which several parameters $v_j$ coincide do not contribute to the sum (11.85). Indeed, the integrand in (11.104) contains two determinants. If we calculate the residue in a solution in which, for example, $v_1 = v_2$, then these determinants vanish.

Another type of spurious solutions occurs for $\zeta = \pi p/q$, where $p$ and $q$ are positive integers and $p < q < N$. Consider a set of parameters $v_1, \dots, v_q$, such that $v_j = w + i\zeta(j-1)$, where $w$ is an arbitrary complex number. Then $\mathcal{Y}_\kappa(v_j|\bar{v}) = 0$ for $j = 1, \dots, q$. Therefore, solutions of the twisted Bethe equations can contain such subsets. However, it is clear that if the integration contours are circles of small radius centered in $i\zeta/2$, they cannot contain all the elements of such a subset. In addition, it is easy to prove that the rows of the matrix $\mathcal{M}^{(\kappa)}(\bar{v}, \bar{u})$ are linearly dependent in this case. Indeed, we have

$$\sum_{j=1}^{q} t(v_j, u_k) = \sum_{j=1}^{q} t(u_k, v_j) = 0, \tag{11.120}$$

for $v_j = w + i\zeta(j-1)$. To prove this statement, it is enough to present the function $t(x, y)$ in the form

$$t(x, y) = i \sin\zeta\big(\coth(x - y - i\zeta) - \coth(x - y)\big), \tag{11.121}$$

and substitute this expression in the sums (11.120) with $v_j = w + i\zeta(j-1)$. Since the entries of the matrix $\mathcal{M}^{(\kappa)}(\bar{v}, \bar{u})$ have the structure (11.97)

$$\mathcal{M}^{(\kappa)}_{jk}(\bar{v}, \bar{u}) = t(v_j, u_k) + V(u_k)t(u_k, v_j), \tag{11.122}$$

we conclude that

$$\sum_{j=1}^{q} \mathcal{M}^{(\kappa)}_{jk}(\bar{v}, \bar{u}) = 0. \tag{11.123}$$

Hence, the determinant of $\mathcal{M}^{(\kappa)}(\bar{v}, \bar{u})$ vanishes for these solutions.

Similarly, one can show that $d(\bar{v}) \det_N \widetilde{\mathcal{M}}^{(\kappa)}(\bar{v}, \bar{u}) = 0$, if the roots $\bar{v}$ contain a pair $\{v_{k_1}, v_{k_2}\}$ such that $v_{k_1} = \xi_\ell$ and $v_{k_2} = \xi_\ell + i\zeta$, $\ell = 1, \dots, L$. Indeed, using $a(\xi_\ell + i\zeta) = 0$ and $d(\xi_\ell) = 0$, we obtain

$$\begin{aligned} d(v_{k_1})\widetilde{\mathcal{M}}^{(\kappa)}_{j,k_1}(\bar{v}, \bar{u}) &= (-1)^{N-1} a(\xi_\ell)\, t(u_j, \xi_\ell) \frac{h(\bar{u}, \xi_\ell)}{h(\xi_\ell, \bar{u})}, \\ d(v_{k_2})\widetilde{\mathcal{M}}^{(\kappa)}_{j,k_2}(\bar{v}, \bar{u}) &= d(\xi_\ell + i\zeta) t(\xi_\ell + i\zeta, u_j) = d(\xi_\ell + i\zeta) t(u_j, \xi_\ell). \end{aligned} \tag{11.124}$$

Thus, the columns $d(v_{k_1})\widetilde{\mathcal{M}}^{(\kappa)}_{j,k_1}$ and $d(v_{k_2})\widetilde{\mathcal{M}}^{(\kappa)}_{j,k_2}$ are proportional to each other. Besides, it is obvious that integration contours cannot contain such solutions either.

We see that the formulas for scalar products turn out to be sufficiently 'smart'. They discard spurious solutions automatically. Therefore, the latter do not contribute to the integral (11.92).

Let us summarize. We see that we can consider multiple integral representation (11.76) for correlation functions as the result of the summation of form factors. It is clear that within this approach, we can obtain other multiple integral representations. Indeed, we took advantage of only one possibility, using different determinant representations for two scalar products. However, we could, for example, express both scalar products in terms of the determinant of the matrix $\mathcal{M}^{(\kappa)}$. Then we would get only second order poles at points $\bar{u}$ in the integrand. By expressing both scalar products in terms of the determinant of the matrix $\widetilde{\mathcal{M}}^{(\kappa)}$, we would get singularities only in inhomogeneities. Both of these possibilities exist, but to date, they have not been practically explored. Perhaps, one of these paths may lead to new multiple integral representations, which will be more convenient for further analysis.

Note also that works [N. Kitanine et al (2009b); K. K. Kozlowski et al (2011a)] used representations (9.118), (9.145) for the scalar products. This approach was found to be appropriate for calculating the long-distance asymptotics of the correlation function of the third spin components in the $XXZ$ chain. However, the method for calculating the asymptotics turned out to be highly technically complex.

Of course, the method of obtaining the multiple integral representations for correlation functions by summing form factors can also be applied to other correlation functions. In this approach, we do not need to calculate the action of the product of transfer matrices on the Bethe vector. All that is required is explicit formulas for form factors. We can apply this method to calculate any two-point correlation function if the form factors of local operators are known. In particular, this method also works for the model of the one-dimensional Bose gas. It leads to a multiple integral representation similar to (11.76) [N. Kitanine et al (2007)]. We can also consider time-dependent correlation functions, although in this case, the multiple integral representations are more complex [N. Kitanine et al (2005d)].

## 11.6 Correlation function of free fermions

The $XXZ$ chain becomes equivalent to the free fermions at $\Delta = 0$ [E. Lieb et al (1961); P. Jordan and E. Wigner (1928)]. This model is often referred to as the $XX$ or $XX0$ chain. It is the isotropic case of the $XY$ chain ($XYZ$ chain at $J_z = 0$), which is also equivalent to free fermions. On the one hand, this model is quite simple. For example, a system of Bethe equations

(10.126) turns into a set of decoupled equations and is explicitly solved.[1] On the other hand, many correlation functions in this model (and in the $XY$ chain) turn out to be non-trivial. Therefore, correlation functions of the $XX$ and $XY$ chains were actively studied [E. Lieb et al (1961); B. M. McCoy (1968); E. Barouch et al (1973); B. M. McCoy et al (1983a,b); Th. Niemeijer (1967); S. Katsura et al (1970); H. G. Vaidya and C. A. Tracy (1978a); J. H. H. Perk and H. W. Capel (1977); T. Tonegawa (1981); M. D'lorio et al (1983a,b); F. Colomo et al (1992, 1993); A. R. Its et al (1993); A. G. Izergin et al (1998)]. In particular, it was shown that the correlation functions in these models are closely related to classical exactly solvable equations [B. M. McCoy and C. A. Tracy (1973); B. M. McCoy et al (1977); T. T. Wu et al (1976); H. G. Vaidya and C. A. Tracy (1978b); J. H. H. Perk (1980); J. H. H. Perk et al (1984); A. R. Its et al (1995, 1996)]. The algebraic Bethe ansatz was applied to the $XX$ chain in [N. Kitanine et al (2002c); F. Göhmann et al (2019, 2020a,b)].

The integral representation for the generating function $Q_n^{(\kappa)}$ significantly simplifies for $\zeta = \pi/2$. This allows us to explicitly calculate the two-point function of the third components of spin (11.15). The simplification of the generating function is based on formula (7.77) for the Cauchy determinant.

We proceed to calculate the generating function for $\zeta = \pi/2$. Substituting this value in the expressions for $\mathcal{M}_{jk}^{(\kappa)}(\bar{z}, \bar{u})$, $\rho(u_j, z_k)$, and $W_m(\bar{z}, \bar{u})$, we obtain

$$\begin{aligned} \mathcal{M}_{jk}^{(\kappa)}(\bar{z}, \bar{u}) &= \frac{2i(\kappa - 1)}{\sinh 2(z_j - u_k)}, \\ \rho(u_j, z_k) &= \frac{i}{\pi \sinh 2(u_j - z_k)}, \\ W_m(\bar{z}, \bar{u}) &= \prod_{j=1}^{m} \prod_{k=1}^{m} \frac{\cosh^2(z_j - u_k)}{\cosh(z_j - z_k)\cosh(u_j - u_k)}. \end{aligned} \tag{11.125}$$

Thus, using (7.77), we obtain

$$\begin{aligned} Q_n^{(\kappa)} = \sum_{m=0}^{n} \frac{(\kappa - 1)^m}{(2\pi)^{2m}(m!)^2} \oint \mathrm{d}\bar{z} \int_{-q}^{q} d\bar{u} \prod_{j=1}^{m} \left( \frac{\varphi(z_j)}{\varphi(u_j)} \right)^n \\ \times \left( \det_m \left[ \frac{1}{\sinh(u_j - z_k)} \right] \right)^2. \end{aligned} \tag{11.126}$$

[1] Recall that the point $\Delta = 0$ corresponds to $\zeta = \pi/2$.

Here integrations over $z_j$-variables are taken along the contours surrounding the point $-i\pi/4$, and

$$\varphi(u) = \frac{\sinh(u - \frac{i\pi}{4})}{\sinh(u + \frac{i\pi}{4})} = -e^{ip_0(u)}, \tag{11.127}$$

where $p_0(u)$ is defined in (10.127).

Let us also introduce a Fermi momentum $k_F = p_0(q)$. It follows from the integral equation (10.132) that $p_0'(u) = 2\pi\rho(u)$ because the kernel $\mathcal{K}(u)$ (10.131) vanishes at $\zeta = \pi/2$. Then we have

$$\frac{N}{L} \equiv D = \int_{-q}^{q} \rho(u)\,\mathrm{d}u = \frac{1}{2\pi}\int_{-q}^{q} p_0'(u)\,\mathrm{d}u = \frac{k_F}{\pi}. \tag{11.128}$$

Recall also that the two-point correlation function of the third components of spin is extracted from the generating function $Q_n^{(\kappa)}$ by (11.20), (11.21)

$$\langle\sigma_1^z\sigma_{n+1}^z\rangle = 2\frac{\mathrm{d}^2}{\mathrm{d}\kappa^2}D_n^2 Q_n^{(\kappa)}\Big|_{\kappa=1} + 2\langle\sigma^z\rangle - 1, \tag{11.129}$$

where the symbol $D_n^2$ denotes the second derivative on the lattice. The latter acts on functions of the discrete argument $\phi(n)$ as follows:

$$D_n^2\phi(n) = \phi(n+1) - 2\phi(n) + \phi(n-1). \tag{11.130}$$

The $m$th term in the sum (11.126) is proportional to $(\kappa - 1)^m$. To calculate the correlation function, we need to take the second derivative with respect to $\kappa$ at $\kappa = 1$. Then it is sufficient to consider only terms with $m = 1$ and $m = 2$ (the term corresponding to $m = 0$ vanishes after taking the derivatives).

Let us consider the contribution to the generating function for $m = 1$:

$$\Lambda_1(n) = \frac{\kappa - 1}{(2\pi)^2}\int_{-q}^{q} \mathrm{d}u \oint \mathrm{d}z \frac{\varphi^n(z)\varphi^{-n}(u) - 1}{\sinh^2(u - z)}. \tag{11.131}$$

We have replaced $\varphi^n(z)\varphi^{-n}(u)$ with $\varphi^n(z)\varphi^{-n}(u) - 1$ since this is equivalent to adding a holomorphic function within the integration contour over $z$. Thanks to this trick, we turned the second order pole at $z = u$ into a simple pole. Now, instead of calculating the integral over $z$ by taking the residue in the $n$th order pole at $z = -i\pi/4$, we can take the residue at the point $z = u$, where we have a simple pole. Indeed, the integrand in (11.131) is $i\pi$-periodic over $z$ and decreases exponentially at $z \to \pm\infty$. Hence, the integral over the boundary of an infinite horizontal strip of the width $i\pi$ vanishes. On the other hand, this integral is equal to the sum of the residues at $z = -i\pi/4$ and $z = u$. Therefore, the integral along the contour around $-i\pi/4$ can be replaced by an integral along the contour around $z = u$ with the opposite sign. Then we immediately find

$$\begin{aligned}\Lambda_1(n) &= -\frac{in(\kappa-1)}{2\pi}\int_{-q}^{q}\frac{\mathrm{d}}{\mathrm{d}u}\log\left(\frac{\sinh(u-\frac{i\pi}{4})}{\sinh(u+\frac{i\pi}{4})}\right)\mathrm{d}u \\ &= \frac{n(\kappa-1)}{2\pi}\int_{-q}^{q}p_0'(u)\,\mathrm{d}u = \frac{n(\kappa-1)k_F}{\pi}. \end{aligned} \tag{11.132}$$

This contribution vanishes after taking the second $\kappa$-derivative and the lattice derivative. However, it gives us a possibility to once again calculate the magnetization $\langle\sigma^z\rangle$ via (11.23):

$$\langle\sigma^z\rangle = 1 - \frac{2k_F}{\pi}. \tag{11.133}$$

It agrees with the result (9.23) since the kernel $\mathcal{K}(u)$ (10.131) vanishes for $\zeta = -i\pi/2$, leading to $k_F = \pi D$.

A contribution to the generating function at $m = 2$ reads

$$\begin{aligned}\Lambda_2(n) = \frac{(\kappa-1)^2}{64\pi^4}&\int_{-q}^{q}\mathrm{d}u_1\,\mathrm{d}u_2\oint\mathrm{d}z_1\,\mathrm{d}z_2\det_2\left(\frac{1}{\sinh(z_k-u_j)}\right) \\ &\times\det_2\left(\frac{\varphi^n(z_k)\varphi^{-n}(u_j)-1}{\sinh(z_k-u_j)}\right). \end{aligned} \tag{11.134}$$

Here we also subtracted 1 from the product $\varphi^n(z_k)\varphi^{-n}(u_j)$ to turn the double poles into simple ones. It is easy to check that such a subtraction gives additional contributions to the integrand, but they turn out to be analytic within the integration contours either in $z_1$, or in $z_2$, or both variables at once. As in the case of $m = 1$, we will integrate over the variables $z_1$ and $z_2$, taking the residues at the points $z_j = u_k$.

We can replace

$$\det_2\left(\frac{1}{\sinh(z_k-u_j)}\right) \longrightarrow \frac{2}{\sinh(z_1-u_1)\sinh(z_2-u_2)}, \tag{11.135}$$

due to the symmetry of the integrand with respect to $z_1$ and $z_2$. Then, the integrand has only simple poles at $z_1-u_1$ and $z_2-u_2$ outside the integration contours. Integrating over the $z$-variables we arrive at

$$\Lambda_2(n) = -\frac{(\kappa-1)^2}{8\pi^2}\int_{-q}^{q}\mathrm{d}u_1\,\mathrm{d}u_2\det_2\left(\frac{\varphi^n(u_k)\varphi^{-n}(u_j)-1}{\sinh(u_k-u_j)}\right). \tag{11.136}$$

It is convenient to represent the integral (11.136) as the sum of two terms: $\Lambda_2(n) = \Lambda_2^{(1)}(n) + \Lambda_2^{(2)}(n)$. Here

$$\Lambda_2^{(1)}(n) = \frac{(\kappa-1)^2 n^2}{8\pi^2} \int_{-q}^{q} p_0'(u_1) p_0'(u_2)\, du_1\, du_2 = \frac{(\kappa-1)^2 n^2 k_F^2}{2\pi^2}, \quad (11.137)$$

and

$$\Lambda_2^{(2)}(n) = \frac{(\kappa-1)^2}{8\pi^2} \int_{-q}^{q} \frac{\varphi^n(u_2)\varphi^{-n}(u_1) + \varphi^n(u_1)\varphi^{-n}(u_2) - 2}{\sinh^2(u_1-u_2)}\, du_1\, du_2. \quad (11.138)$$

The integral (11.138) can be simplified if we immediately proceed to the second lattice derivative. Taking into account the symmetry of the integrand over $u_1$ and $u_2$, we find

$$D_n^2 \Lambda_2^{(2)}(n) = \frac{(\kappa-1)^2}{4\pi^2} \int_{-\infty}^{\infty} \frac{\varphi^{n-1}(u_1)}{\varphi^{n-1}(u_2)} \left( \frac{\varphi(u_1)}{\varphi(u_2)} - 1 \right)^2 \frac{du_1\, du_2}{\sinh^2(u_1-u_2)}. \quad (11.139)$$

It is easy to check that

$$\frac{\varphi(u_1)}{\varphi(u_2)} - 1 = \frac{i \sinh(u_1-u_2)}{\sinh(u_1+\frac{i\pi}{4})\sinh(u_2-\frac{i\pi}{4})}. \quad (11.140)$$

Substituting this expression into (11.139), we obtain

$$\begin{aligned} D_n^2 \Lambda_2^{(2)}(n) &= -\frac{(\kappa-1)^2}{4\pi^2} \left| \int_{-q}^{q} \frac{\sinh^{n-1}(u-\frac{i\pi}{4})}{\sinh^{n+1}(u+\frac{i\pi}{4})}\, du \right|^2 \\ &= -\frac{(\kappa-1)^2}{4\pi^2} \left| \int_{-q}^{q} e^{inp_0(u)}\, dp_0(u) \right|^2 = -\frac{(\kappa-1)^2 \sin^2(nk_F)}{\pi^2 n^2}. \end{aligned} \quad (11.141)$$

Thus, we have

$$2D_n^2 \frac{d^2}{d\kappa^2} \Lambda_2(n) \Big|_{\kappa=1} = \frac{4k_F^2}{\pi^2} - \frac{4\sin^2(nk_F)}{\pi^2 n^2}. \quad (11.142)$$

Substituting all the contributions into (11.129), we finally arrive at

$$\langle \sigma_1^z \sigma_{n+1}^z \rangle = \left( \frac{2k_F}{\pi} - 1 \right)^2 - \frac{4\sin^2(nk_F)}{\pi^2 n^2}. \quad (11.143)$$

This is an isotropic case of the result obtained in [Th. Niemeijer (1967)] for the $\sigma^z$-correlation function in the $XY$ chain.

In the case of zero magnetic field, we have $D = 1/2$ (see (10.149)). This means that $k_F = \pi/2$. Then the correlation function $\langle \sigma_1^z \sigma_{n+1}^z \rangle$ simplifies

$$\langle \sigma_1^z \sigma_{n+1}^z \rangle = 2\frac{(-1)^n - 1}{\pi^2 n^2}. \quad (11.144)$$

Thus, we have obtained an explicit and relatively simple result for the correlation function of the third components of spin in the $XXZ$ chain for

$\Delta = 0$. This simplicity is mainly because the model is equivalent to free fermions. Note, however, that other correlation functions are already more complicated. For example, a two-point function $\langle \sigma_1^+ \sigma_{n+1}^- \rangle$ is expressed in terms of the determinant of an $n \times n$ matrix. If we add the dependence of local spin operators on time, then the answer for such a function is given in the form of a Fredholm determinant (see, e.g., [F. Colomo et al (1993)] and references therein).

## 11.7 Emptiness formation probability

A particular case of the generating function $Q_n^{(\kappa)}$ gives one more correlation function called *emptiness formation probability* (EFP) [V. E. Korepin et al (1994)]. This term is used because, in the model of the one-dimensional Bose gas, an analogous correlation function does have the sense of the probability of the absence of particles in an interval of a fixed length. A similar meaning can be given to this correlation in the $XXZ$ chain if we introduce the notion of a *quasi-particle.*

We assume that there is a quasi-particle in the $k$th site if the spin is directed downward at this site. If the spin is directed upwards, then there is no quasi-particle at this site. Then the vacuum vector $|0\rangle$ is a state without quasi-particles. An on-shell vector $B(\bar{u})|0\rangle$ with $\#\bar{u} = N$ is a state with $N$ quasi-particles. We should only remember that the positions of these quasi-particles are not fixed. The vector $B(\bar{u})|0\rangle$ is a linear combination of states, in each of which the quasi-particles occupy fixed sites $j_1, \dots, j_N$. The sum in this linear combination runs through all possible sites (see (4.38)).

Consider the limit of the operator $\mathcal{Q}_{1,n}^{(\kappa)}$ (11.16) at $\kappa \to 0$. We have

$$\lim_{\kappa \to 0} \mathcal{Q}_{1,n}^{(\kappa)} = \mathcal{Q}_{1,n}^{(0)} = \prod_{k=1}^{n} E_k^{11}. \tag{11.145}$$

In other words, the operator $\mathcal{Q}_{1,n}^{(0)}$ is a projector onto the states in which there are no quasiparticles at sites with numbers $1, \dots, n$. Therefore, if $B(\bar{u})|0\rangle$ is a ground state, then the average value

$$P_n = \mathrm{Lim}\, \frac{\langle 0|C(\bar{u}) \mathcal{Q}_{1,n}^{(0)} B(\bar{u})|0\rangle}{\langle 0|C(\bar{u}) B(\bar{u})|0\rangle}, \tag{11.146}$$

has the meaning of the emptiness formation probability (in the sense of the absence of quasi-particles) in the ground state in an interval of length $n$.

In the standard language, $P_n$ is the probability of finding a cluster of $n$ particles with spins up in the ground state. If the external magnetic field

$\mu = 0$, then the ground state is symmetric with respect to replacing the spins up with the spins down. Hence, the probability of detecting a cluster of $n$ particles with spins down[2] also is equal to $P_n$. Therefore, apart from the limit $\kappa \to 0$, we can also consider the limit $\kappa \to \infty$

$$\lim_{\kappa\to\infty} \kappa^{-n} \mathcal{Q}_{1,n}^{(\kappa)} = \prod_{k=1}^{n} E_k^{22}, \tag{11.147}$$

when the operator $\mathcal{Q}_{1,n}^{(\kappa)}$ turns into a projector onto states with spins down at sites with numbers $1, \dots, n$. Hence, we conclude that the EFP arises in two limiting cases of the generating function $Q_n^{(\kappa)}$:

$$P_n = \lim_{\kappa\to 0} Q_n^{(\kappa)} = \lim_{\kappa\to\infty} \kappa^{-n} Q_n^{(\kappa)}, \qquad \mu = 0. \tag{11.148}$$

We now turn to the integral representation (11.77) for the generating function $Q_n^{(\kappa)}$ at zero magnetic field. In this integral representation, the parameter $\kappa$ enters only the matrix $\mathcal{M}^{(\kappa)}(\bar{z}, \bar{u})$ (11.51). Obviously, for a fixed $m$, the determinant of this matrix is a polynomial in $\kappa$ of degree $m$. Therefore, if we use the second formula (11.148) (where $\kappa \to \infty$), there will be only one term with $m = n$ in the whole sum over $m$. It is easy to see that

$$\lim_{\kappa\to\infty} \kappa^{-n} \det_n \mathcal{M}_{jk}^{(\kappa)}(\bar{z}, \bar{u}) = \left( \prod_{a,b=1}^{n} \frac{\sinh(u_a - z_b - i\zeta)}{\sinh(z_b - u_a - i\zeta)} \right) \det_n t(u_k, z_j). \tag{11.149}$$

Then we obtain

$$\begin{aligned} P_n = \oint \frac{d\bar{z}}{(2\pi \sin\zeta)^n (n!)^2} \int_{-\infty}^{\infty} d\bar{u} \left( \prod_{j=1}^{n} \varphi^n(z_j) \varphi^{-n}(u_j) \right) \det_n t(u_k, z_j) \\ \times \left( \prod_{a,b=1}^{n} \frac{\sinh^2(u_a - z_b - i\zeta)}{\sinh(z_a - z_b - i\zeta) \sinh(u_a - u_b - i\zeta)} \right) \\ \times \det_n \left[ \frac{i}{2\zeta \sinh \frac{\pi}{\zeta}(u_j - z_k)} \right]. \end{aligned} \tag{11.150}$$

In the resulting representation, we can get rid of the integrals over $z$-variables if we artificially introduce inhomogeneities. Namely, we make

[2]In the language of quasi-particles, such a probability could be called the completeness formation probability.

the following change of the functions $\varphi^n(z)$ and $\varphi^{-n}(u)$:

$$\begin{aligned}\varphi^n(z) &= \left(\frac{\sinh(z-\frac{i\zeta}{2})}{\sinh(z+\frac{i\zeta}{2})}\right)^n \to \prod_{k=1}^{n} \frac{\sinh(z-\xi_k-i\zeta)}{\sinh(z-\xi_k)},\\ \varphi^{-n}(u) &= \left(\frac{\sinh(u+\frac{i\zeta}{2})}{\sinh(u-\frac{i\zeta}{2})}\right)^n \to \prod_{k=1}^{n} \frac{\sinh(u-\xi_k)}{\sinh(u-\xi_k-i\zeta)}.\end{aligned} \tag{11.151}$$

Here $\{\xi_1,\dots,\xi_n\}$ is a set of parameters belonging to a neighborhood of the point $-i\zeta/2$. It is clear that the original functions $\varphi^n(z)$ and $\varphi^{-n}(u)$ appear in the limit $\xi_k \to -i\zeta/2$. We denote by $P_n(\bar{\xi})$ the EFP modified in this way. Then

$$\begin{aligned}P_n(\bar{\xi}) &= \oint \frac{d\bar{z}}{(2\pi\sin\zeta)^n(n!)^2} \int_{-\infty}^{\infty} d\bar{u}\, \det_n\left[\frac{i}{2\zeta\sinh\frac{\pi}{\zeta}(u_j-z_k)}\right]\\ &\quad\times \det_n t(u_k,z_j) \prod_{a,b=1}^{n} \frac{\sinh(z_a-\xi_b-i\zeta)\sinh(u_a-\xi_b)}{\sinh(z_a-\xi_b)\sinh(u_a-\xi_b-i\zeta)}\\ &\quad\times \prod_{a,b=1}^{n} \frac{\sinh^2(u_a-z_b-i\zeta)}{\sinh(z_a-z_b-i\zeta)\sinh(u_a-u_b-i\zeta)}.\end{aligned} \tag{11.152}$$

We now can easily calculate the $n$-fold integral over $z_j$. Since the number of integration variables is equal to the number of poles, the integral is equal to the sum of the residues at the points $z_j = \xi_{p(j)}$, where $p$ is any permutation in the set $\{\xi_1,\dots,\xi_n\}$. However, because of the symmetry of the integrand over $z_j$, we can take the residues only at the points $z_j = \xi_j$ and multiply the resulting contribution by $n!$. As a result, we obtain a new representation that no longer contains contour integrals

$$\begin{aligned}P_n(\bar{\xi}) &= \frac{1}{(-i\sin\zeta)^n n!}\left(\prod_{\substack{a,b=1\\a\neq b}}^{n} \frac{1}{\sinh(\xi_a-\xi_b)}\right)\int_{-\infty}^{\infty} d\bar{u}\\ &\quad\times \det_n\left[\frac{i}{2\zeta\sinh\frac{\pi}{\zeta}(u_j-\xi_k)}\right] \det_n\left[\frac{-\sin^2\zeta}{\sinh(u_j-\xi_k)\sinh(u_j-\xi_k-i\zeta)}\right]\\ &\quad\times \prod_{a,b=1}^{n} \frac{\sinh(u_a-\xi_b-i\zeta)\sinh(u_a-\xi_b)}{\sinh(u_a-u_b-i\zeta)}.\end{aligned} \tag{11.153}$$

Thus, we computed the integrals over the $z_j$-variables, but this did not happen for free. Now, to calculate the EFP, we will have to take the limit $\xi_k \to -i\zeta/2$. This procedure is far from trivial.

**Remark 11.4.** One can also introduce the inhomogeneities $\xi_k$ in the integral representation for the generating function $Q_n^{(\kappa)}$. This allows us to compute the integrals over the variables $z_j$. However, the number of integration variables is generally less than the number of poles in this case. Therefore, the result of integration will give a sum over partitions of the set $\bar{\xi}$, but not one term, as in the EFP case.

**Remark 11.5.** The representation for the EFP appeared in the limit of the generating function $Q_n^{(\kappa)}$ at $\kappa \to \infty$. However, the integral representation (11.153) can be derived more simply. Using the formula (11.147) and the formulas of the quantum inverse problem, one can easily convince himself that

$$P_n(\bar{\xi}) = \mathrm{Lim}\left(\prod_{k=1}^{n} \frac{1}{\tau(\xi_k|\bar{u})}\right) \frac{\langle 0|C(\bar{u})\left(\prod_{k=1}^{n} D(\xi_k)\right)B(\bar{u})|0\rangle}{\langle 0|C(\bar{u})B(\bar{u})|0\rangle}. \tag{11.154}$$

A multiple action of the operators $D(\xi_k)$ on a vector Bethe $B(\bar{u})|0\rangle$ is given by (7.63). Using this formula, we can obtain multiple integral representation (11.152) for the thermodynamic limit of the EFP, similarly to how this was with the generating function $Q_{n;N}^{(\kappa)}$. An interested reader can see this for himself.

### 11.7.1 *Emptiness formation probability at* $\Delta = 1/2$

For the value of the anisotropy parameter $\Delta = 1/2$ (that is, $\zeta = \pi/3$), one can explicitly calculate all the integrals over $u_j$ in the representation (11.153) [N. Kitanine et al (2002b)]. Let us rewrite equation (11.153) for $\zeta = \pi/3$:

$$\begin{aligned} P_n(\bar{\xi}) = & \frac{1}{(-i\sin\frac{\pi}{3})^n n!}\left(\prod_{\substack{a,b=1\\ a\neq b}}^{n} \frac{1}{\sinh(\xi_a-\xi_b)}\right)\int_{-\infty}^{\infty} d\bar{u} \\ & \times \det_n\left[\frac{3i}{2\pi\sinh 3(u_j-\xi_k)}\right] \det_n\left[\frac{-\sin^2\frac{\pi}{3}}{\sinh(u_j-\xi_k)\sinh(u_j-\xi_k-\frac{i\pi}{3})}\right] \\ & \times \prod_{a,b=1}^{n} \frac{\sinh(u_a-\xi_b-\frac{i\pi}{3})\sinh(u_a-\xi_b)}{\sinh(u_a-u_b-\frac{i\pi}{3})}. \end{aligned} \tag{11.155}$$

The main simplification of the integrand at $\zeta = \pi/3$ is based on an elementary formula

$$\sinh 3x = 4\sinh x \sinh(x+\tfrac{i\pi}{3})\sinh(x-\tfrac{i\pi}{3}). \tag{11.156}$$

We also use the fact that one of the determinants in (11.155) is explicitly computed via (7.77)

$$\det_n\left[\frac{3i}{2\pi\sinh 3(u_j-\xi_k)}\right]=\left(\frac{3i}{2\pi}\right)^n\frac{\prod_{j>k}^n\sinh 3(u_j-u_k)\sinh 3(\xi_k-\xi_j)}{\prod_{j,k=1}^n\sinh 3(u_j-\xi_k)}. \tag{11.157}$$

Then after simple algebra, the integral representation for $P_n(\bar{\xi})$ takes the form

$$P_n(\bar{\xi})=\alpha_n(\bar{\xi})\int_{-\infty}^{\infty}du_1\dots du_n\ \det_n\left[\frac{1}{\sinh(u_j-\xi_k+\frac{i\pi}{3})}\right]$$
$$\times\det_n\left[\frac{1}{\sinh(u_j-\xi_k)\sinh(u_j-\xi_k-\frac{i\pi}{3})}\right], \tag{11.158}$$

where

$$\alpha_n(\bar{\xi})=\frac{(-1)^{n(n-1)/2}}{n!\ 2^{n^2}}\left(\frac{3i}{4\pi}\right)^n\prod_{\substack{j,k=1\\ j\neq k}}^n\frac{1}{\sinh(\xi_j-\xi_k)}\prod_{j>k}^n\frac{\sinh 3(\xi_k-\xi_j)}{\sinh(\xi_k-\xi_j)}. \tag{11.159}$$

Now the integrals over $u_j$ can be taken using the following lemma.

**Lemma 11.1.** *Let $F_j(u)$ and $G_j(u)$, $j=1,\dots,n$, be functions that are square integrable on an interval $[a,b]$ ($a$ and $b$ can be infinite). Then*

$$\int_a^b\det_n F_k(u_j)\det_n G_k(u_j)\,du_1\dots du_n=n!\det_n\left(\int_a^b F_j(u)G_k(u)\,du\right). \tag{11.160}$$

**Proof.** We have

$$\begin{aligned}
&\int_a^b\det_n F_k(u_j)\det_n G_k(u_j)\,du_1\dots du_n\\
&\quad=\sum_p(-1)^{[p]}\int_a^b\left(\prod_{j=1}^n F_j(u_{p(j)})\,du_j\right)\det_n G_k(u_j)\\
&\quad=\sum_p(-1)^{[p]}\int_a^b\left(\prod_{j=1}^n F_j(u_j)\,du_j\right)\det_n G_k(u_{p^{-1}(j)})\\
&\quad=\sum_p(-1)^{[p]}\left(\prod_{j=1}^n F_j(u_j)\,du_j\right)(-1)^{[p]}\det_n G_k(u_j)\\
&\quad=n!\int_a^b\left(\prod_{j=1}^n F_j(u_j)\,du_j\right)\det_n G_k(u_j).
\end{aligned} \tag{11.161}$$

Let us explain the transformations. First, we write down the first of the two determinants according to the definition. Further, in each term of the sum over permutations, the integration variables were changed $u_{p(j)} \to u_j$. With such a replacement, the rows of the second determinant rearrange. Returning these rows to the initial positions, we get the additional sign $(-1)^{[p]}$, which cancels with the original $(-1)^{[p]}$. Since after this, nothing depends on the permutation, the sum of these permutations gives $n!$. In fact, we proved that each term of the sum over the permutations gives the same contribution as the product of the diagonal terms of the matrix $F_k(u_j)$ since the original integrand was symmetric with respect to the variables $u_1, \dots, u_n$.

Let us continue our transformations:

$$n! \int_a^b \left( \prod_{j=1}^n F_j(u_j)\, du_j \right) \det_n G_k(u_j)$$

$$= n! \int_a^b \det_n \Big( F_j(u_j) G_k(u_j)\, du_j \Big) = n! \det_n \left( \int_a^b F_j(u_j) G_k(u_j)\, du_j \right). \tag{11.162}$$

Here we have multiplied the $j$th row of the matrix $G_k(u_j)$ by the factor $F_j(u_j)\, du_j$. Now each integration variable $u_j$ enters only in the $j$th row. Therefore, one can perform the integration separately in every row. It remains to remove the subscript $j$ from each integration variable, and we arrive at the assertion of the lemma. □

In our case,

$$\begin{aligned} F_k(u_j) &= \frac{1}{\sinh(u_j - \xi_k)\sinh(u_j - \xi_k - \frac{i\pi}{3})}, \\ G_k(u_j) &= \frac{1}{\sinh(u_j - \xi_k + \frac{i\pi}{3})}, \end{aligned} \tag{11.163}$$

and the integration contour $[a, b]$ is the real axis $\mathbb{R}$. Thus, we should calculate the integral

$$J = \int_{-\infty}^{\infty} \frac{du}{\sinh(u - \xi_j)\sinh(u - \xi_j - \frac{i\pi}{3})\sinh(u - \xi_k + \frac{i\pi}{3})}. \tag{11.164}$$

The value of this integral depends on the imaginary parts of the inhomogeneities $\xi_j$. Therefore it is convenient to put $\xi_j = \epsilon_j - \frac{i\pi}{6}$. Then we can assume that $\Im \epsilon_j = 0$ and $\epsilon_j \to 0$ in the homogeneous limit. The integral takes the form

$$J = \int_{-\infty}^{\infty} \frac{du}{\sinh(u - \epsilon_j + \frac{i\pi}{6})\sinh(u - \epsilon_j - \frac{i\pi}{6})\sinh(u - \epsilon_k + \frac{i\pi}{2})}. \tag{11.165}$$

The result of the integration gives us

$$J = -4\pi i \, \frac{\sinh \frac{1}{2}(\epsilon_j - \epsilon_k)}{\sinh \frac{3}{2}(\epsilon_j - \epsilon_k)}, \tag{11.166}$$

and hence,

$$\begin{aligned} P_n(\bar{\epsilon}) = {} & \frac{(-1)^{n(n-1)/2} 3^n}{2^{n^2}} \prod_{j>k}^{n} \frac{\sinh 3(\epsilon_k - \epsilon_j)}{\sinh(\epsilon_k - \epsilon_j)} \\ & \times \prod_{\substack{j,k=1 \\ j \neq k}}^{n} \frac{1}{\sinh(\epsilon_j - \epsilon_k)} \det_n \left[ \frac{\sinh \frac{1}{2}(\epsilon_j - \epsilon_k)}{\sinh \frac{3}{2}(\epsilon_j - \epsilon_k)} \right]. \end{aligned} \tag{11.167}$$

All the integrals are already taken in the resulting representation, and we only need to proceed to the homogeneous limit $\epsilon_j \to 0$. To compute the limit, we could use the following statement.

**Proposition 11.2.** *For any sufficiently smooth function of two variables* $\Phi(x, y)$

$$\begin{aligned} & \lim_{\substack{x_1, \ldots, x_n \to x \\ y_1, \ldots, y_n \to y}} \frac{\det_n \Phi(x_j, y_k)}{\prod_{j<k}^{n} \sinh(x_j - x_k) \sinh(y_j - y_k)} \\ & = \left( \prod_{k=1}^{n-1} \frac{1}{k!} \right)^2 \det_n \left( \frac{\partial^{j+k-2} \Phi(x, y)}{\partial x^{j-1} \partial y^{k-1}} \right). \end{aligned} \tag{11.168}$$

*In this case, the words 'sufficiently smooth' mean the existence of all partial derivatives in the rhs of* (11.168).

We suggest that the reader prove proposition 11.2 or at least check its validity for small $n$ (such checking will suggest the idea of the proof).

We will only use the fact that the limit in (11.168) does not depend on how exactly the parameters $x_j$ and $y_k$ tend to their limit values. Hence, one can choose quite a specific way of taking the limit; the result does not depend on it. Therefore, we first introduce an additional set of variables $\{\epsilon'_1, \ldots, \epsilon'_n\}$. Then

$$\begin{aligned} & \lim_{\epsilon_j \to 0} \prod_{\substack{j,k=1 \\ j \neq k}}^{n} \frac{1}{\sinh(\epsilon_j - \epsilon_k)} \det_n \left[ \frac{\sinh \frac{1}{2}(\epsilon_j - \epsilon_k)}{\sinh \frac{3}{2}(\epsilon_j - \epsilon_k)} \right] \\ & = \lim_{\substack{\epsilon_j \to 0 \\ \epsilon'_k \to 0}} \prod_{j>k}^{n} \frac{1}{\sinh(\epsilon_j - \epsilon_k) \sinh(\epsilon'_k - \epsilon'_j)} \det_n \left[ \frac{\sinh \frac{1}{2}(\epsilon_j - \epsilon'_k)}{\sinh \frac{3}{2}(\epsilon_j - \epsilon'_k)} \right]. \end{aligned} \tag{11.169}$$

Indeed, due to proposition 11.2, the limit in the rhs of (11.169) exists and does not depend on how the parameters $\epsilon_j$ and $\epsilon'_j$ approach their limiting values. Since they all go to zero, we can put $\epsilon_j = \epsilon'_j$ before taking the limit, which gives us the lhs of (11.169).

Using the auxiliary set of variables $\{\epsilon'_1, \dots, \epsilon'_n\}$, we can write down the representation for the EFP as follows:

$$P_n = \frac{3^n}{2^{n^2}} \lim_{\substack{\epsilon_j \to 0 \\ \epsilon'_k \to 0}} \prod_{j>k}^{n} \frac{\sinh 3(\epsilon_k - \epsilon_j)}{\sinh(\epsilon_k - \epsilon_j)} \frac{1}{\sinh(\epsilon_j - \epsilon_k)\sinh(\epsilon'_j - \epsilon'_k)}$$
$$\times \det_n \left[ \frac{\sinh \frac{1}{2}(\epsilon_j - \epsilon'_k)}{\sinh \frac{3}{2}(\epsilon_j - \epsilon'_k)} \right]. \tag{11.170}$$

Now we can set $\epsilon_j = 2\epsilon j$, $\epsilon'_k = 2\epsilon(1-k)$ and consider the limit $\epsilon \to 0$. Then we find

$$P_n = (-1)^{n(n-1)/2} 3^{n(n+1)/2} 2^{-n^2}$$
$$\times \lim_{\epsilon \to 0} \left( \prod_{j>k}^{n} \frac{1}{\sinh^2 2\epsilon(j-k)} \right) \det_n \left[ \frac{\sinh \epsilon(j+k-1)}{\sinh 3\epsilon(j+k-1)} \right]. \tag{11.171}$$

It turns out that the determinant in the rhs of equation (11.171) can be explicitly calculated [G. Kuperberg (1996)]. However, for this, we have to prove an auxiliary lemma.

**Lemma 11.2.** *For any complex $\alpha$ and $\beta$,*

$$\det_n \left( \frac{\sinh \alpha(j+k-1)}{\sinh \beta(j+k-1)} \right)$$
$$= 2^{n^2-n} \left( \prod_{j>k}^{n} \sinh^2 \beta(j-k) \right) \left( \prod_{j,k=1}^{n} \frac{\sinh(\alpha + \beta(j-k))}{\sinh \beta(j+k-1)} \right). \tag{11.172}$$

**Proof.** We transform the determinant in (11.172) as follows:

$$\det_n \left( \frac{\sinh \alpha(j+k-1)}{\sinh \beta(j+k-1)} \right) = \det_n \left( \frac{e^{\alpha(j+k-1)} - e^{-\alpha(j+k-1)}}{2\sinh \beta(j+k-1)} \right)$$
$$= \left( \prod_{j=1}^{n} e^{-\alpha(j-1)} \right) \left( \prod_{k=1}^{n} e^{-\alpha k} \right) \det_n \left( \frac{e^{2\alpha(j+k-1)} - 1}{2\sinh \beta(j+k-1)} \right)$$
$$= e^{-\alpha n^2} \det_n \left( \frac{e^{2\alpha(j+k-1)} - 1}{2\sinh \beta(j+k-1)} \right). \tag{11.173}$$

The determinant in the rhs of (11.173) is a polynomial in $e^{2\alpha}$. To find this polynomial degree, one should neglect the term $-1$ in the combination $e^{2\alpha(j+k-1)}-1$. After this, we immediately find that the polynomial degree is $n^2$. Thus,

$$\det_n\left(\frac{\sinh\alpha(j+k-1)}{\sinh\beta(j+k-1)}\right)=e^{-\alpha n^2}\,\mathfrak{P}_{n^2}\left(e^{2\alpha}\right). \tag{11.174}$$

To find a polynomial, it is enough to find its zeros and the coefficient of the leading term $e^{2\alpha n^2}$. The zeros can be easily found. First, for $\alpha=0$, all matrix elements of the matrix in the lhs of (11.174) vanish. We are dealing with a matrix of zero rank. Therefore, for $\alpha=0$, the determinant has zero of order $n$.

If $\alpha=\beta$, then all the matrix elements are equal to 1, and we obtain a matrix of the rank 1. Therefore, for $\alpha=\beta$, the determinant has zero of order $n-1$. Similarly, the determinant has zero of order $n-1$ for $\alpha=-\beta$.

It is easy to see that for $\alpha=\pm 2\beta$, we obtain a matrix of the rank 2. Therefore, the determinant has zeros of order $n-2$. And so on up to $\alpha=\pm(n-1)\beta$, where the determinant has first order zeros. If we now sum up the orders of all the found zeros, then we obtain $n^2$. Hence, there are no other zeros.

It remains to find the coefficient of the leading term $e^{2\alpha n^2}$. To do this, we send $\alpha\to+\infty$ in (11.174) and calculate the coefficient of $e^{\alpha n^2}$. It is easy to see that this coefficient is

$$2^{-n}\det_n\left(\frac{1}{\sinh\beta(j+k-1)}\right). \tag{11.175}$$

This is a special case of the Cauchy determinant calculated in corollary 7.1. Indeed, setting in (7.77) $x_j=j\beta$ and $y_k=(1-k)\beta$, we arrive at (11.175). Therefore,

$$2^{-n}\det_n\left(\frac{1}{\sinh\beta(j+k-1)}\right)=2^{-n}\frac{\prod_{j>k}^n\sinh^2\beta(j-k)}{\prod_{j,k=1}^n\sinh\beta(j+k-1)}. \tag{11.176}$$

Thus, we arrive at

$$\begin{aligned}\det_n\left(\frac{\sinh\alpha(j+k-1)}{\sinh\beta(j+k-1)}\right)&=e^{-\alpha n^2}\,2^{-n}\frac{\prod_{j>k}^n\sinh^2\beta(j-k)}{\prod_{j,k=1}^n\sinh\beta(j+k-1)}\\&\quad\times\left(e^{2\alpha}-1\right)^n\prod_{m=1}^{n-1}\left(e^{2\alpha}-e^{2m\beta}\right)^{n-m}\\&\quad\times\left(e^{2\alpha}-e^{-2m\beta}\right)^{n-m}.\end{aligned} \tag{11.177}$$

A straightforward calculation shows that this coincides with (11.172). □

We can now apply lemma 11.2 to calculating the determinant (11.171). For this, it is enough to set $\alpha = \epsilon$ and $\beta = 3\epsilon$ in (11.172). We obtain

$$P_n = (-1)^{n(n-1)/2} 3^{n(n+1)/2} 2^{-n} \lim_{\epsilon \to 0} \prod_{j<k}^{n} \frac{\sinh^2 3\epsilon(j-k)}{\sinh^2 2\epsilon(j-k)} \times \prod_{j,k=1}^{n} \frac{\sinh \epsilon(1+3(j-k))}{\sinh 3\epsilon(j+k-1)}. \tag{11.178}$$

The limit $\epsilon \to 0$ becomes trivial, and we arrive at

$$P_n = \frac{(-3)^{\frac{n^2-n}{2}}}{2^{n^2}} \prod_{j,k=1}^{n} \frac{3(j-k)+1}{j+k-1}. \tag{11.179}$$

We provide the reader with the opportunity to complete the calculation and obtain

$$P_n = \frac{\mathcal{A}_n}{2^{n^2}}, \tag{11.180}$$

where

$$\mathcal{A}_n = \prod_{k=0}^{n-1} \frac{(3k+1)!}{(n+k)!}. \tag{11.181}$$

It is noteworthy that the number $\mathcal{A}_n$ is an integer and is well known in combinatorics. This is the number of the *alternating sign matrices* of the size $n \times n$ (see [W. H. Mills et al (1982); D. Zeilberger (1996); D. Bressoud (1999)]). We refer the reader to works [G. Kuperberg (1996); A. V. Razumov and Yu. G. Stroganov (2001); A. V. Razumov, Yu. G. Stroganov (2006a,b); Yu. G. Stroganov (2006); R. Behrend et al (2012)] for a more detailed description of the relationship between alternating sign matrices and the $XXZ$ spin chain at $\Delta = 1/2$. We also suggest that the reader prove that representation (11.180) for the EFP is equivalent to the following formula:

$$P_n = \left(\frac{\sqrt{3}}{2}\right)^{3n^2} \prod_{k=1}^{n} \frac{\Gamma(k+\frac{1}{3})\Gamma(k-\frac{1}{3})}{\Gamma(k+\frac{1}{2})\Gamma(k-\frac{1}{2})}. \tag{11.182}$$

The product of the $\Gamma$-functions is calculated in terms of the Barnes $G$-function (see Section A.2), and we arrive at

$$P_n = \left(\frac{\sqrt{3}}{2}\right)^{3n^2} \frac{G(\frac{3}{2})G(\frac{1}{2})G(n+\frac{4}{3})G(n+\frac{2}{3})}{G(\frac{4}{3})G(\frac{2}{3})G(n+\frac{3}{2})G(n+\frac{1}{2})}. \tag{11.183}$$

Using (A.27) for the asymptotic behavior of the Barnes $G$-function at $n \to \infty$, we obtain the asymptotics of the EFP

$$P_n = \left(\frac{\sqrt{3}}{2}\right)^{3n^2} n^{-5/36} \frac{G(\frac{3}{2})G(\frac{1}{2})}{G(\frac{4}{3})G(\frac{2}{3})}\bigl(1+O(n^{-1})\bigr), \qquad n \to \infty. \quad (11.184)$$

Thus, we obtained an exact result for the EFP at the value of the anisotropy parameter $\Delta = 1/2$. First, this is due to the complete factorization of the $n$-fold integral over the variables $u_j$ in representation (11.153). However, more important is that we were able to compute the homogeneous limit in the resulting expression. As we have already noted, factorization of multiple integral representations in an inhomogeneous chain is possible for any $\Delta$. At the same time, the calculation of the homogeneous limit is not straightforward.

## Chapter 12

# Asymptotics of correlation functions via form factor expansion

In Chapter 11, we already used the method of calculating correlation functions via form factor expansion. For this, we presented the sum over the roots of the Bethe equations as a multiple contour integral. This chapter uses the form factor expansion for calculating long-distance and large-time asymptotics of correlation functions.

The calculation of asymptotics using form factors is widely used in quantum integrable models in which the Hamiltonian has a gap in the spectrum (massive models). Form factors in quantum integrable massive models have been calculated in numerous works [M. Karowski and P. Weisz (1978); M. Karowski (1979); F. A. Smirnov (1990, 1992); A. Fring et al (1990); G. Mussardo (1992); A. Koubek and G. Mussardo (1993); C. Ahn et al (1993); S. Lukyanov (1997); V. Brazhnikov and S. Lukyanov (1998); M. Jimbo and T. Miwa (1995)]. The algebraic Bethe ansatz was used to calculate correlation functions of the $XXZ$ chain in the massive phase [A. G. Izergin et al (1999); M. Dugave et al (2015a,b, 2016); C. Babenko et al (2021a,b)].

In massive models, the correlation functions decrease exponentially. To find the exponential decay rate, it is usually sufficient to calculate the form factor corresponding to the first excited state. However, in the one-dimensional repulsive Bose gas, the situation is different. We saw in Section 10.3.3 that excitations at the Fermi boundaries have zero energy in the thermodynamic limit. Therefore, the one-dimensional Bose gas at zero temperature is critical. The $XXZ$ chain has the same property when $-1 < \Delta \leq 1$. The asymptotic behavior of correlation functions in such models is given by critical exponents, determining the power law decay.

Predictions concerning these exponents were obtained via the Luttinger liquid approach [A. Luther and I. Peschel (1975); F. D. M. Haldane (1980,

1981a,b); V. V. Cheianov and M. Pustilnik (2008); A. Imambekov and L. I. Glazman (2008, 2009a,b)] or the connection to conformal field theory (CFT) [I. Affleck (1985); H. W. J. Blöte et al (1986); J. L. Cardy (1984, 1986); A. M. Polyakov (1970)], together with exact computations of finite-size corrections to the spectrum [H. J. de Vega and F. Woynarovich (1985); C. Destri and H. J. de Vega (1988, 1995); F. Woynarovich (1987); F. Woynarovich et al (1989)]. However, it was impossible to confirm these (indirect) predictions via direct computations based on exact results for an integrable model until recently. The algebraic Bethe ansatz allows us to find the critical exponents via straightforward summation of the form factor series.

In critical models, an infinite number of form factors contribute to the leading asymptotics of the correlation function. However, all these contributions can be taken into account and explicitly summed up. As a result, it turns out that, as in massive models, to find the leading critical exponent, it is enough to calculate the thermodynamic limit of just one form factor. This form factor corresponds to the excited state with minimal energy at a large but finite system size $L$. It turns out that in the thermodynamic limit, this form factor behaves as $L^{-\gamma/2}$, where $\gamma$ is the desired critical exponent.

The thermodynamic limit of form factors in the one-dimensional Bose gas and the $XXZ$ chain at zero temperature was studied in [N. A. Slavnov (1990); N. Kitanine et al (2009c, 2011a); K. K. Kozlowski (2011, 2017)]. Application of the form factor approach to the asymptotic analysis of the ground state correlation functions was performed in [N. Kitanine et al (2011b, 2012); K. K. Kozlowski and V. Terras (2011); N. Kitanine et al (2014); K. K. Kozlowski (2015); K. K. Kozlowski and J. M. Maillet (2015); K. K. Kozlowski (2018a, 2019)]. Form factors at a finite temperature were considered in [M. Dugave et al (2013, 2014, 2015a,b, 2016); F. Göhmann et al (2006)].

We begin this chapter with a general description of the problem in Section 12.1. Section 12.2 defines a special generating function that allows us to calculate the two-point correlation function of densities in the one-dimensional Bose gas. In Section 12.3, we calculate the thermodynamic limit of a special form factor in which the excited state is a twisted ground state. Here we introduce the concept of the smooth and discrete parts of the form factor amplitude. We show that the discrete part behaves as some negative power of the system size $L$. It is this exponent of decreasing that will eventually turn out to be the critical exponent. In Section 12.4, we consider the form factors corresponding to the excited states with particles and holes. We show that the thermodynamic limit of these form factors strongly depends on the positions of the particles and holes with respect to the Fermi boundaries. Here we introduce the concept of critical excited states and the corresponding form factors. We argue that these states make the leading contribution to the asymptotics of the correlation function. In Section 12.5, we calculate the sum of the critical form factors and obtain the asymptotics of the generating function. Then we derive the long-distance asymptotics of the correlation function of densities in Section 12.6. The obtained result coincides with the predictions of the CFT and Luttinger liquid approaches. Section 12.7 is devoted to the correlation function of fields. We find the critical exponents by analyzing the form factors in the thermodynamic limit. Finally, in Section 12.8, we consider the long-distance and large-time asymptotics for a fixed ratio $x/t$. We show that in this case, the asymptotics of the correlation functions may contain contributions that are not predicted by the CFT and Luttinger liquid approaches.

## 12.1 General scheme of the form factor approach

In Section 11.5, we calculated the sum over the excited states, reducing it to a multiple contour integral. After that, we studied the thermodynamic limit of this integral. Now we will try to calculate the thermodynamic limit of individual form factors and then perform their summation. We will try to use the fact that excited states can be described in terms of particles and holes (excitations) in the thermodynamic limit. Thus, we can expect that in the thermodynamic limit, summation over the form factors can be replaced by either integration over the momenta of particles and holes or summation over the corresponding quantum numbers. Note that the description in terms of particles and holes is possible only for excited states whose energy remains finite in the thermodynamic limit. However, we assume that the contribution to the correlation functions from states with infinite energy vanishes in the thermodynamic limit. This assumption seems quite reasonable from a physical point of view.

We consider correlation functions in the one-dimensional Bose gas model. Many calculations of this chapter can also be performed without significant changes for the $XXZ$ ($XXX$) spin chains in the massless phase [N. Kitanine et al (2009c, 2011a,b)]. However, there are bound states in spin chains. The contribution of these states must be taken into account when calculating correlation functions. It turns out that this contribution compensates some additional contributions arising from the summation of particle-hole excitations [K. K. Kozlowski (2018a, 2019)]. There are no bound states in the one-dimensional Bose gas with a positive coupling constant, which simplifies the task.

We have seen that if in the thermodynamic limit of the one-dimensional Bose gas, the momenta of excitations tend to the Fermi boundary, then the energy of this excitation is zero (see (10.88)). This leads to the fact that when calculating the asymptotics of correlation functions, we have to consider the contribution of an infinite number of excited states. Another consequence of the continuity of the Hamiltonian spectrum is the highly non-trivial behavior of form factors in the thermodynamic limit. Let us briefly describe this behavior and the main stages of calculating the asymptotics of correlation functions in the form factor approach framework.

Let $\mathcal{O}(x,t)$ be a local operator that depends on the spatial coordinate $x$ and the time $t$. We consider a two-point correlation function of the operators $\mathcal{O}^\dagger(x_1,t_1)$ and $\mathcal{O}(x_2,t_2)$ and use its form factor expansion (11.7)

$$\begin{aligned}&\langle\!\langle \mathcal{O}^\dagger(x_1,t_1)\mathcal{O}(x_2,t_2)\rangle\!\rangle\\&\quad=\sum_{\bar v,N'}\frac{\langle \boldsymbol{\Phi}_N(\bar u)|\mathcal{O}^\dagger(x_1,t_1)|\boldsymbol{\Phi}_{N'}(\bar v)\rangle\langle \boldsymbol{\Phi}_{N'}(\bar v)|\mathcal{O}(x_2,t_2)|\boldsymbol{\Phi}_N(\bar u)\rangle}{\langle \boldsymbol{\Phi}_N(\bar u)|\boldsymbol{\Phi}_N(\bar u)\rangle\langle \boldsymbol{\Phi}_{N'}(\bar v)|\boldsymbol{\Phi}_{N'}(\bar v)\rangle}.\end{aligned}\tag{12.1}$$

The dependence on the spatial and time coordinates gives a simple phase factor, and we arrive at

$$\langle\!\langle \mathcal{O}^\dagger(x_1,t_1)\mathcal{O}(x_2,t_2)\rangle\!\rangle=\sum_{\bar v,N'}e^{i(x_1-x_2)\mathcal{P}_{\rm ex}-i(t_1-t_2)\mathcal{E}_{\rm ex}}|\boldsymbol{\mathcal{F}}_{\mathcal{O}}|^2,\tag{12.2}$$

where $\mathcal{E}_{\rm ex}$ and $\mathcal{P}_{\rm ex}$ respectively are the excitation energy and momentum. The coefficient $\boldsymbol{\mathcal{F}}_{\mathcal{O}}$ is the normalized form factor of the operator $\mathcal{O}(0,0)$

$$\boldsymbol{\mathcal{F}}_{\mathcal{O}}=\frac{\langle \boldsymbol{\Phi}_{N'}(\bar v)|\mathcal{O}(0,0)|\boldsymbol{\Phi}_N(\bar u)\rangle}{\sqrt{\langle \boldsymbol{\Phi}_N(\bar u)|\boldsymbol{\Phi}_N(\bar u)\rangle\langle \boldsymbol{\Phi}_{N'}(\bar v)|\boldsymbol{\Phi}_{N'}(\bar v)\rangle}}.\tag{12.3}$$

We will call the quantity $|\boldsymbol{\mathcal{F}}_{\mathcal{O}}|^2$ the *amplitude* of the form factor of the operator $\mathcal{O}$.

The question of convergence of the form factor expansion (12.2) is very complex, and it has not been studied to date. We will assume that the series (12.3) absolutely converges if we shift the time $t=t_1-t_2$ to the lower half-plane: $t\to t-i0$. Since $\mathcal{E}_{\rm ex}>0$, this shift makes the phase factor decrease with increasing excitation energy.

Our first task is to calculate the thermodynamic limit of the form factor when $|\boldsymbol{\Phi}_N(\bar u)\rangle$ goes to the ground state and $\langle \boldsymbol{\Phi}_{N'}(\bar v)|$ goes to the excited state with $n$ particles and $n$ holes. For this, we can use explicit determinant formulas obtained via the algebraic Bethe ansatz. It turns out that in the limit $N,L\to\infty$, $N/L=D$, the form factor amplitude $|\boldsymbol{\mathcal{F}}_{\mathcal{O}}|^2$ behaves as a negative power of the system size $L$. More precisely,

$$\mathrm{Lim}\,|\boldsymbol{\mathcal{F}}_{\mathcal{O}}|^2L^{\gamma_{\mathcal{O}}(\{\lambda_p,\lambda_h\})}=\mathfrak{A}^{(\rm fin)}_{\mathcal{O}}(\{\lambda_p,\lambda_h\}|\{p,h\}).\tag{12.4}$$

Function $\mathfrak{A}_{\mathcal{O}}^{(\mathrm{fin})}(\{\lambda_p, \lambda_h\}|\{p, h\})$ has a finite thermodynamic limit. We call it a *finite amplitude* of the form factor. The exponent $\gamma_{\mathcal{O}}$ depends on the particle/hole momenta, and it turns out that $\gamma_{\mathcal{O}} > 0$. Thus, the form factor vanishes in the thermodynamic limit.

This fact should not surprise us. On the contrary, if we expect that the correlation functions in the thermodynamic limit remain finite, then the form factor must be small in this limit. Indeed, when replacing summation over a particle/hole position with integration over the corresponding momentum, the factor $L^{-1}$ is included in the integration measure (see (10.26)). Therefore, for example, the amplitude of a form factor corresponding to an excited state with $n$ particles and $n$ holes must be proportional to $L^{-2n}$. However, as we will see later, $\gamma_{\mathcal{O}} > 2n$. Thus, even when taking into account the integration measure, the form factor remains infinitely small.

The finite amplitude of the form factor $\mathfrak{A}_{\mathcal{O}}^{(\mathrm{fin})}$ depends on the momenta $\{\lambda_p\}$ and $\{\lambda_h\}$. However, in this case, we have pointed out that it also depends on the quantum numbers $\{p\}$ and $\{h\}$. Of course, the assignment of quantum numbers completely fixes the excited state, particularly the values of the momenta $\{\lambda_p\}$ and $\{\lambda_h\}$. By explicitly specifying the dependence on momenta and quantum numbers, we wanted to stress some essential properties of the finite amplitude $\mathfrak{A}_{\mathcal{O}}^{(\mathrm{fin})}$.

Consider, for definiteness, the dependence of the finite amplitude on quantum numbers $\{p\}$ (the dependence on $\{h\}$ is similar). Generically, a finite deviation of the integer $p_a$, say, $p_a \to p_a + k$ with $|k| \ll N$, leads to a deviation of order $L^{-1}$ for the corresponding momentum $\lambda_{p_a}$ and deviations of order $L^{-2}$ for all other momenta. However, the finite amplitude properties for such deviations largely depend on which area is the momentum $\lambda_{p_a}$. If $\lambda_{p_a}$ is separated from the Fermi boundaries by a finite distance, then the change $p_a \to p_a + k$ produces a variation of the finite amplitude of order $L^{-1}$. In this sense, one can say that $\mathfrak{A}_{\mathcal{O}}^{(\mathrm{fin})}$ depends on $\lambda_{p_a}$ rather than on the corresponding quantum number. On the contrary, if in the thermodynamic limit, $\lambda_{p_a} \to \pm q$, then the finite deviation $p_a \to p_a + k$ leads to a finite variation of $\mathfrak{A}_{\mathcal{O}}^{(\mathrm{fin})}$. Hence, in this case, it is natural to say that $\mathfrak{A}_{\mathcal{O}}^{(\mathrm{fin})}$ depends on $p_a$.

These properties of the finite amplitude $\mathfrak{A}_{\mathcal{O}}^{(\mathrm{fin})}$ should be taken into account when we work with the series (12.2). If the quantum number $p_a$ is such that the corresponding momentum $\lambda_{p_a}$ is separated from $\pm q$ by a finite gap, then the sum over $p_a$ can be replaced by the integral over $\lambda_{p_a}$. However, due to a highly complex dependence of $\mathfrak{A}_{\mathcal{O}}^{(\mathrm{fin})}$ on $\{\lambda_p\}$ and $\{\lambda_h\}$, the explicit analytic evaluation of these integrals is hardly possible. The situation is simplified if, for some reason, the integrals under the question are localized around some point (or several points). Then the finite amplitude can be replaced by its constant value in the point of localization. In particular, this localization occurs when calculating the long-distance and/or large-time asymptotics of correlation functions. In this case, we are dealing with integrals of rapidly oscillating functions. Then the main contribution to the asymptotics is given either by the boundaries of the integration domain or by the saddle point.

At the same time, the sum over the quantum numbers $\{p\}$ and $\{h\}$ should be kept for those particles and holes, whose momenta go to the Fermi boundaries in the thermodynamic limit. Note that the excitation energy of such states is zero (more precisely, it is of order $L^{-1}$) in the thermodynamic limit, so we can expect that these states give the main contribution to the correlation function. We will see that this is indeed the case.

Thus, the general strategy for calculating correlation functions using form factor expansion is as follows. We should consider $N$ and $L$ large but finite. This applies not only to the form factor amplitude $|\mathcal{F}_{\mathcal{O}}|^2$ but also to the phase factor $\exp\{i(x_1 - x_2)\mathcal{P}_{\rm ex} - i(t_1 - t_2)\mathcal{E}_{\rm ex}\}$, which depends on the total energy $\mathcal{E}_{\rm ex}$ and the total momentum $\mathcal{P}_{\rm ex}$ of the excited state. This energy and momentum should be calculated, including $1/L$-corrections. When calculating the finite amplitude $\mathfrak{A}_{\mathcal{O}}^{(\rm fin)}$ at $N$ and $L$ large, we have to distinguish between cases when a particle (hole) is separated from the Fermi boundaries and when it is located on the boundary. After calculating the exponent $\gamma_{\mathcal{O}}$ and the finite amplitude $\mathfrak{A}_{\mathcal{O}}^{(\rm fin)}$ for each of these cases, we can proceed to the summation of the form factor series. Here we should first calculate the sum over the quantum numbers $\{p\}$ and $\{h\}$ for excitations lying on the Fermi boundary. In the result obtained, we can already take the limit $N, L \to \infty$ and then proceed to the integration over $\{\lambda_p\}$ and $\{\lambda_h\}$. However, this last step is not required when calculating the asymptotics of correlation functions, since the integrals are localized at the Fermi boundaries $\pm q$ and/or saddle points.

In conclusion of this section, we note that making the form factor expansion method mathematically rigorous is a very difficult task. We have already mentioned the problem of convergence of the form factor expansion. However, in addition to this problem, other issues require special attention. These include the problem of analytical continuation of the Fredholm determinants with respect to the twist parameter, the question of corrections to the thermodynamic limit of form factors, the role of bound states in spin chains, and many others. Detailed coverage of all these issues is beyond the scope of this book. We refer the reader interested in more rigorous mathematical justification of the form factor expansion method to the works [K. K. Kozlowski (2018a, 2019)].

## 12.2 Generating function

As a basic example, we study a two-point correlation function of densities in the one-dimensional Bose gas with chemical potential (10.99). Let

$$\hat{q}(x) = \int_0^x \Psi^\dagger(y)\Psi(y)\,\mathrm{d}y, \qquad \hat{q}(x,t) = e^{iHt}\hat{q}(x)e^{-iHt}. \tag{12.5}$$

The operator $\hat{q}(x)$ can be called the operator of the number of particles in the interval $[0, x]$. Hence, its $x$-derivative makes sense of the density of the particles. Consider a normalized expectation value

$$\langle\!\langle \hat{q}'(x_1,t_1)\hat{q}'(x_2,t_2)\rangle\!\rangle = \frac{\langle \mathbf{\Phi}_N(\bar{u})|\hat{q}'(x_1,t_1)\hat{q}'(x_2,t_2)|\mathbf{\Phi}_N(\bar{u})\rangle}{\langle \mathbf{\Phi}_N(\bar{u})|\mathbf{\Phi}_N(\bar{u})\rangle}. \tag{12.6}$$

Our ultimate aim is to calculate the asymptotic behavior of the correlation function (12.6) for large $x = x_1 - x_2$ and $t = t_1 - t_2$, in the thermodynamic limit $N, L \to \infty$ and $N/L = D$.

We can apply the form factor expansion directly to the correlation function (12.6). However, as for the correlation function of the third components of spin in the $XXZ$ chain, it is more convenient to consider a special generating function. Let us introduce an operator

$$\mathcal{Q}^{(\alpha)} = e^{iHt_1}e^{i\alpha\hat{q}(x_1)}e^{iH_\alpha(t_2-t_1)}e^{-i\alpha\hat{q}(x_2)}e^{-iHt_2}. \tag{12.7}$$

Here $H_\alpha$ is the Hamiltonian of the one-dimensional Bose gas with quasi-periodic boundary conditions. A generating function $Q_N^{(\alpha)}(x,t)$ is then defined as follows:

$$Q_N^{(\alpha)}(x,t) = \langle\!\langle \mathcal{Q}^{(\alpha)}\rangle\!\rangle, \qquad \text{where} \quad x = x_1 - x_2, \quad t = t_1 - t_2. \tag{12.8}$$

It is easy to see that

$$\frac{\partial}{\partial x}\frac{\partial}{\partial \alpha}\, e^{iHt}e^{\pm i\alpha\hat{q}(x)}e^{-iH_\alpha t}\Big|_{\alpha=0} = \pm i\Psi^\dagger(x,t)\Psi(x,t). \tag{12.9}$$

Hence,

$$\frac{1}{2}\frac{\partial^2}{\partial x_1\partial x_2}\frac{\partial^2 Q_N^{(\alpha)}(x,t)}{\partial \alpha^2}\Big|_{\alpha=0} = \langle\!\langle \Psi^\dagger(x_1,t_1)\Psi(x_1,t_1)\Psi^\dagger(x_2,t_2)\Psi(x_2,t_2)\rangle\!\rangle, \tag{12.10}$$

and we arrive at the density correlation function. The Hamiltonian $H_\alpha$ is introduced in order to decompose the generating function $Q_N^{(\alpha)}(x,t)$ over twisted on-shell Bethe vectors that are eigenvectors for $H_\alpha$, provided that $\kappa = e^{-i\alpha}$.

We emphasize once again that the introduction of the function $Q_N^{(\alpha)}(x,t)$ is not mandatory. However, this is convenient, at least for two reasons. First, we have seen that when calculating the form factors of the operator $\hat{q}(x)$, we should distinguish between two cases: when the states coincide and when they are different. The introduction of the twisted on-shell vectors allows us to consider these two cases simultaneously. Second, we will see below that some eigenstates that play an important role in the form factor expansion can be interpreted as shifting the parameter $\alpha$ (see the derivation of equation (10.71)).

Let us decompose the generating function with respect to the complete set of the twisted on-shell Bethe vectors

$$Q_N^{(\alpha)}(x,t) = \sum_{\bar{v}} \frac{\langle \mathbf{\Phi}_N(\bar{u})|e^{iHt_1}e^{i\alpha\hat{q}(x_1)}e^{-iH_\alpha t_1}|\mathbf{\Phi}_N^{(\kappa)}(\bar{v})\rangle}{\langle \mathbf{\Phi}_N(\bar{u})|\mathbf{\Phi}_N(\bar{u})\rangle} \times \frac{\langle \mathbf{\Phi}_N^{(\kappa)}(\bar{v})|e^{iH_\alpha t_2}e^{-i\alpha\hat{q}(x_2)}e^{-iHt_2}|\mathbf{\Phi}_N(\bar{u})\rangle}{\langle \mathbf{\Phi}_N^{(\kappa)}(\bar{v})|\mathbf{\Phi}_N^{(\kappa)}(\bar{v})\rangle}, \tag{12.11}$$

where the sum is taken over solutions of the twisted Bethe equations, and $\kappa = e^{-i\alpha}$. It is clear that we can restrict ourselves only to the excited states depending on $N$ Bethe parameters since otherwise, all matrix elements vanish. Acting with the operators $H$ and $H_\alpha$ on the corresponding eigenvectors, we obtain

$$Q_N^{(\alpha)}(x,t) = \sum_{\bar{v}} e^{-it\mathcal{E}_{\rm ex}} \frac{\langle \mathbf{\Phi}_N(\bar{u})|e^{i\alpha\hat{q}(x_1)}|\mathbf{\Phi}_N^{(\kappa)}(\bar{v})\rangle}{\langle \mathbf{\Phi}_N(\bar{u})|\mathbf{\Phi}_N(\bar{u})\rangle} \times \frac{\langle \mathbf{\Phi}_N^{(\kappa)}(\bar{v})|e^{-i\alpha\hat{q}(x_2)}|\mathbf{\Phi}_N(\bar{u})\rangle}{\langle \mathbf{\Phi}_N^{(\kappa)}(\bar{v})|\mathbf{\Phi}_N^{(\kappa)}(\bar{v})\rangle}, \tag{12.12}$$

where the total excitation energy $\mathcal{E}_{\rm ex}$ is

$$\mathcal{E}_{\rm ex} = \sum_{j=1}^{N}(v_j^2 - u_j^2). \tag{12.13}$$

Using (9.105) and (9.113), we find

$$Q_N^{(\alpha)}(x,t) = \sum_{\bar{v}} e^{-it\mathcal{E}_{\rm ex}+ix\mathcal{P}_{\rm ex}} \frac{(\langle \mathbf{\Phi}_N^{(\kappa)}(\bar{v})|\mathbf{\Phi}_N(\bar{u})\rangle)^2}{\langle \mathbf{\Phi}_N^{(\kappa)}(\bar{v})|\mathbf{\Phi}_N^{(\kappa)}(\bar{v})\rangle\ \langle \mathbf{\Phi}_N(\bar{u})|\mathbf{\Phi}_N(\bar{u})\rangle}, \tag{12.14}$$

where the total excitation momentum $\mathcal{P}_{\rm ex}$ is

$$\mathcal{P}_{\rm ex} = \sum_{j=1}^{N}(v_j - u_j). \tag{12.15}$$

Recall that any solution of the usual Bethe equations in the one-dimensional Bose gas with $c > 0$ is real. If we choose $\alpha$ to be real, then the solutions of the twisted Bethe equations are also real. Then

$$\Big(\langle \mathbf{\Phi}_N^{(\kappa)}(\bar{v})|\mathbf{\Phi}_N(\bar{u})\rangle\Big)^* = e^{i\alpha N}\langle \mathbf{\Phi}_N^{(\kappa)}(\bar{v})|\mathbf{\Phi}_N(\bar{u})\rangle \tag{12.16}$$

for real $\bar{u}$, $\bar{v}$, and $\alpha$. This property follows from representations (6.92) and (6.93) for the function $\boldsymbol{S}_N^{(\kappa)}(\bar{v}|\bar{u})$, which is equal to the scalar product $\langle \mathbf{\Phi}_N^{(\kappa)}(\bar{v})|\mathbf{\Phi}_N(\bar{u})\rangle$. Thus, we arrive at

$$Q_N^{(\alpha)}(x,t) = \sum_{\bar{v}} |\boldsymbol{\mathcal{F}}_{\mathcal{Q}}|^2 e^{ix\mathcal{P}_{\rm ex} - it\mathcal{E}_{\rm ex}}, \tag{12.17}$$

where the form factor amplitude $|\boldsymbol{\mathcal{F}}_{\mathcal{Q}}|^2$ is

$$|\boldsymbol{\mathcal{F}}_{\mathcal{Q}}|^2 = \frac{\Big|\langle \mathbf{\Phi}_N^{(\kappa)}(\bar{v})|\mathbf{\Phi}_N(\bar{u})\rangle\Big|^2 e^{-i\alpha N}}{\langle \mathbf{\Phi}_N^{(\kappa)}(\bar{v})|\mathbf{\Phi}_N^{(\kappa)}(\bar{v})\rangle \; \langle \mathbf{\Phi}_N(\bar{u})|\mathbf{\Phi}_N(\bar{u})\rangle}. \tag{12.18}$$

We see that the form factor amplitude reduces to the normalized scalar product of on-shell and twisted on-shell Bethe vectors.

We expect that the amplitude $|\boldsymbol{\mathcal{F}}_{\mathcal{Q}}|^2$ at large $N$ and $L$ behaves so that

$$\operatorname{Lim} |\boldsymbol{\mathcal{F}}_{\mathcal{Q}}|^2 L^{\gamma_{\mathcal{Q}}(\{\lambda_p,\lambda_h\})} = \mathfrak{A}_{\mathcal{Q}}^{(\rm fin)}(\{\lambda_p,\lambda_h\}|\{p,h\}). \tag{12.19}$$

Our first task is to find the exponent $\gamma_{\mathcal{Q}}$ and the finite amplitude $\mathfrak{A}_{\mathcal{Q}}^{(\rm fin)}$. To do this, we will use determinant representations for the scalar products of on-shell and twisted on-shell Bethe vectors.

The norm of the ground state is given by (10.180)–(10.184):

$$\langle \mathbf{\Phi}_N(\bar{u})|\mathbf{\Phi}_N(\bar{u})\rangle = c^N\left(\prod_{\substack{a,b=1\\a\neq b}}^{N} \frac{u_a - u_b - ic}{u_a - u_b}\right)\left(\prod_{a=1}^{N} 2\pi L\hat{\rho}_u(u_a)\right)$$
$$\times \det_N\left(\delta_{jk} - \frac{\mathcal{K}(u_j - u_k)}{2\pi L\hat{\rho}_u(u_k)}\right), \tag{12.20}$$

where $\hat{\rho}_u(z) = \hat{\xi}_u'(z)$ is the ground state distribution density. The norm of the excited state is given by (10.189)–(10.191):

$$e^{i\alpha N}\langle \mathbf{\Phi}_N^{(\kappa)}(\bar{v})|\mathbf{\Phi}_N^{(\kappa)}(\bar{v})\rangle = c^N\left(\prod_{\substack{a,b=1\\a\neq b}}^{N} \frac{v_a - v_b - ic}{v_a - v_b}\right)\left(\prod_{a=1}^{N} 2\pi L\hat{\rho}_v(v_a)\right)$$
$$\times \det_N\left(\delta_{jk} - \frac{\mathcal{K}(u_j - u_k)}{2\pi L\hat{\rho}_v(v_k)}\right). \tag{12.21}$$

where $\hat{\rho}_v(z) = \hat{\xi}_v'(z)$ is the distribution density of the excited state.

We also have several determinant representations for the scalar product $\langle\mathbf{\Phi}_N^{(\kappa)}(\bar v)|\mathbf{\Phi}_N(\bar u)\rangle$. We will use equation (9.118). Being specified for the one-dimensional Bose gas, it has the following form:

$$\langle\mathbf{\Phi}_N^{(\kappa)}(\bar v)|\mathbf{\Phi}_N(\bar u)\rangle = \frac{(e^{-i\alpha}-1)}{V^+(\omega)-e^{-i\alpha}V^-(\omega)}\left(\prod_{a,b=1}^{N}\frac{v_a-u_b-ic}{v_a-u_b}\right) \times\left(\prod_{a=1}^{N}\left(1-e^{-2\pi i\hat F(u_a)}\right)\right)\det_N\left(\delta_{jk}-U_{jk}\right). \tag{12.22}$$

Recall the notation used in (12.22). Functions $V^\pm(z)$ are given by the products

$$V^\pm(z)=\prod_{l=1}^{N}\frac{z-v_l\pm ic}{z-u_l\pm ic}. \tag{12.23}$$

Function $\hat F(z)$ is the shift function describing the excited state $|\mathbf{\Phi}_N^{(\kappa)}(\bar v)\rangle$. Using representation (10.62) for $\hat F(z)$, we can easily show that

$$e^{2\pi i\hat F(z)}=e^{i\alpha}\frac{V^+(z)}{V^-(z)}. \tag{12.24}$$

The entries $U_{jk}$ have the following form:

$$U_{jk}=\frac{\prod_{l=1}^{N}(u_k-v_l)}{\prod_{l=1;\,l\neq k}^{N}(u_k-u_l)}\left(\prod_{l=1}^{N}\frac{u_k-u_l+ic}{u_k-v_l+ic}\right)\frac{\mathcal{K}_\alpha(u_k,u_j)-\mathcal{K}_\alpha(\omega,u_j)}{1-e^{-2\pi i\hat F(u_k)}}, \tag{12.25}$$

where

$$\mathcal{K}_\alpha(u,v)=\frac{1}{u-v+ic}+\frac{e^{-i\alpha}}{v-u+ic}. \tag{12.26}$$

Finally, $\omega$ is an arbitrary complex number. We showed in Section 9.4 that combination (12.22) does not depend on $\omega$. In what follows, it will be convenient for us to assume that $\omega$ is real.

We now can write down an explicit representation for the form factor amplitude (12.18). However, before doing this, we present the amplitude $|\mathcal{F}_{\mathcal{Q}}|^2$ as the product of two factors:

$$|\mathcal{F}_{\mathcal{Q}}|^2=\mathfrak{D}(\bar u,\bar v,\alpha)\mathfrak{S}(\bar u,\bar v,\alpha). \tag{12.27}$$

We call $\mathfrak{D}(\bar u,\bar v,\alpha)$ a *discrete part* of the amplitude. It is given by

$$\mathfrak{D}(\bar u,\bar v,\alpha)=\frac{\prod_{j>k}^{N}(u_j-u_k)^2(v_j-v_k)^2}{\prod_{j,k=1}^{N}(u_j-v_k)^2}\prod_{j=1}^{N}\frac{\sin^2\pi F(u_j)}{\pi^2L^2\hat\rho_u(u_j)\hat\rho_v(v_j)}. \tag{12.28}$$

The factor $\mathfrak{S}(\bar{u},\bar{v},\alpha)$ is called a *smooth part* of the amplitude:

$$\mathfrak{S}(\bar{u},\bar{v},\alpha)=\left(\frac{\sin\frac{\alpha}{2}}{\sin\pi\hat{F}(\omega)}\right)^2\frac{W_N(\bar{u},\bar{v})}{V^+(\omega)V^-(\omega)}$$
$$\times\frac{\left|\det_N\left(\delta_{jk}-U_{jk}\right)\right|^2}{\det_N\left(\delta_{jk}-\frac{\mathcal{K}(u_j-u_k)}{2\pi L\hat{\rho}_u(u_k)}\right)\det_N\left(\delta_{jk}-\frac{\mathcal{K}(v_j-v_k)}{2\pi L\hat{\rho}_v(v_k)}\right)}. \tag{12.29}$$

Here

$$W_N(\bar{u},\bar{v})=\prod_{j,k=1}^{N}\frac{(u_j-v_k-ic)(v_j-u_k-ic)}{(u_j-u_k-ic)(v_j-v_k-ic)}. \tag{12.30}$$

The division into discrete and smooth parts is related to their different behavior in the thermodynamic limit. The smooth part has a well-defined thermodynamic limit in which it depends on the momenta of excitations and is a functional of the shift function. In contrast, the discrete part does not have a uniform thermodynamic limit. It is the discrete part for large $L$ and $N$ that gives the factor $L^{-\gamma_{\mathcal{Q}}}$, which we mentioned above. The thermodynamic limit of the discrete part also strongly depends on the positions of the particles and holes relative to the Fermi boundary.

## 12.3 Special form factor

We start with the case where the vector $|\mathbf{\Phi}_N^{(\kappa)}(\bar{v})\rangle$ is a twisted ground state. Then $v_j=u_j+\alpha/L$. The shift function for this state is proportional to the dressed charge $F(u)=\alpha Z(u)/2\pi$ (see (10.138)). Recall that the dressed charge is proportional to the ground state density $Z(u)=2\pi\rho(u)$ in the one-dimensional Bose gas.

### 12.3.1 *Smooth part*

#### 12.3.1.1 *Analytical properties of the scalar product*

To calculate the limit in the smooth part, we need to replace the determinant of the matrix $\delta_{jk}-U_{jk}$ with a Fredholm determinant using the method described in Section 10.7.4. To do this, we first establish some analytical properties of the scalar product

$$S_N^{(\kappa)}(\bar{v}|\bar{u})=\langle\mathbf{\Phi}_N^{(\kappa)}(\bar{v})|\mathbf{\Phi}_N(\bar{u})\rangle,\qquad \kappa=e^{-i\alpha}. \tag{12.31}$$

Although we consider real $\alpha$, the relation $v_j=u_j+\alpha/L$ is also preserved for any complex $\alpha$. Therefore, we can investigate the analytical properties of the scalar product $S_N^{(\kappa)}(\bar{v}|\bar{u})$ as a function of $\alpha$.

In addition to representation (12.22), we also have equation (6.92) for $S_N^{(\kappa)}(\bar v|\bar u)$. Specifying it for the one-dimensional Bose gas, we obtain

$$S_N^{(\kappa)}(\bar v|\bar u) = \frac{\prod_{a,b=1}^N (v_a - u_b - ic)}{\prod_{a>b}^N (u_a - u_b)(v_b - v_a)} \det_N \mathcal{M}^{(\kappa)}, \tag{12.32}$$

where

$$\mathcal{M}_{jk}^{(\kappa)} = \frac{-ic}{(v_j - u_k)(v_j - u_k - ic)} + \frac{-ice^{-i\alpha}}{(u_k - v_j)(u_k - v_j - ic)} \frac{V^-(u_k)}{V^+(u_k)}, \tag{12.33}$$

and $V^{\pm}(z)$ are given by (12.23). Using (12.24) we recast the matrix element $\mathcal{M}_{jk}^{(\kappa)}$ as follows:

$$\mathcal{M}_{jk}^{(\kappa)} = \frac{-ic}{(v_j - u_k)(v_j - u_k - ic)} + \frac{-ic\, e^{-2\pi i \hat F(u_k)}}{(u_k - v_j)(u_k - v_j - ic)}. \tag{12.34}$$

The parameters $\bar v$ depend on $\alpha$: $v_j = v_j(\alpha)$, $j = 1, \dots, N$. Therefore, we may have $v_j(\alpha) = u_k$ for some $j$, $k$, and $\alpha$. At first glance, it seems that the matrix element $\mathcal{M}_{jk}^{(\kappa)}$ becomes singular. Let us show that this is not the case.

Let $v_j = u_k$ for some $j$ and $k$. Then due to the definition of the shift function (10.59), we obtain

$$\hat F(u_k) = L(\hat\xi_u(u_k) - \hat\xi_v(u_k)) = L(\hat\xi_u(u_k) - \hat\xi_v(v_j)) = k - j. \tag{12.35}$$

Hence, $e^{-2\pi i \hat F(u_k)} = 1$. We see that the residue of the matrix element $\mathcal{M}_{jk}^{(\kappa)}$ (12.34) at $v_j = u_k$ vanishes. Thus, $\det \mathcal{M}^{(\kappa)}$ is an analytical function of $\alpha$ in the strip $\Im\alpha < c$. Therefore, when calculating this determinant, we can restrict ourselves to sufficiently small real values of $\alpha$ and then use the analytical continuation.

We studied matrices of the form (12.25) in Section 10.7.4. The determinant of the matrix $\delta_{jk} - U_{jk}$ can be replaced by the Fredholm determinant

$$\det_N(\delta_{jk} - U_{jk}) = \det\left(\mathbf{I} - \tfrac{1}{2\pi i}\hat{\mathbf{U}}\right), \tag{12.36}$$

with the kernel

$$\hat U(z, z') = \left(\prod_{l=1}^N \frac{(z' - v_l)(z' - u_l + ic)}{(z' - u_l)(z' - v_l + ic)}\right) \frac{\mathcal{K}_\alpha(z', z) - \mathcal{K}_\alpha(\omega, z)}{1 - e^{-2\pi i \hat F(z')}}. \tag{12.37}$$

The integral operator $\mathbf{I} + \frac{1}{2\pi i}\hat{\mathbf{U}}$ acts on a contour $\Gamma([-q, q])$, which surrounds the Fermi zone $[-q, q]$ (see Fig. 10.1).

Replacement (12.36) is possible even at finite $N$, but it is necessary that the only singularities of the kernel $\hat U(z, z')$ within the contour $\Gamma([-q, q])$ are the points $u_l$, $l = 1, \dots, N$. In our case, singularities can also occur at points of the interval $[-q, q]$ where the shift function $\hat{F}(z')$ takes integer values.

In the case under consideration, the shift function is proportional to $\alpha$. Therefore, we can always choose $\alpha$ so small as to provide the condition $|\hat{F}(z)| < 1$ for $z \in [-q, q]$. After calculating the thermodynamic limit of the scalar product for such $\alpha$, we use the analytic continuation.

We can, however, relax these conditions. Suppose that $\hat{F}(u_k) = m$, where $m$ is an integer, and $u_k$ is separated from the Fermi boundaries. This means that $1 \ll k \ll N$. Then it follows from (10.59) that

$$\hat{\xi}_v(u_k) = \frac{k - m}{L}. \tag{12.38}$$

Since $m$ is finite (does not depend on $N$), we conclude that $k - m \in \{1, \dots, N\}$. Hence, $u_k = v_{k-m}$, which implies $\prod_{l=1}^{N}(u_k - v_l) = 0$. Thus, the matrix elements $U_{jk}$ in the $k$th column are not singular, despite the shift function taking an integer value.

If it turns out that $k - m \notin \{1, \dots, N\}$, then the singularity compensation does not occur, and the pole remains. However, this is only possible if $u_k$ goes to one of the Fermi boundaries $\pm q$ in the thermodynamic limit. To avoid this situation, we assume that the shift function at the Fermi boundaries $\hat{F}(\pm q)$ is not equal to integers. Further on, when studying form factors with particles and holes, we will impose the same conditions for the shift function. This will enable us to make replacement (12.36).

#### 12.3.1.2 *Thermodynamic limit of the smooth part*

Using formulas of Section 10.7, we easily compute the thermodynamic limit of the smooth part. We start with the kernel of the integral operator $\hat{U}(z, z')$. In fact, we only need to calculate the limit of the product over $l$ in (12.37). Using equation (10.174) at $n = 0$, we obtain

$$\mathrm{Lim} \prod_{l=1}^{N} \frac{(z' - v_l)(z' - u_l + ic)}{(z' - u_l)(z' - v_l + ic)} = e^{\tilde{F}(z'+ic) - \tilde{F}(z')}, \tag{12.39}$$

where $\tilde{F}(z')$ is the Cauchy transform of the shift function:

$$\tilde{F}(z') = \int_{-q}^{q} \frac{F(u)\,\mathrm{d}u}{z' - u}. \tag{12.40}$$

Here we used the fact that $z' \in \Gamma([-q, q])$, while $u \in [-q, q]$. Therefore, $|z' - u| > 0$.

Thus, we arrive at

$$U(z,z') = \operatorname{Lim} \hat{U}(z,z') = e^{\tilde{F}(z'+ic)-\tilde{F}(z')} \frac{\mathcal{K}_\alpha(z',z) - \mathcal{K}_\alpha(\omega,z)}{1-e^{-2\pi i F(z')}}. \tag{12.41}$$

The limits of other products in the smooth part are also simple. All of them were computed in Section 10.7.1. For instance,

$$\operatorname{Lim} V^+(\omega)V^-(\omega) = e^{-\tilde{F}(\omega-ic)-\tilde{F}(\omega+ic)}, \tag{12.42}$$

and

$$\operatorname{Lim} \log W_N(\bar{u},\bar{v}) = C_0[F], \tag{12.43}$$

where

$$C_0[F] = -\int_{-q}^{q} \frac{F(u)F(v)}{(u-v-ic)^2}\,du\,dv. \tag{12.44}$$

Finally, the limits of two determinants in the denominator of (12.29) were computed in Section 10.7.3 (see (10.188), (10.194)).

Substituting everything in (12.29) and assuming for simplicity that $\omega$ is real, we obtain

$$\begin{aligned} \operatorname{Lim} \mathfrak{S}(\bar{u},\bar{v},\alpha) &= \mathfrak{S}[F] \\ &= e^{C_0[F]} \left| e^{\tilde{F}(\omega-ic)} \frac{\sin\frac{\alpha}{2}}{\sin \pi F(\omega)} \frac{\det\left(\mathbf{I} - \frac{1}{2\pi i}\mathbf{U}\right)}{\det\left(\mathbf{I} - \frac{1}{2\pi}\mathbf{K}\right)} \right|^2. \end{aligned} \tag{12.45}$$

It remains to substitute the shift function $F(u) = \alpha Z(u)/2\pi$ in these formulas. However, we draw the reader's attention that we did not use this specific form of the shift function when calculating the thermodynamic limit. Therefore, in the case of an excited state with particles and holes, the calculation of the limit of the smooth part of the form factor amplitude is quite similar.

We also note that we neglected corrections of order $O(L^{-2})$ when replacing sums with integrals.

### 12.3.2 *Thermodynamic limit of the discrete part*

#### 12.3.2.1 *Discrete part for impenetrable bosons*

Computing the discrete part becomes especially easy in the impenetrable bosons model ($c \to +\infty$). This simple example clearly shows why the form factor is proportional to the fractional power of the system size in the thermodynamic limit. Note also that the smooth part of the form factor amplitude (12.29) goes to 1 at $c \to +\infty$. We recommend that the reader independently verify this statement. Thus, the form factor is reduced to the discrete part for impenetrable bosons.

We have for $c \to +\infty$

$$u_j = \frac{2\pi}{L}\left(j - \frac{N+1}{2}\right), \qquad v_j = \frac{2\pi}{L}\left(j - \frac{N+1}{2} + \frac{\alpha}{2\pi}\right). \tag{12.46}$$

It is also easy to see that

$$\hat{\rho}_u(z) = \hat{\rho}_v(z) = \frac{1}{2\pi}, \qquad \hat{F}(z) = \frac{\alpha}{2\pi}. \tag{12.47}$$

Then the discrete part (12.28) takes the following form:

$$\mathfrak{D}(\bar{u}, \bar{v}, \alpha) = \left(\frac{\sin\frac{\alpha}{2}}{\pi}\right)^{2N} \frac{\prod_{j>k}^{N}(j-k)^4}{\prod_{j,k=1}^{N}(j-k-\frac{\alpha}{2\pi})^2}. \tag{12.48}$$

Calculating the product over $k$, we obtain

$$\mathfrak{D}(\bar{u}, \bar{v}, \alpha) = \prod_{j=1}^{N} \frac{\Gamma^4(j)}{\Gamma^2(j-\frac{\alpha}{2\pi})\Gamma^2(j+\frac{\alpha}{2\pi})}, \tag{12.49}$$

where we used

$$\Gamma(z)\Gamma(1-z) = \frac{\pi}{\sin \pi z}. \tag{12.50}$$

The remaining product is expressed in terms of the Barnes $G$-function due to (A.28)

$$\mathfrak{D}(\bar{u}, \bar{v}, \alpha) = \frac{G^2(1-\frac{\alpha}{2\pi})G^2(1+\frac{\alpha}{2\pi})G^4(N+1)}{G^2(N+1-\frac{\alpha}{2\pi})G^2(N+1+\frac{\alpha}{2\pi})}. \tag{12.51}$$

Using the asymptotics of the Barnes $G$-function (A.27) at $N \to \infty$, we easily find

$$\mathfrak{D}(\bar{u}, \bar{v}, \alpha) = N^{-\alpha^2/2\pi^2}\Big(1 + O(N^{-1})\Big), \qquad N \to \infty. \tag{12.52}$$

So, the form factor behaves as $L^{-\alpha^2/2\pi^2}$ in the limit $N, L \to \infty$, $N/L = D$. This behavior is entirely due to the discrete part.

#### 12.3.2.2 *Discrete part in the general case*

Let us now proceed to calculate the limit of the discrete part (12.28) in the general case $c < \infty$. In fact, all further calculations in this section are a reduction of the general case to impenetrable bosons.

From the condition $v_j = u_j + \alpha/L$, in particular, it follows that $v_j - v_k = u_j - u_k$ for any $j, k = 1, \dots, N$. One can use this to simplify some calculations. However, we will calculate the thermodynamic limit of the discrete part in the most general case, when the shift of variables $v_j$ relative to $u_j$ is determined by a shift function $\hat{F}(u_j)$ of a generic type. We will also not use the fact that the shift function satisfies some integral equation. First, this approach will allow us to easily calculate the thermodynamic limit of the discrete part of the form factor amplitude when the excited state contains particles and holes. Second, we will establish some general properties of the discrete part that do not depend on a specific model and specific excited state.

Let us introduce two functions

$$\phi_u(x,y) = \frac{x-y}{\hat{\xi}_u(x)-\hat{\xi}_u(y)}, \qquad \phi_v(x,y) = \frac{x-y}{\hat{\xi}_v(x)-\hat{\xi}_v(y)}. \tag{12.53}$$

Obviously, $\phi_u(x,y)$ and $\phi_v(x,y)$ are symmetric functions of $x$ and $y$. It is also obvious that

$$\phi_u(x,x) = 1/\hat{\rho}_u(x), \qquad \phi_v(x,x) = 1/\hat{\rho}_v(x). \tag{12.54}$$

Let us represent the discrete part (12.28) as a product of four factors:

$$\mathfrak{D}(\bar{u},\bar{v},\alpha) = \prod_{l=1}^{4} \mathfrak{D}_N^{(l)}. \tag{12.55}$$

Here

$$\mathfrak{D}_N^{(1)} = \prod_{k=1}^{N} \frac{\hat{\rho}_v(u_k)}{\hat{\rho}_u(u_k)}, \tag{12.56}$$

$$\mathfrak{D}_N^{(2)} = \frac{\prod_{j>k}^{N} \phi_v^2(u_j,u_k)\phi_v^2(v_j,v_k)}{\prod_{j,k=1}^{N} \phi_v^2(u_j,v_k)} \prod_{k=1}^{N} \frac{1}{\hat{\rho}_v(v_k)\hat{\rho}_v(u_k)}, \tag{12.57}$$

$$\mathfrak{D}_N^{(3)} = \left(\prod_{j>k}^{N} \frac{\hat{\xi}_v(u_j)-\hat{\xi}_v(u_k)}{\hat{\xi}_u(u_j)-\hat{\xi}_u(u_k)}\right)^2 \prod_{j=1}^{N} e^{2\psi(j)\left(\hat{F}(u_j)-\hat{F}(u_{N+1-j})\right)}, \tag{12.58}$$

and

$$\mathfrak{D}_N^{(4)} = \frac{\prod_{j>k}^{N}(\hat{\xi}_u(u_j)-\hat{\xi}_u(u_k))^2(\hat{\xi}_v(v_j)-\hat{\xi}_v(v_k))^2}{\prod_{j,k=1}^{N}(\hat{\xi}_v(u_j)-\hat{\xi}_v(v_k))^2} \times \prod_{j=1}^{N}\left(\frac{\sin\pi F(u_j)}{\pi L} e^{\psi(j)\left(\hat{F}(u_{N+1-j})-\hat{F}(u_j)\right)}\right)^2. \tag{12.59}$$

In these formulas, we introduced a logarithmic derivative of the $\Gamma$-function

$$\psi(z) = \frac{\mathrm{d}}{\mathrm{d}z}\log\Gamma(z). \tag{12.60}$$

The limits of the first three products are relatively simple. Differentiating relation (10.59) we find

$$\hat{\rho}_v(z) = \hat{\rho}_u(z) - \frac{\hat{F}'(z)}{L}. \tag{12.61}$$

Hence,

$$\log \mathfrak{D}_N^{(1)} = \sum_{k=1}^{N} \log\left(1 - \frac{\hat{F}'(u_k)}{L\hat{\rho}_u(u_k)}\right) = -\sum_{k=1}^{N}\left(\frac{\hat{F}'(u_k)}{L\hat{\rho}_u(u_k)} + O(L^{-2})\right). \tag{12.62}$$

From this, we immediately find

$$\operatorname{Lim} \log \mathfrak{D}_N^{(1)} = -\int_{-q}^{q} F'(u)\,\mathrm{d}u = F_- - F_+, \tag{12.63}$$

where $F_\pm = F(\pm q)$.

It is also easy to see that

$$\mathfrak{D}_N^{(2)} = \prod_{j,k=1}^{N} \frac{\phi_v(u_j, u_k)\phi_v(v_j, v_k)}{\phi_v(v_j, u_k)\phi_v(u_j, v_k)}, \tag{12.64}$$

where we used (12.54) and the symmetry of the function $\phi_v(x,y)$. Then $\log \mathfrak{D}_N^{(2)}$ has the form (10.163), and we can use (10.165):

$$\operatorname{Lim} \log \mathfrak{D}_N^{(2)} = -\int_{-q}^{q} F(u)F(v)\frac{\partial^2}{\partial u \partial v} \log\left(\frac{\xi(u) - \xi(v)}{u - v}\right) \mathrm{d}u\,\mathrm{d}v. \tag{12.65}$$

To compute the limit of $\mathfrak{D}_N^{(3)}$, we use (10.59) again

$$\frac{\hat{\xi}_v(u_j) - \hat{\xi}_v(u_k)}{\hat{\xi}_u(u_j) - \hat{\xi}_u(u_k)} = 1 - \frac{\hat{F}(u_j) - \hat{F}(u_k)}{L(\hat{\xi}_u(u_j) - \hat{\xi}_u(u_k))} = 1 - \frac{\hat{F}(u_j) - \hat{F}(u_k)}{j - k}. \tag{12.66}$$

Then

$$\begin{aligned}\log \mathfrak{D}_N^{(3)} &= 2\sum_{j=1}^{N} \psi(j)\big(\hat{F}(u_j) - \hat{F}(u_{N+1-j})\big) \\ &\quad - \sum_{j>k}^{N}\left\{2\frac{\hat{F}(u_j) - \hat{F}(u_k)}{j - k} + \frac{(\hat{F}(u_j) - \hat{F}(u_k))^2}{L^2(\hat{\xi}_u(u_j) - \hat{\xi}_u(u_k))^2} + O(L^{-3})\right\}.\end{aligned} \tag{12.67}$$

One of the sums in (12.67) can be partly calculated via (A.3):

$$\begin{aligned}
&\sum_{j>k}^{N} \frac{\hat{F}(u_j) - \hat{F}(u_k)}{j-k} \\
&\quad = \sum_{j=2}^{N}\sum_{k=1}^{j-1} \frac{\hat{F}(u_j)}{j-k} - \sum_{k=1}^{N-1}\sum_{j=k+1}^{N} \frac{\hat{F}(u_k)}{j-k} \\
&\quad = \sum_{j=1}^{N} \hat{F}(u_j)(\psi(j) - \psi(1)) - \sum_{k=1}^{N} \hat{F}(u_k)(\psi(N+1-k) - \psi(1)) \\
&\quad = \sum_{j=1}^{N} \psi(j)(\hat{F}(u_j) - \hat{F}(u_{N+1-j})). \qquad (12.68)
\end{aligned}$$

We see that this contribution cancels with the sum in the first line of (12.67). The remaining sum in (12.67) has a well-defined thermodynamic limit, and we obtain

$$\begin{aligned}
\log \mathfrak{D}_N^{(3)} &= -\int_{-q}^{q} \rho(u)\,\mathrm{d}u \int_{-q}^{u} \rho(v)\,\mathrm{d}v \left(\frac{F(u)-F(v)}{\xi(u)-\xi(v)}\right)^2 \\
&= -\frac{1}{2}\int_{-q}^{q} \left(\frac{F(u)-F(v)}{\xi(u)-\xi(v)}\right)^2 \rho(u)\rho(v)\,\mathrm{d}u\,\mathrm{d}v. \qquad (12.69)
\end{aligned}$$

It is easy to check that when calculating the limits of products $\mathfrak{D}_N^{(l)}$, $l = 1, 2, 3$, we neglected corrections of order at most $O(L^{-1})$. Indeed, when developing the logarithms in series, we neglected the terms that ultimately give contributions of order $O(L^{-1})$. At the same time, when replacing sums with integrals, corrections of order $O(L^{-2})$ occur.

Let us now proceed to the calculation of the limit of $\mathfrak{D}_N^{(4)}$ (12.59).

**Proposition 12.1.** *For $L$ and $N$ large enough and $N/L = D$, the product $\mathfrak{D}_N^{(4)}$ behaves as*

$$\mathfrak{D}_N^{(4)} = \mathfrak{D}_{\mathrm{fin}}^{(4)} L^{-(F_-^2+F_+^2)}\left(1 + O\left(\tfrac{\log(DL)}{L}\right)\right). \qquad (12.70)$$

*Here the finite part $\mathfrak{D}_{\mathrm{fin}}^{(4)}$ has the following representation:*

$$\begin{aligned}
\mathfrak{D}_{\mathrm{fin}}^{(4)} &= G^2(1-F_-)G^2(1+F_+)(2\pi)^{F_- - F_+} e^{F_+ - F_-} \\
&\quad \times \exp\left\{\int_{-q}^{q}\left(\frac{F^2(q)-F^2(u)}{\xi(q)-\xi(u)} - \frac{F^2(-q)-F^2(u)}{\xi(-q)-\xi(u)}\right)\rho(u)\,\mathrm{d}u\right\}, \qquad (12.71)
\end{aligned}$$

*and $G(z)$ is the Barnes $G$-function.*

**Proof.** First of all, we note that

$$\frac{\prod\limits_{k<j}^{N}(\hat{\xi}_u(u_j)-\hat{\xi}_u(u_k))^2(\hat{\xi}_v(v_j)-\hat{\xi}_v(v_k))^2}{\prod\limits_{j,k=1}^{N}(\hat{\xi}_v(u_j)-\hat{\xi}_v(v_k))^2}=\frac{L^{2N}\prod\limits_{k<j}^{N}(j-k)^4}{\prod\limits_{j,k=1}^{N}(L\hat{\xi}_v(u_j)-k)^2}. \tag{12.72}$$

Presenting $L\hat{\xi}_v(u_j)=j-\hat{F}(u_j)$ and calculating the product over $k$, we obtain

$$\frac{\prod\limits_{k<j}^{N}(\hat{\xi}_u(u_j)-\hat{\xi}_u(u_k))^2(\hat{\xi}_v(v_j)-\hat{\xi}_v(v_k))^2}{\prod\limits_{j,k=1}^{N}(\hat{\xi}_v(u_j)-\hat{\xi}_v(v_k))^2}=L^{2N}\prod_{j=1}^{N}\frac{\Gamma^4(j)\Gamma^2(1-j+\hat{F}(u_j))}{\Gamma^2(N+1-j+\hat{F}(u_j))}. \tag{12.73}$$

Using (12.50), we arrive at

$$\mathfrak{D}_N^{(4)}=\prod_{j=1}^{N}\frac{\Gamma^4(j)\exp\left\{2\psi(j)\big(\hat{F}(u_{N+1-j})-\hat{F}(u_j)\big)\right\}}{\Gamma^2(j+\hat{F}(u_{N+1-j}))\Gamma^2(j-\hat{F}(u_j))}. \tag{12.74}$$

We see that $\mathfrak{D}_N^{(4)}$ can be factorized as follows:

$$\mathfrak{D}_N^{(4)}=H_1[\hat{F}]H_2[\hat{F}], \tag{12.75}$$

where

$$H_1[\hat{F}]=\prod_{k=1}^{N}\frac{\Gamma^2(j)\exp\left\{-2\psi(j)\hat{F}(u_j)\right\}}{\Gamma^2(j-\hat{F}(u_j))}, \tag{12.76}$$

and $H_2[\hat{F}]$ being obtained from (12.76) by the replacement $\hat{F}(u_j)\to -\hat{F}(u_{N+1-j})$.

We first consider $H_1[\hat{F}]$. We use the fact that the shift function $\hat{F}(z)$ is bounded uniformly for any value of $N$. Let $|\hat{F}(u_j)|<\nu$ for some $\nu\in\mathbb{N}$ and $j=1,\dots,N$. Then we present $H_1[\hat{F}]$ as

$$H_1[\hat{F}]=H_1^{(1)}[F_-]\cdot H_1^{(2)}[\hat{F}]\cdot H_1^{(3)}[\hat{F}], \tag{12.77}$$

where

$$H_1^{(1)}[F_-]=\prod_{j=1}^{N}\frac{\Gamma^2(j)e^{-2\psi(j)F_-}}{\Gamma^2(j-F_-)}, \tag{12.78}$$

$$H_1^{(2)}[\hat{F}]=\prod_{j=1}^{\nu-1}\frac{\Gamma^2(j-F_-)e^{-2\psi(j)(\hat{F}(u_j)-F_-)}}{\Gamma^2(j-\hat{F}(u_j))}, \tag{12.79}$$

and

$$H_1^{(3)}[\hat{F}] = \prod_{j=\nu}^{N} \frac{\Gamma^2(j-F_-)e^{-2\psi(j)(\hat{F}(u_j)-F_-)}}{\Gamma^2(j-\hat{F}(u_j))}. \tag{12.80}$$

The product (12.78) can be calculated explicitly via (A.11), (A.28). We obtain

$$H_1^{(1)}[F_-] = \frac{G^2(N+1)G^2(1-F_-)e^{-2(N\psi(N)+1-N)F_-}}{G^2(N+1-F_-)}. \tag{12.81}$$

Using asymptotic expansion for the Barnes $G$-function (A.27) and the $\psi$-function (A.4), we find

$$H_1^{(1)}[F_-] = (DL)^{-F_-^2}\, G^2(1-F_-)e^{(\log 2\pi-1)F_-}\big(1+O(L^{-1})\big). \tag{12.82}$$

Thus, this product vanishes in the thermodynamic limit. The remaining two products $H_1^{(2)}$ and $H_1^{(3)}$ have finite limits.

The limit of (12.79) is equal to 1 due to $\hat{F}(u_j) - F_- = O(L^{-1})$ for $j = 1, \dots, \nu-1$. Finally, to compute the limit of (12.80), we can use Taylor integral representation for the logarithm of the $\Gamma$-function

$$\log\frac{\Gamma(j+x)}{\Gamma(j)} = x\psi(j) + \frac{x^2}{2}\psi'(j) + \frac{x^3}{2}\int_0^1 (1-t)^2\psi''(j+tx)\,\mathrm{d}t, \tag{12.83}$$

provided $|x| < j$. Recall that $|\hat{F}(u_j)| < \nu$; hence, we can apply (12.83) to $\Gamma(j-F_-)$ and $\Gamma(j-\hat{F}(u_j))$. Then

$$\begin{aligned}\log H_1^{(3)}[\hat{F}] &= \sum_{j=\nu}^{N}(F_-^2 - \hat{F}^2(u_j))\psi'(j) - \sum_{j=\nu}^{N}\int_0^1 (1-t)^2 \\ &\quad\times\Big(F_-^3\psi''(j+tF_-) - \hat{F}^3(u_j)\psi''(j+t\hat{F}(u_j))\Big)\,\mathrm{d}t.\end{aligned} \tag{12.84}$$

The limit of the sum in the first line of (12.84) is obtained by applying lemma A.2 and corollary A.1. Indeed, using $u_j = \hat{\xi}_u^{-1}(\frac{j}{L})$ we find

$$\begin{aligned}&\sum_{j=\nu}^{N}\Big(F_-^2 - \hat{F}^2(\hat{\xi}_u^{-1}(\tfrac{j}{L}))\Big)\psi'(j) \\ &\qquad = \int_0^D \frac{F_-^2 - F^2(\xi^{-1}(x))}{x}\,\mathrm{d}x + O\Big(\tfrac{\log(DL)}{L}\Big).\end{aligned} \tag{12.85}$$

Setting $\xi^{-1}(x) = u$ we obtain

$$\begin{aligned}&\sum_{j=\nu}^{N}\Big(F_-^2 - \hat{F}^2(\hat{\xi}_u^{-1}(\tfrac{j}{L}))\Big)\psi'(j) \\ &\qquad = \int_{-q}^{q} \frac{F^2(-q) - F^2(u)}{\xi(u)-\xi(-q)}\rho(u)\,\mathrm{d}u + O\Big(\tfrac{\log(DL)}{L}\Big).\end{aligned} \tag{12.86}$$

The thermodynamic limit of the sum in the second line of (12.84) vanishes due to lemma A.3.

Combining (12.82) and (12.86) we arrive at

$$H_1[\hat{F}] = (DL)^{-F_-^2}\, G^2(1-F_-)(2\pi)^{F_-}e^{-F_-}$$

$$\times \exp\left\{-\int_{-q}^{q}\frac{F^2(-q)-F^2(u)}{\xi(-q)-\xi(u)}\rho(u)\,\mathrm{d}u\right\}\left(1+O\left(\tfrac{\log(DL)}{L}\right)\right). \quad (12.87)$$

The thermodynamic limit of $H_2[\hat{F}]$ in (12.75) is calculated in the same way. We obtain

$$H_2[\hat{F}] = (DL)^{-F_+^2}G^2(1+F_+)(2\pi)^{-F_+}e^{F_+}$$

$$\times \exp\left\{\int_{-q}^{q}\frac{F^2(q)-F^2(u)}{\xi(q)-\xi(u)}\rho(u)\,\mathrm{d}u\right\}\left(1+O\left(\tfrac{\log(DL)}{L}\right)\right). \quad (12.88)$$

Substituting (12.87) and (12.88) in (12.75), we arrive at the statement of the proposition. □

Thus, combining all the results obtained, we arrive at the following representation for the thermodynamic limit of the discrete part $\mathfrak{D}(\bar{u},\bar{v},\alpha)$:

$$\mathfrak{D}(\bar{u},\bar{v},\alpha) = \mathfrak{D}_{\mathrm{fin}}L^{-(F_-^2+F_+^2)}\left(1+O\left(\tfrac{\log(DL)}{L}\right)\right), \quad (12.89)$$

where

$$\mathfrak{D}_{\mathrm{fin}} = \frac{G^2(1+F_+)G^2(1-F_-)}{\left(2q\rho(q)\right)^{F_+^2+F_-^2}}(2\pi)^{F_--F_+}e^{C_1[F]}. \quad (12.90)$$

The functional $C_1[F]$ has the form:

$$C_1[F] = \frac{1}{2}\int_{-q}^{q}\frac{F'(u)F(v)-F(u)F'(v)}{u-v}\,\mathrm{d}u\,\mathrm{d}v + F_+J_+[F] - F_-J_-[F], \quad (12.91)$$

where

$$J_\pm[F] = \int_{-q}^{q}\frac{F(u)-F_\pm}{u\mp q}\,\mathrm{d}u. \quad (12.92)$$

It results from the combination of integrals (12.65), (12.69), (12.87), and (12.88).

To find the exponent $\gamma_{\mathcal{Q}}$ and the finite amplitude $\mathfrak{A}_{\mathcal{Q}}^{(\mathrm{fin})}$ in (12.19), we need to set $F(u) = \alpha Z(u)/2\pi$. We have

$$\gamma_{\mathcal{Q}} = \frac{\alpha^2\mathcal{Z}^2}{2\pi^2}, \quad (12.93)$$

where $\mathcal{Z} = Z(q) = Z(-q)$. The finite amplitude $\mathfrak{A}_{\mathcal{Q}}^{(\mathrm{fin})}$ is

$$\mathfrak{A}_{\mathcal{Q}}^{(\mathrm{fin})} = \mathfrak{S}\big[\tfrac{\alpha Z}{2\pi}\big] \frac{G^2\big(1, \frac{\alpha \mathcal{Z}}{2\pi}\big)}{\big(2q\rho(q)\big)^{\gamma_{\mathcal{Q}}}} e^{C_1\left[\frac{\alpha Z}{2\pi}\right]}, \tag{12.94}$$

where $G(1, z) = G(1 - z)G(1 + z)$. The smooth part $\mathfrak{S}\big[\frac{\alpha Z}{2\pi}\big]$ is given by (12.45), in which we should set $F(u) = \alpha Z(u)/2\pi$.

In the considered case, the excited state has no particles and holes. Therefore, the exponent $\gamma_{\mathcal{Q}}$ and the finite amplitude $\mathfrak{A}_{\mathcal{Q}}^{(\mathrm{fin})}$ do not depend on the particle/hole momenta and the corresponding quantum numbers. They are only functionals of the shift function:

$$\gamma_{\mathcal{Q}} = \gamma_{\mathcal{Q}}\big[\tfrac{\alpha Z}{2\pi}\big], \qquad \mathfrak{A}_{\mathcal{Q}}^{(\mathrm{fin})} = \mathfrak{A}_{\mathcal{Q}}^{(\mathrm{fin})}\big[\tfrac{\alpha Z}{2\pi}\big]. \tag{12.95}$$

In the conclusion of this section, we draw the reader's attention that when calculating the limit of the discrete part of the form factor amplitude, we used only the fact that the distributions of the parameters $\bar{u}$ and $\bar{v}$ were given respectively by some counting functions $\hat{\xi}_u(z)$ and $\hat{\xi}_v(z)$, which, in turn, determined some shift function. No other properties of the counting functions were used. Therefore, calculating the behavior of the discrete part for large $N$ and $L$ is quite universal. In particular, the same method can be used to analyze the discrete part of the field form factors.

## 12.4 Form factor with particles and holes

Let us now consider a general form factor. Let the excited state be a twisted state containing $n$ particles and $n$ holes. We assume that the particles and holes are determined by a set of quantum numbers $\{p_j, h_j\}$ and a set of momenta $\{\lambda_{p_j}, \lambda_{h_j}\}$, so that $\hat{\xi}_v(\lambda_{p_j}) = p_j/L$ and $\hat{\xi}_v(\lambda_{h_j}) = h_j/L$ for $j = 1, \dots, n$. Here $\hat{\xi}_v(z)$ is the counting function corresponding to the roots $\bar{v}$ of the twisted Bethe equations. To calculate various thermodynamic limits, we will use the standard method of passing to variables $\bar{v}'$, such that $\hat{\xi}_v(v'_j) = j/L$, $j = 1, \dots, N$. The shift of these variables relative to the ground state parameters $\bar{u}$ is given by the shift function $F(z)$ (10.59).

### 12.4.1 *Smooth part*

Calculating the limit of the smooth part of the form factor amplitude (12.29) is still rather simple. It completely repeats the calculation done in

Section 12.3.1.2. We only need to replace the shift function with a modified function $F_{\rm mod}(u)$ (10.168). The final result reads

$$
\begin{aligned}
\operatorname{Lim} \mathfrak{S}(\bar{u}, \bar{v}, \alpha) &= \mathfrak{S}(\{\lambda_{p_j}, \lambda_{h_j}\}, [F]) \\
&= e^{C_0} \left( \prod_{j,k=1}^{n} \frac{(\lambda_{p_j} - \lambda_{h_k} - ic)(\lambda_{h_j} - \lambda_{p_k} - ic)}{(\lambda_{p_j} - \lambda_{p_k} - ic)(\lambda_{h_j} - \lambda_{h_k} - ic)} \right) \left| \prod_{l=1}^{n} \frac{\omega - \lambda_{h_l} - ic}{\omega - \lambda_{p_l} - ic} \right|^2 \\
&\times \left| e^{\tilde{F}(\omega - ic) + \sum_{j=1}^{n} (\tilde{F}(\lambda_{p_j} - ic) - \tilde{F}(\lambda_{h_j} - ic))} \frac{\sin \frac{\alpha}{2}}{\sin \pi F(\omega)} \frac{\det\left(\mathbf{I} - \frac{1}{2\pi i}\mathbf{U}\right)}{\det\left(\mathbf{I} - \frac{1}{2\pi}\mathbf{K}\right)} \right|^2,
\end{aligned}
\tag{12.96}
$$

where $C_0$ is given by (12.44). The kernel of the integral operator $\mathbf{I} - \frac{1}{2\pi i}\mathbf{U}$ has the following form:

$$
\begin{aligned}
U(z, z') = e^{\tilde{F}(z' + ic) - \tilde{F}(z')} \prod_{l=1}^{n} \frac{(z' - \lambda_{p_l})(z' - \lambda_{h_l} + ic)}{(z' - \lambda_{h_l})(z' - \lambda_{p_l} + ic)} \\
\times \frac{\mathcal{K}_\alpha(z', z) - \mathcal{K}_\alpha(\omega, z)}{1 - e^{-2\pi i F(z')}}.
\end{aligned}
\tag{12.97}
$$

The smooth part depends on the momenta of excitations but does not depend on the corresponding quantum numbers. Therefore, this limit is stable when any quantum number obtains a finite deviation. For example, if we shift $p_j \to p_j + m$, where $m$ remains finite in the thermodynamic limit, the corresponding momentum $\lambda_{p_j}$ does not change. Hence, the smooth part (12.96) also does not change.

### 12.4.2 *Discrete part*

The discrete part (12.28) can now be written as follows:

$$
\begin{aligned}
\mathfrak{D}(\bar{u}, \bar{v}, \alpha) = \widetilde{\mathfrak{D}}(\bar{u}, \bar{v}', \alpha) \left( \prod_{j=1}^{n} P(\lambda_{p_j}, p_j) H(\lambda_{h_j}, h_j) \right) \\
\times \frac{\prod_{j>k}^{n} (\lambda_{p_j} - \lambda_{p_k})^2 (\lambda_{h_j} - \lambda_{h_k})^2}{\prod_{j,k=1}^{n} (\lambda_{p_j} - \lambda_{h_k})^2}.
\end{aligned}
\tag{12.98}
$$

Here

$$
\widetilde{\mathfrak{D}}(\bar{u}, \bar{v}', \alpha) = \left( \prod_{j=1}^{N} \frac{\sin^2 \pi F(u_j)}{\pi^2 L^2 \hat{\rho}_u(u_j) \hat{\rho}_v(v'_j)} \right) \frac{\prod_{j>k}^{N} (u_j - u_k)^2 (v'_j - v'_k)^2}{\prod_{j,k=1}^{N} (u_j - v'_k)^2},
\tag{12.99}
$$

where the set $\bar{v}'$ is defined in (10.56). The functions $P(\lambda_{p_j}, p_j)$ and $H(\lambda_{h_j}, h_j)$ are given by

$$P(\lambda_{p_j}, p_j) = \frac{1}{\hat{\rho}_v(\lambda_{p_j})} \left( \prod_{k=1}^{N} \frac{v'_k - \lambda_{p_j}}{u_k - \lambda_{p_j}} \right)^2, \tag{12.100}$$

$$H(\lambda_{h_j}, h_j) = \hat{\rho}_v(\lambda_{h_j}) \frac{\prod_{k=1}^{N}(u_k - \lambda_{h_j})^2}{\prod_{k=1;k\neq h_j}^{N}(v'_k - \lambda_{h_j})^2}. \tag{12.101}$$

The limit of $\widetilde{\mathfrak{D}}(\bar{u}, \bar{v}', \alpha)$ is computed in the same way as we described in the previous section. It is given by (12.89), (12.90), in which the shift function corresponds to the general excited state with particles and holes (see (10.67)).

Presenting the products $P(\lambda_{p_j}, p_j)$ and $H(\lambda_{h_j}, h_j)$ in the form

$$P(\lambda_{p_j}, p_j) = \frac{1}{\hat{\rho}_v(\lambda_{p_j})} \left( \prod_{k=1}^{N} \frac{\phi_v(v'_k, \lambda_{p_j})}{\phi_u(u_k, \lambda_{p_j})} \frac{k - p_j}{k - p_j - \hat{F}(\lambda_{p_j})} \right)^2, \tag{12.102}$$

$$H(\lambda_{h_j}, h_j) = \frac{1}{L^2 \hat{\rho}_v(\lambda_{h_j})} \left( \prod_{k=1}^{N} \frac{\phi_u(u_k, \lambda_{h_j})}{\phi_v(v'_k, \lambda_{h_j})} \right)^2 \frac{\prod_{k=1}^{N}(k - h_j - \hat{F}(\lambda_{h_j}))^2}{\prod_{k=1;k\neq h_j}^{N}(k - h_j)^2}, \tag{12.103}$$

we can partly calculate their thermodynamic limits by the method developed in the previous section. Let

$$J[F](w) = \int_{-q}^{q} \frac{F(u) - F(w)}{u - w}\, du. \tag{12.104}$$

Then for $L$ and $N$ large enough, we obtain

$$\begin{aligned} P(\lambda_{p_j}, p_j) = {} & \frac{1}{\rho(\lambda_{p_j})} \left( \frac{(\lambda_{p_j} - q)(\xi(\lambda_{p_j}) - \xi(-q))}{(\lambda_{p_j} + q)(\xi(\lambda_{p_j}) - \xi(q))} \right)^{2F(\lambda_{p_j})} \\ & \times e^{2J[F](\lambda_{p_j})} \left( \frac{\Gamma(p_j)\Gamma(p_j - N + F(\lambda_{p_j}))}{\Gamma(p_j - N)\Gamma(p_j + F(\lambda_{p_j}))} \right)^2 \left(1 + O(L^{-1})\right), \end{aligned} \tag{12.105}$$

and

$$\begin{aligned} H(\lambda_{h_j}, h_j) = {} & \frac{\sin^2 \pi F(\lambda_{h_j})}{L^2 \pi^2 \rho(\lambda_{h_j})} \left( \frac{(\lambda_{h_j} + q)(\xi(q) - \xi(\lambda_{h_j}))}{(q - \lambda_{h_j})(\xi(\lambda_{h_j}) - \xi(-q))} \right)^{2F(\lambda_{h_j})} \\ & \times e^{-2J[F](\lambda_{h_j})} \left( \frac{\Gamma(h_j + F(\lambda_{h_j}))\Gamma(N + 1 - h_j - F(\lambda_{h_j}))}{\Gamma(h_j)\Gamma(N + 1 - h_j)} \right)^2 \\ & \times \left(1 + O(L^{-1})\right). \end{aligned} \tag{12.106}$$

**Remark 12.1.** We assume in (12.105) that $p_j > N$. If $p_j < 1$, then the $\Gamma$-functions of non-positive integer arguments should be understood as a limit:

$$\frac{\Gamma(p_j)}{\Gamma(p_j - N)} = \lim_{\epsilon \to 0} \frac{\Gamma(p_j + \epsilon)}{\Gamma(p_j - N + \epsilon)} = (-1)^N \frac{\Gamma(N + 1 - p_j)}{\Gamma(1 - p_j)}. \tag{12.107}$$

This rule should be used in all further formulas.

Further calculation of the limits of $P(\lambda_{p_j}, p_j)$ and $H(\lambda_{h_j}, h_j)$ strongly depends on the position of the particle and hole momenta relative to the Fermi boundaries.

### 12.4.3 *Excitations away from the Fermi boundaries*

Suppose that, in the thermodynamic limit, particles and holes are separated from the Fermi boundaries: $\lambda_{p_j} \neq \pm q$ and $\lambda_{h_j} \neq \pm q$. This means that either $p_j - N \gg 1$ or $-p_j \gg 1$ and either $N - h_j \gg 1$ or $h_j \gg 1$. Then the arguments of the $\Gamma$-functions in (12.105) and (12.106) are large positive numbers. Hence, we can apply Stirling's formula for their asymptotics

$$\frac{\Gamma(z + a)}{\Gamma(z)} = z^a \big(1 + O(z^{-1})\big), \qquad z \to +\infty. \tag{12.108}$$

We obtain

$$\begin{aligned} \operatorname{Lim} \frac{\Gamma(p_j)\Gamma(p_j - N + F(\lambda_{p_j}))}{\Gamma(p_j - N)\Gamma(p_j + F(\lambda_{p_j}))} &= \left(\frac{p_j - N}{p_j}\right)^{F(\lambda_{p_j})} \\ &= \left(\frac{\xi(\lambda_{p_j}) - \xi(q)}{\xi(\lambda_{p_j}) - \xi(-q)}\right)^{F(\lambda_{p_j})}, \end{aligned} \tag{12.109}$$

$$\begin{aligned} &\operatorname{Lim} \frac{\Gamma(h_j + F(\lambda_{h_j}))\Gamma(N + 1 - h_j - F(\lambda_{h_j}))}{\Gamma(h_j)\Gamma(N + 1 - h_j)} \\ &= \left(\frac{h_j}{N - h_j}\right)^{F(\lambda_{h_j})} = \left(\frac{\xi(\lambda_{h_j}) - \xi(-q)}{\xi(q) - \xi(\lambda_{h_j})}\right)^{F(\lambda_{h_j})}. \end{aligned} \tag{12.110}$$

Here we used $\xi(q) = D$ and $\xi(-q) = 0$. Substituting these results in (12.105) and (12.106) we find

$$\operatorname{Lim} P(\lambda_{p_j}, p_j) = \frac{1}{\rho(\lambda_{p_j})} \exp\left\{2 \int_{-q}^{q} \frac{F(u)\,\mathrm{d}u}{u - \lambda_{p_j}}\right\}, \tag{12.111}$$

and

$$\operatorname{Lim} L^2 H(\lambda_{h_j}, h_j) = \frac{\sin^2 \pi F(\lambda_{h_j})}{\pi^2 \rho(\lambda_{h_j})} \exp\left\{-2 V.P. \int_{-q}^{q} \frac{F(u)\,\mathrm{d}u}{u - \lambda_{h_j}}\right\}. \tag{12.112}$$

Here $V.P. \int$ means the principal value of the integral.

We see that the limiting values of $P(\lambda_{p_j}, p_j)$ and $L^2 H(\lambda_{h_j}, h_j)$ do not depend on the quantum numbers $p_j$ and $h_j$. They are smooth functions of the momenta $\lambda_{p_j}$ and $\lambda_{h_j}$.

Thus, if the particle and hole momenta are not on the Fermi boundaries, the discrete part behaves as follows:

$$\begin{aligned}
&\mathfrak{D}(\bar{u},\bar{v},\alpha)\\
&\quad=\frac{G^2(1+F_+)G^2(1-F_-)}{(2q\rho(q)L)^{F_+^2+F_-^2}}(2\pi)^{F_- - F_+}e^{C_1[F]}\\
&\quad\times\left(\prod_{j=1}^{n}\frac{\sin^2\pi F(\lambda_{h_j})}{L^2\pi^2\rho(\lambda_{p_j})\rho(\lambda_{h_j})}\exp\left\{2V.P.\int_{-q}^{q}\frac{(\lambda_{p_j}-\lambda_{h_j})F(u)\,du}{(u-\lambda_{p_j})(u-\lambda_{h_j})}\right\}\right)\\
&\quad\times\frac{\prod_{j>k}^{n}(\lambda_{p_j}-\lambda_{p_k})^2(\lambda_{h_j}-\lambda_{h_k})^2}{\prod_{j,k=1}^{n}(\lambda_{p_j}-\lambda_{h_k})^2}\Big(1+O\Big(\tfrac{\log(DL)}{L}\Big)\Big).
\end{aligned}\tag{12.113}$$

Recall that $F(u)$ is the shift function (10.67) that also depends on the momenta of particles and holes: $F(u) = F(u|\{\lambda_{p_j}, \lambda_{h_j}\})$. We see that the discrete part does not depend on the quantum numbers $\{p_j, h_j\}$. Given that the smooth part is also independent of quantum numbers, we conclude that the entire form factor amplitude depends only on the particles and holes momenta. We also see that the form factor amplitude is proportional to the fractional power of the system size $L^{-2n-F_+^2-F_-^2}$. When integrating over the momenta of particles and holes, the coefficient $L^{-2n}$ becomes a part of the integration measure. However, the $L^{-F_+^2-F_-^2}$ factor remains. At the same time, we see that the result obtained is valid only if $\lambda_{p_j} \ne \pm q$ and $\lambda_{h_j} \ne \pm q$, $j = 1, \dots, n$. Indeed, otherwise, the Cauchy determinant in the last line of (12.113) has singularities, and the integral in the third line diverges.

### 12.4.4 *Critical states and critical form factors*

First of all, we show with a simple example what happens to the discrete part of the form factor amplitude when the momenta of particles and/or holes fall on the Fermi boundary in the thermodynamic limit. Let $\lambda_{p_j} \to q$ for some $j \in \{1, \dots, n\}$ in the thermodynamic limit. This means that the corresponding quantum number $p_j$ has the form $p_j = N + p_j^+$, where $p_j^+/N \to 0$ for $N \to \infty$. Substituting this $\lambda_{p_j}$ and $p_j$ in (12.105) we obtain

$$P(q,p_j^+) = \frac{1}{\rho(q)}\left(\frac{D}{2q\rho(q)}\right)^{2F_+} \times \left(\frac{\Gamma(p_j^+ + F_+)\Gamma(N+p_j^+)}{\Gamma(p_j^+)\Gamma(N+p_j^+ + F_+)}\right)^2 e^{2J_+[F]}\Big(1+O(L^{-1})\Big). \tag{12.114}$$

It is clear that now we cannot apply Stirling's formula to the ratio of $\Gamma$-functions $\Gamma(p_j^+ + F_+)/\Gamma(p_j^+)$ since $p_j^+$ can be finite. The arguments of the second pair of $\Gamma$-functions go to infinity. Therefore, taking the limit, we arrive at

$$P(q,p_j^+) = \frac{e^{2J_+[F]}}{\rho(q)\Big(2q\rho(q)L\Big)^{2F_+}}\left(\frac{\Gamma(p_j^+ + F_+)}{\Gamma(p_j^+)}\right)^2 \Big(1+O(L^{-1})\Big). \tag{12.115}$$

First of all, we see that $P(q,p_j^+)$ tends to zero or infinity, depending on the sign of $F_+$. Besides, the finite part of $P(q,p_j^+)$ depends on the quantum number $p_j^+$. Moreover, it changes by a finite value with a finite change in the quantum number, say $p_j^+ \to p_j^+ + 1$. At the same time, a finite change in $p_j^+$ yields an infinitely small change in the corresponding momentum $\lambda_{p_j}$ (of order $L^{-1}$). Therefore, we obviously cannot replace the sum with an integral when summing form factors over such excited states.

The example above shows that we need to study the form factors more carefully when the particle and hole momenta are at the Fermi boundary. Of course, there are states in which some of the momenta tend to $\pm q$, and the rest of the momenta are separated from these points. We will consider these states and the corresponding form factors in Section 12.8. Now we will consider the case when all particles and holes momenta coincide with the Fermi boundaries in the thermodynamic limit.

**Definition 12.1.** Let an excited state contain $n$ particles and $n$ holes. It is called a *critical excited state* if all the momenta $\lambda_{p_j}, \lambda_{h_j} = \pm q$, $j = 1,\dots,n$, in the thermodynamic limit. The corresponding form factor is called a *critical form factor*.

The excitation energy of any critical state is equal to zero in the thermodynamic limit (more precisely, it behaves as $1/L$). Therefore, it is natural to expect that the critical states produce the main contribution to the long-distance asymptotics of correlation functions. There is also one more argument to support this assumption.

Consider the long-distance asymptotics of the generating function $Q_N^{(\alpha)}(x,t)$, while the time $t$ remains finite. The dependence on the distance $x$ is included in the form factor series as a phase factor $\exp\{ix\mathcal{P}_{\rm ex}\}$ (see (12.17)). This factor becomes rapidly oscillating at $x$ large. Then the main contribution to the asymptotics occurs either when summing in the vicinity of the saddle point or at the boundaries of the summation region. In the thermodynamic limit, the momentum of the excited state $\mathcal{P}_{\rm ex}$ is (see (10.95)):

$$\mathcal{P}_{\rm ex} = \alpha D + \sum_{j=1}^{n}(k(\lambda_{p_j}) - k(\lambda_{h_j})), \tag{12.116}$$

where the dressed momentum $k(u)$ is given by (10.94). Due to (10.94), we have $k'(u) = 2\pi\rho(u) > 0$ for arbitrary $u \in \mathbb{R}$. Therefore, the oscillating exponent has no saddle point. Hence, the main contribution to the asymptotics of the correlation function is produced by the neighborhoods of the points $\pm q$, which are the summation boundaries for the particle and hole momenta values.

The above considerations allow us to assume that we can only limit ourselves to the sum of critical form factors for calculating the long-distance asymptotics of correlation functions. We will see that this assumption agrees with predictions arising from the Luttinger liquid and conformal field theory approaches [I. Affleck (1985); H. W. J. Blöte et al (1986); J. L. Cardy (1984, 1986); F. D. M. Haldane (1980, 1981a,b); A. Luther and I. Peschel (1975)]. We will also see that it is the summation over quantum numbers in the vicinity of the Fermi boundaries that allows us to get rid of the infinitely small prefactor $L^{-\gamma_{\mathcal{Q}}}$ and ultimately ensures finiteness of the generating function $Q_N^{(\alpha)}(x,t)$.

Different critical excited states can be characterized by the specific distribution of particles and holes on the Fermi boundaries $\pm q$. Let $n_p^{\pm}$ momenta of particles be equal to $\pm q$ and $n_h^{\pm}$ momenta of holes be equal to $\pm q$ in the thermodynamic limit. Obviously,

$$n_p^+ + n_p^- = n_h^+ + n_h^- = n. \tag{12.117}$$

**Definition 12.2.** Let in a given critical excited state, the distribution of particles and holes on the Fermi boundaries be such that

$$n_p^+ - n_h^+ = n_h^- - n_p^- = \ell, \qquad -n \le \ell \le n. \tag{12.118}$$

Then we say that this critical excited state belongs to the $\mathbf{P}_\ell$ class. The corresponding critical form factor also will be said to belong to the $\mathbf{P}_\ell$ class.

The separation of critical states into different classes is carried out according to their excitation momenta in the thermodynamic limit. Indeed, it follows from (12.116) that for all the states of $\mathbf{P}_\ell$ class, it is equal to $2\ell k_F + \alpha D$, where $k_F = \pi D$ is the Fermi momentum. Different critical excited states belonging to the $\mathbf{P}_\ell$ class depend on different sets of quantum numbers $\{p\}$ and $\{h\}$.

For $L$ large but finite, the momentum of every individual critical state has a deviation from $2\ell k_F + \alpha D$ of order $1/L$. It follows from (10.89)–(10.92) that

$$\mathcal{P}_{\text{ex}} = \alpha D + \frac{2\pi}{L} \sum_{k=1}^{n} (p_k - h_k), \tag{12.119}$$

where $p_k$ and $h_k$ are the quantum numbers of particles and holes. It is convenient to re-parameterize these integers for the critical excited states as follows:

$$\begin{aligned} p_j &= p_j^+ + N, && \text{if } \lambda_{p_j} = q\,, \\ p_j &= 1 - p_j^-\,, && \text{if } \lambda_{p_j} = -q\,, \\ h_j &= N + 1 - h_j^+, && \text{if } \lambda_{h_j} = q\,, \\ h_j &= h_j^-\,, && \text{if } \lambda_{h_j} = -q\,. \end{aligned} \tag{12.120}$$

All the integers $\{p^\pm\}$ and $\{h^\pm\}$ introduced above are positive and vary in a range such that

$$\lim_{N\to\infty} \frac{\sum p_j^\pm}{N} = \lim_{N\to\infty} \frac{\sum h_j^\pm}{N} = 0\,. \tag{12.121}$$

Hence, all the particles and holes collapse to $\pm q$ in the thermodynamic limit. Then the expression for the excitation momentum of critical states belonging to the $\mathbf{P}_\ell$ class becomes

$$\mathcal{P}_{\text{ex}} = \frac{\alpha_\ell k_F}{\pi} + \frac{2\pi}{L}(E^+ - E^- + \ell), \tag{12.122}$$

where $\alpha_\ell = \alpha + 2\pi\ell$ and

$$E^\pm = \sum_{j=1}^{n_p^\pm} (p_j^\pm - 1) + \sum_{j=1}^{n_h^\pm} h_j^\pm. \tag{12.123}$$

All critical states of the $\mathbf{P}_\ell$ class are described by the same shift function in the thermodynamic limit. Let us denote it by $F_\ell(u)$. Then it follows from (10.67) that

$$F_\ell(u) = \frac{\alpha}{2\pi} Z(u) + \ell\big(\phi(u, -q) - \phi(u, q)\big). \tag{12.124}$$

Due to (10.71), we obtain

$$F_\ell(u) = \frac{\alpha_\ell}{2\pi} Z(u) - \ell. \tag{12.125}$$

It is easy to check that the limits of the sums (10.167), (10.171) then take the form

$$\operatorname{Lim} \mathcal{J}_1 = \frac{\alpha_\ell}{2\pi} \int_{-q}^{q} f'(u) Z(u) \, du, \tag{12.126}$$

$$\operatorname{Lim} \mathcal{J}_2 = -\left(\frac{\alpha_\ell}{2\pi}\right)^2 \int_{-q}^{q} \frac{\partial^2 f(u,v)}{\partial u \partial v} Z(u) Z(v) \, du \, dv. \tag{12.127}$$

At the same time, to calculate the smooth part of the amplitude of the form factor corresponding to the twisted ground state, we used formulas (10.162), (10.165), in which we should put $F(u) = \alpha Z(u)/2\pi$. We see that (12.126), (12.127) differ from (10.162), (10.165) only by a shift of the auxiliary parameter $\alpha \to \alpha_\ell = \alpha + 2\pi\ell$. Therefore, the smooth part of the amplitude of the $\mathbf{P}_\ell$ class form factors is the same as that of the form factors corresponding to the $\alpha_\ell$-twisted ground state. This coincidence is not accidental.

The class $\mathbf{P}_\ell$ contains an infinite number of states. Indeed, knowing that a particle (resp. hole) momentum equals $\pm q$ does not allow to fix the corresponding integers $p_k$ (resp. $h_k$) unambiguously. On the other hand, quantities with well-defined thermodynamic limits (such as the smooth part of the form factor amplitude) must only depend on the macroscopic realization $\{\lambda_p\}$ and $\{\lambda_h\}$ of the excited state and not on the microscopic quantum numbers $\{p\}$ and $\{h\}$ leading to such a macroscopic configuration of the momenta. Therefore, we can calculate the thermodynamic limit by choosing any representative of the $\mathbf{P}_\ell$ class. It is easy to see that one of these representatives is just the $\alpha_\ell$-twisted ground state.

### 12.4.5 *Discrete part of critical form factor amplitude*

In Section 12.4.4, we have already started calculating the limit of the discrete part of the form factor amplitude when the momenta of particles and holes tend to the Fermi boundary. Let us complete this calculation. Assume that we are dealing with a critical form factor of the $\mathbf{P}_\ell$ class. The distribution of particles and holes at the Fermi boundaries is given by equation (12.118), and the quantum numbers are given by formulas (12.120). The shift function $F_\ell(u)$ has the form (12.125), and $F_{\ell,\pm} = F_\ell(\pm q)$.

Then we have (see (12.115))

$$P(\lambda_{p_j}, p_j^+) = \frac{e^{2J_+[F_\ell]}}{\rho(q)\Big(2q\rho(q)L\Big)^{2F_{\ell,+}}} \left(\frac{\Gamma(p_j^+ + F_{\ell,+})}{\Gamma(p_j^+)}\right)^2 \Big(1 + O(L^{-1})\Big), \tag{12.128}$$

for $\lambda_{p_j} \to q$. Similarly, using (12.107) we find

$$P(\lambda_{p_j}, p_j^-) = \frac{e^{2J_-[F_\ell]}\Big(2q\rho(q)L\Big)^{2F_{\ell,-}}}{\rho(q)} \left(\frac{\Gamma(p_j^- - F_{\ell,-})}{\Gamma(p_j^-)}\right)^2 \Big(1 + O(L^{-1})\Big), \tag{12.129}$$

for $\lambda_{p_j} \to -q$.

The limit $H(\lambda_{h_j}, h_j)$ (12.106) is calculated in the same way. We have

$$\begin{aligned} H(\lambda_{h_j}, h_j^+) &= \frac{e^{-2J_+[F_\ell]}\sin^2 \pi F_{\ell,+}\Big(2q\rho(q)L\Big)^{2F_{\ell,+}}}{(\pi L)^2\rho(q)} \\ &\quad\times \left(\frac{\Gamma(h_j^+ - F_{\ell,+})}{\Gamma(h_j^+)}\right)^2 \Big(1 + O(L^{-1})\Big), \end{aligned} \tag{12.130}$$

for $\lambda_{h_j} \to q$, and

$$\begin{aligned} H(\lambda_{h_j}, h_j^-) &= \frac{e^{-2J_-[F_\ell]}\sin^2 \pi F_{\ell,-}}{(\pi L)^2\rho(q)\Big(2q\rho(q)L\Big)^{2F_{\ell,-}}} \\ &\quad\times \left(\frac{\Gamma(h_j^- + F_{\ell,-})}{\Gamma(h_j^-)}\right)^2 \Big(1 + O(L^{-1})\Big), \end{aligned} \tag{12.131}$$

for $\lambda_{h_j} \to -q$.

Thus, the product of the factors $P(\lambda_{p_j}, p_j)$ and $H(\lambda_{h_j}, h_j)$ in (12.98) has the following form:

$$\begin{aligned} &\prod_{j=1}^{n} P(\lambda_{p_j}, p_j) H(\lambda_{h_j}, h_j) \\ &\quad = e^{2\ell(J_+[F_\ell] - J_-[F_\ell])} \\ &\qquad \times\Gamma^2\left(\begin{matrix}\{p^+ + F_{\ell,+}\}, \{p^- - F_{\ell,-}\}, \{h^+ - F_{\ell,+}\}, \{h^- + F_{\ell,-}\} \\ \{p^+\}, \{p^-\}, \{h^+\}, \{h^-\}\end{matrix}\right) \\ &\qquad \times \frac{(\sin \pi F_{\ell,+})^{2n_h^+}(\sin \pi F_{\ell,-})^{2n_h^-}}{(\pi L\rho(q))^{2n}\Big(2q\rho(q)L\Big)^{2\ell(F_{\ell,+} + F_{\ell,-})}} \Big(1 + O(L^{-1})\Big). \end{aligned} \tag{12.132}$$

Here we have used the standard hypergeometric type notation for ratios of Γ-functions:

$$\Gamma\begin{pmatrix} a_1 , & \dots , & a_r \\ b_1 , & \dots , & b_s \end{pmatrix} = \frac{\prod_{k=1}^{r} \Gamma(a_k)}{\prod_{k=1}^{s} \Gamma(b_k)}. \tag{12.133}$$

We have only to calculate the thermodynamic limit of the Cauchy determinant in equation (12.98). This is a relatively simple task. For example, if two particles with momenta $\lambda_{p_j}$ and $\lambda_{p_k}$ are located at different boundaries of the Fermi zone (say, $\lambda_{p_j} \to q$ and $\lambda_{p_k} \to -q$), then obviously

$$\lambda_{p_j} - \lambda_{p_k} = 2q + O(L^{-1}). \tag{12.134}$$

A similar estimate holds for the momentum difference of two holes or the momentum difference of a particle and a hole. If the momenta tend to the same boundary (say, $\lambda_{p_j} \to q$ and $\lambda_{p_k} \to q$), then

$$\begin{aligned}\lambda_{p_j} - \lambda_{p_k} &= \hat{\xi}_v^{-1}\left(\tfrac{N+p_j^+}{L}\right) - \hat{\xi}_v^{-1}\left(\tfrac{N+p_k^+}{L}\right) \\ &= \frac{p_j^+ - p_k^+}{L}\left(\hat{\xi}_v^{-1}\right)'\left(\tfrac{N}{L}\right) + O(L^{-2}) = \frac{p_j^+ - p_k^+}{L\rho(q)} + O(L^{-2}).\end{aligned} \tag{12.135}$$

We do the same with the other differences and eventually come to the following estimate:

$$\begin{aligned}\frac{\prod_{j>k}^{n}(\lambda_{p_j} - \lambda_{p_k})^2(\lambda_{h_j} - \lambda_{h_k})^2}{\prod_{j,k=1}^{n}(\lambda_{p_j} - \lambda_{h_k})^2} &= \left(L\rho(q)\right)^{2n-2\ell^2}\left(2q\right)^{-2\ell^2} L_{n_p^+,n_h^+}(\{p^+\}, \\ &\qquad \{h^+\})L_{n_p^-,n_h^-}(\{p^-\},\{h^-\})\Big(1 + O(L^{-1})\Big),\end{aligned} \tag{12.136}$$

where

$$L_{n,n'}(\{p\},\{h\}) = \frac{\prod_{j>k}^{n}(p_j - p_k)^2 \prod_{j>k}^{n'}(h_j - h_k)^2}{\prod_{j=1}^{n}\prod_{k=1}^{n'}(p_j + h_k - 1)^2}. \tag{12.137}$$

Gathering all the results obtained, we find the exponent $\gamma_{\mathcal{Q}}$ and the finite amplitude $\mathfrak{A}_{\mathcal{Q}}^{(\mathrm{fin})}$ in the case of the critical form factor of the $\mathbf{P}_\ell$ class. We have

$$\gamma_{\mathcal{Q}} = (F_{\ell,+} + \ell)^2 + (F_{\ell,-} + \ell)^2. \tag{12.138}$$

Due to (12.125), we obtain

$$\gamma_{\mathcal{Q}} = \frac{\alpha_\ell^2}{2\pi^2}\mathcal{Z}^2. \tag{12.139}$$

Thus, the exponent $\gamma_{\mathcal{Q}}$ is the same as the exponent of the form factor corresponding to the $\alpha_\ell$-twisted ground state. This is to be expected since $\gamma_{\mathcal{Q}}$ does not depend on a particular representative of the $\mathbf{P}_\ell$ class.

Note that the exponent $\gamma_{\mathcal{Q}}$ is determined by the values of the shift function at the Fermi boundaries for critical form factors. It can also be written as

$$\gamma_{\mathcal{Q}} = \operatorname{Lim} L^2\Big(\hat{\rho}_u^2(u_{\min})(v_{\min}-u_{\min})^2+\hat{\rho}_u^2(u_{\max})(v_{\max}-u_{\max})^2\Big). \quad (12.140)$$

Indeed, it is easy to see that equations (12.138) and (12.140) are equivalent. However, representation (12.140) is the most general for the power-law behavior of the form factor amplitude. It holds not only for the amplitude $|\mathcal{F}_{\mathcal{Q}}|^2$ but also for the amplitude of the field form factor.

As we already explained, the smooth part does not depend on a particular representative of the $\mathbf{P}_\ell$ class. Then the finite amplitude $\mathfrak{A}_{\mathcal{Q}}^{(\text{fin})}$ is

$$\begin{aligned}
&\mathfrak{A}_{\mathcal{Q}}^{(\text{fin})}(\{p^\pm\},\{h^\pm\}|[F_\ell])\\
&= \mathfrak{S}\big[F_\ell+\ell\big]\frac{G^2(1+F_{\ell,+})G^2(1-F_{\ell,-})}{\big(2q\rho(q)\big)^{(F_{\ell,+}+\ell)^2+(F_{\ell,-}+\ell)^2}}\\
&\times(2\pi)^{F_{\ell,-}-F_{\ell,+}}e^{C_1[F_\ell+\ell]}\left(\frac{\sin\pi F_{\ell,+}}{\pi}\right)^{2n_h^+}\left(\frac{\sin\pi F_{\ell,-}}{\pi}\right)^{2n_h^-}\\
&\times R_{n_p^+,n_h^+}(\{p^+\},\{h^+\}|F_{\ell,+})R_{n_p^-,n_h^-}(\{p^-\},\{h^-\}|-F_{\ell,-}),
\end{aligned} \quad (12.141)$$

where

$$R_{n,n'}(\{p\},\{h\}|\nu) = L_{n,n'}(\{p\},\{h\})\Gamma^2\left(\begin{matrix}\{p+\nu\},\,\{h-\nu\}\\ \{p\},\,\{h\}\end{matrix}\right). \quad (12.142)$$

We also used a property of the functional $C_1[F]$

$$C_1[F+a] = C_1[F]+2a(J_+[F]-J_-[F]), \quad (12.143)$$

for any constant $a$.

We see that the finite amplitude $\mathfrak{A}_{\mathcal{Q}}^{(\text{fin})}$ depends on the quantum numbers $\{p^\pm\},\{h^\pm\}$ and functionally depends on the shift function $F_\ell$.

To conclude this section, we discuss a question of corrections to the results obtained. The finite amplitude was calculated up to the corrections of order $O\big(\log(DL)/L\big)$. It comes from the evaluation of the product $\widetilde{\mathfrak{D}}(\bar{u},\bar{v}',\alpha)$ (12.99). Therefore, it does not depend on quantum numbers. However, other corrections that we neglected when calculating the large $L$ behavior of the discrete part may depend on $\{p^\pm\}$ and $\{h^\pm\}$. For example, such corrections obviously occur in formulas (12.134), (12.135). Strictly speaking, such corrections can give a finite contribution to the correlation function when summing over quantum numbers. There are, however, arguments that allow us to state that these contributions are subleading in the asymptotic expansion of the correlation function for large $x$. We will discuss these arguments later. In the meantime, we assume that we can neglect all corrections to the finite amplitude when calculating the large $x$ asymptotic behavior of the correlation function.

## 12.5 Summation of critical form factors

In this section, we calculate the sum of the critical form factors. It takes place in two stages. At the first stage, we sum up the form factors within a $\mathbf{P}_\ell$ class, and at the next stage, we take the sum over different classes. The latter turns into the sum over the integer $\ell$ that characterizes each class. At the first stage, we need to calculate the sum over quantum numbers $\{p^\pm, h^\pm\}$.

Let us denote by $Q^{(\alpha)}_{\rm cr}(x,t)$ the part of the generating function $Q^{(\alpha)}_N(x,t)$ that arises by summing only the critical form factors. Then

$$Q^{(\alpha)}_{\rm cr}(x,t) = \sum_{\ell\in\mathbb{Z}}\sum_{\{p^\pm,h^\pm\}} e^{ix\mathcal{P}_{\rm ex}-it\mathcal{E}_{\rm ex}}|\boldsymbol{\mathcal{F}}_{\mathcal{Q}}(\{p^\pm\},\{h^\pm\}|\mathbf{P}_\ell)|^2, \tag{12.144}$$

where $|\boldsymbol{\mathcal{F}}_{\mathcal{Q}}(\{p^\pm\},\{h^\pm\}|\mathbf{P}_\ell)|^2$ is the amplitude of the critical form factor of the $\mathbf{P}_\ell$ class.

We have already calculated the momentum $\mathcal{P}_{\rm ex}$ of the excited state from the $\mathbf{P}_\ell$ class, taking into account the $1/L$-corrections (see equation (12.122)). To sum up critical form factors within a fixed $\mathbf{P}_\ell$ class, we also need to know the energy, including $1/L$-corrections. This calculation is much more complex than calculating the momentum corrections. Therefore, we only give the final result here, and the details of the calculations are given in appendix B.

Let a critical excited state be characterized by the quantum numbers (12.120). Then the $1/L$-correction to the energy of this state has the following form:

$$\mathcal{E}_{\rm ex} = \frac{2\pi v_F}{L}(E^+ + E^-) + \frac{1}{L}\Delta\mathcal{E}(\mathbf{P}_\ell) + O(L^{-2}). \tag{12.145}$$

Here $E^\pm$ are given by (12.123), and $v_F$ is called the *Fermi velocity*

$$v_F = \left.\frac{\partial\varepsilon(u)}{\partial k(u)}\right|_{u=q} = \frac{\varepsilon'(q)}{k'(q)}. \tag{12.146}$$

This value makes sense of the velocity of sound in the gas at the Fermi boundary. The term $\Delta\mathcal{E}(\mathbf{P}_\ell)$ is common for all the states of the $\mathbf{P}_\ell$ class. It does not depend on the specific representative of this class.

According to the assumption formulated at the end of the previous section, we can replace the amplitude $|\boldsymbol{\mathcal{F}}_{\mathcal{Q}}(\{p^\pm\},\{h^\pm\}|\mathbf{P}_\ell)|^2$ in (12.144) with the product of $L^{-\gamma_{\mathcal{Q}}}$ and the finite amplitude $\mathfrak{A}^{(\rm fin)}_{\mathcal{Q}}$. We can ignore all corrections to the finite amplitude. Given that the exponent $\gamma_{\mathcal{Q}}$ is independent of quantum numbers, we obtain

$$Q_{\rm cr}^{(\alpha)}(x,t) = \sum_{\ell\in\mathbb{Z}} e^{ix\alpha_\ell k_F/\pi} e^{(ix\Delta\mathcal{P}(\mathbf{P}_\ell)-it\Delta\mathcal{E}(\mathbf{P}_\ell))/L} L^{-(F_{\ell,+}+\ell)^2-(F_{\ell,-}+\ell)^2} \\ \times \sum_{\{p^\pm,h^\pm\}} e^{2\pi i[(x-v_Ft)E^+-(x+v_Ft)E^-]/L}\, \mathfrak{A}_{\mathcal{Q}}^{(\rm fin)}(\{p^\pm\},\{h^\pm\}|[F_\ell]). \tag{12.147}$$

In this formula, corrections to the excitation momentum and energy that do not depend on quantum numbers are respectively indicated by symbols $\Delta\mathcal{P}(\mathbf{P}_\ell)$ and $\Delta\mathcal{E}(\mathbf{P}_\ell)$. We also intentionally do not use expression (12.125) for $F_\ell(u)$ yet. We do this to perform calculations in the most general way, without using any additional information about the shift function.

It is easy to see that the internal sum over the quantum numbers in (12.147) factorizes into the product of the sum over $\{p^+,h^+\}$ and the sum over $\{p^-,h^-\}$:

$$\sum_{\{p^\pm,h^\pm\}} e^{2\pi i[(x-v_Ft)E^+-(x+v_Ft)E^-]/L}\, \mathfrak{A}_{\mathcal{Q}}^{(\rm fin)}(\{p^\pm\},\{h^\pm\}|[F_\ell]) \\ = \mathfrak{S}\big[F_\ell+\ell\big]\frac{e^{C_1[F_\ell+\ell]}(2\pi)^{F_{\ell,-}-F_{\ell,+}}G^2(1+F_{\ell,+})G^2(1-F_{\ell,-})}{(2q\rho(q))^{(F_{\ell,+}+\ell)^2+(F_{\ell,-}+\ell)^2}}\Sigma^+\Sigma^-, \tag{12.148}$$

where

$$\Sigma^\pm = \sum_{\substack{n_p,n_h\ge 0\\ n_p-n_h=\pm\ell}}\ \sum_{\substack{1\le p_1<\cdots<p_{n_p}<\infty\\ 1\le h_1<\cdots<h_{n_h}<\infty}} \left(\frac{\sin\pi F_{\ell,\pm}}{\pi}\right)^{2n_h}\left(\prod_{k=1}^{n_h} e^{\pm 2\pi i(x\mp v_Ft)h_k/L}\right) \\ \times \left(\prod_{j=1}^{n_p} e^{\pm 2\pi i(x\mp v_Ft)(p_j-1)/L}\right) R_{n_p,n_h}(\{p\},\{h\}|\pm F_{\ell,\pm}). \tag{12.149}$$

Remarkably, this sum can be explicitly calculated. Similar sums (for $\ell=0$) were studied earlier in the works devoted to $Z$-measures on partitions [S. Kerov et al (1993); S. Kerov (2000); A. Borodin and G. Olshanski (2000a,b); A. Okounkov (2001); A. Borodin and G. Olshanski (2001)].

**Proposition 12.2.** *Let $|z|<1$. For any integer $\ell$ and an arbitrary complex $\nu$, we define a function $f_\ell(\nu,z)$ as a series*

$$f_\ell(\nu,z) = \sum_{\substack{n,n'\ge 0\\ n-n'=\ell}}\ \sum_{\substack{1\le p_1<\cdots<p_n<\infty\\ 1\le h_1<\cdots<h_{n'}<\infty}} \left(\frac{\sin\pi\nu}{\pi}\right)^{2n'} \\ \times\left(\prod_{j=1}^{n} z^{p_j-1}\right)\left(\prod_{k=1}^{n'} z^{h_k}\right) R_{n,n'}(\{p\},\{h\}|\nu). \tag{12.150}$$

*Then the series is absolutely convergent and*

$$f_\ell(\nu, z) = \frac{G^2(1+\ell+\nu)}{G^2(1+\nu)} \frac{z^{\ell(\ell-1)}}{(1-z)^{(\nu+\ell)^2}}. \tag{12.151}$$

In our case, $\nu = F_{\ell,\pm}$ and $z = \exp\{\pm 2\pi i(x \mp v_F t)/L\}$. Therefore, to make the series (12.149) convergent, we make a shift $t \to t - i0$. The original proof of proposition 12.2 can be found in [A. Borodin and G. Olshanski (2000a)]. In paper [N. Kitanine et al (2011b)], the proof was given using the representation for the sum (12.150) as the determinant of an infinite matrix. A nice method for calculating the sum (12.150) using free fermions can also be found in [K. K. Kozlowski and J. M. Maillet (2015)] (see also [A. Alexandrov and A. Zabrodin (2013)]).

Using equation (12.150), we immediately compute the sums $\Sigma^\pm$ (12.149), and thus, the sum over quantum numbers in (12.148). We have

$$\begin{aligned}\sum_{\{p^\pm, h^\pm\}} & e^{2\pi i[(x-v_F t)E^+ - (x+v_F t)E^-]/L}\, \mathfrak{A}_Q^{(\mathrm{fin})}(\{p^\pm\},\{h^\pm\}|[F_\ell]) \\ &= \mathfrak{G}[F_\ell+\ell] \frac{e^{C_1[F_\ell+\ell]}(2\pi)^{F_{\ell,-}-F_{\ell,+}}}{(2q\rho(q))^{(F_{\ell,+}+\ell)^2+(F_{\ell,-}+\ell)^2}} \\ &\times \frac{e^{-4i\pi\ell(x+\ell v_F t)/L} G^2(1+F_{\ell,+}+\ell) G^2(1-F_{\ell,-}-\ell)}{\left(1-e^{2i\pi(x-v_F t)/L}\right)^{(F_{\ell,+}+\ell)^2} \left(1-e^{-2i\pi(x+v_F t)/L}\right)^{(F_{\ell,-}+\ell)^2}}. \end{aligned} \tag{12.152}$$

Now we can easily take the limit $L \to \infty$:

$$\begin{aligned}\lim_{L\to\infty} & \frac{L^{-(F_{\ell,+}+\ell)^2-(F_{\ell,-}+\ell)^2} e^{-4i\pi\ell(x+\ell v_F t)/L}}{\left(1-e^{2i\pi(x-v_F t)/L}\right)^{(F_{\ell,+}+\ell)^2} \left(1-e^{-2i\pi(x+v_F t)/L}\right)^{(F_{\ell,-}+\ell)^2}} \\ &= \frac{e^{\frac{i\pi}{2}\delta_\ell}(2\pi)^{-(F_{\ell,+}+\ell)^2-(F_{\ell,-}+\ell)^2}}{(x-v_F t)^{(F_{\ell,+}+\ell)^2}(x+v_F t)^{(F_{\ell,-}+\ell)^2}}, \end{aligned} \tag{12.153}$$

where

$$\delta_\ell = (F_{\ell,+}+\ell)^2 - (F_{\ell,-}+\ell)^2. \tag{12.154}$$

Thus, we can proceed to the thermodynamic limit in (12.147)

$$\begin{aligned}\operatorname{Lim} Q_{\mathrm{cr}}^{(\alpha)}(x,t) = \sum_{\ell\in\mathbb{Z}} & e^{ix\alpha_\ell k_F/\pi} \frac{e^{C_1[F_\ell+\ell]}(2\pi)^{F_{\ell,-}-F_{\ell,+}}}{(4\pi q\rho(q))^{(F_{\ell,+}+\ell)^2+(F_{\ell,-}+\ell)^2}} \\ &\times \frac{e^{\frac{i\pi}{2}\delta_\ell}\mathfrak{G}[F_\ell+\ell] G^2(1+F_{\ell,+}+\ell) G^2(1-F_{\ell,-}-\ell)}{(x-v_F t)^{(F_{\ell,+}+\ell)^2}(x+v_F t)^{(F_{\ell,-}+\ell)^2}}. \end{aligned} \tag{12.155}$$

So far, we have not used any information about the shift function $F_\ell(u)$ of the $\mathbf{P}_\ell$ class. Now we use (12.125). This leads us to

$$\operatorname{Lim} Q_{\mathrm{cr}}^{(\alpha)}(x,t) = \sum_{\ell\in\mathbb{Z}} \left\{ \mathfrak{S}\left[\tfrac{\alpha_\ell Z}{2\pi}\right] G^2(1,\tfrac{\alpha_\ell}{2\pi}\mathcal{Z}) e^{C_1[\frac{\alpha_\ell}{2\pi}Z]} \big(2q\rho(q)\big)^{-\frac{\alpha_\ell^2 \mathcal{Z}^2}{2\pi^2}} \right\} \times \frac{e^{ix\alpha_\ell k_F/\pi}}{\big(4\pi^2(x^2-v_F^2t^2)\big)^{\frac{\alpha_\ell^2\mathcal{Z}^2}{4\pi^2}}}, \tag{12.156}$$

and we recall that $\mathcal{Z} = Z(\pm q)$ and $G(1,z) = G(1+z)G(1-z)$. Note that the expression in braces is nothing more than the finite amplitude of the form factor corresponding to the $\alpha_\ell$-twisted ground state. Indeed, comparing this expression with (12.94), we conclude that

$$\mathfrak{S}\left[\tfrac{\alpha_\ell Z}{2\pi}\right] G^2(1,\tfrac{\alpha_\ell}{2\pi}\mathcal{Z}) e^{C_1[\frac{\alpha_\ell}{2\pi}Z]} \big(2q\rho(q)\big)^{-\frac{\alpha_\ell^2 \mathcal{Z}^2}{2\pi^2}} = \mathfrak{A}_{\mathcal{Q}}^{(\mathrm{fin})}\left[\tfrac{\alpha_\ell Z}{2\pi}\right]. \tag{12.157}$$

Thus, we arrive at

$$\operatorname{Lim} Q_{\mathrm{cr}}^{(\alpha)}(x,t) = \sum_{\ell\in\mathbb{Z}} \frac{e^{ix\alpha_\ell k_F/\pi}}{\big(4\pi^2(x^2-v_F^2t^2)\big)^{\frac{\alpha_\ell^2\mathcal{Z}^2}{4\pi^2}}} \mathfrak{A}_{\mathcal{Q}}^{(\mathrm{fin})}\left[\tfrac{\alpha_\ell Z}{2\pi}\right]. \tag{12.158}$$

## 12.6 Long-distance asymptotics of the correlation function

The resulting representation for the generating function makes it easy to calculate the long-distance asymptotic behavior of the density correlation function. Recall that we considered the case $x \to \infty$, while the time $t$ remained finite. Therefore, from the point of view of asymptotic analysis,

$$x^2 - v_F^2 t^2 = x^2 + O(1). \tag{12.159}$$

Thus, the dependence on time gives only subleading contributions to every term of the sum (12.158). Actually, we only needed this dependence for regularization of the sum of the form (12.149). In the asymptotic regime, we can put $t = 0$. Then expansion (12.158) takes the form

$$\operatorname{Lim} Q_{\mathrm{cr}}^{(\alpha)}(x) = \sum_{\ell\in\mathbb{Z}} \mathfrak{A}_{\mathcal{Q}}^{(\mathrm{fin})}\left[\tfrac{\alpha_\ell Z}{2\pi}\right] (2\pi x)^{-\frac{\alpha_\ell^2\mathcal{Z}^2}{2\pi^2}} e^{ix\alpha_\ell k_F/\pi}. \tag{12.160}$$

Each term of this series has a simple physical meaning. The exponential factor contains the momentum of the $\alpha_\ell$-twisted ground state. The pre-exponential factor is nothing more than the amplitude of the form factor (12.18) between the ground state and the $\alpha_\ell$-twisted ground state up to the replacement $L \to 2\pi x$. On the other hand, each term of series (12.160) is the result of the summation of critical form factors of the $\mathbf{P}_\ell$ class, one of the representatives of which is the $\alpha_\ell$-twisted ground state. Thus, the summation of the critical form factors within the class leads to the rescaling $L \to 2\pi x$.

From a physical point of view, this rescaling is quite natural. Indeed, our calculations show that each form factor in the thermodynamic limit behaves as $L^{-\gamma_{\mathcal{Q}}}$. On the other hand, we expect that the correlation function is finite in the thermodynamic limit. Hence, the factor $L^{-\gamma_{\mathcal{Q}}}$ must somehow disappear after summing the form factors. However, we have seen that when summing critical form factors, the entire dependence on $L$ is contained only in the phase factor, where $L$ is included only in the form of a combination of $2\pi x/L$. Therefore, if the compensation of the factor $L^{-\gamma_{\mathcal{Q}}}$ does occur, then the factor $(2\pi x)^{-\gamma_{\mathcal{Q}}}$ must inevitably occur instead. We made sure of this using an explicit calculation. However, we could also predict this rescaling in advance if we postulate that the correlation functions are finite.

Here it is appropriate to return to the question of corrections, which we discussed at the end of Section 12.4.5 (see page 311). Recall that calculating the finite amplitude of the form factor, we neglected all the corrections in the $1/L$-expansion. At the same time, at least some of these corrections depend on quantum numbers. Therefore, when summing form factors within a specific class, they can give a finite contribution. However, we can now fully expect that these corrections turn into subleading terms in the $1/x$-expansion due to the rescaling mechanism described above. In any case, they do not contribute to the leading term of the asymptotics corresponding to form factors of the fixed $\mathbf{P}_\ell$ class.

Thus, we conclude that the coefficients in the expansion (12.160) are actually functions of $x$: $\mathfrak{A}^{(\text{fin})}_{\mathcal{Q}} = \mathfrak{A}^{(\text{fin})}_{\mathcal{Q}}(x)$. In turn, these functions are decomposed into asymptotic series with respect to the negative powers of $x$. In formula (12.160), we took into account only the leading terms of these expansions. Calculating corrections to these leading contributions is a too complex technical task, which is hardly feasible within the form factor decomposition method. Therefore, we will not touch on this issue at all.

Summing up the above, we will consider expansion (12.160) to be the final answer for the generating function's asymptotics. At the same time, we remember that the coefficients $\mathfrak{A}^{(\text{fin})}_{\mathcal{Q}}$ are some functions of $x$. Within our method, we can only calculate the limiting values of these functions at $x \to \infty$. These limiting values are given in formula (12.160).

According to equation (12.10), to calculate the asymptotics of the density correlation function $\langle \hat{q}'(x_1)\hat{q}'(x_2)\rangle$, it is sufficient to take the second

$\alpha$-derivative of (12.160) at $\alpha = 0$ and then to differentiate the result twice over $x$ (recall that $x = x_1 - x_2$). We obtain

$$\langle \hat{q}'(x_1)\hat{q}'(x_2)\rangle = -\frac{1}{2}\frac{\partial^2}{\partial x^2}\frac{\partial^2}{\partial \alpha^2}\sum_{\ell\in\mathbb{Z}} \frac{\mathfrak{A}^{(\mathrm{fin})}_{\mathcal{Q}}\left[\frac{\alpha_\ell Z}{2\pi}\right]}{(2\pi x)^{\frac{\alpha_\ell^2 z^2}{2\pi^2}}}\, e^{ix\alpha_\ell k_F/\pi}\Big|_{\alpha=0}. \tag{12.161}$$

Two cases should be distinguished: $\ell = 0$ and $\ell \neq 0$. Let us start with the second case.

### 12.6.1 *The case $\ell \neq 0$*

For $\ell \neq 0$, the second $\alpha$-derivative acts only on the finite amplitude $\mathfrak{A}^{(\mathrm{fin})}_{\mathcal{Q}}\left[\frac{\alpha_\ell Z}{2\pi}\right]$ because the latter is proportional to $\alpha^2$. Indeed, the finite amplitude is given by (12.157). The smooth part of the amplitude is given by (12.45), in which one should replace $\alpha$ with $\alpha_\ell = \alpha + 2\pi\ell$:

$$\mathfrak{S}\left[\tfrac{\alpha_\ell Z}{2\pi}\right] = e^{C_0\left[\frac{\alpha_\ell Z}{2\pi}\right]}\left| e^{\frac{\alpha_\ell}{2\pi}\tilde{Z}(\omega - ic)}\frac{\sin\frac{\alpha}{2}}{\sin\left(\frac{\alpha_\ell}{2}Z(\omega)\right)}\frac{\det\left(\mathbf{I} - \frac{1}{2\pi i}\mathbf{U}\right)}{\det\left(\mathbf{I} - \frac{1}{2\pi}\mathbf{K}\right)}\right|^2, \tag{12.162}$$

where $\tilde{Z}(\omega)$ is the Cauchy transform of the dressed charge $Z(u)$

$$\tilde{Z}(\omega) = \int_{-q}^{q}\frac{Z(u)\,\mathrm{d}u}{\omega - u}. \tag{12.163}$$

Recall that although $\omega$ is an arbitrary real number, the smooth part $\mathfrak{S}$ does not depend on it (see appendix C). Then we can put, for example, $\omega = q$. Then $\sin\left(\pi\ell Z(\omega)\right) \neq 0$, since we assume that the shift function is not an integer at the Fermi boundaries.

We see that when calculating the second $\alpha$-derivative at $\alpha = 0$, we must differentiate only $\sin^2(\alpha/2)$; otherwise, we get zero. In the remaining part of the finite amplitude, we just have to put $\alpha = 0$, that is, $\alpha_\ell = 2\pi\ell$. Thus, we arrive at

$$\frac{\partial^2}{\partial\alpha^2}\frac{\mathfrak{A}^{(\mathrm{fin})}_{\mathcal{Q}}\left[\frac{\alpha_\ell Z}{2\pi}\right]}{(2\pi x)^{\frac{\alpha_\ell^2 z^2}{2\pi^2}}}\, e^{ix\alpha_\ell k_F/\pi}\Big|_{\alpha=0} = \frac{\mathfrak{A}^{(\mathrm{fin})}_{\hat{q}}[\ell Z]}{(2\pi x)^{2\ell^2 Z^2}}\, e^{2ix\ell k_F}, \qquad \ell \neq 0, \tag{12.164}$$

where

$$\mathfrak{A}^{(\mathrm{fin})}_{\hat{q}}[\ell Z] = \frac{\partial^2 \mathfrak{A}^{(\mathrm{fin})}_{\mathcal{Q}}\left[\frac{\alpha_\ell Z}{2\pi}\right]}{\partial\alpha^2}\Big|_{\alpha=0}. \tag{12.165}$$

In fact, $\mathfrak{A}^{(\mathrm{fin})}_{\hat{q}}[\ell Z]$ is the finite amplitude of the form factor of the operator $\hat{q}$ between the ground state and the $2\pi\ell$-twisted ground state.

When calculating the derivative with respect to $x$, we must differentiate only the exponential factor $e^{2ix\ell k_F}$ since the derivative of the power factor gives corrections to the leading asymptotics. Thus, the leading asymptotic contribution for $\ell \neq 0$ has the following form:

$$-\frac{1}{2}\frac{\partial^2}{\partial x^2}\frac{\partial^2}{\partial \alpha^2}\frac{\mathfrak{A}_{\mathcal{Q}}^{(\text{fin})}\left[\frac{\alpha_\ell Z}{2\pi}\right]}{(2\pi x)^{\frac{\alpha_\ell^2 \mathcal{Z}^2}{2\pi^2}}} e^{ix\alpha_\ell k_F/\pi}\bigg|_{\alpha=0}$$

$$= 2\ell^2 k_F^2 \frac{\mathfrak{A}_{\hat{q}}^{(\text{fin})}[\ell Z]}{(2\pi x)^{2\ell^2\mathcal{Z}^2}} e^{2ix\ell k_F} + O(x^{-2\ell^2\mathcal{Z}^2-1}), \qquad \ell \neq 0. \qquad (12.166)$$

### 12.6.2 *The case* $\ell = 0$

In this case, the second $\alpha$-derivative should only be applied to the exponential and power factor. It is easy to see that if at least one derivative over $\alpha$ acts on the finite amplitude, then the subsequent differentiation over $x$ will give zero. Thus, we obtain

$$-\frac{1}{2}\frac{\partial^2}{\partial x^2}\frac{\partial^2}{\partial \alpha^2}\frac{\mathfrak{A}_{\mathcal{Q}}^{(\text{fin})}\left[\frac{\alpha Z}{2\pi}\right]}{(2\pi x)^{\frac{\alpha^2 \mathcal{Z}^2}{2\pi^2}}} e^{ix\alpha k_F/\pi}\bigg|_{\alpha=0}$$

$$= \left(D^2 - \frac{\mathcal{Z}^2}{2\pi^2 x^2}\right)\lim_{\alpha\to 0}\mathfrak{A}_{\mathcal{Q}}^{(\text{fin})}\left[\tfrac{\alpha Z}{2\pi}\right], \qquad \ell = 0, \qquad (12.167)$$

and we recall that the average density $D$ is related to the Fermi momentum by $D = k_F/\pi$.

We need only to calculate the limit of the finite amplitude at $\alpha = 0$. It follows from equations (12.157) and (12.162) that

$$\lim_{\alpha\to 0}\mathfrak{A}_{\mathcal{Q}}^{(\text{fin})}\left[\tfrac{\alpha Z}{2\pi}\right] = \frac{1}{Z^2(\omega)}\lim_{\alpha\to 0}\left|\frac{\det\left(\mathbf{I} - \frac{1}{2\pi i}\mathbf{U}\right)}{\det\left(\mathbf{I} - \frac{1}{2\pi}\mathbf{K}\right)}\right|^2. \qquad (12.168)$$

Consider how the kernel $U(z, z')$ behaves at $\alpha \to 0$. It is given by (12.41)

$$U_\Gamma(z, z') = e^{\tilde{F}(z'+ic)-\tilde{F}(z')}\frac{\mathcal{K}_\alpha(z', z) - \mathcal{K}_\alpha(\omega, z)}{1 - e^{-2\pi i F(z')}}, \qquad (12.169)$$

in which one should set $F(z) = \frac{\alpha}{2\pi}Z(z)$. We added the subscript $\Gamma$ to the kernel $U(z, z')$ to emphasize that the integral operator acts on the contour $\Gamma([-q, q])$ surrounding the interval $[-q, q]$. We can replace the kernel (12.169) with a new kernel $U_{[-q,q]}(z, z')$ that will act directly on the interval $[-q, q]$:

$$U_\Gamma(z, z') \to U_{[-q,q]}(z, z') = U_\Gamma(z, z' - i0) - U_\Gamma(z, z' + i0). \qquad (12.170)$$

The rhs of (12.170) is non-vanishing due to the factor $e^{-\tilde{F}(z')}$, which has a cut on the interval $[-q, q]$. Then

$$\det\left(\mathbf{I} - \tfrac{1}{2\pi i}\mathbf{U}_\Gamma\right) = \det\left(\mathbf{I} - \tfrac{1}{2\pi i}\mathbf{U}_{[-q,q]}\right). \tag{12.171}$$

Using

$$\tilde{F}(z - i0) - \tilde{F}(z + i0) = 2\pi i F(z), \tag{12.172}$$

we obtain

$$U_{[-q,q]}(z, z') = -e^{\tilde{F}(z'+ic)-\tilde{F}(z'+i0)}\big(\mathcal{K}_\alpha(z', z) - \mathcal{K}_\alpha(\omega, z)\big). \tag{12.173}$$

In this expression, we can safely set $F(z) = \frac{\alpha}{2\pi} Z(z)$ and take the limit $\alpha \to 0$:

$$\frac{1}{2\pi i} U_{[-q,q]}(z, z')\Big|_{\alpha=0} = \frac{1}{2\pi}\big(\mathcal{K}(z - z') - \mathcal{K}(z - \omega)\big), \tag{12.174}$$

where $\mathcal{K}(z)$ is given by (10.10). Thus,

$$\det\left(\mathbf{I} - \tfrac{1}{2\pi i}\mathbf{U}_\Gamma\right)\Big|_{\alpha=0} = \det\left(\mathbf{I} - \tfrac{1}{2\pi}(\mathbf{K} - \mathbf{K}^{(\omega)})\right), \tag{12.175}$$

where the kernel of the operator $\mathbf{K}^{(\omega)}$ is

$$\mathcal{K}^{(\omega)}(z, z') = \mathcal{K}(z - \omega), \tag{12.176}$$

and the integral operator in the rhs of (12.175) acts on $[-q, q]$. Finally, we note that due to corollary C.1

$$\det\left(\mathbf{I} - \tfrac{1}{2\pi}(\mathbf{K} - \mathbf{K}^{(\omega)})\right) = Z(\omega)\det\left(\mathbf{I} - \tfrac{1}{2\pi}\mathbf{K}\right), \tag{12.177}$$

and hence,

$$\lim_{\alpha \to 0} \mathfrak{A}_{\mathcal{Q}}^{(\mathrm{fin})}\left[\tfrac{\alpha Z}{2\pi}\right] = 1. \tag{12.178}$$

Thus, we arrive at

$$-\frac{1}{2}\frac{\partial^2}{\partial x^2}\frac{\partial^2}{\partial \alpha^2}\frac{\mathfrak{A}_{\mathcal{Q}}^{(\mathrm{fin})}\left[\frac{\alpha Z}{2\pi}\right]}{(2\pi x)^{\frac{\alpha^2 \mathcal{Z}^2}{2\pi^2}}}\, e^{ix\alpha k_F/\pi}\Big|_{\alpha=0} = D^2 - \frac{\mathcal{Z}^2}{2\pi^2 x^2}. \tag{12.179}$$

As a result, the asymptotic expansion for the two-point density correlation function has the form

$$\langle \hat{q}'(x_1)\hat{q}'(x_2)\rangle = D^2 - \frac{\mathcal{Z}^2}{2\pi^2 x^2} + 2k_F^2 \sum_{\ell \in \mathbb{Z};\, \ell \neq 0} \frac{\ell^2 \mathfrak{A}_{\hat{q}}^{(\mathrm{fin})}[\ell Z]}{(2\pi x)^{2\ell^2 \mathcal{Z}^2}}\, e^{2ix\ell k_F}. \tag{12.180}$$

In this expression, the oscillating terms (with $\ell \neq 0$) arise when summing form factors corresponding to excited states with non-zero total momentum. The non-oscillating term corresponds to the sum of form factors of the class $\mathbf{P}_0$. If we used ordinary (non-twisted) excited states in the form factor expansion, then the ground state would also fall into this class. It is easy to see that the term $D^2$ in (12.180) is just the contribution from the ground state because

$$D^2 = \langle \hat{q}'(x) \rangle^2 = \big(\mathrm{Lim} \langle\!\langle \Psi^\dagger(x)\Psi(x) \rangle\!\rangle\big)^2. \tag{12.181}$$

All other states of the $\mathbf{P}_0$ class generate the term $-\mathcal{Z}^2/2\pi^2 x^2$. Therefore, when summing over ordinary (non-twisted) excited states, we would face the problem of separating the ground state from the class $\mathbf{P}_0$. We succeeded in avoiding this difficulty by introducing the auxiliary twist parameter $\alpha$ from the very beginning. Due to this twist parameter, we have eliminated the possibility that one of the excited states coincides with the ground state. Thus, we unified the procedure for summing critical form factors in any class $\mathbf{P}_\ell$, regardless of $\ell$. In particular, this explains the fact that all the terms in the asymptotics of the generating function (12.160) have the same structure for any $\ell$. In contrast, the term $\ell = 0$ is distinguished in the asymptotics of the correlation function (12.180).

## 12.7 Correlation function of fields

The same method is used to calculate the two-point correlation function of the fields

$$\langle \Psi^\dagger(x_1, t_1)\Psi(x_2, t_2) \rangle = \mathrm{Lim} \, \frac{\langle \mathbf{\Phi}_N(\bar{u}) | \Psi^\dagger(x_1, t_1)\Psi(x_2, t_2) | \mathbf{\Phi}_N(\bar{u}) \rangle}{\langle \mathbf{\Phi}_N(\bar{u}) | \mathbf{\Phi}_N(\bar{u}) \rangle}. \tag{12.182}$$

There are differences, but they are not fundamental. Therefore, we do not give all the details of the calculations but briefly discuss some of the nuances.

The main difference from the density correlation function is that the number of Bethe parameters in excited states for the field form factors differs from the number of Bethe parameters in the ground state. For example, the field $\Psi(x, t)$ form factor does not vanish if the number of Bethe parameters in the excited state is $N - 1$ (see (9.45)). Therefore, the excited states never coincide with the ground state. As a result, there is no need to introduce an auxiliary twist. We can construct a form factor decomposition based on ordinary excited states rather than on twisted ones. On the other hand, we can also keep the auxiliary twist parameter. In this case, the transition between classes $\mathbf{P}_0$ and $\mathbf{P}_\ell$ can be interpreted as shifting this parameter $\alpha \to \alpha_\ell = \alpha + 2\pi\ell$.

For $N$ and $L$ finite, the correlation function of the fields is given by the following form factor expansion:

$$\langle\!\langle \Psi^\dagger(x_1,t_1)\Psi(x_2,t_2)\rangle\!\rangle = \sum_{\bar v} |\boldsymbol{\mathcal{F}}_\Psi|^2 e^{ix\mathcal{P}_{\rm ex}-it\mathcal{E}_{\rm ex}}, \tag{12.183}$$

where $x = x_1 - x_2$, $t = t_1 - t_2$. As usual, $\mathcal{P}_{\rm ex}$ and $\mathcal{E}_{\rm ex}$ respectively are the excitation momentum and energy. The form factor amplitude $|\boldsymbol{\mathcal{F}}_\Psi|^2$ is

$$|\boldsymbol{\mathcal{F}}_\Psi|^2 = \frac{|\langle \boldsymbol{\Phi}_{N-1}(\bar v)|\Psi(0,0)|\boldsymbol{\Phi}_N(\bar u)\rangle|^2}{\langle \boldsymbol{\Phi}_{N-1}(\bar v)|\boldsymbol{\Phi}_{N-1}(\bar v)\rangle\ \langle \boldsymbol{\Phi}_N(\bar u)|\boldsymbol{\Phi}_N(\bar u)\rangle}. \tag{12.184}$$

It is convenient to use equation (9.147) for calculating the thermodynamic limit of the amplitude. As in the case discussed above, the amplitude $|\boldsymbol{\mathcal{F}}_\Psi|^2$ at large $N$ and $L$ behaves so that

$$\operatorname{Lim} |\boldsymbol{\mathcal{F}}_\Psi|^2 L^{\gamma_\Psi(\{\lambda_p,\lambda_h\})} = \mathfrak{A}_\Psi^{(\rm fin)}(\{\lambda_p,\lambda_h\}|\{p,h\}), \tag{12.185}$$

where $\mathfrak{A}_\Psi^{(\rm fin)}$ is a finite amplitude.

For critical form factors of the $\mathbf{P}_\ell$ class, the exponent $\gamma_\Psi$ is given by the universal formula (12.140). In the case of the field form factor, it reads as follows:

$$\gamma_\Psi = (F_{\ell,+} + \ell - 1)^2 + (F_{\ell,-} + \ell)^2, \tag{12.186}$$

where $F_{\ell,\pm} = F_\ell(\pm q)$, and $F_\ell(u)$ is the shift function of the critical states of the $\mathbf{P}_\ell$ class (that is, the shift function of the $2\pi\ell$-twisted ground state). We see that it is slightly different from (12.138) because $F_{\ell,+}$ has an additional shift of $-1$. This shift is because the number of parameters $\bar v$ in the excited state $\langle \boldsymbol{\Phi}_{N-1}(\bar v)|$ is $N-1$. Indeed, due to (12.140), we have

$$\begin{aligned}\operatorname{Lim} L\hat\rho_u(u_{\max})(v_{\max} - u_{\max}) &= \operatorname{Lim} L\hat\rho_u(u_N)(v_{N-1} - u_N)\\ &= \operatorname{Lim} L\hat\rho_u(u_N)(v_{N-1} - v'_N + v'_N - u_N),\end{aligned} \tag{12.187}$$

where $\hat\xi_v(v'_N) = N/L$. It is easy to see that

$$\begin{aligned}\operatorname{Lim} L\hat\rho_u(u_N)(v'_N - u_N) &= F_\ell(q) + \ell,\\ \operatorname{Lim} L\hat\rho_u(u_N)(v_{N-1} - v'_N) &= -1.\end{aligned} \tag{12.188}$$

The shift function for the $\mathbf{P}_\ell$ class is found from equation (10.103) with $\Delta N = -1$ and $\alpha = 0$

$$F_\ell(u) = \frac{1}{2}Z(u) + \phi(u,q) + \ell\big(\phi(u,-q) - \phi(u,q)\big). \tag{12.189}$$

Using equations (10.119)–(10.121) we find

$$F_{\ell,-} = \frac{1}{2\mathcal{Z}} + \ell(\mathcal{Z}-1), \qquad F_{\ell,+} = 1 - \frac{1}{2\mathcal{Z}} + \ell(\mathcal{Z}-1), \tag{12.190}$$

and hence,

$$\gamma_\Psi = \frac{1}{2\mathcal{Z}^2} + 2\ell^2\mathcal{Z}^2. \tag{12.191}$$

To calculate the long-distance asymptotic behavior (at a finite time) of the correlation function of the fields, we again restrict ourselves to the critical form factors. The summation of the latter still reduces to the formula (12.150). Since we consider time $t$ to be finite at this stage, we can ignore the time dependence in the asymptotic formula or even put $t = 0$ at all.[1] Then the summation of the critical form factors results in rescaling $L \to 2\pi x$. As a result, we arrive at the following asymptotic expansion:

$$\langle \Psi^\dagger(x_1)\Psi(x_2)\rangle = \frac{1}{(2\pi x)^{\frac{1}{2\mathcal{Z}^2}}}\sum_{\ell\in\mathbb{Z}}(-1)^\ell \frac{\mathfrak{A}_\Psi^{(\mathrm{fin})}[F_\ell]}{(2\pi x)^{2\ell^2\mathcal{Z}^2}}\, e^{2ix\ell k_F}. \tag{12.192}$$

The alternating sign $(-1)^\ell$ appears due to the phase factor $\exp\{\frac{i\pi}{2}\delta_\ell\}$ (see (12.153), (12.154)). This phase was vanishing in the case of the generating function. Now it is

$$\delta_\ell = (F_{\ell,+} + \ell - 1)^2 - (F_{\ell,-} + \ell)^2 = -2\ell. \tag{12.193}$$

As with the density correlation function, the finite amplitudes $\mathfrak{A}_\Psi^{(\mathrm{fin})}[F_\ell]$ actually depend on the distance $x$. Formula (12.192) takes into account only the limiting values of these functions at $x \to \infty$.

## 12.8 Large-time asymptotics

So far, we have considered the asymptotics of correlation functions at $x \to +\infty$ and $x \gg v_F t$. In this case, the main asymptotic contribution is produced only by the states with particles and holes close to the Fermi boundary. Let us now consider the case when $x \to +\infty$, $t \to +\infty$, and $x/t = const$. We will assume that $x/t \neq v_F$. Then the situation is slightly different.

Let us turn back to the form factor expansion (12.2). At this stage, the explicit form of the operator $\mathcal{O}(x,t)$ is not essential. In the thermodynamic limit, the oscillating phase can be presented in the form

$$-it\mathcal{E}_{\mathrm{ex}} + ix\mathcal{P}_{\mathrm{ex}} = -it\sum_{a=1}^{n}\big(\omega(\lambda_{p_a}) - \omega(\lambda_{h_a})\big), \tag{12.194}$$

where

$$\omega(u) = \varepsilon(u) - \frac{x}{t}k(u). \tag{12.195}$$

[1]The time dependence is only needed for the regularization of sums (12.149).

Here $k(u)$ and $\varepsilon(u)$ are the dressed momentum (10.94) and the dressed energy (10.76). Recall also that $x = x_1 - x_2$ and $t = t_1 - t_2$. In the limit $x \to +\infty$, $t \to +\infty$, and $x/t = const$, the phase factor rapidly oscillates and has a saddle point. In addition to the contributions from the Fermi boundaries $\pm q$, we now also have to consider the contributions of the saddle point.

The existence and the uniqueness of the saddle point of the phase $\omega(u)$ is closely related to the properties of the velocity of sound $v(u)$ in the gas

$$v(u) = \frac{\partial \varepsilon}{\partial k} = \frac{\varepsilon'(u)}{k'(u)}. \tag{12.196}$$

Under certain constraints between the coupling constant $c$ and the chemical potential $\mu$, it can be proved that this function is strictly monotonic: $v'(u) > 0$. Numerical analysis of the equations (10.94) and (10.76) confirms that $v'(u) > 0$ for arbitrary $c$ and $\mu$. Therefore, below we assume that $v(u)$ is a strictly monotonic function. It follows also from equations (10.94) and (10.76) that $v(u) \to \pm\infty$ as $u \to \pm\infty$. This ensures that the phase $\omega(u)$ has a unique saddle point $u_0$ determining by the equation $v(u_0) = x/t$. Moreover, it follows from (10.94) that $k'(u) > 0$. Thus,

$$\omega''(u_0) = \varepsilon''(u_0) - v(u_0)k''(u_0) = v'(u_0)k'(u_0) > 0. \tag{12.197}$$

There exist two asymptotic regimes. In the space-like regime ($x/t > v_F$) the saddle point is outside the Fermi zone $|u_0| > q$, while in the time-like regime ($x/t < v_F$) it is inside $|u_0| < q$. The contribution of the saddle point to the asymptotics of correlation functions occurs when integrating over the momenta of excitations in a neighborhood of the saddle point. In the space-like regime, this is integration over the momenta of particles; in the time-like regime, this is integration over the momenta of holes. Thus, we deal with the localization of the integrals over $\{\lambda_p\}$ (resp. $\{\lambda_h\}$) in a small vicinity of $u_0$. Of course, these integral contributions should be 'dressed' by the excitations on the Fermi boundaries, as we did in the previous sections.

Let us now describe the distributions of particles and holes in the form factors giving rise to the leading asymptotics of the correlation functions. As before, we denote by $n$ the total number of particles (holes) in a given excited state. Consider for definiteness the space-like regime. Then among the particles, we chose $n_0$ of them with momenta $\bar{\lambda}_p^0 = \{\lambda_{p_1}^0, \dots \lambda_{p_{n_0}}^0\}$ lying in a small vicinity $B(u_0)$ of the saddle point $u_0$. Other $n - n_0$ particles are at the Fermi boundaries $\pm q$. Let $n_p^+$ particles have the momentum $q$ and $n_p^-$ particles have the momentum $-q$. As for the holes, all of them are on the Fermi surface in the space-like regime. Let $n_h^\pm$ refer to the numbers of holes with momenta $\pm q$, respectively. As before, the excitations in the vicinity of the Fermi boundaries are described by the quantum numbers $\{p^{\pm}\}$ and $\{h^{\pm}\}$ (12.120).

We use the same integers to describe the particle/holes distributions in the time-like regime, but $n_0$ denotes the number of holes with momenta close to the saddle point. It is convenient to introduce a parameter $\tau$ such that $\tau = 1$ in the space-like regime and $\tau = -1$ in the time-like regime. Then, in both cases, we have

$$n_p^+ + n_p^- + \tau n_0 = n_h^+ + n_h^-. \tag{12.198}$$

Let $\ell$ refer to the right particle/hole discrepancy number $\ell = n_p^+ - n_h^+$. Then the left particle/hole discrepancy number can be written as $n_p^- - n_h^- = -\tau n_0 - \ell$. Below we call the excited states described above the *quasi-critical excited states*. They can be characterized by two numbers $\tau n_0$ and $\ell$. Thus, they can be combined into the classes, denoted below by $\mathbf{P}_{\tau n_0,\ell}$. We will also say that the corresponding form factors (*quasi-critical form factors*) belong to the $\mathbf{P}_{\tau n_0,\ell}$ class. In the thermodynamic limit and after the localization $\lambda^0_{p_a} = u_0$ $(a = 1,\dots,n_0)$, all the states of the $\mathbf{P}_{\tau n_0,\ell}$ class have the excitation momentum $\mathcal{P}_{\rm ex} = (2\ell + \tau n_0)k_F + \tau n_0 k(u_0)$ and the excitation energy $\mathcal{E}_{\rm ex} = \tau n_0 \varepsilon(u_0)$. These values are independent of the quantum numbers $p^{\pm}$ and $h^{\pm}$. However, we already know that to sum up the quasi-critical form factors, it is essential to consider the finite-size corrections to the excitation energy and momentum.

Calculating the exponent $\gamma_{\mathcal{O}}$ and the finite amplitude of the form factors of the $\mathbf{P}_{\tau n_0,\ell}$ class is completely similar to what has already been done. Therefore, we do not give details of this calculation but will focus only on a small difference. Consider for definiteness the space-like regime. The time-like regime can be treated in the same way.

In the case of localization of the integrals around some $u_0$, one can set the corresponding $\lambda^0_{p_a} = u_0$, $a = 1,\dots,n_0$, in the arguments of the smooth part of the form factor amplitude. However, we can not do this in the discrete part. Indeed, due to (12.98), the discrete part is proportional to the square of the Vandermonde determinant[2]

$$\mathfrak{D}(\{\lambda_p\},\{\lambda_h\}|\{p\},\{h\}) = \mathcal{V}^2(\bar{\lambda}^0_p)\,\widetilde{\mathfrak{D}}(\{\lambda_p\},\{\lambda_h\}|\{p\},\{h\}), \tag{12.199}$$

[2]This property is characteristic not only for the density correlation function in the one-dimensional Bose gas but also for other correlation functions in models in which the ground state is a Fermi zone.

where

$$\mathcal{V}(\bar{\lambda}_p^0) = \prod_{j>k}^{n_0} (\lambda_{p_j}^0 - \lambda_{p_k}^0), \tag{12.200}$$

and $\widetilde{\mathfrak{D}}$ has a non-vanishing limit at $\lambda_{p_a}^0 = u_0$, $a = 1, \dots, n_0$. Therefore, when calculating the finite amplitude $\mathfrak{A}_{\mathcal{O}}^{(\mathrm{fin})}$, we must first extract the square of the Vandermonde determinant from $\mathfrak{A}_{\mathcal{O}}^{(\mathrm{fin})}$:

$$\mathfrak{A}_{\mathcal{O}}^{(\mathrm{fin})} = \mathcal{V}^2(\bar{\lambda}_p^0)\, \widetilde{\mathfrak{A}}_{\mathcal{O}}^{(\mathrm{fin})}. \tag{12.201}$$

We call $\widetilde{\mathfrak{A}}_{\mathcal{O}}^{(\mathrm{fin})}$ a *reduced finite amplitude*. Calculating the integrals over $\bar{\lambda}_p^0$, we can set all the integration variables equal to $u_0$ in the arguments of the reduced finite amplitude, but we should keep the prefactor $\mathcal{V}^2(\bar{\lambda}_p^0)$ as it is.

This is the only significant difference that should be taken into account when calculating the thermodynamic limit of quasi-critical form factors. We have already done all the other calculations in the previous sections, so we will just list the main results.

The momentum and the energy of quasi-critical states have the following form (taking into account first-order corrections):

$$\mathcal{P}_{\mathrm{ex}} = \sum_{a=1}^{n_0} k(\lambda_{p_a}) + (2\ell + n_0)k_F + \frac{1}{L}\mathcal{P}_{n_0,\ell} + \frac{2\pi}{L}\left(E^+ - E^-\right), \tag{12.202}$$

$$\mathcal{E}_{\mathrm{ex}} = \sum_{a=1}^{n_0} \varepsilon(\lambda_{p_a}) + \frac{1}{L}\mathcal{E}_{n_0,\ell} + \frac{2\pi v_F}{L}\left(E^+ + E^-\right). \tag{12.203}$$

Here $\mathcal{P}_{n_0,\ell}$ and $\mathcal{E}_{n_0,\ell}$ are constants for all the states belonging to a given $\mathbf{P}_{n_0,\ell}$ class, and $E^\pm$ are given by (12.123).

The oscillating phase near the saddle point can be written up to the second order term

$$\omega(\lambda_{p_a}) = \omega(u_0) + \frac{\omega''(u_0)}{2}(\lambda_{p_a} - u_0)^2 + o\left((\lambda_{p_a} - u_0)^2\right). \tag{12.204}$$

The shift function of the $\mathbf{P}_{n_0,\ell}$ class quasi-critical states is

$$\begin{aligned} F_{n_0,\ell}(u) = &\left(\frac{\alpha}{2\pi} - \frac{\Delta N}{2}\right) Z(u) - (\ell + \Delta N)\phi(u,q) \\ &+ (\ell + n_0)\phi(u,-q) - n_0\phi(u,u_0), \end{aligned} \tag{12.205}$$

where $\Delta N = N' - N$ is 0 for the density correlation function and $\pm 1$ for the correlation functions of the fields. The exponent $\gamma_{\mathcal{O}}$ of the corresponding quasi-critical form factors can be computed from the boundary values of the shift function (see (12.140))

$$\gamma_{\mathcal{O}} = (F_{n_0,\ell}^+ + \ell)^2 + (F_{n_0,\ell}^- + \ell + n_0)^2 + n_0, \tag{12.206}$$

where

$$F^-_{n_0,\ell} = F_{n_0,\ell}(-q), \quad F^+_{n_0,\ell} = F_{n_0,\ell}(q) + \Delta N. \tag{12.207}$$

The additional constant $n_0$ in (12.206) occurs because $n_0$ particles are separated from the Fermi boundaries.

To describe the reduced finite amplitude of quasi-critical form factors, we consider a special representative of the $\mathbf{P}_{n_0,\ell}$ class. This quasi-critical state is characterized by quantum numbers $\{p^\pm, h^\pm\}$ (12.120) satisfying the following conditions:

$$\begin{aligned}
&\{p^+\} = \{1,\dots,\ell\} \quad \text{for} \quad \ell > 0; && \{p^+\} = \emptyset \quad \text{for} \quad \ell \le 0;\\
&\{h^+\} = \{1,\dots,|\ell|\} \quad \text{for} \quad \ell < 0; && \{h^+\} = \emptyset \quad \text{for} \quad \ell \ge 0;\\
&\{p^-\} = \{1,\dots,|\ell| - n_0\} \quad \text{for} \quad \ell \le -n_0; && \{p^-\} = \emptyset \quad \text{for} \quad \ell > -n_0;\\
&\{h^-\} = \{1,\dots,\ell + n_0\} \quad \text{for} \quad \ell > -n_0; && \{h^-\} = \emptyset \quad \text{for} \quad \ell \le -n_0.
\end{aligned} \tag{12.208}$$

In all the cases listed above, $\lambda^0_{p_a} = u_0$, $a = 1,\dots,n_0$. It is easy to see that this state is analogous to the $\alpha_\ell$-twisted ground state for $n_0 > 0$.

Let $\widetilde{\mathfrak{A}}^{(\mathrm{fin})}_{\mathcal{O}}(u_0, [F_{n_0,\ell}])$ denote the reduced finite amplitude of the form factor that corresponds to the excited state described above. Then the reduced finite amplitudes of all other form factors of the $\mathbf{P}_{n_0,\ell}$ class have the following form:

$$\begin{aligned}
\widetilde{\mathfrak{A}}^{(\mathrm{fin})}_{\mathcal{O}}(u_0, \{p^\pm\}, \{h^\pm\}) &= \frac{\widetilde{\mathfrak{A}}^{(\mathrm{fin})}_{\mathcal{O}}(u_0, [F_{n_0,\ell}])\, G^2(1 + F^+_{n_0,\ell}) G^2(1 - F^-_{n_0,\ell})}{G^2(1 + \ell + F^+_{n_0,\ell}) G^2(1 - \ell - n_0 - F^-_{n_0,\ell})}\\
&\quad \times R_{n_p^+, n_h^+}(\{p^+\}, \{h^+\} | F^+_{n_0,\ell})\\
&\quad \times R_{n_p^-, n_h^-}(\{p^-\}, \{h^-\} | - F^-_{n_0,\ell}),
\end{aligned} \tag{12.209}$$

where the functions $R_{n,n'}$ are given by equation (12.142). The derivation of relation (12.209) and the explicit representation for the reduced finite amplitude $\widetilde{\mathfrak{A}}^{(\mathrm{fin})}_{\mathcal{O}}(u_0, [F_{n_0,\ell}])$ is completely analogous to the derivation described above. Therefore, we do not provide details.

The contribution of the quasi-critical form factors into the form factor expansion reads

$$\begin{aligned}
\langle\!\langle \mathcal{O}^\dagger(x,t)\mathcal{O}(0,0)\rangle\!\rangle_{\mathrm{qcr}} &= \sum_{n_0=0}^{\infty} \sum_{\ell\in\mathbb{Z}} L^{-\gamma_{\mathcal{O}}}\, e^{ix(2\ell+n_0)k_F - itn_0\omega(u_0)} \sum_{\{\lambda_{p_a}\}\{p^\pm\},\{h^\pm\}}\\
&\quad \times e^{i\left(x\mathcal{P}^{(d)}_{\mathrm{ex}} - t\mathcal{E}^{(d)}_{\mathrm{ex}}\right)} \mathcal{V}^2(\{\lambda_{p_a}\}) \widetilde{\mathfrak{A}}^{(\mathrm{fin})}_{\mathcal{O}}(u_0, \{p^\pm\}, \{h^\pm\}).
\end{aligned} \tag{12.210}$$

Here the oscillating phase is

$$x\mathcal{P}_{\mathrm{ex}}^{(d)} - t\mathcal{E}_{\mathrm{ex}}^{(d)} = \frac{x}{L}\mathcal{P}_{n_0,\ell} - \frac{t}{L}\mathcal{E}_{n_0,\ell} - t\frac{\omega''(u_0)}{2}\sum_{a=1}^{n_0}(\lambda_{p_a} - u_0)^2 + \frac{2\pi}{L}(x - v_F t)E^+ - \frac{2\pi}{L}(x + v_F t)E^-. \tag{12.211}$$

It is easy to see that the sum of form factors (12.210) factorizes into three separate sums taken over the positions of particles and holes close to the left Fermi boundary $-q$, particles and holes close to the right Fermi boundary $q$, and particles close to the saddle point $u_0$. The sums around the Fermi boundaries are computed via (12.150). We obtain

$$\begin{aligned}&\langle \mathcal{O}^\dagger(x,t)\mathcal{O}(0,0)\rangle_{\mathrm{qcr}} \\ &= \operatorname{Lim}\sum_{n_0=0}^{\infty}\sum_{\ell\in\mathbb{Z}}\sum_{\bar\lambda_p^0\in B(u_0)} \mathcal{V}^2(\bar\lambda_p^0)\,\widetilde{\mathfrak{A}}_{\mathcal{O}}^{(\mathrm{fin})}(u_0,[F_{n_0,\ell}]) \\ &\times \frac{L^{-\gamma_{\mathcal{O}}} e^{ix(2\ell+n_0)k_F - itn_0\omega(u_0)}\prod_{a=1}^{n_0} e^{-it\frac{\omega''(u_0)}{2}(\lambda_{p_a}-u_0)^2}}{\left(1-e^{2\pi i(x-v_F t)/L}\right)^{(F_{n_0,\ell}^+ +\ell)^2}\left(1-e^{-2\pi i(x+v_F t)/L}\right)^{(F_{n_0,\ell}^- +\ell+n_0)^2}}.\end{aligned} \tag{12.212}$$

Here we have already partially taken the limit $L \to \infty$, putting all non-essential $1/L$-corrections equal to zero. The remaining limits are also easily calculated. Using the explicit form of the exponent $\gamma_{\mathcal{O}}$ (12.206), we find

$$\begin{aligned}&\langle \mathcal{O}^\dagger(x,t)\mathcal{O}(0,0)\rangle_{\mathrm{qcr}} \\ &= \operatorname{Lim}\sum_{n_0=0}^{\infty}\sum_{\ell\in\mathbb{Z}}\sum_{\bar\lambda_p^0\in B(u_0)} \mathcal{V}^2(\bar\lambda_p^0)\,\widetilde{\mathfrak{A}}_{\mathcal{O}}^{(\mathrm{fin})}(u_0,[F_{n_0,\ell}]) \\ &\times \frac{L^{-n_0} e^{ix(2\ell+n_0)k_F - itn_0\omega(u_0) + i\frac{\pi}{2}\delta_{\ell,n_0}}\prod_{a=1}^{n_0} e^{-it\frac{\omega''(u_0)}{2}(\lambda_{p_a}-u_0)^2}}{\left(2\pi(x-v_F t)\right)^{(F_{n_0,\ell}^+ +\ell)^2}\left(2\pi(x+v_F t)\right)^{(F_{n_0,\ell}^- +\ell+n_0)^2}},\end{aligned} \tag{12.213}$$

where

$$\delta_{\ell,n_0} = (F_{n_0,\ell}^+ + \ell)^2 - (F_{n_0,\ell}^- + \ell + n_0)^2. \tag{12.214}$$

Finally, the sum over the momenta $\lambda_{p_a}^0$ turns into an integral[3]

$$\begin{aligned}&\operatorname{Lim}\sum_{\bar\lambda_p^0\in B(u_0)} L^{-n_0}\mathcal{V}^2(\{\lambda_{p_a}\})\prod_{a=1}^{n_0} e^{-it\frac{\omega''(u_0)}{2}(\lambda_{p_a}-u_0)^2} \\ &= \frac{1}{n_0!}\int_{B(u_0)} \mathcal{V}^2(\bar\lambda_p^0)\prod_{a=1}^{n_0}\rho(\lambda_{p_a})e^{-\frac{it\omega''(u_0)}{2}(\lambda_{p_a}-u_0)^2}\,\mathrm{d}\lambda_{p_a}.\end{aligned} \tag{12.215}$$

[3]See [K. K. Kozlowski (2018a)] for more rigorous analysis of the saddle point contribution.

The factor $1/n_0!$ occurs because any permutations in the set $\bar{\lambda}_p^0$ correspond to the same state. Using the standard saddle point considerations and the fact that $\omega''(u_0) > 0$, we reduce integral (12.215), in the limit $t \to \infty$, $x \to \infty$, to Mehta's integral [M. Gaudin and M. L. Mehta (1960); M. L. Mehta and F. J. Dyson (1963); M. L. Mehta (2004)] (see also Appendix D)

$$\begin{aligned}
&\frac{1}{n_0!}\int_{B(u_0)} \mathcal{V}^2(\bar{\lambda}_p^0)\prod_{a=1}^{n_0}\rho(\lambda_{p_a})e^{-\frac{it\omega''(u_0)}{2}(\lambda_{p_a}-u_0)^2}\mathrm{d}\lambda_{p_a}\\
&\quad=\frac{\rho^{n_0}(u_0)e^{-i\pi n_0^2/4}}{n_0!}\int_{\mathbb{R}}\mathcal{V}^2(\bar{\lambda}_p^0)\prod_{a=1}^{n_0}e^{-\frac{t\omega''(u_0)}{2}\lambda_{p_a}^2}\mathrm{d}\lambda_{p_a}\cdot\left(1+O(t^{-1})\right)\\
&\quad=e^{-i\pi n_0^2/4}\frac{(\sqrt{2\pi}\rho(u_0))^{n_0}G(n_0+1)}{\left(t\omega''(u_0)\right)^{n_0^2/2}}\cdot\left(1+O(t^{-1})\right).
\end{aligned} \tag{12.216}$$

Substituting (12.216) into (12.213), we obtain the leading asymptotic terms for all the oscillating harmonics in the space-like regime

$$\begin{aligned}
\langle\mathcal{O}^\dagger(x,t)\mathcal{O}(0,0)\rangle_{\mathrm{qcr}} &= \sum_{n_0=0}^{\infty}\sum_{\ell\in\mathbb{Z}}\frac{(\sqrt{2\pi}\rho(u_0))^{n_0}G(1+n_0)}{\left(t\omega''(u_0)\right)^{n_0^2/2}}\\
&\quad\times\frac{\widetilde{\mathfrak{A}}_{\mathcal{O}}^{(\mathrm{fin})}(u_0,[F_{n_0,\ell}])e^{ix(2\ell+n_0)k_F-in_0t\omega(u_0)+i\frac{\pi}{2}\delta_+(\ell,n_0)}}{\left(2\pi(x-v_Ft)\right)^{(F^+_{n_0,\ell}+\ell)^2}\left(2\pi(x+v_Ft)\right)^{(F^-_{n_0,\ell}+\ell+n_0)^2}},
\end{aligned} \tag{12.217}$$

where

$$\delta_+(\ell,n_0) = (F^+_{n_0,\ell}+\ell)^2 - (F^-_{n_0,\ell}+\ell+n_0)^2 - n_0^2/2. \tag{12.218}$$

Similar computations can be done in the time-like regime. The final result is only slightly different

$$\begin{aligned}
\langle\mathcal{O}^\dagger(x,t)\mathcal{O}(0,0)\rangle_{\mathrm{qcr}} &= \sum_{n_0=0}^{\infty}\sum_{\ell\in\mathbb{Z}}\frac{(\sqrt{2\pi}\rho(u_0))^{n_0}G(1+n_0)}{\left(t\omega''(u_0)\right)^{n_0^2/2}}\\
&\quad\times\frac{\widetilde{\mathfrak{A}}_{\mathcal{O}}^{(\mathrm{fin})}(u_0,[F_{-n_0,\ell}])\,e^{ix(2\ell-n_0)k_F+in_0t\omega(u_0)+i\frac{\pi}{2}\delta_-(\ell,n_0)}}{|2\pi(x-v_Ft)|^{(F^+_{-n_0,\ell}+\ell)^2}|2\pi(x+v_Ft)|^{(F^-_{-n_0,\ell}+\ell-n_0)^2}},
\end{aligned} \tag{12.219}$$

where

$$\delta_-(\ell,n_0) = (F^+_{-n_0,\ell}+\ell)^2 + (F^-_{-n_0,\ell}+\ell-n_0)^2 + n_0^2/2. \tag{12.220}$$

Thus, we deal with a double asymptotic series. For $n_0 = 0$, we reproduce a series obtained by considering the contributions of only critical form factors. However, if $n_0 > 0$, additional contributions of the saddle point occur.

Evidently, the terms with $n_0 > 1$ and $|\ell| > 1$ are subdominant, and often they decay quicker than corrections to the first terms. We give, however, the results to these sub-leading contributions as they correspond to different oscillating harmonics.

### 12.8.1 *Correlation function of fields*

We can now use the results obtained for the asymptotic analysis of specific correlation functions. If we restrict ourselves to calculating only the power-law behavior of correlation functions, we only need to find the shift functions corresponding to the states of the $\mathbf{P}_{\tau n_0,\ell}$ classes.

Consider the two-point function $\langle \Psi^\dagger(x,t)\Psi(0,0)\rangle$. The shift function of any state of the $\mathbf{P}_{\tau n_0,\ell}$ class is given by (12.205) with $\alpha = 0$, $\Delta N = -1$, and $n_0$ replaced with $\tau n_0$. Then we find

$$\begin{aligned} F^+_{\tau n_0,\ell} + \ell &= \ell \mathcal{Z} - \frac{1}{2\mathcal{Z}} + \tau n_0\, \Phi_+, \\ F^-_{\tau n_0,\ell} + \ell + \tau n_0 &= \ell \mathcal{Z} + \frac{1}{2\mathcal{Z}} + \tau n_0\, \Phi_-, \end{aligned} \tag{12.221}$$

where

$$\begin{aligned} \Phi_+ &= \phi(q,-q) - \phi(q,u_0) = -\phi(q,u_0) + \frac{\mathcal{Z}-\mathcal{Z}^{-1}}{2}, \\ \Phi_- &= 1 + \phi(-q,-q) - \phi(-q,u_0) = -\phi(-q,u_0) + \frac{\mathcal{Z}+\mathcal{Z}^{-1}}{2}. \end{aligned} \tag{12.222}$$

Then we obtain the following asymptotic expansion of the correlation function:

$$\begin{aligned} &\langle \Psi^\dagger(x,t)\Psi(0,0)\rangle \\ &= \sum_{n_0=0}^{\infty} \sum_{\ell\in\mathbb{Z}} \frac{(\sqrt{2\pi}\rho(u_0))^{n_0}\, G(1+n_0)}{|t\omega''(u_0)|^{n_0^2/2}} \\ &\quad\times \frac{\widetilde{\mathfrak{A}}^{(\mathrm{fin})}_{\Psi}(u_0,[F_{\tau n_0,\ell}])\, e^{ix(2\ell+\tau n_0)k_F - i\tau n_0 t\omega(u_0) + i\frac{\pi}{2}\delta_\tau(\ell,n_0)}}{|2\pi(x-v_F t)|^{(\ell\mathcal{Z}-\frac{1}{2\mathcal{Z}}+\tau n_0\,\Phi_+)^2}\, |2\pi(x+v_F t)|^{(\ell\mathcal{Z}+\frac{1}{2\mathcal{Z}}+\tau n_0\,\Phi_-)^2}}, \end{aligned} \tag{12.223}$$

where

$$\delta_\tau(\ell,n_0) = \Big(\ell\mathcal{Z} + \frac{1}{2\mathcal{Z}} + \tau n_0\,\Phi_-\Big)^2 + \Big(\ell\mathcal{Z} - \frac{1}{2\mathcal{Z}} + \tau n_0\,\Phi_+\Big)^2 - \frac{\tau n_0^2}{2}. \tag{12.224}$$

To get some insight into this asymptotic behavior, let us look at the possible values of the exponents $\theta_{n_0,\ell,\tau,u_0}$ associated with the terms behaving as $|t|^{-\theta_{n_0,\ell,\tau,u_0}}$ in (12.223). They are positive numbers given as the sum of three squares

$$\theta_{n_0,\ell,\tau,u_0} = \frac{n_0^2}{2} + \left(\ell\mathcal{Z} - \frac{1}{2\mathcal{Z}} + \tau n_0 \Phi_+\right)^2 + \left(\ell\mathcal{Z} + \frac{1}{2\mathcal{Z}} + \tau n_0 \Phi_-\right)^2. \quad (12.225)$$

For $n_0 = 0$, the exponents do not depend on $\tau$ and $u_0$. We get

$$\theta_{0,\ell} = 2\ell^2\mathcal{Z}^2 + \frac{1}{2\mathcal{Z}^2}. \quad (12.226)$$

The dominant term is given by $\ell = 0$ with critical exponent $1/2\mathcal{Z}^2 \leq 1/2$, since $\mathcal{Z} \geq 1$ in the one-dimensional Bose gas with finite positive coupling constant ($\mathcal{Z} = 1$ only in the limit $c \to +\infty$). This estimate follows from (10.40) and $\mathcal{Z} = 2\pi\rho(q)$.

The rest of the series, namely the terms with $n_0 \geq 1$, corresponds to the contribution of the saddle point located in the non-linear part of the spectrum (separated from the Fermi boundary). Note that the Luttinger liquid theory and the CFT approaches fail to predict such contributions and the corresponding oscillatory harmonics, which explicitly depend on the saddle point $u_0$ (hidden in $\tau\Phi_\pm$).

To see the possible relevance of these additional terms, let us now discuss the relative values of the different critical exponents. For $n_0 \geq 1$, $\theta_{n_0,\ell,\tau,u_0} \geq 1/2$, whereas $1/2\mathcal{Z}^2 < 1/2$. Hence, the term $n_0 = 0$, $\ell = 0$ gives the dominant contribution to the asymptotic series (12.223), confirming the Luttinger liquid and the CFT prediction.

The origin of the first subleading term of the series (12.223) is more involved. In the CFT and Luttinger liquid theory, it is given by $n_0 = 0$, $\ell = \pm 1$. The corresponding critical exponent $2\mathcal{Z}^2 + 1/2\mathcal{Z}^2$ is greater than or equal to $5/2$. It so happens that the leading saddle point contribution can be dominant with respect to this first subleading CFT contribution.

To see this, let us consider the impenetrable bosons ($c \to +\infty$), where $\mathcal{Z} = \Phi_- = 1$ and $\Phi_+ = 0$. In this case, all the dependence of the critical exponents $\theta_{n_0,\ell,\tau,u_0}$ on $u_0$ reduces to the value of $\tau = \pm 1$. Namely, it just indicates whether we are in the space-like ($\tau = 1$) or in the time-like ($\tau = -1$) asymptotic regimes. Then

$$\theta_{n_0,\ell,\tau} = \frac{n_0^2}{2} + \left(\ell - \frac{1}{2}\right)^2 + \left(\ell + \frac{1}{2} + \tau n_0\right)^2. \quad (12.227)$$

For a given $n_0$, the exponent $\theta_{n_0,\ell,\tau}$ would reach its minimal value at $\ell = -\tau n_0/2$ if we consider $\ell$ as a real variable. Hence, for $n_0 = 1$, the dominant term is realized with the nearest integer values $\ell = 0, -\tau$ with corresponding critical exponents $\theta_{1,0,\tau}$ and $\theta_{1,-\tau,\tau}$.

In the time-like regime, $\tau = -1$. The leading critical exponents coming from the saddle point are $\theta_{1,0,-1} = 1$ and $\theta_{1,1,-1} = 1$, which produce two dominant terms in (12.223) with respect to the first subleading CFT critical exponents $\theta_{0,\pm 1,1} = 5/2$. Hence, in this case, the saddle point provides the first oscillatory subleading contribution to the asymptotic behavior of the series (12.223).

In the space-like regime, $\tau = 1$. The leading critical exponents coming from the saddle point are $\theta_{1,0,1} = 3$ and $\theta_{1,-1,1} = 3$. Hence, they are subdominant with respect to $\theta_{0,\pm 1,-1} = 5/2$.

### 12.8.2 *Density correlation function*

Let us now discuss the correlation function of densities $\langle \hat{q}'(x,t)\hat{q}'(0,0)\rangle$. As usual, we start with the generating function $\langle Q_N^{(\alpha)}(x,t)\rangle$. Then we should put $\Delta N = 0$ in the expression for the shift function (12.205). Then we obtain for states of the $\mathbf{P}_{\tau n_0,\ell}$ class

$$\begin{aligned} F^+_{\tau n_0,\ell} + \ell &= \frac{\alpha_\ell}{2\pi}\mathcal{Z} + \tau n_0\, \Phi_+, \\ F^-_{\tau n_0,\ell} + \ell + \tau n_0 &= \frac{\alpha_\ell}{2\pi}\mathcal{Z} + \tau n_0\, \Phi_-. \end{aligned} \tag{12.228}$$

Substituting these expressions in (12.217) and (12.219), we find

$$\begin{aligned} &Q^{(\alpha)}_{\rm qcr}(x,t) \\ &\quad = \sum_{n_0=0}^{\infty}\sum_{\ell\in\mathbb{Z}} \frac{(\sqrt{2\pi}\rho(u_0))^{n_0}\, G(1+n_0)}{|t\omega''(u_0)|^{n_0^2/2}} \\ &\qquad \times \frac{\widetilde{\mathfrak{A}}^{(\rm fin)}_{\mathcal{Q}}(u_0,[F_{\tau n_0,\ell}])e^{ix(\alpha_\ell+\pi\tau n_0)k_F/\pi - i\tau n_0 t\omega(u_0) + i\frac{\pi}{2}\delta_\tau(n_0,\ell,\alpha)}}{\left(2\pi|x-v_F t|\right)^{(\alpha_\ell\mathcal{Z}/2\pi + n_0\tau\Phi_+)^2}\left(2\pi|x+v_F t|\right)^{(\alpha_\ell\mathcal{Z}/2\pi+n_0\tau\Phi_-)^2}}, \end{aligned} \tag{12.229}$$

where $Q^{(\alpha)}_{\rm qcr}(x,t)$ is the contribution of the quasi-critical form factors to the generating function, and

$$\delta_\tau(n_0,\ell,\alpha) = \left(\frac{\alpha_\ell}{2\pi}\mathcal{Z} + n_0\tau\Phi_+\right)^2 - \left(\frac{\alpha_\ell}{2\pi}\mathcal{Z} + n_0\tau\Phi_-\right)^2 - \frac{\tau n_0^2}{2}. \tag{12.230}$$

It then remains to differentiate (12.229) with respect to $\alpha$ and $x$ to get the asymptotic behavior of the two-point function. Just as in the time-independent correlation function, we should consider the case $\ell = n_0 = 0$ separately. Using the results of Sections 12.6.1 and 12.6.2, we obtain

$$\begin{aligned}
&\langle \hat{q}'(x,t)\hat{q}'(0,0)\rangle \\
&= D^2 - \frac{\mathcal{Z}^2}{2\pi^2}\frac{x^2+v_F^2t^2}{(x^2-v_F^2t^2)^2} \\
&\quad + \sum_{n_0=0}^{\infty}\sum_{\substack{\ell\in\mathbb{Z}\\ |\ell|+n_0>0}} \frac{(\sqrt{2\pi}\rho(u_0))^{n_0}\, G(1+n_0)}{2|t\omega''(u_0)|^{n_0^2/2}}\widetilde{\mathfrak{A}}_{\hat{q}}^{(\text{fin})}(u_0,[F_{\tau n_0,\ell}]) \\
&\quad \times \frac{\big(2\ell+\tau n_0(k_F+k(u_0))\big)^2 e^{ix(2\ell+\tau n_0)k_F - i\tau n_0 t\omega(u_0) + i\frac{\pi}{2}\delta_\tau(n_0,\ell,0)}}{\big(2\pi|x-v_Ft|\big)^{(\ell\mathcal{Z}+\tau n_0\Phi_+)^2}\big(2\pi|x+v_Ft|\big)^{(\ell\mathcal{Z}+\tau n_0\Phi_-)^2}}.
\end{aligned} \tag{12.231}$$

Here

$$\mathfrak{A}_{\hat{q}}^{(\text{fin})}(u_0,[F_{\tau n_0,\ell}]) = \frac{\partial^2\mathfrak{A}_{\mathcal{Q}}^{(\text{fin})}(u_0,[F_{\tau n_0,\ell}])}{\partial\alpha^2}\bigg|_{\alpha=0}, \qquad |\ell|+n_0>0. \tag{12.232}$$

Let us look now at the possible values of the critical exponents $\theta_{n_0,\ell,\tau,u_0}$ associated with the terms behaving as $|t|^{-\theta_{n_0,\ell,\tau,u_0}}$ in (12.231). For oscillating terms in (12.231), the critical exponents are given by the sum of three squares

$$\theta_{n_0,\ell,\tau,u_0} = \frac{n_0^2}{2} + \big(\ell\mathcal{Z}+\tau n_0\Phi_+\big)^2 + \big(\ell\mathcal{Z}+\tau n_0\Phi_-\big)^2. \tag{12.233}$$

For $n_0=0$, they do not depend on $\tau$ and $u_0$. We have

$$\theta_{0,\ell} = 2\ell^2\mathcal{Z}^2. \tag{12.234}$$

Hence, as for the field correlation functions, the series over $\ell$ at $n_0=0$ reproduces the CFT and Luttinger liquid predictions, with the exact values of the amplitudes given in terms of specific (renormalized) form factors of the density. The dominant term of the asymptotic expansion (12.231) is evidently the constant term. The origin of the first subleading term is more involved in general.

The terms with $n_0 > 1$ are subdominant with respect to the non-oscillating contribution, which has a critical exponent equal to 2. Hence, after the constant term, the leading behavior is produced either by the $n_0 = 0$ and $n_0 = 1$ terms or by the non-oscillating term. For $n_0 = 0$, the leading terms correspond to $\ell = \pm 1$ with critical exponent $2\mathcal{Z}^2 \geq 2$. The terms with $n_0 = 0$ and $\ell > 1$ are evidently subleading. It is more complicated to select the leading contribution for $n_0 = 1$ and compare it to 2 or $2\mathcal{Z}^2$ since it depends on the regime and the values of the constants $\Phi_\pm$.

To see at least the possible relevance of these different terms, let us consider the example of the impenetrable bosons again. There, the leading non-constant terms correspond to $n_0 = 1$, $\ell = 0$ and $n_0 = 1$, $\ell = -\tau$. Both of them have the critical exponent 3/2, which is dominant with respect to the non-oscillating term and the first subleading oscillating term predicted from CFT (both with critical exponent 2). In this example, the saddle point provides the main contribution after the constant term as soon as $x/t$ takes a finite value in the asymptotic regime. Hence, in this case, we again obtain a contribution to the asymptotics that is not predicted by the Luttinger liquid theory and CFT.

Let us summarize. We have shown that the form factor expansion is well suited for calculating the asymptotics of correlation functions in critical models. We have considered the correlation functions of the one-dimensional bosons as an example. However, completely analogous calculations can be done in the massless spin chains [N. Kitanine et al (2009c, 2011a,b, 2014); K. K. Kozlowski (2018a, 2019)]. Explicit representations for form factors in finite-size systems play a crucial role in these calculations. In turn, we obtained these explicit formulas using the algebraic Bethe ansatz.

# Appendix A

# The $\psi$-function and the Barnes $G$-function

In this appendix, we consider some properties of the $\psi$-function and the Barnes $G$-function.

## A.1 The $\psi$-function

The $\psi$-function (the digamma function) is defined as the logarithmic derivative of the $\Gamma$-function

$$\psi(z) = \frac{\mathrm{d}}{\mathrm{d}z} \log \Gamma(z). \tag{A.1}$$

The multiplication property of the $\Gamma$-function implies

$$\psi^{(n)}(z+1) - \psi^{(n)}(z) = \frac{(-1)^n n!}{z^{n+1}}, \quad \text{where} \quad \psi^{(n)}(z) = \frac{\mathrm{d}^n}{\mathrm{d}z^n}\psi(z). \tag{A.2}$$

It follows from (A.2) that

$$\sum_{k=0}^{N-1} \frac{1}{(k+a)^{n+1}} = \frac{(-1)^n}{n!}\Big(\psi^{(n)}(N+a) - \psi^{(n)}(a)\Big). \tag{A.3}$$

Here $a \neq 0, -1, -2, \dots$.

Let $|z| \to \infty$ with $-\pi < \arg(z) < \pi$. Then

$$\begin{aligned} \psi(z) &= \log z - \frac{1}{2z} + O\left(\frac{1}{z^2}\right), \\ \psi^{(n)}(z) &= \frac{(-1)^{n-1}(n-1)!}{z^n} + O\left(\frac{1}{z^{n+1}}\right), \qquad n \geq 1. \end{aligned} \tag{A.4}$$

For $n \geq 1$, the $n$th-derivative of the $\psi$-function admits the following integral representation:

$$\psi^{(n)}(z) = -\int_0^1 \frac{x^{z-1} \log^n x}{1-x}\,\mathrm{d}x. \tag{A.5}$$

The latter implies, in particular, that

$$(-1)^{n-1}\psi^{(n)}(z) > 0, \qquad \text{for} \quad z > 0, \quad n \geq 1. \tag{A.6}$$

It is also easy to see that

$$\psi'(z) - \frac{1}{z} = \int_0^1 x^{z-1}\left(\frac{\log x}{x-1} - 1\right) dx > 0, \qquad \text{for} \quad z > 0. \tag{A.7}$$

**Lemma A.1.** *For $n \geq 0$, $a \neq 0, -1, -2, \ldots$, and an arbitrary complex $\alpha$, one has the following identity:*

$$\sum_{k=1}^{N} \psi^{(n)}(k+a)e^{\alpha k} = \frac{1}{e^\alpha - 1}\Bigg[\psi^{(n)}(N+a)e^{\alpha(N+1)} - \psi^{(n)}(a)e^\alpha \\ - (-1)^n n! \sum_{k=0}^{N-1} \frac{e^{\alpha(k+1)}}{(k+a)^{n+1}}\Bigg]. \tag{A.8}$$

**Proof.** Denote the lhs of (A.8) as $f(\alpha)$,

$$f(\alpha) = \sum_{k=1}^{N} \psi^{(n)}(k+a)\, e^{\alpha k}. \tag{A.9}$$

Shifting $k$ to $k+1$ and using (A.2), we obtain

$$\begin{aligned} f(\alpha) &= e^\alpha \sum_{k=0}^{N-1} \psi^{(n)}(k+a+1)\, e^{\alpha k} \\ &= e^\alpha \sum_{k=0}^{N-1} \psi^{(n)}(k+a)\, e^{\alpha k} + (-1)^n n! \sum_{k=0}^{N-1} \frac{e^{\alpha(k+1)}}{(k+a)^{n+1}} \\ &= e^\alpha f(\alpha) + e^\alpha \psi^{(n)}(a) - e^{\alpha(N+1)}\psi^{(n)}(N+a) \\ &\quad + (-1)^n n! \sum_{k=0}^{N-1} \frac{e^{\alpha(k+1)}}{(k+a)^{n+1}}. \end{aligned} \tag{A.10}$$

Then (A.8) follows immediately. □

Taking derivatives of (A.8) with respect to $\alpha$ at $\alpha = 0$, we obtain formulas for the sums of the type $\psi^{(n)}(k+a)k^p$. Let us explicitly give several of them. We have

$$\sum_{k=1}^{N} \psi(k+a) = (N+a)\psi(N+a) - a\psi(a) - N, \tag{A.11}$$

for $p=0$, $n=0$, and

$$\sum_{k=1}^{N}\psi^{(n)}(k+a)=(N+a)\psi^{(n)}(N+a)-a\psi^{(n)}(a+1)$$
$$+n\Big(\psi^{(n-1)}(N+a)-\psi^{(n-1)}(a+1)\Big), \quad (A.12)$$

for $p=0$, $n\geq 1$. We also have

$$\sum_{k=1}^{N}k\psi'(k+a)=\tfrac{(N+a)(N+1-a)}{2}\psi'(N+a)+\tfrac{a(a-1)}{2}\psi'(a+1)$$
$$-\tfrac{2a-1}{2}\Big(\psi(N+a)-\psi(a+1)\Big)+\tfrac{N-1}{2}, \quad (A.13)$$

for $p=1$, $n=1$, and

$$\sum_{k=1}^{N}k\psi^{(n)}(k+a)=\tfrac{(N+a)(N+1-a)}{2}\psi^{(n)}(N+a)+\tfrac{a(a-1)}{2}\psi^{(n)}(a+1)$$
$$-\tfrac{n(2a-1)}{2}\Big(\psi^{(n-1)}(N+a)-\psi^{(n-1)}(a+1)\Big)$$
$$-\tfrac{n(n-1)}{2}\Big(\psi^{(n-2)}(N+a)-\psi^{(n-2)}(a+1)\Big), \quad (A.14)$$

for $p=1$, $n\geq 2$.

The following two lemmas allow us to calculate the thermodynamic limits of sums containing the $\psi$-function.

**Lemma A.2.** *Let $f(x)$ be a smooth function on an interval $[0,D]$, and let*

$$S_1[f]=\sum_{k=1}^{N}\left[f\left(\tfrac{k}{L}\right)-f(0)\right]\psi'(k). \quad (A.15)$$

*Then the sum $S_1[f]$ behaves as*

$$S_1[f]=\int_0^D\frac{f(y)-f(0)}{y}\,\mathrm{d}y+O\Big(\tfrac{\log(DL)}{L}\Big), \quad (A.16)$$

*in the thermodynamic limit ($N,L\to\infty$, $N/L=D$).*

**Proof.** We present $S_1[f]$ as follows:

$$S_1^{(N,L)}[f]=\sum_{k=1}^{N}\left(\psi'(k)-\frac{1}{k}\right)\frac{k}{L}\int_0^1 f'(\tfrac{kt}{L})\,\mathrm{d}t+\sum_{k=1}^{N}\frac{1}{L}\int_0^1 f'(\tfrac{kt}{L})\,\mathrm{d}t. \quad (A.17)$$

It is easy to see that the first sum vanishes in the limit considered. Indeed, we have

$$\left|\sum_{k=1}^{N}\left(\psi'(k)-\frac{1}{k}\right)\frac{k}{L}\int_0^1 f'(\tfrac{kt}{L})\,\mathrm{d}t\right|\leq\frac{\|f'\|}{L}\sum_{k=1}^{N}\left(k\psi'(k)-1\right),$$

where $\|f'\| = \sup_{x\in[0,D]} |f'(x)|$. Using (A.13), we obtain

$$\sum_{k=1}^{N} (k\psi'(k) - 1) = \frac{1}{2}\left[N(N+1)\psi'(N) + \psi(N) - \psi(1) - N - 1\right]. \quad \text{(A.18)}$$

Using (A.4), we conclude that this contribution is of order $O(\log N)$.

Thus, the limit of $S_1[f]$ reduces to the limit of the second term in (A.17), which is quite straightforward due to the Euler–Maclaurin summation formula

$$\begin{aligned}\sum_{k=1}^{N} \frac{1}{L}\int_0^1 f'\left(\tfrac{kt}{L}\right) \mathrm{d}t &= \int_0^D \mathrm{d}y \int_0^1 f'(yt)\,\mathrm{d}t + O(L^{-1}) \\ &= \int_0^D \frac{f(y)-f(0)}{y}\,\mathrm{d}y + O(L^{-1}). \end{aligned} \quad \text{(A.19)}$$

Combining the estimates obtained, we arrive at the statement of lemma A.2. □

**Corollary A.1.** *Let $\nu$ be a finite positive integer. Then under the conditions of lemma A.2,*

$$\sum_{k=\nu}^{N} \left[f\left(\tfrac{k}{L}\right) - f(0)\right]\psi'(k) = \int_0^D \frac{f(y)-f(0)}{y}\,\mathrm{d}y + O\left(\tfrac{\log(DL)}{L}\right), \quad \text{(A.20)}$$

*in the thermodynamic limit.*

**Proof.** The difference between the sum $S_1[f]$ (A.15) and the sum (A.20) is

$$S_1[f] - \sum_{k=\nu}^{N} \left[f\left(\tfrac{k}{L}\right) - f(0)\right]\psi'(k) = \sum_{k=1}^{\nu-1} \left[f\left(\tfrac{k}{L}\right) - f(0)\right]\psi'(k). \quad \text{(A.21)}$$

Since $f(x)$ is a smooth function, each term in this finite sum is of order $O(L^{-1})$. □

**Lemma A.3.** *Let $F(x)$ and $f(x)$ be smooth functions on an interval $[0, D]$ such that $\|F\| = \sup_{x\in[0,D]} |F(x)| < \nu$, where $\nu$ is a positive integer. Then, for $m > 1$, the sum*

$$S_m[f,F] = \sum_{k=\nu}^{N} \left[f\left(\tfrac{k}{L}\right)\psi^{(m)}\left(F\left(\tfrac{k}{L}\right) + k\right) - f(0)\,\psi^{(m)}(F(0)+k)\right], \quad \text{(A.22)}$$

*vanishes in the thermodynamic limit.*

**Proof.** We present $S_m[f, F]$ as follows:

$$S_m[f,F] = \sum_{k=\nu}^{N} \frac{k}{L} \int_0^1 \mathrm{d}t \Big[ f'\left(\tfrac{tk}{L}\right) \psi^{(m)}\left(F\left(\tfrac{tk}{L}\right) + k\right) + f\left(\tfrac{tk}{L}\right) F'\left(\tfrac{tk}{L}\right) \psi^{(m+1)}\left(F\left(\tfrac{tk}{L}\right) + k\right)\Big]. \quad \text{(A.23)}$$

Due to (A.6) and (A.7), $(-1)^{m-1}\psi^{(m)}(z)$ is a positive monotonically decreasing function for $z > 0$. Then one has the following estimate:

$$|S_m[f,F]| \leq \frac{\|F'\|\|f\| + \|f'\|}{L} \sum_{k=\nu}^{N} k\Big[(-1)^m \psi^{(m+1)}(k - \|F\|) + (-1)^{m-1}\psi^{(m)}(k - \|F\|)\Big]. \quad \text{(A.24)}$$

The sum in (A.24) is readily computed via (A.14). It is easy to see that it is of order $O(\log N)$ at most. □

## A.2 The Barnes G-function

The Barnes $G$-function $G(z)$ is defined as the unique solution of a difference equation

$$\begin{aligned} &G(1+z) = \Gamma(z)G(z), \\ &\frac{\mathrm{d}^3}{\mathrm{d}z^3} \log G(1+z) \geq 0, \quad \text{for} \quad z \geq 0, \end{aligned} \quad \text{(A.25)}$$

with the initial condition $G(1) = 1$. It has the following integral representation

$$\log G(1+z) = \frac{z(1-z)}{2} + \frac{z}{2}\log 2\pi + \int_0^z x\psi(x)\,\mathrm{d}x, \qquad \Re z > -1. \quad \text{(A.26)}$$

The asymptotics of the Barnes $G$-function at $|z| \to \infty$, $-\pi < \arg(z) < \pi$ is

$$\log G(1+z) = \left(\frac{z^2}{2} - \frac{1}{12}\right)\log z - \frac{3z^2}{4} + \frac{z}{2}\log 2\pi + \zeta'(-1) + O(z^{-1}), \quad \text{(A.27)}$$

where $\zeta'(z)$ is the derivative of the Riemann $\zeta$-function.

It is easy to see that due to (A.25),

$$\prod_{k=0}^{N} \Gamma(k+a) = \frac{G(N+1+a)}{G(a)}. \quad \text{(A.28)}$$

# Appendix B

# Finite-size corrections to the excitation energy

Finite-size corrections to the excitation energy of critical states can be divided into two types:

$$\mathcal{E}_{\rm ex} = \frac{1}{L}(\Delta\mathcal{E}(\mathbf{P}_\ell) + \Delta\mathcal{E}_{\rm ph}) + O(L^{-2}). \tag{B.1}$$

Here $\Delta\mathcal{E}(\mathbf{P}_\ell)$ depends on the class $\mathbf{P}_\ell$ but not on the specific representative of this class. In other words, $\Delta\mathcal{E}(\mathbf{P}_\ell)$ does not depend on the quantum numbers $p_j^\pm$ and $h_j^\pm$. On the contrary, the term $\Delta\mathcal{E}_{\rm ph}$ depends on these quantum numbers. The term $\Delta\mathcal{E}(\mathbf{P}_\ell)$ remains constant when we take the sum of form factors within a specific $\mathbf{P}_\ell$ class. Therefore, it remains proportional to $1/L$ and vanishes in the thermodynamic limit. Thus, when calculating $1/L$-corrections to the excitation energy, we should find the term $\Delta\mathcal{E}_{\rm ph}$ only. Any contribution to the term $\Delta\mathcal{E}(\mathbf{P}_\ell)$ can be neglected.

To calculate corrections to the energy, we have to find $1/L$-corrections to the shift function. They can also be divided into two types

$$\hat{F}(z) = F_\ell(z) + \frac{1}{L}(F(z;\mathbf{P}_\ell) + F_{\rm ph}(z)) + O(L^{-2}). \tag{B.2}$$

Here $F_\ell(z)$ is the thermodynamic limit of the shift function of the $\mathbf{P}_\ell$ class (12.125). The term $F(z;\mathbf{P}_\ell)$ is common for all the states of the $\mathbf{P}_\ell$ class, while the term $F_{\rm ph}(z)$ depends on the quantum numbers $p_j^\pm$ and $h_j^\pm$. We are only interested in this latter term of the shift function.

## B.1 Corrections to the shift function

Let us now find $1/L$-corrections to the shift function $\hat{F}(z)$. Recall that $\hat{F}(z)$ is defined as

$$\hat{F}(z) - \frac{1}{2\pi}\sum_{k=1}^{N}(\theta(z-u_k) - \theta(z-v_k')) = \frac{\alpha + \omega_n(z)}{2\pi}, \tag{B.3}$$

(see (10.63)). Here $v'_k$ are defined by (10.56), and

$$\omega_n(z) = \sum_{j=1}^{n} \big(\theta(z - \lambda_{h_j}) - \theta(z - \lambda_{p_j})\big). \tag{B.4}$$

We consider a state from the $\mathbf{P}_\ell$ class. Then, in the thermodynamic limit,

$$\omega_n(z) = \ell\big(\theta(z + q) - \theta(z - q)\big). \tag{B.5}$$

This ultimately leads to (12.125) for the $O(1)$ contribution to the shift function.

Consider now corrections to the function $\omega_n(z)$. Due to the shift function (10.59) definition, we have

$$\hat{\xi}_v(q) = \hat{\xi}_u(q) - \frac{\hat{F}(q)}{L} = \frac{N + 1/2 - \hat{F}(q)}{L}. \tag{B.6}$$

Here we used (10.48). This implies

$$q = \hat{\xi}_v^{-1}\Big(\tfrac{N+1/2-\hat{F}(q)}{L}\Big). \tag{B.7}$$

From this, we find for any finite $m$

$$\hat{\xi}_v^{-1}\Big(\tfrac{N+m}{L}\Big) = q + \frac{m - 1/2 + F_\ell(q)}{L\rho(q)} + O(L^{-2}). \tag{B.8}$$

Similarly,

$$\hat{\xi}_v^{-1}\Big(\tfrac{m}{L}\Big) = -q + \frac{m - 1/2 + F_\ell(-q)}{L\rho(q)} + O(L^{-2}). \tag{B.9}$$

Let quantum numbers of particles and holes be defined by (12.120). Then

$$\begin{aligned}\omega_n(z) = {}&\sum_{j=1}^{n_p^+} \theta\Big(\hat{\xi}_v^{-1}\big(\tfrac{N+p_j^+}{L}\big) - z\Big) - \sum_{j=1}^{n_h^+} \theta\Big(\hat{\xi}_v^{-1}\big(\tfrac{N+1-h_j^+}{L}\big) - z\Big)\\ &+ \sum_{j=1}^{n_p^-} \theta\Big(\hat{\xi}_v^{-1}\big(\tfrac{1-p_j^-}{L}\big) - z\Big) - \sum_{j=1}^{n_h^-} \theta\Big(\hat{\xi}_v^{-1}\big(\tfrac{h_j^-}{L}\big) - z\Big).\end{aligned} \tag{B.10}$$

Using (B.8), (B.9), we obtain

$$\begin{aligned}\omega_n(z) = {}&\ell\big(\theta(z + q) - \theta(z - q)\big) + \frac{1}{L\rho(q)}\Big(\mathcal{K}(z - q)E^+ - \mathcal{K}(z + q)E^-\Big)\\ &+ \frac{\ell}{L\rho(q)}\Big(\mathcal{K}(z - q)(F_\ell(q) + 1/2) - \mathcal{K}(z + q)(F_\ell(-q) - 1/2)\Big)\\ &+ O(L^{-2}),\end{aligned} \tag{B.11}$$

where $E^{\pm}$ are given by (12.123). Note that the term in the second line of (B.11) is common to all states of the $\mathbf{P}_\ell$ class. Therefore, it only contributes to the function $F(z;\mathbf{P}_\ell)$ but not to the function $F_{\rm ph}(z)$.

Consider now corrections to the lhs of (B.3). There are two types of corrections. First, we have to develop the difference $\theta(z-u_k)-\theta(z-v'_k)$ up to the second order in $v'_k-u_k$:

$$\theta(z-u_k)-\theta(z-v'_k)=(v'_k-u_k)\mathcal{K}(z-u_k)+\frac{(v'_k-u_k)^2}{2}\mathcal{K}'(z-u_k)+O(L^{-3}). \tag{B.12}$$

Secondly, we should more precisely estimate the shift of $v'_k$ with respect to $u_k$:

$$v'_k-u_k=\frac{a_k^{(1)}}{L}+\frac{a_k^{(2)}}{L^2}+O(L^{-3}). \tag{B.13}$$

On the one hand,

$$\frac{\hat{F}(v'_k)}{L}=\frac{\hat{F}(u_k)}{L}+\frac{a_k^{(1)}\hat{F}'(u_k)}{L^2}+O(L^{-3}). \tag{B.14}$$

On the other hand,

$$\begin{aligned}\frac{\hat{F}(v'_k)}{L}&=\hat{\xi}_u(v'_k)-\hat{\xi}_v(v'_k)\\&=\frac{a_k^{(1)}}{L}\hat{\rho}_u(u_k)+\frac{1}{L^2}\left(a_k^{(2)}\hat{\rho}_u(u_k)+\frac{(a_k^{(1)})^2}{2}\hat{\rho}'_u(u_k)\right)+O(L^{-3}).\end{aligned} \tag{B.15}$$

Comparing (B.14) and (B.15), we find

$$a_k^{(1)}=\frac{\hat{F}(u_k)}{\hat{\rho}_u(u_k)},\qquad a_k^{(2)}=\frac{1}{2\hat{\rho}_u(u_k)}\left(\frac{\hat{F}^2(u_k)}{\hat{\rho}_u(u_k)}\right)'. \tag{B.16}$$

We see that the coefficient $a_k^{(2)}$ is completely determined by the thermodynamic limit of the shift function $F(u_k)$. The latter, in turn, is given by (12.125). Thus, this correction to the shift $v'_k-u_k$ depends only on the $\mathbf{P}_\ell$ class but not on the specific representative of this class. Therefore, it does not contribute to the function $F_{\rm ph}(z)$. The same obviously applies to the corrections in equation (B.12).

Thus, replacing the sum with an integral in (B.3), we arrive at the following integral equation:

$$\begin{aligned}\hat{F}(z)-\frac{1}{2\pi}\int_{-q}^{q}\mathcal{K}(z-v)\hat{F}(v)\,\mathrm{d}v&=\frac{\alpha}{2\pi}+\frac{\ell}{2\pi}\big(\theta(z+q)-\theta(z-q)\big)+\frac{1}{2\pi L\rho(q)}\\&\qquad\Big(\mathcal{K}(z-q)E^{+}-\mathcal{K}(z+q)E^{-}\Big)+\frac{1}{L}\mathcal{G}_\ell(z).\end{aligned} \tag{B.17}$$

Here $\mathcal{G}_\ell(z)$ denotes all the terms which do not depend on the quantum numbers $p_j^\pm$ and $h_j^\pm$. Substituting here (B.2), we obtain an integral equation for the function $F_{\rm ph}(z)$

$$F_{\rm ph}(z) - \frac{1}{2\pi}\int_{-q}^{q} \mathcal{K}(z-v)F_{\rm ph}(v)\,{\rm d}v = \frac{1}{2\pi\rho(q)}\Big(\mathcal{K}(z-q)E^+ - \mathcal{K}(z+q)E^-\Big), \tag{B.18}$$

with obvious solution

$$F_{\rm ph}(z) = \frac{1}{\rho(q)}\Big(\mathcal{R}(z,q)E^+ - \mathcal{R}(z,-q)E^-\Big). \tag{B.19}$$

Here $\mathcal{R}(u,v)$ is the resolvent of the integral operator $\mathbf{I} - \frac{1}{2\pi}\mathbf{K}$ (10.108).

## B.2 Excitation energy

We now can easily find the excitation energy

$$\mathcal{E}_{\rm ex} = \sum_{k=1}^{N}(v_k^2 - u_k^2) = \sum_{k=1}^{N}({v'_k}^2 - u_k^2) + \sum_{j=1}^{n}(\lambda_{p_j}^2 - \lambda_{h_j}^2). \tag{B.20}$$

If the quantum numbers of particles and holes are defined by (12.120), then

$$\sum_{j=1}^{n}(\lambda_{p_j}^2 - \lambda_{h_j}^2) = \sum_{j=1}^{n_p^+}\Big(\hat\xi_v^{-1}\big(\tfrac{N+p_j^+}{L}\big)\Big)^2 - \sum_{j=1}^{n_h^+}\Big(\hat\xi_v^{-1}\big(\tfrac{N+1-h_j^+}{L}\big)\Big)^2 + \sum_{j=1}^{n_p^-}\Big(\hat\xi_v^{-1}\big(\tfrac{1-p_j^-}{L}\big)\Big)^2 - \sum_{j=1}^{n_h^-}\Big(\hat\xi_v^{-1}\big(\tfrac{h_j^-}{L}\big)\Big)^2. \tag{B.21}$$

Using (B.8), (B.9), we find

$$\sum_{j=1}^{n}(\lambda_{p_j}^2 - \lambda_{h_j}^2) = \frac{2q}{L\rho(q)}\big(E^+ + E^- + \ell(F_\ell(q) + F_\ell(-q))\big). \tag{B.22}$$

The term $\ell(F_\ell(q) + F_\ell(-q))$ contributes to the correction $\Delta\mathcal{E}(\mathbf{P}_\ell)$ only.

We also have

$$\sum_{k=1}^{N}({v'_k}^2 - u_k^2) = \sum_{k=1}^{N}2u_k(v'_k - u_k) + \sum_{k=1}^{N}(v'_k - u_k)^2. \tag{B.23}$$

The second term can be neglected as it also contributes only to $\Delta\mathcal{E}(\mathbf{P}_\ell)$. Thus, replacing the remaining sum with an integral, we obtain

$$\mathcal{E}_{\rm ex} = \int_{-q}^{q} 2u\left(F_\ell(u) + \frac{1}{L}F_{\rm ph}(u)\right){\rm d}u + \frac{2q}{L\rho(q)}(E^+ + E^-) + \frac{\Delta\mathcal{E}(\mathbf{P}_\ell)}{L}. \tag{B.24}$$

We suggest that the reader verify that the leading contribution to the excitation energy vanishes, as it should be. The correction $\Delta\mathcal{E}_{\text{ph}}$ is given by

$$\Delta\mathcal{E}_{\text{ph}} = \int_{-q}^{q} 2uF_{\text{ph}}(u)\,\mathrm{d}u + \frac{2q}{\rho(q)}(E^+ + E^-). \tag{B.25}$$

It follows from (10.77) that

$$2v + \int_{-q}^{q} 2u\mathcal{R}(u,v)\,\mathrm{d}u = \varepsilon'(v). \tag{B.26}$$

Then we obtain

$$\begin{aligned}\int_{-q}^{q} 2uF_{\text{ph}}(u)\,\mathrm{d}u &= \frac{1}{\rho(q)}\int_{-q}^{q} 2u(\mathcal{R}(u,q)E^+ - \mathcal{R}(u,-q)E^-)\,\mathrm{d}u\\ &= \frac{\varepsilon'(q) - 2q}{\rho(q)}(E^+ + E^-),\end{aligned} \tag{B.27}$$

and substituting this result in (B.25), we finally arrive at

$$\Delta\mathcal{E}_{\text{ph}} = \frac{\varepsilon'(q)}{\rho(q)}(E^+ + E^-) = 2\pi v_F(E^+ + E^-), \tag{B.28}$$

where $v_F$ is the Fermi velocity (12.146).

## Appendix C

# Identities for Fredholm determinants

In this section, we prove that the smooth part of the form factor amplitude does not depend on an arbitrary parameter $\omega$. To do this, we will establish some general properties of Fredholm determinants. Let $\mathcal{C}$ be a contour in the complex plane. Consider an invertible integral operator $\mathbf{I}+\mathbf{V}$ acting on functions from $L^2(\mathcal{C})$ as follows:

$$[f(\mathbf{I}+\mathbf{V})](z') = f(z') + \int_{\mathcal{C}} V(z,z')f(z)\,\mathrm{d}z. \tag{C.1}$$

We also introduce the inverse operator $\mathbf{I}-\mathbf{R} = (\mathbf{I}+\mathbf{V})^{-1}$ so that the kernel of the resolvent $\mathcal{R}(z,z')$ satisfies the following equations:

$$\begin{aligned} \mathcal{R}(z,z') + \int_{\mathcal{C}} \mathcal{R}(z,w)V(w,z')\,\mathrm{d}w = V(z,z'),\\ \mathcal{R}(z,z') + \int_{\mathcal{C}} V(z,w)\mathcal{R}(w,z')\,\mathrm{d}w = V(z,z'). \end{aligned} \tag{C.2}$$

**Proposition C.1.** *Let $f(z) \in L^2(\mathcal{C})$ and let*

$$g(z') = [f(\mathbf{I}+\mathbf{V})](z'). \tag{C.3}$$

*Then for arbitrary $z_0 \in \mathcal{C}$ the following identity holds:*

$$\det(\mathbf{I}+\mathbf{V}) = \frac{g(z_0)}{f(z_0)}\,\det(\mathbf{I}+\mathbf{V}-\mathbf{V}_g^0). \tag{C.4}$$

*Here the kernel of the operator $\mathbf{V}_g^0$ is*

$$V_g^0(z,z') = \frac{g(z')}{g(z_0)}V(z,z_0). \tag{C.5}$$

**Remark C.1.** We can consider identity (C.4) as an analog of Cramer's rule to solve a system of linear equations.

**Proof.** We transform the determinant in the rhs of (C.4) as follows:

$$\det(\mathbf{I}+\mathbf{V}-\mathbf{V}_g^0) = \det(\mathbf{I}+\mathbf{V})\det\Big((\mathbf{I}-\mathbf{R})(\mathbf{I}+\mathbf{V}-\mathbf{V}_g^0)\Big)$$
$$= \det(\mathbf{I}+\mathbf{V})\det\Big(\mathbf{I}-(\mathbf{I}-\mathbf{R})\mathbf{V}_g^0\Big). \tag{C.6}$$

Due to (C.2), we find

$$V_g^0(z,z') - \int_{\mathcal{C}} \mathcal{R}(z,w)V_g^0(w,z')\,\mathrm{d}w = \mathcal{R}_g^0(z,z'), \tag{C.7}$$

where

$$\mathcal{R}_g^0(z,z') = \frac{g(z')}{g(z_0)}\mathcal{R}(z,z_0). \tag{C.8}$$

Thus,

$$\det(\mathbf{I}+\mathbf{V}-\mathbf{V}_g^0) = \det(\mathbf{I}+\mathbf{V})\det(\mathbf{I}-\mathbf{R}_g^0). \tag{C.9}$$

Since the kernel $\mathcal{R}_g^0(z,z')$ is degenerate, the determinant $\det(\mathbf{I}-\mathbf{R}_g^0)$ is easily calculated

$$\det(\mathbf{I}-\mathbf{R}_g^0) = 1 - \int_{\mathcal{C}} \frac{g(z)}{g(z_0)}\mathcal{R}(z,z_0)\,\mathrm{d}z. \tag{C.10}$$

To see this, it is enough to substitute the kernel $\mathcal{R}_g^0(z,z')$ in the definition of the Fredholm determinant (10.176). Due to the degeneracy of the kernel $\mathcal{R}_g^0(z,z')$, all the terms of the series with $n>1$ are zero.

Due to (C.3), we find

$$1 - \int_{\mathcal{C}} \frac{g(z)}{g(z_0)}\mathcal{R}(z,z_0)\,\mathrm{d}z = \frac{f(z_0)}{g(z_0)}, \tag{C.11}$$

which completes the proof. □

**Corollary C.1.** *Let under the condition of proposition C.1 $f(z)$ solve an integral equation*

$$[f(\mathbf{I}+\mathbf{V})](z') = 1. \tag{C.12}$$

*Then for arbitrary $z_0 \in \mathcal{C}$,*

$$f(z_0) = \frac{\det(\mathbf{I}+\mathbf{V}-\mathbf{V}^0)}{\det(\mathbf{I}+\mathbf{V})}, \tag{C.13}$$

*where $V^0(z,z') = V(z,z_0)$.*

**Proof.** This is a particular case of proposition C.1 for $g(z)=1$. □

**Remark C.2.** If the kernel $V(z,z')$ is analytic within some domain $\mathcal{D}$ containing $\mathcal{C}$, then identity (C.4) holds for $z_0 \in \mathcal{D}$. This remark also applies to corollary C.1.

Let now $\mathcal{C} = \Gamma([-q,q])$ on which the operator $\mathbf{U}$ with the kernel $U(z,z')$ (12.41) acts (see Fig. 10.1). Let also $f(z)$ be a holomorphic function outside $\Gamma([-q,q])$ and bounded at $|z| \to \pm\infty$. Then

$$\begin{aligned}\frac{1}{2\pi i}\int_{\Gamma([-q,q])} &\big(\mathcal{K}_\alpha(z',z) - \mathcal{K}_\alpha(\omega,z)\big) f(z)\,\mathrm{d}z \\ &= e^{-i\alpha} f(\omega - ic) - f(\omega + ic) - e^{-i\alpha} f(z' - ic) + f(z' + ic). \end{aligned} \tag{C.14}$$

Here $\mathcal{K}_\alpha(z)$ is given by (12.26), and the integral has been computed by the residues lying outside the contour $\Gamma([-q,q])$. Compose a kernel $V_\omega(z,z')$ as

$$V_\omega(z,z') = \frac{f(z')}{2\pi i}\frac{\mathcal{K}_\alpha(z',z) - \mathcal{K}_\alpha(\omega,z)}{e^{-i\alpha} f(z' - ic) - f(z' + ic)}. \tag{C.15}$$

Then due to (C.14), we obtain

$$[f(\mathbf{I} + \mathbf{V}_\omega)](z') = g(z'), \tag{C.16}$$

where

$$g(z) = f(z)\frac{e^{-i\alpha} f(\omega - ic) - f(\omega + ic)}{e^{-i\alpha} f(z - ic) - f(z + ic)}. \tag{C.17}$$

Let us apply (C.4) to the operator $\mathbf{I} + \mathbf{V}_\omega$ with $\omega \in \Gamma([-q,q])$. We obtain

$$\det(\mathbf{I} + \mathbf{V}_\omega) = \frac{e^{-i\alpha} f(\omega - ic) - f(\omega + ic)}{e^{-i\alpha} f(z_0 - ic) - f(z_0 + ic)} \det(\mathbf{I} + \mathbf{V}_\omega - \mathbf{V}^0_{\omega,g}), \tag{C.18}$$

where

$$V^0_{\omega,g}(z,z') = \frac{g(z')}{g(z_0)} V_\omega(z,z_0). \tag{C.19}$$

A straightforward calculation shows that

$$\begin{aligned} V_\omega(z,z') &- \frac{g(z')}{g(z_0)} V_\omega(z,z_0) \\ &= \frac{f(z')}{2\pi i}\frac{\mathcal{K}_\alpha(z',z) - \mathcal{K}_\alpha(z_0,z)}{e^{-i\alpha} f(z' - ic) - f(z' + ic)} = V_{z_0}(z,z'). \end{aligned} \tag{C.20}$$

Thus, (C.18) yields

$$\frac{\det(\mathbf{I} + \mathbf{V}_\omega)}{e^{-i\alpha} f(\omega - ic) - f(\omega + ic)} = \frac{\det(\mathbf{I} + \mathbf{V}_{z_0})}{e^{-i\alpha} f(z_0 - ic) - f(z_0 + ic)}, \tag{C.21}$$

for arbitrary $z_0$ such that $|\Im(z_0 - z)| < c$, $z \in \Gamma([-q, q])$. Hence, the combination

$$\frac{\det(\mathbf{I} + \mathbf{V}_\omega)}{e^{-i\alpha} f(\omega - ic) - f(\omega + ic)}, \tag{C.22}$$

is $\omega$-independent.

In particular, we can take

$$f(z) = e^{-\tilde{F}(z)} \prod_{l=1}^{n} \frac{z - \lambda_{p_l}}{z - \lambda_{h_l}}. \tag{C.23}$$

Clearly, this $f(z)$ is holomorphic for $z \notin [-q, q]$ and $f(z) \to 1$ at $|z| \to \infty$. Then

$$\begin{aligned} V_\omega(z, z') = &\frac{e^{\tilde{F}(z'+ic) - \tilde{F}(z')}}{2\pi i(\mathfrak{X}(z') - 1)} \prod_{l=1}^{n} \frac{(z' - \lambda_{p_l})(z' - \lambda_{h_l} + ic)}{(z' - \lambda_{h_l})(z' - \lambda_{p_l} + ic)} \\ &\times \left(\mathcal{K}_\alpha(z', z) - \mathcal{K}_\alpha(\omega, z)\right), \end{aligned} \tag{C.24}$$

where

$$\mathfrak{X}(z') = e^{\tilde{F}(z'+ic) - \tilde{F}(z'-ic) - i\alpha} \prod_{l=1}^{n} \frac{(z' - \lambda_{p_l} - ic)(z' - \lambda_{h_l} + ic)}{(z' - \lambda_{h_l} - ic)(z' - \lambda_{p_l} + ic)}. \tag{C.25}$$

Observe that

$$\tilde{F}(z' + ic) - \tilde{F}(z' - ic) = -i \int_{-q}^{q} \mathcal{K}(z - u) F(u)\, du. \tag{C.26}$$

Then using the integral equation (10.65) for the shift function, we obtain

$$\mathfrak{X}(z') = e^{-2\pi i F(z')}, \tag{C.27}$$

and thus,

$$V_\omega(z, z') = -\frac{1}{2\pi i} U(z, z'), \tag{C.28}$$

where $U(z, z')$ is given by (12.97).

Similar calculation gives us

$$e^{-i\alpha} f(\omega - ic) - f(\omega + ic) = \left(e^{-2\pi i F(\omega)} - 1\right) e^{-\tilde{F}(\omega+ic)} \prod_{l=1}^{n} \frac{\omega - \lambda_{p_l} + ic}{\omega - \lambda_{h_l} + ic}. \tag{C.29}$$

Then by virtue of (C.22), we conclude that the combination

$$\left|\det(\mathbf{I} - \tfrac{1}{2\pi i}\mathbf{U})\right|^2 \frac{e^{\tilde{F}(\omega+ic) + \tilde{F}(\omega-ic)}}{\sin^2(\pi F(\omega))} \prod_{l=1}^{n} \frac{(\omega - \lambda_{h_l})^2 + c^2}{(\omega - \lambda_{p_l})^2 + c^2}, \tag{C.30}$$

does not depend on $\omega$.

# Appendix D

# Integrals with Vandermonde determinant

Let $\{p_n(x)\}$, $n = 0, 1, \ldots$, be a system of polynomials normalized by the condition

$$p_n(x) = x^n + o(x^n). \tag{D.1}$$

Let these polynomials be orthogonal on an interval $[a, b]$ with some weight $e^{-w(x)}$

$$\int_a^b p_n(x)p_m(x)e^{-w(x)}\,\mathrm{d}x = \delta_{nm}h_n, \tag{D.2}$$

where $w(x)$ is a real-valued function.

**Remark D.1.** The interval $[a, b]$ is not necessarily finite. We can also consider infinite intervals, for instance, the real axis. In this case, we require that $w(x) \to +\infty$ if $|x| \to \infty$.

**Proposition D.1.** *Let $\bar{x} = \{x_1, \ldots, x_n\}$ and let $\mathcal{V}(\bar{x})$ be the Vandermonde determinant of $\bar{x}$. Then*

$$\int_a^b \mathcal{V}^2(\bar{x}) \prod_{j=1}^{n} e^{-w(x_j)}\,\mathrm{d}x_j = n! \prod_{k=0}^{n-1} h_k, \tag{D.3}$$

*where $h_k$ are given by* (D.2).

**Proof.** First of all, we note that

$$\mathcal{V}(\bar{x}) = \det_n p_{k-1}(x_j). \tag{D.4}$$

Indeed, let us represent the polynomials $p_{k-1}(x)$ as

$$p_{k-1}(x) = \sum_{l=1}^{n} p_{k-1}^{(l-1)} x^{l-1}, \tag{D.5}$$

where $p_{k-1}^{(l-1)} = 0$ for $l > k$, and $p_{k-1}^{(k-1)} = 1$. Then

$$\det_n p_{k-1}(x_j) = \det_n \left( \sum_{l=1}^{n} p_{k-1}^{(l-1)} x_j^{l-1} \right) = \det_n \left( x_j^{k-1} \right) \det_n \left( p_{k-1}^{(j-1)} \right). \tag{D.6}$$

It is obvious now that

$$\det_n \left( x_j^{k-1} \right) = \mathcal{V}(\bar{x}), \qquad \det_n \left( p_{k-1}^{(j-1)} \right) = 1, \tag{D.7}$$

because the matrix $p_{k-1}^{(j-1)}$ is upper-triangular and $p_{k-1}^{(k-1)} = 1$.

Thus, we have

$$\int_a^b \mathcal{V}^2(\bar{x}) \prod_{j=1}^{n} e^{-w(x_j)} \, \mathrm{d}x_j = \int_a^b \left( \det_n p_{k-1}(x_j) \right)^2 \prod_{j=1}^{n} e^{-w(x_j)} \, \mathrm{d}x_j. \tag{D.8}$$

Due to lemma 11.1, we obtain

$$\int_a^b \left( \det_n p_{k-1}(x_j) \right)^2 \prod_{j=1}^{n} e^{-w(x_j)} \, \mathrm{d}x_j$$
$$= n! \det_n \left( \int_a^b p_{k-1}(x) p_{j-1}(x) e^{-w(x)} \, \mathrm{d}x \right). \tag{D.9}$$

Due to orthogonality (D.2), we obtain a diagonal matrix with the elements $\delta_{jk} h_{k-1}$. This implies (D.3). □

Let now $w(x) = \nu x^2$ with $\nu > 0$, and the interval $[a, b]$ is the real axis. Then obviously, the polynomials $p_n(x)$ are related to the Hermite polynomials:

$$p_n(x) = \left( \frac{-1}{2\nu} \right)^n e^{\nu x^2} \frac{d^n}{dx^n} e^{-\nu x^2}. \tag{D.10}$$

Substituting (D.10) in (D.2) we easily find

$$h_n = \sqrt{\frac{\pi}{\nu}} \frac{n!}{(2\nu)^n}. \tag{D.11}$$

Hence,

$$\prod_{k=0}^{n-1} h_k = \left( \frac{\pi}{\nu} \right)^{n/2} \frac{G(n+1)}{(2\nu)^{n(n-1)/2}}. \tag{D.12}$$

# Bibliography


Affleck I., Critical behavior of two-dimensional systems with continuous symmetries, *Phys. Rev. Lett.* **55** (1985) 1355–1358.

Ahn C., Delfino G., and Mussardo G., Mapping between the sinh-Gordon and ising models, *Phys. Lett. B* **317** (1993) 573–580, `arXiv:hep-th/9306103`.

Alcaraz F. C., Barber M. N., Batchelor M. T., Baxter R. J., and Quispel G. R. W., Surface exponents of the quantum $XXZ$, Ashkin-Teller and Potts models, *J. Phys. A: Math. Gen.* **20** (1987) 6397–6410.

Alexandrov A., and Zabrodin A., Free fermions and tau-functions, *J. Geom. Phys.* **67** (2013) 37–80, `arXiv:math-ph/1212.6049`

Asakawa H., and Suzuki M., Finite-size correction in the $XXZ$ model and the Hubbard model with boundary fields, *J. Phys. A: Math, Gen.* **29** (1996) 225–246.

Avan J., Belliard S., Grosjean N., and Pimenta R., Modified algebraic Bethe ansatz for $XXZ$ chain on the segment - III - proof, *Nucl. Phys. B* **899** (2015) 229–246, `arXiv:1506.02147`.

Babenko C., Göhmann F., Kozlowski K. K., and Suzuki J., A thermal form factor series for the longitudinal two-point function of the Heisenberg-Ising chain in the antiferromagnetic massive regime, *J. Math. Phys.* **62** (2021) 041901, `arXiv:2011.12752`.

Babenko C., Göhmann F., Kozlowski K. K., Sirker J., and Suzuki J., Exact real-time longitudinal correlation functions of the massive $XXZ$ chain, *Phys. Rev. Lett.* **126** (2021) 210602, `arXiv:2012.07378`.

Barouch E., McCoy B. M., and Wu T. T., Zero-Field susceptibility of the two-dimensional ising model near $Tc$, *Phys. Rev. Lett.* **31** (1973) 1409–1411.

Baxter R. J., Partition function of the eight vertex lattice model, *Annals Phys.* **70** (1972) 193–228.

Baxter R. J., One-dimensional anisotropic Heisenberg chain, *Annals Phys.* **70** (1972) 323–327.

Baxter R. J., Eight-vertex model in lattice statistics and one-dimensional anisotropic Heisenberg chain I, II, III, *Annals Phys.* **76** (1973) 1–24, 25–47, 48–71.

Baxter R. J., Solvable eight-vertex model on an arbitrary planar lattice, *Phil. Trans. Royal Soc. London* **289** (1978) 315–346.

Baxter R. J., *Exactly solved models in statistical mechanics* (Academic Press, 1982).

Baxter R. J., Completeness of the Bethe ansatz for the six and eight-vertex models, *J. Statist. Phys.* **108**:1–2 (2002) 1–48, `arXiv:cond-mat/0111188`.

Bazhanov V. V., Integrable quantum systems and classical Lie algebras, *Comm. Math. Phys.* **113** (1987) 471–503.

Bazhanov V. V., and Stroganov Yu. G., A new class of factorized $S$-matrices and triangle equations, *Phys. Lett. B* **105** (1981) 278–280.

Behrend R., Di Francesco P., and Zinn-Justin P., On the weighted enumeration of alternating sign matrices and descending plane partitions, *J. Combin. Theory Ser. A* **119**:2 (2012) 331–363, `arXiv:1103.1176`.

Biegel D., Karbach M., and Müller G., Transition rates via Bethe ansatz for the spin-1/2 Heisenberg chain, *Europhys. Lett.* **59**:6 (2002) 882–888, `arXiv:cond-mat/0205430`.

Biegel D., Karbach M., and Müller G., Transition rates via Bethe ansatz for the spin-1/2 planar $XXZ$ antiferromagnet, *J. Phys. A: Math. Gen.* **36** (2003) 5361–5368, `arXiv:cond-mat/0303147`.

Belavin A. A., Dynamical symmetry of integrable quantum systems, *Nucl. Phys. B* **180** (1981) 189–200.

Belavin A. A., Discrete groups and the integrability of quantum systems, *Funct. Anal. Appl.* **14** (1981) 260–267.

Belliard S., Modified algebraic Bethe ansatz for $XXZ$ chain on the segment - I: Triangular cases, *Nucl. Phys. B* **892** (2015) 1–20, `arXiv:1408.4840`.

Belliard S., and Crampé N., Heisenberg $XXX$ model with general boundaries: Eigenvectors from Algebraic Bethe ansatz, *SIGMA* **9** (2013) 072, `arXiv:1309.6165`.

Belliard S., Pakuliak S., Ragoucy E., and Slavnov N. A., Bethe vectors of $GL(3)$-invariant integrable models, *J. Stat. Mech.* (2013) P02020, `arXiv:1210.0768`.

Belliard S., Pakuliak S., Ragoucy E., and Slavnov N. A., Bethe vectors of quantum integrable models with $GL(3)$ trigonometric $R$-matrix, *SIGMA* **9** (2013) 058, `arXiv:1304.7602`.

Belliard S., and Pimenta R., Slavnov and Gaudin–Korepin Formulas for Models without $U(1)$ Symmetry: The Twisted $XXX$ Chain, *SIGMA* **11** (2015) 099, `arXiv:1506.06550`.

Belliard S., and Pimenta R., Modified algebraic Bethe ansatz for $XXZ$ chain on the segment - II -general cases, *Nuclear Phys. B* **894** (2015) 527–552, `arXiv:1412.7511`.

Belliard S., and Pimenta R., Slavnov and Gaudin–Korepin Formulas for Models without $U(1)$ Symmetry: The $XXX$ chain on the segment, *J. Phys. A: Math. Theor.* **49** (2016) 17LT01, `arXiv:1507.03242`.

Belliard S., and Ragoucy E., The nested Bethe ansatz for "all" closed spin chains, *J. Phys. A: Math. Theor.* **41** (2008) 295202, `arXiv:0804.2822`.

Belliard S., and Slavnov N. A., Why scalar products in the algebraic Bethe ansatz have determinant representation, *J. High Energy Phys.* **10** (2019) 103, `arXiv: 1908.00032`.

Belliard S., Slavnov N. A., and Vallet B., Scalar product of twisted $XXX$ modified Bethe vectors, *J. Stat. Mech.*, **9** (2018) 093103, `arXiv:1805.11323`.

Belliard S., Slavnov N. A., and Vallet B., Modified algebraic Bethe Ansatz: Twisted $XXX$ case, *SIGMA* **14** (2018) 054, `arXiv:1804.00597`.

Berezin F. A., Pokhil G. P., and Finkelberg V. M., The Schrödinger equation for a system of one-dimensional particles with point interactions, *Vestn. Mosk. Gos. Univ., Mathematics and Mechanics* **1** (1964) 21–28, (in Russian).

Berg B., Karowski M., Weicz P., and Kurak V., Factorized $U(n)$ symmetric $S$-matrices in two dimensions, *Nucl. Phys. B* **134** (1978) 125–132.

Bethe H., (1931) Eigenwerte und Eigenfunktionen der linearen Atomkette, *Zeitschrift für Physik* **71** (1931) 205–226.

Blöte H. W. J., Cardy J. L., and Nightingale M. P., Conformal invariance, the central charge, and universal finite-size amplitudes at criticality, *Phys. Rev. Lett.* **56** (1986) 742–745.

Bogolyubov N. M., and Izergin A. G., Lattice completely integrable regularization of the sine-Gordon model for small coupling constants, *Theor. Math. Phys.* **59**:2 (1984) 441–452.

Bogolyubov N. M., and Izergin A. G., Lattice sine-Gordon model with local Hamiltonian, *Theor. Math. Phys.* **61**:3 (1984) 1195–1204.

Bogolyubov N. M., Izergin A. G., and Korepin V. E., Structure of the vacuum in the quantum sine-Gordon model, *Phys. Lett. B* **159** (1985) 345–347.

Bogolyubov N. M., Izergin A. G., and Korepin V. E., Quantum inverse scattering method and correlation functions, in Shastry B. S., Jha S. S., and Singh V. (eds.), *Exactly Solvable Problems in Condensed Matter and Relativistic Field Theory, Lecture Notes in Physics* **242** (1985) 220–316, Berlin: Springer.

Bogolyubov N. M., Izergin A. G., and Korepin V. E., Critical exponents in completely integrable models of quantum statistical physics, *Theor. Math. Phys.* **70**:1 (1987) 94–102.

Bogolyubov N. M., and Korepin V. E., Correlation length of the one-dimensional Bose gas, *Nucl. Phys. B* **257** (1985) 766–778.

Bogolyubov N. M., and Korepin V. E., Quantum nonlinear Schrödinger equation on a lattice, *Theor. Math. Phys.* **66**:3 (1986) 300–305.

Boos H. E., Damerau J., Göhmann F., Klümper A., Suzuki J., and Weisse A., Short-distance thermal correlations in the XXZ chain, *J. Stat. Mech.* (2008) P08010, `arXiv:0806.3953`.

Boos H. E., and Göhmann F., On the physical part of the factorized correlation functions of the $XXZ$ chain, *J. Phys. A: Math. Theor.* **42** (2009) 315001, `arXiv:0903.5043`.

Boos H. E., Göhmann F., Klümper A., and Suzuki J., Factorization of multiple integrals representing the density matrix of a finite segment of the Heisenberg spin chain, *J. Stat. Mech.* (2006) P04001, `arXiv:hep-th/0603064`.

Boos H. E., Göhmann F., Klümper A., and Suzuki J., Factorization of the finite temperature correlation functions of the $XXZ$ chain in a magnetic field, *J. Phys. A: Math. Theor.* **40** (2007) 10699, `arXiv:0705.2716`.

Boos H., Jimbo M., Miwa T., Smirnov F., and Takeyama Y., A recursion formula for the correlation functions of an inhomogeneous $XXX$ model, *St. Petersburg Math. J.* **17**:1 (2006) 85–117, `arXiv:hep-th/0405044`.

Boos H., Jimbo M., Miwa T., Smirnov F., and Takeyama Y., Reduced qKZ equation and correlation functions of the $XXZ$ model, *Comm. Math. Phys.* **261** (2006) 245–276, arXiv:hep-th/0412191.

Boos H., Jimbo M., Miwa T., Smirnov F., and Takeyama Y., Hidden Grassmann structure in the $XXZ$ model, *Comm. Math. Phys.* **272** (2007) 263–281, arXiv:hep-th/0606280.

Boos H., Jimbo M., Miwa T., Smirnov F., and Takeyama Y., Hidden Grassmann structure in the $XXZ$ model II: Creation Operators, *Comm. Math. Phys.* **286** (2009) 875–932, arXiv:0801.1176.

Boos H. E., and Korepin V. E., Quantum spin chains and Riemann zeta function with odd arguments, *J. Phys. A: Math. Gen.* **34** (2001) 5311–5316, arXiv:hep-th/0104008.

Boos H. E., and Korepin V. E., Evaluation of integrals representing correlations in the $XXX$ Heisenberg spin chain, in Kashiwara M., and Miwa T. (eds.), *MathPhys Odyssey 2001. Progress in Mathematical Physics*, **23**. Boston, MA: Birkhäuser, arXiv:hep-th/0105144.

Boos H. E., Korepin V. E., and Smirnov F. A., Emptiness formation probability and quantum Knizhnik–Zamolodchikov equation, *Nucl. Phys. B* **658** (2003) 417–439, arXiv:hep-th/0209246.

Borodin A., and Olshanski, G. Harmonic functions on multiplicative graphs and interpolation polynomials, *Electron. J. Combin.* **7** (2000) R28, arXiv:math/9912124.

Borodin A., and Olshanski G., Distributions on partitions, point processes, and the hypergeometric kernel, *Comm. Math. Phys.* **211** (2000) 335–358, arXiv:math/9904010.

Borodin A., and Olshanski G., $Z$-Measures on partitions, Robinson–Schensted–Knuth correspondence, and $\beta = 2$ ensembles, in Bleher P. M., and Its A. R. (eds.), *Random Matrix Models and their Applications*, MSRI Publications **40**, Cambridge Univ. Press, 2001, 71–94, arXiv:math/9905189.

Bortz M., and Sirker J., Boundary susceptibility in the open $XXZ$-chain, *J. Phys. A: Math. Gen.* **38** (2005) 5957–5978, arXiv:cond-mat/0501559.

Brazhnikov V., and Lukyanov S., Angular quantization and form-factors in massive integrable models, *Nucl. Phys. B* **512** (1998) 616–636, arXiv:hep-th/9707091.

Bressoud D., *Proofs and Confirmations: The Story of the Alternating Sign Matrix Conjecture*, Cambridge, England: Cambridge University Press, 1999.

Bullough R. K., Chen Y.-Z., Timonen J., Tognetti V., and Vaia R., Classical thermodynamics of the Heisenberg chain in a field by generalized Bethe ansatz method, *Phys. Lett. A* **145** (1990) 154–158.

Calabrese P., and Caux J.-S., Correlation functions of the one-dimensional attractive Bose gas, *Phys. Rev. Lett.* **98** (2007) 150403, arXiv:cond-mat/0612192.

Cao J., Yang W. L., Shi K., and Wang Y., Off-diagonal Bethe ansatz and exact solution a topological spin ring, *Phys. Rev. Lett.* **111** (2013) 137201, arXiv:1305.7328.

Cao J., Yang W. L., Shi K., and Wang Y., Off-diagonal Bethe ansatz solution of the $XXX$ spin-chain with arbitrary boundary conditions, *Nucl. Phys. B* **875** (2013) 152–165, arXiv:1306.1742.

Cardy J. L., Conformal invariance and universality in finite-size scaling, *J. Phys. A: Math. Gen.* **17** (1984) L385–L388.

Cardy J. L., Operator content of two-dimensional conformally invariant theories, *Nucl. Phys. B* **270** (1986) 186–204.

Case K. M., Polynomial constants for the quantized NLS equation, *J. Math. Phys.* **25** (1984) 2306–2314.

Caux J.-S., and Calabrese P., Dynamical density-density correlations in the one-dimensional Bose gas, *Phys. Rev. A* **74** (2006) 031605, `arXiv:cond-mat/0603654`.

Caux J.-S., Calabrese P., and Slavnov N. A., One-particle dynamical correlations in the one-dimensional Bose gas, *J. Stat. Mech.* (2007) P01008, `arXiv:cond-mat/0611321`.

Caux J.-S., Essler F. H. L., and Loew U., Dynamical structure factor of the anisotropic Heisenberg chain in a transverse field, *Phys. Rev. B* **68** (2003) 134431, `arXiv:cond-mat/0303587`.

Caux J.-S., Hagemans R., and Maillet J.-M., Computation of dynamical correlation functions of Heisenberg chains: the gapless anisotropic regime, *J. Stat. Mech.* (2005) P09003, `arXiv:cond-mat/0506698`.

Caux J.-S., and Maillet J.-M., Computation of dynamical correlation functions of Heisenberg chains in a field, *Phys. Rev. Lett.* **95** (2005) 077201, `arXiv:cond-mat/0502365`.

Caux J.-S., and Mossel J., Remarks on the notion of quantum integrability, *J. Stat. Mech.* (2011) P02023, `arXiv:1012.3587`.

Cheianov V. V., and Pustilnik M., Threshold singularities in the dynamic response of gapless integrable models, *Phys. Rev. Lett.* **100** (2008) 126403, `arXiv:0710.3589`.

Cherednik I. V., On 'quantum' deformation of irreducible finite-dimensional representations of $\mathfrak{gl}_N$, *Sov. Math. Dokl.* **33** (1986) 507–510.

Clémente-Gallardo J., and Marmo G., Towards a definition of quantum integrability, *Int. J. Geom. Meth. Mod. Phys.* **6** (2009) 129–172, `arXiv:0808.3819`.

Colomo F., Izergin A. G., Korepin V. E., and Tognetti V., Correlators in the Heisenberg XXO chain as Fredholm determinants, *Phys. Lett. A* **169** (1992) 243–247.

Colomo F., Izergin A. G., Korepin V. E., and Tognetti V., Temperature correlation functions in the XXO Heisenberg chain, *Theor. Math. Phys.* **194** (1993) 11–38.

Crampé N., Algebraic Bethe ansatz for the totally asymmetric simple exclusion process with boundaries, *J. Phys. A: Math. Theor.* **48** (2015) 08FT01, `arXiv:1411.7954`.

Creamer D. B., Thacker H. B., and Wilkinson D., Gel'fand–Levitan method for operator fields, *Phys. Rev. D* **21** (1980) 1523–1528.

Damerau J., Göhmann F., Hasenclever N. P., and Klümper A., Density matrices for finite segments of Heisenberg chains of arbitrary length, *J. Phys. A: Math. Theor.* **40** (2007) 4439–4453, `arXiv:cond-mat/0701463`.

Date E., Jimbo M., Kuniba A., Miwa T., and Okado M., Exactly solvable SOS models. II: Proof of the star-triangle relation and combinatorial identities, *Adv. Stud. Pure Math. Phys.* **16** (1988) 17–122.

Davies B., The quantum Gelfand–Levitan equation and nonlinear Schrödinger equation, *Inverse Problems* **4** (1988) 47–58.

Davies B., Higher conservation laws for the quantum non-linear Schrödinger equation, *Physica A* **167**:2 433–456.

Davies B., and Kieu T. D., The quantum inverse method and the non-linear Schrödinger equation, *Inverse Problems* **2** (1986) 141–160.

Davies B., and Korepin V. E., Higher conservation laws for the quantum non-linear Schrödinger equation, `arXiv:1109.6604`.

Destri C., and de Vega H. J., Integrable quantum field theories and conformal field theories from lattice models in the light-cone approach, *Phys. Lett. B* **201** (1988) 261–268.

Destri C., and de Vega H. J., Unified approach to thermodynamic Bethe ansatz and finite size corrections for lattice models and field theories, *Nucl. Phys. B* **438** (1995) 413–454, `arXiv:hep-th/9407117`.

D'lorio M., Armstrong R. L., and Taylor D. R., Longitudinal and transverse spin dynamics of a one-dimensional $XY$ system studied by chlorine nuclear relaxation in $PrCl_3$, *Phys. Rev. B* **27** (1983) 1664–1673.

D'lorio M., Glaus U., and Stoll E., Transverse spin dynamics of a one-dimensional $XY$ system: A fit to spin-spin relaxation data, *Solid State Comm.* **47**:5 (1983) 313–315.

Dorlas T. C., Orthogonality and completeness of the Bethe ansatz eigenstates of the nonlinear Schrödinger model, *Comm. Math. Phys.* **154**:2 (1993) 347–376.

Dorlas T. C., and Samsonov M., *On the thermodynamic limit of the 6-vertex model*, `arXiv:0903.2657`.

Drinfeld V. G., Hamiltonian structures on Lie groups, Lie bialgebras and the geometric meaning of the classical Yang–Baxter equations, *Sov. Math. Dokl.* **268**:2 (1983) 285–287.

Dugave M., Göhmann F., and Kozlowski K. K., Thermal form factors of the $XXZ$ chain and the large-distance asymptotics of its temperature dependent correlation functions, *J. Stat. Mech.* (2013) P07010, `arXiv:1305.0118`.

Dugave M., Göhmann F., and Kozlowski K. K., Low-temperature large-distance asymptotics of the transversal two-point functions of the $XXZ$ chain, *J. Stat. Mech.* (2014) P04012, `arXiv:1401.4132`.

Dugave M., Göhmann F., Kozlowski K. K., and Suzuki J., On form-factor expansions for the $XXZ$ chain in the massive regime, *J. Stat. Mech.* (2015) P05037, `arXiv:1412.8217`.

Dugave M., Göhmann F., Kozlowski K. K., and Suzuki J., Low-temperature spectrum of correlation lengths of the $XXZ$ chain in the antiferromagnetic massive regime, *J. Phys. A: Math. Theor.* **48** (2015) 334001, `arXiv:1504.07923`.

Dugave M., Göhmann F., Kozlowski K. K., and Suzuki J., Thermal form factor approach to the ground-state correlation functions of the $XXZ$ chain in the antiferromagnetic massive regime, *J. Phys. A: Math. Theor.* **49** (2016) 394001, `arXiv:1605.07968`.

Duminil-Copin H., Kozlowski K. K., Krachun D., Manolescu I., and Tikhonovskaya T., *On the six-vertex model's free energy*, `arXiv:2012.11675`.

Eckle H.-P., *Models of Quantum Matter. A First Course on Integrability and the Bethe Ansatz*, Oxford: Oxford University Press, 2019.

Enriquez B., Khoroshkin S., and Pakuliak S., Weight functions and Drinfeld currents, *Comm. Math. Phys.* **276** (2007) 691–725, `arXiv:math/0610398`.

Essler F. H. L., Frahm H., Göhmann F., Klümper A., and Korepin V. E., *The One-Dimensional Hubbard Model*, Cambridge University Press, 2005.

Essler F. H. L., Frahm H., Izergin A. G., and Korepin V. E., Determinant Representations for Correlation Functions of Spin-1/2 $XXX$ and $XXZ$ Heisenberg Magnets, *Comm. Math. Phys.* **174** (1995) 191–214, `arXiv:hep-th/9406133`.

Essler F. H. L., Frahm H., Its A. R., and Korepin V. E., Integro-Difference Equation for a correlation function of the spin-1/2 Heisenberg $XXZ$ chain, *Nucl. Phys. B* **446** (1995) 448–460, `arXiv:cond-mat/9503142`.

Essler F. H. L., Frahm H., Its A. R., and Korepin V. E., Determinant representation for a quantum correlation function of the lattice sine-Gordon model, *J. Phys. A: Math. Gen.* **30** (1997) 219–244, `arXiv:hep-th/9512090`.

Faddeev L. D., Integrable models in $(1 + 1)$-dimensional quantum field theory, in Zuher J.-B., and Stora R. (eds.), *Recent Advances in Field Theory and Statistical Mechanics* (Les Houches, 1982), North-Holland, Amsterdam, 1984, 561–608.

Faddeev L. D., How algebraic Bethe ansatz works for integrable models, in Connes A., Gawedzki K., and Zinn-Justin J. (eds.), *Quantum Symmetries (Les Houches, 1995)*, North-Holland, Amsterdam, 1998, 149–219, `arXiv:hep-th/9605187`.

Faddeev L. D., Sklyanin E. K., and Takhtajan L. A., Quantum Inverse Problem. I, *Theor. Math. Phys.* **40** (1979) 688–706.

Faddeev L. D., and Takhtajan L. A., The quantum method of the inverse problem and the Heisenberg $XYZ$ model, *Russian Math. Surveys* **34**:5 (1979) 11–68.

Faddeev L. D., and Takhtajan L. A., Spectrum and scattering of excitations in the one-dimensional isotropic Heisenberg model, *J. Math. Sci.* **24** (1984) 241–267.

Faddeev L. D., and Takhtajan L. A., *Hamiltonian Method in the Theory of Solitons*, Amsterdam: Springer, 1987.

Fateev V. A., and Zamolodchikov A. B., Self-dual solutions of the star-triangle relations in $Z_N$ models, *Phys. Lett. A* **92** (1982) 37–39.

Fowler M., and Zotos X., Quantum sine-Gordon thermodynamics: The Bethe ansatz method, *Phys. Rev. B* **24** (1981) 2634–2639.

Frahm H., and Zvyagin A. A., The open spin chain with impurity: An exact solution, *J. Phys.: Condens. Matter* **9** (1997) 9939–9946.

Franchini F., An introduction to integrable techniques for one-dimensional quantum systems, *Lecture Notes in Physics* **940** (2017) Springer, Cham., `arXiv:1609.02100`.

Frappat L., Khoroshkin S., Pakuliak S., and Ragoucy E., Bethe ansatz for the universal weight function, *Ann. Henri Poincaré* **10** (2009) 513–548, `arXiv:0810.3135`.

Fring A., Mussardo G., and Simonetti P., Form factors for integrable lagrangian field theories, the sinh-Gordon model, *Nucl. Phys. B* **393** (1993) 413–441, `arXiv:hep-th/9211053`.

Fujimoto S., and Eggert S., Boundary susceptibility in the spin-1/2 chain: Curie like behavior without magnetic impurities, *Phys. Rev. Lett.* **92** (2004) 037206, `arXiv:cond-mat/0310230`.

Furusaki A., and Hikihara T., Boundary contributions to specific heat and susceptibility in the spin-1/2 $XXZ$ chain, *Phys. Rev. B* **69** (2004) 094429, `arXiv:cond-mat/0310517`.

Gardner C. S., Greene J. M., Kruskal M. D., and Miura R. M., Method for Solving the Korteweg-de-Vries Equation, *Phys. Rev. Lett.* **19** (1967) 1095–1097.

Gaudin M., *Modèles exacts in mécanique statistique: la méthode de Bethe et ses généralisations*, Note CEA 1559 (1972) (1).

Gaudin M., Diagonalisation d'une classe d'hamiltoniens de spin, *Journal de Physique* **37**:10 (1976) 1087–1098.

Gaudin M., McCoy B. M., and Wu T. T., Normalization sum for the Bethe's hypothesis wave functions of the Heisenberg–Ising chain, *Phys. Rev. D* **23** (1981) 417–419.

Gaudin M., *La Fonction d'Onde de Bethe*, Paris: Masson, 1983.

Gaudin M., and Mehta M. L., On the density of eigenvalues of a random matrix, *Nucl. Phys.* **18** (1960) 420–427.

Girardeau M., Relationship between systems of impenetrable Bosons and Fermions in one dimension, *J. Math. Phys.* **1**:6 (1960) 516–523.

Girardeau M. D, Wright E. M., and Triscari J. M., Ground-state properties of a one-dimensional system of hard-core bosons in a harmonic trap, *Phys. Rev. A.* **63**:3 (2001) 033601, `arXiv:cond-mat/0008480`.

Göhmann F., Bortz M., and Frahm H., Surface free energy for systems with integrable boundary conditions, *J. Phys. A: Math. Gen.* **38** (2005) 10879–10892, `arXiv:cond-mat/0508377`.

Göhmann F., Hasenclever N. P., and Seel A., Finite temperature density matrix and two-point correlations in the antiferromagnetic $XXZ$ chain, *J. Stat. Mech.* (2005) P10015, `arXiv:cond-mat/0509765`.

Göhmann F., Karbach M., Klümper A., Kozlowski K. K., and Suzuki J., Thermal form-factor approach to dynamical correlation functions of integrable lattice models, *J. Stat. Mech.* (2017) 113106, `arXiv:1708.04062`.

Göhmann F., Klümper A., and Seel A., Integral representations for correlation functions of the $XXZ$ chain at finite temperature, *J. Phys. A: Math. Gen.* **37** (2004) 7625–7652, `arXiv:hep-th/0405089`.

Göhmann F., Klümper A., and Seel A., Emptiness formation probability at finite temperature for the isotropic Heisenberg chain, *Physica B* **359–361** (2005) 807–809, `arXiv:cond-mat/0406611`.

Göhmann F., Klümper A., and Seel A., Integral representation of the density matrix of the $XXZ$ chain at finite temperatures, *J. Phys. A: Math. Gen.* **38** (2005) 1833–1842, `arXiv:cond-mat/0412062`.

Göhmann F., and Korepin V. E., Solution of the quantum inverse problem, *J. Phys. A: Math. Gen.* **33** (2000) 1199–1220, `arXiv:hep-th/9910253`.

Göhmann F., Kozlowski K. K., Sirker J., and Suzuki J., Equilibrium dynamics of the $XX$ chain, *Phys. Rev. B* **100** (2019) 155428, `arXiv:1906.03143`.

Göhmann F., Kozlowski K. K., and Suzuki J., High-temperature analysis of the transverse dynamical two-point correlation function of the $XX$ quantum-spin chain, *J. Math. Phys.* **61** (2020) 013301, `arXiv:1905.04922`.

Göhmann F., Kozlowski K. K., and Suzuki J., Long-time large-distance asymptotics of the transversal correlation functions of the XX chain in the space-like regime, *Lett. Math. Phys.* **110** (2020) 1783–1797, `arXiv:1908.11555`.

Göhmann F., Seel A., and Suzuki J., Correlation functions of the integrable isotropic spin-1 chain at finite temperature, *J. Stat. Mech.* (2010) P11011, `arXiv:1008.4440`.

Göhmann F., and Suzuki J., Quantum spin chains at finite temperatures, in Feigin B., Jimbo M., and Okado M. (eds.), *New Trends in Quantum Integrable Systems*, World Scientific, 2010 81-100, `arXiv:1002.3194`.

Gromov N., Levkovich-Maslyuk F., and Sizov G., New construction of eigenstates and separation of variables for $SU(N)$ quantum spin chains, *J. High Energy Phys.* **09** (2017) 111, `arXiv:1610.08032`.

Gusev E. V., Weak convergence of wave values in the Heisenberg model, *Rep. Math. Phys.* **18**:3 (1980) 399–410.

Gusev E. V., Thermodynamics and excited states of the Heisenberg model, *Theor. Math. Phys.* **63**:2 (1985) 527–532.

Gusev E. V., Bethe method: Thermodynamics and limit states, *Theor. Math. Phys.* **67**:3 (1986) 606–613.

Hagemans R., Caux J.-S., and Maillet J.-M., How to calculate correlation functions of Heisenberg chains, *AIP Conference Proceedings* **846** (2006) 245–254, `arXiv:cond-mat/0611467`.

Haldane F. D. M., General Relation of Correlation Exponents and Spectral Properties of One-Dimensional Fermi Systems: Application to the Anisotropic $S = \frac{1}{2}$ Heisenberg Chain, *Phys. Rev. Lett.* **45** (1980) 1358–1362.

Haldane F. D. M., Demonstration of the "Luttinger liquid" character of Bethe-ansatz-soluble models of 1-$D$ quantum fluids, *Phys. Lett. A* **81** (1981) 153–155.

Haldane F. D. M., 'Luttinger liquid theory' of one-dimensional quantum fluids. I. Properties of the Luttinger model and their extension to the general 1$D$ interacting spinless Fermi gas, *J. Phys. C: Solid State Phys.* **14** (1981) 2585.

Heisenberg W., Zur Theorie des Ferromagnetismus, *Zeitschrift für Physik* **49** (1928) 619–636.

Holstein T., and Primakoff H., Field dependence of the intrinsic domain magnetization of a ferromagnet, *Phys. Rev.* **58**:12 (1940) 1098–1113.

Hultén L., Über das Austauschproblem eines Kristalles, *Arkiv. Mat. Astron Fys. A* **26** (1939) 1–106.

Hutsalyuk A., Liashyk A., Pakuliak S. Z., Ragoucy E., and Slavnov N. A., Scalar products of Bethe vectors in the models with $\mathfrak{gl}(m|n)$ symmetry, *Nucl. Phys. B* **923** (2017) 277–311, `arXiv:1704.08173`.

Hutsalyuk A., Liashyk A., Pakuliak S. Z., Ragoucy E., and Slavnov N. A., Current presentation for the super-Yangian double $DY(\mathfrak{gl}(m|n))$ and Bethe vectors, *Russian Math. Surveys* **72**:1 (2017) 33–99, `arXiv:1611.09620`.

Imambekov A., and Glazman L. I., Exact exponents of edge singularities in dynamic correlation functions of $1D$ Bose gas, *Phys. Rev. Lett.* **100** (2008) 206805, `arXiv:0711.2697`.

Imambekov A., and Glazman L. I., Universal theory of nonlinear Luttinger liquids, *Science* **323** (2009) 228–231, `arXiv:0806.4779`.

Imambekov A., and Glazman L. I., Phenomenology of one-dimensional quantum liquids beyond the low-energy limit, *Phys. Rev. Lett.* **102** (2009) 126405, `arXiv:0812.1046`.

Its A. R., Izergin A. G., Korepin V. E., and Slavnov N. A., Temperature correlations of quantum spins, *Phys. Rev. Lett.* **70** (1993) 1704–1706, `arXiv:cond-mat/9209034`.

Its A. R., Izergin A. G., Korepin V. E., and Slavnov N. A., The matrix Riemann–Hilbert problem and differential equations for correlation functions of the XXO Heisenberg chain, *St. Petersburg Math. J.* **6**:2 (1995) 315–326.

Its A. R., Izergin A. G., Korepin V. E., and Slavnov N. A., Integrable differential equations for temperature correlation functions of the Heisenberg XXO chain, *J. Math. Sci.* **80** (1996) 1747–1759.

Its A. R., and Slavnov N. A., On the Riemann–Hilbert approach to asymptotic analysis of the correlation functions of the quantum nonlinear Schrödinger equation: Interacting fermion case, *Theor. Math. Phys.* **119**:2 (1999) 541–593, `arXiv:math-ph/9811009`.

Izergin A. G., Partition function of the six-vertex model in a finite volume, *Sov. Phys. Dokl.* **32** (1987) 878–879.

Izergin A. G., *Introduction to the Bethe Ansatz Solvable Models*, Lectures given in 1998–1999, Università di Firenze, Dipartimento di Fisica, `http://theory.fi.infn.it/colomo/others.html`.

Izergin A. G., Coker D. A., and Korepin V. E., Determinant formula for the six-vertex model, *J. Phys. A: Math. Gen.* **25** (1992) 4315–4334.

Izergin A. G., Kitanine N., Maillet J. M., and Terras V., Spontaneous magnetization of the $XXZ$ Heisenberg spin-1/2 chain, *Nucl. Phys. B* **554** (1999) 679–696, `arXiv:solv-int/9812021`.

Izergin A. G., Kitanin N. A., and Slavnov N. A., On correlation functions of the $XY$ model, *J. Math. Sci.* **88**:2 (1998) 224–232.

Izergin A. G., and Korepin V. E., A lattice model related with a nonlinear Schrödinger equation, *Sov. Phys. Dokl.* **26** (1981a) 653–654, `arXiv:0910.0295`.

Izergin A. G., and Korepin V. E., The inverse scattering method approach to the quantum Shabat–Mikhailov model, *Comm. Math. Phys.* **79** (1981b) 303–316.

Izergin A. G., and Korepin V. E., The lattice quantum sine-Gordon model, *Lett. Math. Phys.* **5** (1981c) 199–205.

Izergin A. G., and Korepin V. E., The quantum inverse scattering method approach to correlation functions, *Comm. Math. Phys.* **94** (1984) 67–92.

Izergin A. G., and Korepin V. E., Problem of describing all $L$-operators of the $XXX$ and $XXZ$ models, *J. Sov. Math.* **30** (1984) 2292–2298.

Izergin A. G., Korepin V. E., and Reshetikhin N. Yu., Correlation functions in a one-dimensional Bose gas, *J. Phys. A: Math. Gen.* **20** (1987) 4799–4822.

Izergin A. G., Korepin V. E., and Smirnov F. A., Trace formulas for the quantum nonlinear Schrödinger equation, *Theor. Math. Phys.* **48**:3 (1981) 773–776.

Jimbo M., Quantum $R$-matrix for the generalized Toda system, *Comm. Math. Phys.* **102** (1986) 537–547.

Jimbo M. (ed.), *Yang–Baxter Equation in Integrable Systems*, Advanced Series in Mathematical Physics: **10**, Singapore: World Scientific, 1990.

Jimbo M., Miki K., Miwa T., and Nakayashiki A., Correlation functions of the $XXZ$ model for $\Delta < -1$, *Phys. Lett. A* **168**:4 (1992) 256–263, `arXiv:hep-th/9205055`.

Jimbo M., and Miwa T., Algebraic Analysis of Solvable Lattice Models, *CBMS Regional Conference Series in Mathematics*, **85**, AMS, Providence, Rhode Island 1995.

Jimbo M., and Miwa T., Quantum KZ equation with $|q| = 1$ and correlation functions of the $XXZ$ model in the gapless regime, *J. Phys. A: Math. Gen.* **29** (1996) 2923–2958, `arXiv:hep-th/9601135`.

Jimbo M., Miwa T., and Okado M., Solvable lattice models related to the vector representation of classical simple Lie algebras, *Comm. Math. Phys.* **116** (1988) 507–525.

Jimbo M., Miwa T., and Smirnov F., Hidden Grassmann structure in the $XXZ$ model III: Introducing the Matsubara direction, *J. Phys. A: Math. Theor.* **42** (2009) 304018, `arXiv:0811.0439`.

Jordan P., and Wigner E., *Über das Paulische Äquivalenzverbot, Zeitschrift für Physik* **47**:9 (1928) 631–651.

Karbach M., Biegel D., and Müller G., Quasiparticles governing the zero-temperature dynamics of the one-dimensional spin-1/2 Heisenberg antiferromagnet in a magnetic field, *Phys. Rev. B* **66** (2002) 054405, `arXiv:cond-mat/0205142`.

Karowski M., Exact $S$-matrices and form factors in 1+1 dimensional field theoretic models with soliton behaviour, *Phys. Rep.* **49** (1979) 229–237.

Karowski M., and Weisz P., Exact form factors in (1+1)-dimensional field theoretic models with soliton behaviour, *Nucl. Phys. B* **139** (1978) 455–476.

Kato G., Shiroishi M., Takahashi M., and Sakai K., Next-nearest-neighbour correlation functions of the spin-1/2 $XXZ$ chain at the critical region, *J. Phys. A: Math. Gen.* **36** (2003) L337–L344, `arXiv:cond-mat/0304475`.

Kato G., Shiroishi M., Takahashi M., and Sakai K., Third-neighbour and other four-point correlation functions of spin-1/2 $XXZ$ chain, *J. Phys. A: Math. Gen.* **37** (2004) 5097–5124, `arXiv:cond-mat/0402625`.

Katsura S., Horiguchi T., and Suzuki M., Dynamical properties of the isotropic $XY$ model, *Physica* **46**:1 (1970) 67–86.

Kerov S., Anisotropic Young diagrams and Jack symmetric functions, *Funct. Anal. Appl.* **34**:1 (2000) 41–51, `arXiv:math/9712267`.

Kerov S., Olshanski G., and Vershik A., Harmonic analysis on the infinite symmetric group. A deformation of the regular representation, *Comptes Rend. Acad. Sci. Paris, Ser. I* **316** (1993) 773–778.

Khoroshkin S., and Pakuliak S., A computation of universal weight function for quantum affine algebra $U_q(\widehat{\mathfrak{gl}}_N)$, *J. Math. Kyoto Univ.* **48** (2008) 277–321, `arXiv:0711.2819`.

Khoroshkin S., and Pakuliak S., Generating series for nested Bethe vectors, *SIGMA* **4** (2008) 081, `arXiv:0810.3131`.

Khoroshkin S., Pakuliak S., and Tarasov V., Off-shell Bethe vectors and Drinfeld currents, *J. Geom. Phys.* **57** (2007) 1713–1732, `arXiv:math.QA/0610517`.

Kirillov A. N., and Reshetikhin N. Yu., Exact solution of the integrable $XXZ$ Heisenberg model with arbitrary spin: I. The ground state and the excitation spectrum, *J. Phys. A: Math. Gen.* **20** (1987) 1565–1585.

Kirillov A. N., and Smirnov F. A., Solution of some combinatorial problems which arise in calculating correlators in exactly solvable models, *J. Sov. Mat.* **47** (1989) 2413–2422.

Kitanine N., Kozlowski K. K., Maillet J. M., Niccoli G, Slavnov N. A., and Terras V., Correlation functions of the open $XXZ$ chain I, *J. Stat. Mech.* **0710** (2007) P10009, `arXiv:0707.1995`.

Kitanine N., Kozlowski K. K., Maillet J. M., Niccoli G, Slavnov N. A., and Terras V., Correlation functions of the open $XXZ$ chain II, *J. Stat. Mech.* **0807** (2008) P07010, `arXiv:0803.3305`.

Kitanine N., Kozlowski K. K., Maillet J. M., Slavnov N. A., and Terras V., On correlation functions of integrable models associated to the six-vertex $R$-matrix, *J. Stat. Mech.* (2007) P01022, `arXiv:hep-th/0611142`.

Kitanine N., Kozlowski K. K., Maillet J. M., Slavnov N. A., and Terras V., Riemann–Hilbert approach to a generalized sine kernel and applications, *Comm. Math. Phys.* **291**:3 (2009) 691–761, `arXiv:0805.4586`.

Kitanine N., Kozlowski K. K., Maillet J. M., Slavnov N. A., and Terras V., Algebraic Bethe ansatz approach to the asymptotic behavior of correlation functions, *J. Stat. Mech.* (2009) P04003, `arXiv:0808.0227`.

Kitanine N., Kozlowski K. K., Maillet J. M., Slavnov N. A., and Terras V., On the thermodynamic limit of form factors in the massless $XXZ$ Heisenberg chain, *J. Math. Phys.* **50**:9 (2009) 095209, `arXiv:0903.2916`.

Kitanine N., Kozlowski K. K., Maillet J. M., Slavnov N. A., and Terras V., The thermodynamic limit of particle-hole form factors in the massless $XXZ$ Heisenberg chain, *J. Stat. Mech.* (2011) P05028, `arXiv:1003.4557`.

Kitanine N., Kozlowski K. K., Maillet J. M., Slavnov N. A., and Terras V., A form factor approach to the asymptotic behavior of correlation functions, *J. Stat. Mech.* (2011) P12010, `arXiv:hep-th/1110.0803`.

Kitanine N., Kozlowski K. K., Maillet J. M., Slavnov N. A., and Terras V., Form factor approach to dynamical correlation functions in critical models, *J. Stat. Mech.* (2012) P09001, `arXiv:1206.2630`.

Kitanine N., Kozlowski K. K., Maillet J. M., and Terras V., Long-distance asymptotic behaviour of multi-point correlation functions in massless quantum models, *J. Stat. Mech.* (2014) P05011, `arXiv:1312.5089`.

Kitanine N., Maillet J. M., Slavnov N. A., and Terras V., Spin-spin correlation functions of the $XXZ - 1/2$ Heisenberg chain in a magnetic field, *Nucl. Phys. B* **641** (2002) 487–518, `arXiv:hep-th/0201045`.

Kitanine N., Maillet J. M., Slavnov N. A., and Terras V., Emptiness formation probability of the XXZ spin-1/2 Heisenberg chain at $\Delta = 1/2$, *J. Phys. A: Math. Gen.* **35**:27 (2002) L385–L388, `arXiv:hep-th/0201134`.

Kitanine N., Maillet J. M., Slavnov N. A., and Terras V., Correlation functions of the $XXZ$ spin-1/2 Heisenberg chain at the free fermion point from their multiple integral representations, *Nucl. Phys. B* **642**:3 (2002) 433–455, arXiv:hep-th/0203169.

Kitanine N., Maillet J. M., Slavnov N. A., and Terras V., Large distance asymptotic behavior of the emptiness formation probability of the $XXZ$ spin-1/2 Heisenberg chain, *J. Phys. A: Math. Gen.* **35**:49 (2002) L753–L758, arXiv:hep-th/0210019.

Kitanine N., Maillet J. M., Slavnov N. A., and Terras V., Correlation functions of the $XXZ$ spin-1/2 Heisenberg chain: Recent advances, Proceedings of 6th International Workshop on Conformal Field Theory and Integrable Models, *Int. J. Modern Phys. A* **19** (2004) 248–266.

Kitanine N., Maillet J. M., Slavnov N. A., and Terras V., Exact results for the $\sigma^z$ two-point function of the $XXZ$ chain at $\Delta = 1/2$, *J. Stat. Mech.* (2005) L09002, arXiv:hep-th/0506114.

Kitanine N., Maillet J. M., Slavnov N. A., and Terras V., On the spin-spin correlation functions of the $XXZ$ spin-1/2 infinite chain, *J. Phys. A: Math. Gen.* **38**:34 (2005) 7441–7460, arXiv:hep-th/0407223.

Kitanine N., Maillet J. M., Slavnov N. A., and Terras V., Master equation for spin-spin correlation functions of the XXZ chain, *Nucl. Phys. B* **712**:3 (2005) 600–622, arXiv:hep-th/0406190.

Kitanine N., Maillet J. M., Slavnov N. A., and Terras V., Dynamical correlation functions of the $XXZ$ spin-1/2 chain, *Nucl. Phys. B* **729**:3 (2005) 558–580, arXiv:hep-th/0407108.

Kitanine N., Maillet J. M., Slavnov N. A., and Terras V., On the algebraic Bethe Ansatz approach to the correlation functions of the $XXZ$ spin-1/2 Heisenberg chain, *Solvable Lattice Models*, RIMS, Kyoto, (2005) 14–48, arXiv:hep-th/0505006.

Kitanine N., Maillet J. M., and Terras V., Form factors of the $XXZ$ Heisenberg spin-1/2 finite chain, *Nucl. Phys. B* **554** (1999) 647–678, arXiv:math-ph/9807020.

Kitanine N., Maillet J. M., and Terras V., Correlation functions of the $XXZ$ Heisenberg spin-1/2 chain in a magnetic field, *Nucl. Phys. B* **567** (2000) 554–582, arXiv:math-ph/9907019.

Klümper A., Thermodynamics of the Anisotropic Spin-1/2 Heisenberg Chain and Related Quantum Chains, *Zeitschrift für Physik B Condensed Matter* **91**:4 (1993) 507–519, arXiv:cond-mat/9306019.

Klümper A., Integrability of quantum chains: Theory and applications to the spin-1/2 $XXZ$ chain, *Lecture Notes in Physics* **645** (2004) 349–379, arXiv:cond-mat/0502431.

Kojima T., Korepin V. E., and Slavnov N. A., Determinant representation for dynamical correlation function of the quantum Nonlinear Schrödinger equation, *Comm. Math. Phys.* **188**:3 (1997) 657–689, arXiv:hep-th/9611216.

Kojima T., Korepin V. E., and Slavnov N. A., Completely integrable equation for the quantum correlation function of nonlinear Schrödinger equation, *Comm. Math. Phys.* **189**:3 (1997) 709–728, arXiv:hep-th/9612252.

Korepin V. E., Analysis of bilinear relation of a six-vertex model, *Sov. Phys. Dokl.* **27** (1982) 612–613.

Korepin V. E., Calculation of norms of Bethe wave functions, *Comm. Math. Phys.* **86** (1982) 391–418.

Korepin V. E., Correlation functions of the one-demensional Bose gas in the repulsive case, *Comm. Math. Phys.* **94** (1984) 93–113.

Korepin V. E., Dual field formulation of quantum integrable models, *Comm. Math. Phys.* **113** (1987) 177–190.

Korepin V. E., Generating functional of correlation functions for the nonlinear Schrödinger equation, *Funct. Anal. Appl.* **23**:1 (1989) 12–19.

Korepin V. E., Bogoliubov N. M., and Izergin A. G., *Quantum Inverse Scattering Method and Correlation Functions*, Cambridge: Cambridge Univeristy Press, 1993.

Korepin V. E., Izergin A. G., Essler F., and Uglov D. B., Correlation function of the spin-1/2 XXX antiferromagnet, *Phys. Lett. A* **190** (1994) 182–184, `arXiv:cond-mat/9403066`.

Korepin V. E., and Slavnov N. A., Correlation function of fields in one-dimensional Bose gas, *Comm. Math. Phys.* **136** (1991) 633–644.

Korepin V. E., and Slavnov N. A., The Riemann–Hilbert problem associated with the quantum Nonlinear Schrödinger equation, *J. Phys. A: Math. Gen.* **30** (1997) 8241–8256, `arXiv:hep-th/9706147`.

Korepin V. E., and Slavnov N. A., Normal ordering in the theory of correlation functions of exactly solvable models, *J. Phys. A: Math. Gen.* **30** (1997) 8623–8634, `arXiv:hep-th/9709204`.

Korepin V. E., and Slavnov N. A., The determinant representation for quantum correlation functions of the sinh-Gordon model, *J. Phys. A: Math. Gen.* **31** (1998) 9283–9295, `arXiv:hep-th/9801046`.

Koubek A., and Mussardo G., On the operator content of the sinh-Gordon model, *Phys. Lett. B* **311** (1993) 193–201, `arXiv:hep-th/9306044`.

Kozlowski K. K., On Form Factors of the conjugated field in the non-linear Schrödinger model, *J. Math. Phys.* **52** (2011) 083302, `arXiv:1105.1052`.

Kozlowski K. K., Large-distance and long-time asymptotic behavior of the reduced density matrix in the non-linear Schrödinger model, *Ann. Henri Poincaré* **16** (2015) 437–534, `arXiv:1101.1626`.

Kozlowski K. K., Form factors of bound states in the $XXZ$ chain, *J. Phys. A: Math. Theor.* **50** (2017) 184002, `arXiv:1611.01489`.

Kozlowski K. K., On the thermodynamic limit of form factor expansions of dynamical correlation functions in the massless regime of the $XXZ$ spin-1/2 chain, *J. Math. Phys.* **59** (2018) 091408, `arXiv:1706.09459`.

Kozlowski K. K., On condensation properties of Bethe roots associated with the $XXZ$ chain, *Comm. Math. Phys.* **357** (2018) 1009–1069, `arXiv:1508.05741`.

Kozlowski K. K., Long-distance and large-time asymptotic behaviour of dynamic correlation functions in the massless regime of the $XXZ$ spin-1/2 chain, *J. Math. Phys.* **60** (2019) 073303, `arXiv:1903.00207`.

Kozlowski K. K., and Maillet J. M., Microscopic approach to a class of 1*D* quantum critical models, *J. Phys. A: Math. Theor.* **48** (2015) 484004, `arXiv:1501.07711`.

Kozlowski K. K., Maillet J. M., and Slavnov N. A., Long-distance behavior of temperature correlation functions in the one-dimensional Bose gas, *J. Stat. Mech.* (2011) P03018, `arXiv:1011.3149`.

Kozlowski K. K., Maillet J. M., and Slavnov N. A., Correlation functions of one-dimensional bosons at low temperature, *J. Stat. Mech.* (2011) P03019, `arXiv:1102.1858`.

Kozlowski K. K., and Terras V., Long-time and large-distance asymptotic behavior of the current–current correlators in the non-linear Schrödinger model, *J. Stat. Mech.* (2011) P09013, `arXiv:1101.0844`.

Krantz S. G., and Parks H. R., *The Implicit Function Theorem: History, Theory, and Applications*, Boston, MA: Birkhäuser, 2012.

Krichever I. M., and Sklyanin E. K., Baxter's equation and algebraic geometry, *Funct. Anal. Appl.* **15** (1981) 92–103.

Kulish P. P., and Reshetikhin N. Yu., Generalized Heisenberg ferromagnet and the Gross–Neveu model, *Sov. Phys. JETP* **53** (1981) 108–114.

Kulish P. P., and Reshetikhin N. Yu., GL(3)-invariant solutions of the Yang-Baxter equation and associated quantum systems, *J. Sov. Math.* **34** (1986) 1948–1971.

Kulish P. P., and Reshetikhin N. Yu., Diagonalization of $GL(N)$ invariant transfer matrices and quantum $N$-wave system (Lee model), *J. Phys. A* **16** (1983) L591–L596.

Kulish P. P., Reshetikhin N. Yu., and Sklyanin E. K., Yang–Baxter equation and representation theory I, *Lett. Math. Phys.* **5** (1981) 393–403.

Kulish P. P., and Sklyanin E. K., Quantum spectral transform method: Recent developments, in *Integrable Quantum Field Theories*, Lecture Notes in Physics **151**, Berlin–Heidelberg: Springer, 1982, 61–119.

Kulish P. P., and Sklyanin E. K., Solutions of the Yang–Baxter equation, *J. Sov. Math.* **19** (1982) 1596–1620.

Kuperberg G., Another proof of the alternating sign matrix conjecture, *Internat. Math. Res. Notices*, **1996**:3 (1996) 139–150, `arXiv:math/9712207`.

Levkovich-Maslyuk F., The Bethe Ansatz, *J. Phys. A: Math. Theor.* **49** (2016) 323004, `arXiv:1606.02950`.

Liashyk A., *New approach to scalar products of Bethe vectors*, `arXiv:1907.11875`.

Liashyk A., and Slavnov N. A., On Bethe vectors in $\mathfrak{gl}_3$-invariant integrable models, *J. High Energy Phys.* **06** (2018) 018, `arXiv:1803.07628`.

Lieb E. H., and Liniger W., Exact Analysis of an Interacting Bose Gas. I. The General Solution and the Ground State, *Phys. Rev.* **130** (1963) 1605–1616.

Lieb E. H., Exact Analysis of an Interacting Bose Gas. II. The Excitation Spectrum, *Phys. Rev.* **130** (1963) 1616–1624.

Lieb E., and Mattis D. (eds.), *Mathematical Physics in one dimension*, New York-London: Academic Press, 1966.

Lieb E., Schultz T., and Mattis D., Two soluble models of an antiferromagnetic chain, *Ann. Phys.* **16** (1961) 407–466.

Lukyanov S., Form-factors of exponential fields in the affine $A_{N-1}{}^{(1)}$ Toda model, *Phys. Lett. B* **408** (1997) 192–200, `arXiv:hep-th/9704213`.

Luther A., and Peschel I, Calculation of critical exponents in two dimensions from quantum field theory in one dimension, *Phys. Rev. B* **12** (1975) 3908–3917.

Maillet J. M., and Niccoli G., On quantum separation of variables, *J. Math. Phys.* **59** (2018) 091417, `arXiv:1807.11572`.

Maillet J. M., and Terras V., On the quantum inverse scattering problem, *Nucl. Phys. B* **575** (2000) 627–644, `arXiv:hep-th/9911030`.

McCoy B. M., Spin correlation functions of the $X - Y$ model, *Phys. Rev.* **173** (1968) 531–541.

McCoy B. M., Perk J. H. H., and Schrock R. E., Time-dependent correlation functions of the transverse Ising chain at the critical magnetic field, *Nucl. Phys. B* **220** (1983) 35–47.

McCoy B. M., Perk J. H. H., and Schrock R. E., Correlation functions of the transverse Ising chain at the critical field for large temporal and spatial separations, *Nucl. Phys. B* **220** (1983) 269–282.

McCoy B. M., and Tracy C. A., Neutron scattering and the correlation functions of the Ising model near $Tc$, *Phys. Rev. Lett.* **31** (1973) 1500–1504.

McCoy B. M., Tracy C. A., and Wu T. T., Painleve functions of the third kind, *J. Math. Phys.* **18** (1977) 1058–1092.

McGuire J. B., Study of exactly solvable one-dimensional $N$-body problems, *J. Math. Phys.* **5** (1964) 622–636.

McGuire J. B., Interacting fermions in one dimension: I Repulsive potential, *J. Math. Phys.* **6** (1965) 432–439.

McGuire J. B., Interacting fermions in one dimension: II Attractive potential, *J. Math. Phys.* **7** (1966) 123–132.

Mehta M. L., and Dyson F. J., Statistical theory of the energy levels of complex systems. V, *J. Math. Phys.* **4** (1963) 713–719.

Mehta M. L., *Random matrices*, Pure and Applied Mathematics (Amsterdam), **142** (3rd ed.), Amsterdam: Elsevier/Academic Press, 2004.

Mills W. H., Robbins D. P., and Rumsey H., Proof of the Macdonald conjecture, *Invent. Math.* **66** (1982) 73–87.

Minahan J. A., and Zarembo K., The Bethe-Ansatz for $\mathcal{N} = 4$ Super Yang–Mills, *J. High Energy Phys.* **2003**:03 (2003) 013, `arXiv:hep-th/0212208`.

Molev A., Nazarov M., and G. Olshanski G., Yangians and classical Lie algebras, *Russian Math. Surveys* **51**:2 (1996) 205–282, `arXiv:hep-th/9409025`.

Molev A., *Yangians and Classical Lie Algebras.* Mathematical Surveys and Monographs, 143. American Mathematical Society, Providence, RI, 2007.

Mukhin E., Tarasov V., and Varchenko A., Bethe Algebra of Homogeneous XXX Heisenberg Model Has Simple Spectrum, *Comm. Math. Phys.* **288** (2009) 1–42, `arXiv:0706.0688`.

Mukhin E., Tarasov V., and Varchenko A., Spaces of quasi-exponentials and representations of the Yangian $Y(\mathfrak{gl}_N)$, *Transformation Groups* **19** (2014) 861–885, `arXiv:1303.1578`.

Mussardo G., Off-critical statistical models: Factorized scattering theories and bootstrap program, *Phys. Rep.* **218** (1992) 215–379.

Nepomechie R. I., and Wang C., Twisting singular solutions of Bethe's equations, *J. Phys. A: Math. Theor.* **47** (2014) 505004, arXiv:1409.7382.

Niemeijer Th., Some exact calculations on a chain of spins 12, *Physica* **36**:3 (1967) 377–419.

Okounkov A., SL(2) and $z$-measures, in Bleher P. M., and Its A. R. (eds.), *Random Matrix Models and their Applications*, MSRI Publications **40**, Cambridge University Press, 2001, 407–420, arXiv:math/0002135.

Oota T., Quantum projectors and local operators in lattice integrable models, *J. Phys. A: Math. Gen.* **37** (2004) 441–452, arXiv:hep-th/0304205.

Os'kin A., Pakuliak S., and Silantyev A., On the universal weight function for the quantum affine algebra $U_q(\mathfrak{gl}_N)$, *St. Petersburg Math. J.* **21** (2010) 651–680, arXiv:0711.2821.

Pakuliak S., Ragoucy E., and Slavnov N. A., Zero modes method and form factors in quantum integrable models, *Nucl. Phys. B* **893** (2015) 459–481, arXiv:1412.6037.

Pereira R. G., Sirker J., Caux J.-S., Hagemans R., Maillet J. M., White S. R., and Affleck I., The dynamical spin structure factor for the anisotropic spin-1/2 Heisenberg chain, *Phys. Rev. Lett.* **96** (2006) 257202, arXiv:cond-mat/0603681.

Pereira R. G., Sirker J., Caux J.-S., Hagemans R., Maillet J. M., White S. R., and Affleck I., Dynamical structure factor at small $q$ for the $XXZ$ spin-1/2 chain, *J. Stat. Mech.* (2007) P08022, arXiv:0706.4327.

Perk J. H. H., Equations of motion for the transverse correlations of the one-dimensional $XY$-model at finite temperature, *Phys. Lett. A* **79** (1980) 1–2.

Perk J. H. H., and Capel H. W., Time-dependent $xx$-correlation functions in the one-dimensional $XY$-model, *Physica A* **89** (1977) 265–303.

Perk J. H. H., Capel H. W., Quispel G. R. W., and Nijhoff F. W., Finite-temperature correlations for the Ising chain in a transverse field, *Physica A* **123** (1984) 1–49.

Perk J. H. H., and Schultz C. L., New families of commuting transfer matrices in $q$-state vertex model, *Phys. Lett. A* **84**:8 (1981) 407–410.

Perk J. H. H., and Au-Yang H., Yang–Baxter equations, in Françoise, J.-P., Naber G. L., and Tsou S. T. (eds.), *Encyclopedia of Mathematical Physics*, **5**, Elsevier, ScienceDirect, 2006, 465–473, arXiv:math-ph/0606053.

Polyakov A. M., Conformal symmetry of critical fluctuations, *JETP Lett.* **12** (1970) 381–383.

Plunkett J., *An Introduction to the Algebraic Bethe Ansatz*, VDM Verlag, 2009.

Razumov A. V., and Stroganov Yu. G., Spin chains and combinatorics, *J. Phys. A: Math. Gen.* **34** (2001) 3185–3190, arXiv:cond-mat/0012141.

Razumov A. V., and Stroganov Yu. G., Enumerations of half-turn-symmetric alternating-sign matrices of odd-order, *Theor. Math. Phys.* **148**:3 (2006) 1174–1198, arXiv:math-ph/0504022.

Razumov A. V., and Stroganov Yu. G., Enumeration of quarter-turn symmetric alternating-sign matrices of odd-order, *Theor. Math. Phys.* **149**:3 (2006) 1639–1650, arXiv:math-ph/0507003.

Reshetikhin N. Yu., A method of functional equations in the theory of exactly solvable quantum systems, *Lett. Math. Phys.* **7** (1983) 205–213.

Reshetikhin N. Yu., and Wiegmann P., Towards the classification of completely integrable quantum field theories (the Bethe Ansatz associated with Dynkin diagrams and their automorphisms), *Phys. Lett. B* **189** (1987) 125–131.

Ruelle D., *Statistical Mechanics: Rigorous Results*, New York: W. A. Benjamin, Inc., 1969.

Ryan P., and Volin D., Separated variables and wave functions for rational $\mathfrak{gl}(N)$ spin chains in the companion twist frame, *J. Math. Phys.* **60** (2019) 032701, `arXiv:1810.10996`.

de Sa P. A., and Tsvelik A. M., Anisotropic spin-1/2 Heisenberg chain with open boundary conditions, *Phys. Rev. B* **52** (1995) 3067–3070, `arXiv:cond-mat/9503031`.

Sakai K., Shiroishi M., Nishiyama Y., and Takahashi M., Third-neighbor correlators of a one-dimensional spin-1/2 Heisenberg antiferromagnet, *Phys. Rev. E* **67** (2003) 065101, `arXiv:cond-mat/0302564`.

Sato J., Shiroishi M., and Takahashi M., Evaluation of Dynamic spin structure factor for the spin-1/2 $XXZ$ chain in a magnetic field, *J. Phys. Soc. Jpn.* **73** (2004) 3008–3014, `arXiv:cond-mat/0410102`.

Seel A., Bhattacharyya T., Göhmann F., and Klümper A., A note on the spin-1/2 $XXZ$ chain concerning its relation to the Bose gas, *J. Stat. Mech.* (2007) P08030, `arXiv:0705.3569`.

Sirker J., Pereira R. G., Caux J.-S., Hagemans R., Maillet J. M., White S. R., and Affleck I., Boson decay and the dynamical structure factor for the $XXZ$ chain at finite magnetic field, *Physica B* **403** (2008) 1520–1522, `arXiv:0705.3312`.

Sklyanin E. K., Quantum version of the method of inverse scattering problem, *J. Math. Sci.* **19**:5 (1982) 1546–1596.

Sklyanin E. K., Some algebraic structures connected with the Yang–Baxter equation, *Funct. Anal. Appl.* **16**:4 (1982) 263–270.

Sklyanin E. K., Some algebraic structures connected with the Yang–Baxter equation. Representations of quantum algebras, *Funct. Anal. Appl.* **17**:4 (1983) 273–284.

Sklyanin E. K., Boundary conditions for integrable quantum systems, *J. Phys. A: Math. Gen.* **21** (1988) 2375–2389.

Sklyanin E. K., Separation of variables in the Gaudin model, *J. Sov. Math.* **47** (1989) 2473–2488.

Sklyanin E. K., Quantization of integrable models in the infinite volume. The nonlinear Schrödinger equation, *St. Petersburg Math. J.* **1**:2 (1990) 515–534.

Sklyanin E. K., Separation of variables in the quantum integrable models related to the Yangian $Y[sl(3)]$, *J. Math. Sci.* **80**:3 (1996) 1861–1871, `arXiv:hep-th/9212076`.

Slavnov N. A., Calculation of scalar products of wave functions and form factors in the framework of the algebraic Bethe ansatz, *Theor. Math. Phys.* **79**:2 (1989) 502–508.

Slavnov N. A., Nonequal-time current correlation function in a one-dimensional Bose gas, *Theor. Math. Phys.* **82**:3 (1990) 273–282.

Slavnov N. A., Integral equations for correlation functions of a quantum one-dimensional Bose gas, *Theor. Math. Phys.* **121**:1 (1999) 1358–1376, `arXiv:solv-int/9812028`.

Slavnov N. A., Asymptotics of the Fredholm determinant associated with the correlation functions of the quantum Nonlinear Schrödinger equation, *J. Math. Sci.* **104**:3 (2001) 1135–1143, `arXiv:solv-int/9810016`.

Slavnov N. A., On an identity for dual fields, *J. Math. Sci.* **100**:2 (2000) 2181–2188.

Slavnov N. A., The algebraic Bethe ansatz and quantum integrable systems, *Russian Math. Surveys* **62**:4 (2007) 727–766.

Slavnov N. A., Integral operators with the generalized sine kernel on the real axis, *Theor. Math. Phys.* **165**:1 (2010) 1262–1274, `arXiv:1005.5047`.

Slavnov N. A., Introduction to the theory of quantum integrable systems. Quantum nonlinear Schrödinger equation, *Steklov Math. Inst. Lect. Notes* **18** Moscow, 2011 (in Russian).

Slavnov N. A., Algebraic Bethe ansatz, *Steklov Math. Inst. Lect. Notes* **27** Moscow, 2017, `arXiv:1804.07350`.

Slavnov N. A., *Introduction to the nested algebraic Bethe ansatz*, in: *Les Houches Summer School Lecture Notes Series, Integrability in Atomic and Condensed Matter Physics*, SciPost Phys. Lect. Notes **19** (2020), `arXiv:1911.12811`.

Smirnov F. A., Reductions of the sine-Gordon model as a perturbation of minimal models of conformal field theory, *Nucl. Phys. B* **337** (1990) 156–180.

Smirnov F. A., *Form Factors in Completely Integrable Models of Quantum Field Theory*, Advanced Series in Mathematical Physics: **14**, Singapore: World Scientific, 1992.

Smirnov V. I., *A Course of Higher Mathematics*, Oxford, New York: Pergamon Press, 1964.

Stroganov Yu. G., Izergin–Korepin determinant at a third root of unity, *Theor. Math. Phys.* **146**:1 (2006) 53–62, `arXiv:math-ph/0204042`.

Sutherland B., Further results for the many-body problem in one dimension, *Phys. Rev. Lett.* **20**:3 (1968) 98–100.

Sutherland B., A general model for multicomponent quantum systems, *Phys. Rev. B* **12** (1975) 3795–3805.

Takahashi M., One-dimensional Heisenberg model at finite temperature, *Progr. Theor. Phys.* **46**:2 (1971) 401–415.

Takahashi M., Low-temperature specific heat of spin-1/2 anisotropic Heisenberg ring, *Progr. Theor. Phys.* **50**:5 (1973) 1519–1536.

Takahashi M., and Suzuki M., One-dimensional anisotropic Heisenberg model at finite temperatures, *Progr. Theor. Phys.* **48**:6 (1972) 2187–2209.

Takhtajan L. A., Introduction to algebraic Bethe ansatz, in Shastry B. S., Jha S. S., and Singh V. (eds.), *Exactly Solvable Problems in Condensed Matter and Relativistic Field Theory. Lecture Notes in Physics*, **242**, Berlin, Heidelberg: Springer, 1985.

Tarasov V., Structure of quantum $L$-operators for the $R$-matrix of $XXZ$ model, *Theor. Math. Phys.* **61**:2 (1984) 1065–1072.

Tarasov V., and Varchenko A., *Bases of Bethe Vectors and Difference Equations with Regular Singular Points*, `arXiv:q-alg/9504011`.

Tarasov V., and Varchenko A., Jackson integral representations for solutions of the Knizhnik–Zamolodchikov quantum equation, *St. Petersburg Math. J.* **6** (1995) 275–313, `arXiv:hep-th/9311040`.

Tarasov V., and Varchenko A., Combinatorial formulae for nested Bethe vector, *SIGMA* **9** (2013) 048, `arXiv:math/0702277`.

Thacker H. B., Polynomial conservation laws in $(1+1)$-dimensional classical and quantum field theory, *Phys. Rev. D* **17** (1978) 1031–1040.

Thacker H. B., Exact integrability in quantum field theory and statistical systems, *Rev. Mod. Phys.* **53** (1981) 253–286.

Tolstoy V. N., and Khoroshkin S. M., The universal $R$-matrix for quantum untwisted affine Lie algebras, *Funct. Anal. Appl.* **26** (1992) 69–71.

Tonegawa T., Transverse spin correlation function of the one-dimensional spin-12 $XY$ model, *Solid State Comm.* **40**:11 (1981) 983–986.

Tonks L., The complete equation of state of one, two and three-dimensional gases of hard elastic spheres, *Phys. Rev.* **50** (1936) 955–963.

Vaidya H. G., and Tracy C. A., Crossover scaling function for the one-dimensional $XY$ model at zero temperature, *Phys. Lett. A* **68** (1978) 378–380.

Vaidya H. G., and Tracy C. A., Transverse time-dependent spin correlation functions for the one-dimensional $XY$ model at zero temperature, *Physica A* **92** (1978) 1–41.

de Vega H. J., and Woynarovich F., Method for calculating finite size corrections in Bethe ansatz systems: Heisenberg chain and 6-vertex model, *Nucl. Phys. B* **251** (1985) 439–456.

Vichirko V. I., and Reshetikhin N. Yu., Excitation spectrum of the anisotropic generalization of an $SU_3$ magnet, *Theor. Math. Phys.* **56** (1984) 805–812.

Vladimirov A. A., Proof of the invariance of the Bethe-Ansatz solutions under complex conjugation, *Theor. Math. Phys.* **66**:1 (1986) 102–105.

Wang Y.-S., The reconstruction of local quantum operators for the boundary $XXZ$ spin-1/2 Heisenberg chain, *J. Phys. A: Math. Gen.* **33** (2000) 4009–4016.

Wang Y.-S., The scalar products and the norm of Bethe eigenstates for the boundary $XXX$ Heisenberg spin-1/2 finite chain, *Nucl. Phys. B* **622** (2002) 633–649.

Wang Y., Yang W. L., Cao J., and Shi K., The Algebraic Bethe Ansatz, in *Off-Diagonal Bethe Ansatz for Exactly Solvable Models*, Berlin, Heidelberg: Springer, 2015.

Weigert S., The problem of quantum integrability, *Physica D: Nonlinear Phenomena* **56**:1 (1992) 107–119.

Woynarovich F., Excitation spectrum of the spin-1/2 Heisenberg chain and conformal invariance, *Phys. Rev. Lett.* **59** (1987) 259–261.

Woynarovich F., Eckle H. P., and Truong T. T., Non-analytic finite-size corrections in the one-dimensional Bose gas and Heisenberg chain, *J. Phys. A: Math. Gen.* **22** (1989) 4027–4043.

Wu T. T., McCoy B. M., Tracy C. A., and Barouch E., Spin-spin correlation functions for the two-dimensional Ising model: Exact theory in the scaling region, *Phys. Rev. B* **13** (1976) 316–374.

Yang C. N., Some exact results for the many-body problem in one dimension with repulsive delta-function interaction, *Phys. Rev. Lett.* **19**:23 (1967) 1312–1315.

Yang C. N., and Yang C. P., One-dimensional chain of anisotropic spin-spin interactions. I. Proof of Bethe's hypothesis for ground state in a finite system, *Phys. Rev.* **150** (1966) 321–327.

Yang C. N., and Yang C. P., One-dimensional chain of anisotropic spin-spin interactions. II. Properties of the ground state energy per lattice site in infinite system, *Phys. Rev.* **150** (1966) 327–339.

Yang C. N., and Yang C. P., Thermodynamics of a one-dimensional system of bosons with repulsive delta-function interaction, *J. Math. Phys.* **10** (1969) 1115–1122.

Yang C. N., and Ge M. L. (eds.), Braid group, knot theory and statistical mechanics, *Advanced Series in Mathematical Physics*: **9**, Singapore: World Scientific, 1991.

Zamolodchikov A. B., Exact two-particle $S$-matrices of quantum solitons of the sine-Gordon model, *Comm. Math. Phys.* **55** (1977) 183–186.

Zamolodchikov A. B., and Zamolodchikov Al. B., Realistic factorized $S$-matrix in two-dimensional space-time with isotropic symmetry, *JEPT Lett.* **26** (1977) 457–460.

Zamolodchikov A. B., and Zamolodchikov Al. B., Exact $S$-matrix of Gross–Neveu "elementary" fermions, *Phys. Lett. B* **72** (1978) 481–483.

Zamolodchikov A. B., and Zamolodchikov Al. B., Factorized $S$-matrices in two dimensions as the exact solutions of certain relativistic quantum field theory models, *Ann. Phys.* **120** (1979) 253–291.

Zeilberger D., Proof of the Alternating Sign Matrix Conjecture, *Elec. J. Comb.* **3**:2 (1996) R13, `arXiv:math/9407211`.

# Index

action
  completely indirect, 130–131, 134–135, 138
    recursion, 137, 139
  direct, 130–131, 140
  indirect, 130–131, 142
alternating sign matrices, 276
anisotropy parameter, 214, 232, 270, 277
asymptotics, 219, 251, 262, 277, 279, 281–282, 284, 303, 306, 316, 320, 323, 332, 339
  expansion, 298, 311, 319, 322, 329, 332
  large-time, vii, 279, 281, 284, 322
  leading, 280, 316, 318, 323
  long-distance, vii, 232, 262, 279, 281, 284, 305–306, 315
  series, 316, 328, 330
auxiliary linear space, 5, 15, 17, 62, 65–67

Barnes $G$-function, 276, 293, 296, 298, 335, 339
Bernoulli numbers, 199
Bethe, v, 1–2, 4, 78
  ansatz
    algebraic, vi–vii, ix, 1–2, 4, 21, 23, 25, 31, 33, 44, 69, 71, 85–88, 91, 105, 107, 150, 173, 231–236, 263, 279–280, 282, 333
  coordinate, vi, 2, 4, 8, 21, 69, 191
equations, 31–32, 37–42, 46–57, 69, 81–83, 92, 119, 120, 126–127, 132, 153–154, 161, 175, 189–191, 194–196, 198, 200, 204, 208, 214–215, 219, 221, 234–236, 240, 244, 246, 248, 256, 262, 279, 287
  logarithmic form, 122, 192, 215
  twisted, 45, 50, 52–57, 82–83, 114–116, 118–124, 127, 132, 134–135, 153, 168–169, 177–178, 181, 191, 200–201, 204, 221, 251–256, 260–261, 286–287, 300
parameters, 33, 76, 107, 131, 182, 191, 200, 208–209, 226, 232, 234, 244–246, 256, 286, 320
vector, vi, 31, 33–38, 40–42, 45, 48–50, 53, 56, 69, 72, 76, 78, 79, 81–83, 85, 87–88, 92, 96, 104–105, 108, 113, 126–127, 129–131, 135, 138, 140–143, 153, 173, 251, 253, 262
  dual, 32, 56–57, 139
  norm, 49, 107
  off-shell, 38, 85–89, 129–131, 139, 141–142, 150, 234, 237–238, 246

on-shell, 38, 40–41, 56, 82, 86, 105, 107–108, 123, 127, 132, 153, 157, 169, 174, 182, 191, 225, 234–236, 238, 252
partial, 72, 76–77, 177
total, 72, 76, 94, 176
Bose field, 19, 20, 171–172, 233
bosons, 194
impenetrable, 194–195, 201, 292–293, 330, 332
one-dimensional, vii, 190, 210, 213, 215
bound state, 281, 285
boundary conditions, 3, 8
domain wall, 103
periodic, 8, 15, 169, 191, 238
quasi-periodic, 45, 191, 285

Cauchy
contour integral, 232, 241, 252
determinant, 144–145, 182–183, 187, 263, 275, 304, 310
transform, 229, 291, 317
chemical potential, 208, 285, 323
classical inverse scattering method, 2, 5
commutation relations, 2, 5, 8, 10, 18, 20, 26–29, 33, 36, 39, 42, 59, 85, 87, 91, 107, 142, 159
bilinear, 6
canonical, 19
multiple, 85, 104, 141
commutation scheme
first, 35–36, 173
second, 35–36
composite model, vi, 69–71, 75, 88, 94, 170, 172, 176
conformal field theory, 280–281, 306, 330, 332
continuous limit, 20, 26, 40, 58, 78–81, 172
counting function, 198–202, 204, 208, 209, 215, 217, 300
critical
exponent, 279–281, 329, 330, 332
model, 279, 332

density
average, 189, 193, 195–196, 208–209, 318
distribution, 195–196, 198–199, 201, 209–210, 215–217, 219, 244, 251, 287–288
integral equation, 189–190, 196, 199, 216, 264
inhomogeneous, 217, 219, 248, 250
integral equation, 249
matrix, 231
inhomogeneous, 231
digamma function, 335
dressed charge, 190, 210, 212, 217, 219–220, 289, 317
integral equation, 210, 217, 220
dual fields, vii, 86, 235

emptiness formation probability, 232, 267–270, 274, 276, 277
energy, 39, 81, 195, 205–209, 233–234, 279–280, 312, 325
dressed, 323
excitation, 207, 282, 284, 286, 305, 313, 321, 324, 341, 344–345
Euler–Maclaurin formula, 199, 221, 338
excited state, 190, 198, 200–202, 204–209, 214–215, 217, 222, 226–227, 279–284, 286, 288, 292, 294, 300, 302, 305, 308, 319–321, 323–324, 326
critical, 281, 305–307, 312, 321, 341
norm, 287
quasi-critical, 324–326

Faddeev, v
Fermi
boundary, 190, 199, 204, 206–207, 210, 212, 219–220, 250, 279–280, 282–284, 289, 291, 303–306, 308, 311–312, 317, 322–324, 326–327, 330
momentum, 264, 318
zone, 227, 232, 290, 310, 323
Fermi velocity, 312, 345

Fermi zone, 195
fermions, 19, 194
  free, 194, 220, 232, 262, 266, 314
finite amplitude, 283–284, 287, 299–300, 310–312, 315–318, 321–322, 324–325
  reduced, 325–326
finite-size corrections, 198, 280, 324, 341
Fock space, 19, 235
form factor, vii, 125, 144, 182, 202, 226, 262, 279–280, 292, 333
  amplitude, 280, 282–284, 287–288, 304, 311–312, 315, 321
    discrete part, 280, 288–289, 292–294, 299–301, 304, 308, 311, 324
    smooth part, 280, 289, 291–292, 300–301, 304, 308, 311, 317, 324, 347
  critical, 281, 304–306, 308, 310–312, 316, 320–322
  density, 176
    generating function, 176, 181
  expansion, 189, 232, 236, 251, 256, 260, 279, 282, 284–285, 320–322, 326, 332
  field, 187, 208, 300, 311, 320–321
  local operator, 72, 75, 165–166, 168, 170, 181, 223, 234, 236, 262
  monodromy matrix element, 153–156, 158, 165–168
  quasi-critical, 324–326, 331
  universal, 158–159, 162–164, 168, 170, 175, 181
Fourier transform, 218–219
Fredholm determinant, 223–224, 226–227, 266, 284, 289–290, 347–348
free fermions, 262

Gaudin, 103
  hypothesis, 86, 107
  model, 18
generalized model, 57, 86, 91–93, 95–96, 125–127, 131–132
grand canonical ensemble, 208
ground state, 3, 46, 189–190, 195, 198–202, 206, 208, 214–216, 219, 221, 225, 227, 233, 236, 238, 240–241, 244, 248, 251, 255–256, 267, 280, 282, 300, 315, 317, 320, 324
  antiferromagnetic, 214
  ferromagnetic, 214
  norm, 256, 287
  twisted, 201–202, 204, 221, 280, 289, 308, 310, 315, 317, 321, 326

Heisenberg spin chain, v–vi, 2, 31, 48, 53, 67, 69, 71, 76, 83, 87, 92, 165–166, 189, 214, 232–233, 235
  XX, 262–263
  XX0, 262
  XXX, 1–3, 14–15, 17, 20, 22–24, 33, 39–42, 46, 50–51, 56, 59, 61, 66, 78–79, 105, 165–166, 181, 187, 251, 281
    homogeneous, 18, 40
    inhomogeneous, 17–18, 25, 57, 76, 94, 105, 125
  XXZ, vii, 2, 22, 24, 39–42, 46, 50–51, 56, 61, 66, 69, 78, 82–83, 108, 165–166, 176, 187, 190, 210, 213–217, 231–232, 236, 240, 262, 266–267, 276, 279–281, 285
    homogeneous, 168
    inhomogeneous, 23, 25, 53–54, 61–62, 125, 236, 238, 240
  XY, 262–263
  XYZ, 2, 61, 66, 262
highest coefficient, 86, 93, 96–97, 103, 104, 107, 113, 141, 150, 179
  conjugated, 93, 96, 97
  recursion, 96–98, 102–103
Hilbert space, 1, 2, 5, 7, 12, 23, 33, 35, 56, 165
  dual, 56, 105
hole, 200–201, 203–207, 209, 217, 226, 280–284, 289, 291–292, 294, 300,

302–303, 305–308, 322–324, 327, 342, 344
momentum, 204, 206, 222, 226–227, 229, 281, 303–306, 308, 310, 324
Holstein–Primakov transformation, 59
hyperplane, 21, 81

inhomogeneity, 18, 54, 62, 67, 77, 87, 92, 94, 105, 123–125, 232, 237, 238, 240–241, 243, 251, 256, 258, 260, 262, 268–269, 272
integrability, vi
quantum, 7
integrable system, 1, 4–5, 7, 27, 231
integral equation, 203, 210, 212–213, 344, 348
integrals of motion, vi, 7, 8, 18, 20, 31, 38
generating function, 8, 31

Jacobian, 55, 108, 115, 148, 253

Knizhnik–Zamolodchikov equation, 231

L-operator, 18, 23–24, 29, 32, 57, 59, 61, 67, 70–72, 77
lattice QNLSE model, 19, 28, 57, 78, 171
XXX chain, 14–15, 17–18, 49, 56, 59
inhomogeneous, 18, 76
XXZ chain, 22–23
lattice QNLSE model, 20, 23, 26, 28, 40, 57–58, 67, 78, 172
Lieb–Liniger model, 21
Luttinger liquid, 279, 281, 306, 330, 332

magnetic field, 190, 213–214, 216, 218–220, 231, 250, 266–268
critical, 214
magnetization, 166, 168–169, 181, 189, 214, 219, 239, 250, 265
massive
models, 279–280
phase, 214, 279
Thirring model, 210
massless phase, 214, 281
matrix
Cauchy, 144, 147
Pauli, 3, 236
permutation, 10, 16–18, 62–63, 65–66
positive definite, 150, 192–193
standard basis, 5–6, 10, 18, 61, 67, 166, 236
Mehta's integral, 328
momentum, 198, 204, 207, 306–307, 312, 315, 323, 325
bare, 215
dressed, 207–208, 306, 323
excitation, 207, 282, 284, 287, 307, 313, 321, 324
operator, 20, 171
monodromy matrix, vi, 1, 5, 7, 12–15, 18–19, 23–24, 29, 39, 44–45, 56–58, 61–63, 66, 70, 76, 88, 91, 95, 159, 172
partial, 71, 76
total, 71, 76, 176
twisted, 43, 45
multi-composite model, 69, 76–77

natural order, 108–109, 112

one-dimensional Bose gas, 21, 31, 33, 39, 46, 48, 69, 78, 81, 103, 119, 165–166, 170–171, 175, 181, 187, 189, 191, 213–214, 217, 225, 232, 235, 262, 267, 279–282, 285, 287–290, 324, 329, 332
operator
integral, 196, 203, 206, 210, 216, 223–224, 227, 229, 290, 318–319, 347
kernel, 205, 210, 212, 216, 218, 220, 224, 264, 265, 290, 291, 301, 318, 319, 347–349

resolvent, 210, 344, 347
self-adjoint, 7, 123, 150

partial fraction decomposition, 100, 137
particle, 2–3, 8, 21, 46, 194, 200–201, 204–205, 207, 209, 214, 217, 226, 267, 280–284, 289, 291–292, 294, 300, 302–308, 310, 322–323, 326–327, 342, 344
momentum, 203, 206, 222, 227, 229, 281, 283, 301, 303–306, 308, 310, 323
particles number operator, 20, 177, 190, 208, 285
partition, 69–70, 72–77, 79–81, 86, 90–91, 95–96, 99–105, 107–108, 110–112, 114, 123, 125–126, 130–131, 138, 140–142, 177–180, 232, 237, 241, 243–245, 258, 269, 313
parity, 109, 111
partition function, 103, 233
plane wave, 21
polydisc, 253
power law, 279
pseudovacuum, 23

quantum
determinant, 2, 29
nonlinear Schrödinger equation, 18, 171, 190, 208
number, 200, 202, 204, 209, 281, 283, 284, 300, 301, 304–308, 311–314, 316, 324, 326, 341, 342, 344
space, 5, 17, 67, 233
local, 15, 61, 67, 165, 234
quantum inverse problem, vi, 61, 67, 69, 105, 165–166, 170, 181, 236, 239, 270
quantum inverse scattering method, v–vi, 1–2, 4–5, 7, 171
quantum system, v, vii, 2, 4, 7, 46
quasi-momentum, 217
quasi-particle, 267

R-matrix, vi, 1–2, 5–6, 8–10, 12–15, 17, 19, 22, 43, 57, 62–64, 66–67, 71, 91, 93, 171
elliptic, 12
rational, viii, 2, 12, 22, 26–28, 32, 44, 47, 57, 59, 86, 92, 97, 99, 108, 113, 116, 120, 122, 129, 146, 158–159, 162, 182, 187
trigonometric, viii, 2, 12, 21–22, 27, 45, 86, 92–93, 97–99, 108, 113, 115–116, 121, 129, 146, 159, 162, 183, 216
rapidity, 217
recursion, 133
reference state, 23
Riemann
$\zeta$-function, 339
–Hilbert problem, 220
integral, 196
RTT
relation, 1, 5–6, 8–9, 13–14, 18–19, 22, 28–29, 31–32, 39, 44, 59, 71, 91, 129, 150, 158
algebra, 87, 107, 181

S-matrix, 7
saddle point, 284, 306, 323–325, 327–328, 330, 332
scalar product, vi–viii, 27, 49, 56, 69, 72, 85–87, 89, 91–96, 103–105, 107–108, 113, 115–116, 119, 124–127, 129, 131–135, 141–143, 149–150, 153–159, 162, 174, 177–178, 180–183, 187, 234, 237, 240, 246–247, 252, 254–256, 261–262, 287–289, 291
scattering data, 5
shift function, 190, 198, 201–204, 206, 209–210, 213, 217, 221–222, 229, 288–292, 294, 297, 300–302, 304, 307–308, 311, 313, 315, 317, 321, 325, 329, 331, 341–343, 350
integral equation, 202–203, 209, 217, 343, 350
similarity transformations, 213, 216
sine-Gordon model, 22–23, 67

six-vertex model, 103
space-like regime, 323–324, 328, 330
spectral parameter, 5, 41, 44
Stirling's formula, 303, 305
symmetrization, 78

temperature
  finite, 190, 231–233, 235, 280
  zero, 231, 233, 235, 279–280
tensor product, 3, 5–6, 8, 10, 14–15, 62–63
theorem
  implicit function, 55
  Liouville, 102
  spin-statistics, 194
thermodynamic limit, vii, 40, 144, 166, 182, 189–191, 195, 198–199, 202, 204, 206, 208, 210, 213–217, 221, 223, 225–229, 232–237, 240–241, 244–245, 248–249, 260, 270, 279–285, 289, 291–293, 296, 298–299, 301, 303–308, 310, 314, 316, 321–322, 324–325, 338, 341–343
time-like regime, 323–324, 328, 330
Tonks–Girardeau gas, 194
transfer matrix, vi, 1, 7, 15–20, 23, 31, 33, 37–39, 46, 56, 61–62, 66, 129, 139, 150, 155, 158
  eigenvalue, 37–39, 108, 148, 158
  quantum, vii, 235
  twisted, vi, 32, 45, 50–52, 54, 113, 116, 120, 123, 129–130, 135, 141–143, 154, 232, 239, 251
    eigenvalue, 115, 120, 130, 134–135, 143, 168–169
twist parameter, 45, 55, 82–83, 108, 113, 123, 153, 176, 190, 200, 239, 284, 320
two-point correlation function, vii, 231–233, 262, 282
  densities, 280, 285, 319, 331
    generating function, 280–281, 285–286, 306, 312, 315–316, 320, 322, 331
  fields, 320, 329
  local spin operators, 236, 238, 263–264, 266
    generating function, 232, 238–240, 250–253, 260, 263–265, 267–270

ultracold atoms, v

vacuum, 19, 23, 33–34, 36, 39, 45, 58, 72, 89, 213–214
  dual, 56, 71, 89, 91, 105
  physical, 200
  vector, 2, 23, 34, 41, 44, 56, 59, 77, 88, 91, 105, 214, 267
Vandermonde determinant, 109, 324–325, 351
velocity of sound, 312, 323

Yang–Baxter equation, 8–10, 12, 22, 43, 63

Lightning Source UK Ltd.
Milton Keynes UK
UKHW022150260522
403581UK00003B/304

9 789811 254253